Cottage Industry of Biocontrol Agents and Their Applications

Nabil El-Wakeil · Mahmoud Saleh ·
Mohamed Abu-hashim

Editors

Cottage Industry
of Biocontrol Agents
and Their Applications

Practical Aspects to Deal Biologically
with Pests and Stresses Facing Strategic Crops

 Springer

Editors
Nabil El-Wakeil
National Research Centre
Dokki, Giza, Egypt

Mahmoud Saleh
National Research Centre
Dokki, Giza, Egypt

Mohamed Abu-hashim
Faculty of Agriculture
Zagazig University
Zagazig, Egypt

ISBN 978-3-030-33163-4   ISBN 978-3-030-33161-0   (eBook)
https://doi.org/10.1007/978-3-030-33161-0

This Springer imprint is published by the registered company Springer Nature Switzerland AG
The registered company address is: Gewerbestrasse 11, 6330 Cham, Switzerland

Preface

This book entitled *Cottage Industry of Biocontrol Agents and Their Applications* divided into five main parts comprising 15 chapters written by more than 20 professors, researchers, and scientists from different Egyptian entities.

The late Dr. Doug Waterhouse authored or co-authored 12 books on the biological control that drawn relevant information available from many sources for enabling students and research workers to locate easily, most or all of the information on pests in different countries. The books are essential for planning future biological control projects in different regions over the world. These books will be referred to for years to come, guiding new initiatives and recording part of the history of safely and successfully controlling insects by biological control in the developing countries.

As an introducible chapter to this book, the editors begin with the following title "The Roadmap of the Book." It presents generally different biological control strategies for altered pests (insects, mites, and plant diseases as well as micronutrient deficiency) by using diverse biocontrol agent groups to apply the effectiveness of modern field applications in agriculture which have been given a significant role for the enhancement of agricultural production. These applications comprise parasitoids, predators; microorganisms and biocontrol products for plant diseases. In the last chapter, there is another problem discussed; it is dealing with deficiencies in micronutrients, which affect the plant production. Our idea in this book is presenting a brief about low-cost biological control industry; therefore, all co-authors have tried, each specialist has done his/her best in their branches to present a summary of their ideas, information and research publications to obtain this practical book for our scientists, colleagues and students.

Cairo, Egypt Prof. Dr. Nabil El-Wakeil
Cairo, Egypt Prof. Dr. Mahmoud Saleh
Zagazig, Egypt Dr. Mohamed Abu-hashim
July 2019

Acknowledgements

Many colleagues within National Research Centre and other research institutions as well as other colleagues in Egyptian Universities (Entomologist, Acarologists, Bacterlogist, Virologist, Mycologist, Nematologist and Plant pathologist and Botanists) have provided valuable information from their experience as well as results of different international projects and publications in their book chapters. Although it is not possible to acknowledge all of the many individuals involved, the editors want to thank to all contributors either scientists or the technical members.

The editors would like to express their special thanks to all those who contributed in one way or another to make this high-quality book a real source of information on biological control industry in Egypt supported by the latest findings in the field summarized to support graduate students, researchers, scientists, and decision/policy-makers in Egypt and everywhere who are interested over the world.

Much appreciation and great thanks are due to all the authors who contributed to this book. Without the efforts and patience of all the contributors in writing, reviewing, and revising the different versions of the chapters, it would not have been possible to produce this unique book titled *Cottage Industry of Biocontrol Agents and Their Application*. Also, special thanks are due to the Springer team and editors, who largely supported the authors and editors during the production of this book. The book editor would be happy to receive any comments to improve future editions. Comments, feedback, suggestions for improvement, or new chapters for next editions are welcome and should be sent directly to the book editor.

Cairo, Egypt   Prof. Dr. Nabil El-Wakeil
Cairo, Egypt   Prof. Dr. Mahmoud Saleh
Zagazig, Egypt   Dr. Mohamed Abu-hashim
July 2019

Introduction: The Roadmap of the Book

Many insects that are of tiny or no economic significance in their home country may become main pests when they have been introduced to another country without their own natural enemies. Areas of the world where much of today's agriculture is based on introduced crops are distinguished for two facts [1]. The first fact is that a high part of their significant arthropod insects is exotic; the second is that classical biological control has obtained many important success cases. Worldwide, biological control either applied alone or as one of many strategies of integrated pest management is attracting increasing interest to reduce dependence on insecticides. Therefore, concentrating on biological control industry is very significant issue for given economy and environment for both developed and developing countries.

First Part: Parasitoids, Predacious Insects and Mites for Managing Agricultural Pests

It consists of five chapters dealing mainly with parasitoids and predators for insects and mites. To explain the different ways, especially in the early step of biological control starting with egg parasitoids, which help to kill the egg (first development stage of any insect), by this strategy, you make the way of insect management shorter.

First Chapter: Egg Parasitoids

They are used successfully for many years as inundative biocontrol agents against a wide range of economically important crops, vegetables, fruit orchards, and forest insect pests; they are currently the most widely produced and released natural enemies in biological control over the world [2, 3]. *Trichogramma* species were applied in more than 50 countries, and commercial releases occur in approximately 32 million hectares every year [4]. Egg parasitoids have been used to control many of insect pests which infest crops such as cotton, sorghum, soybean, sugarcane, tomato, and grape [5, 6].

Second Chapter: Larval Parasitoids

The ecto/endo-parasitoids which develop outside/inside their larval hosts are called larval parasitoids. The larval parasitoids lay their eggs into or outside the hosts, which hatch and begin feeding on their larvae in order to complete their growth. When fully fed the parasitoid, larva spins a cocoon and pupates either externally or internally and then emerges as an adult to feed, mate, and continue the life cycle [7]. Most of adult parasitoids feed on flower nectar, pollen, and honeydew. The authors aimed in this chapter to discuss potential of larval parasitoids in sustainable agriculture through monitoring the abundance of natural enemies, the mass production, and the field application [9] either alone or combining with egg parasitoids [8].

Third Chapter: Aphid Parasitoids

Aphids are one of insect groups whose economic importance increases with agriculture development [10]. Natural enemies include parasitoids and others, which maintain aphid population at a low population size [11, 12]. In this chapter, the authors are interested to present the following aspects:

1. Survey and seasonal abundance of certain aphids and estimation of the parasitism percentages of some aphid species during four seasons.
2. Effects of some weather factors (temperature, rainfall, and relative humidity) on the aphid parasitoid population and its aphid hosts.
3. Host suitability and host stage preference of certain aphid species for *Diaeretiella rapae*.
4. Evaluating the role of the parasitoid, *D. rapae* in controlling the cabbage and cauliflower aphids under field conditions.
5. Mass production of *D. rapae* and study of its efficacy in controlling the cabbage and cauliflower aphid species in greenhouses and in the field.

Fourth Chapter: Predacious Insects

The authors in this chapter entitled "Predacious insects and their efficiency in suppressing insect pests" mentioned that there are many predaceous insects in nature. These insect predators help to avoid some insect pests from population outbreaks. These predators positively eat/prey their prey. One of the vital goals for authors is the developing methods which suppress insect species without harming the environment or other organisms. Augmentative releases of insect predators are released into profitable crops with aphid infestations [13–15]. The authors aimed in this chapter to discuss potential of using insect predators to sustain their population through mass production and the field application for suppressing various insect pests in the numerous target crops, as well as against many aphid species in different crops.

Fifth Chapter: Predatory Mites

This chapter entitled "Mass Production of Predatory Mites and Their Efficacy for Controlling Pests". Mites are common found in different habitats as in soil, stored product, ornamentals, and field crops. Mites of the family Tetranychidae are strictly phytophagous and are represented in all regions of Egypt. Spider mites can undoubtedly cause severe crop loss [16]. The Mesostigmata predators are encountered in soil and litter, on aerial parts of plants, nests or galleries of insects, mammals, and birds, where they feed on small insects, nematodes, Collembola, or on phytophagous and mycophagous mites [17, 18]. Prostigmata predators are also found in diverse habitats, including soils and overlaying litter layers as well as aerial parts of plants. Predators of the genus *Agistemus* play an important role in biological control of economic pests [19, 20].

The objective of this chapter is to give a highlight about the most dominant predatory mites associated with fruit trees, vegetables, and crops. The authors concentrated also on selected predatory species from various groups of mites for augmentation release for rapid biological control. The mass production and their potential of biological control in Egypt are also included. The role of these mites in attacking various pests by natural and factitious prey is reviewed.

Second Part: Microorganisms for Controlling Insect Pests

The second part comprised four chapters, which dealing with using microorganisms such as bacteria, fungi, nematodes, and viruses to manage different insect pests.

Sixth Chapter: *Bacillus thuringiensis*

In this chapter entitled "Production and Application of *Bacillus thuringiensis* for Pest Control in Egypt", the authors reported that *Bacillus thuringiensis (B.t)* accounts 80–90% the total world microbial insecticide market. *B.t.* appeared to be a simple, spore-forming infectious bacterium, easy to grow on laboratory media. Steinhaus [21] encouraged the commercialization of *B.t.* leading to the production of Thuricide in 1957.

The leading role in research and development of *B.t.* has been taken by the National Research Centre (NRC), Egypt, to develop research in this area and to combat the key lepidopteran pests of field, oilseed, and vegetable crops. This chapter covers the expanding knowledge of the production, the mode of action, and the application of this pathogen in Egypt. Emphasis has been given to Lepidoptera, the most destructive group of plant pests [22–25].

Seventh Chapter: Entomopathogenic Fungi

In this chapter, the authors tried to discuss isolation, mass production, and application of entomopathogenic fungi for insect control. Entomopathogenic fungi play an exclusively important role in the history of microbial control of insects [26–28]. Entomopathogenic fungi were the first to be recognized as microbial diseases in insects [29]. The authors aimed to discuss the following objectives:

1. Isolation and abundance of the entomopathogenic fungi.
2. Mass production of entomopathogenic fungi.
3. Field applications of entomopathogenic fungi against some insect pests.

Eighth Chapter: Entomopathogenic Nematodes

Commercialization of biopesticides based on entomopathogenic nematodes (EPNs) is the very important objective of this chapter as mentioned by authors. EPNs are considered as an excellent insect biocontrol agents as they can kill their insect host within 48 hours; have a wide host range; have the ability to move searching for hosts; can be in vivo or in vitro mass produced; and present no hazard vertebrates and most non-target invertebrates [30–34]. This chapter focuses on isolation, mass production, formulation, as well as field application of EPNs for the management of insect pests in different crops, vegetables and fruit orchards.

Ninth Chapter: Insect Viruses

In this chapter, the authors discussed using insect viruses as biocontrol agent; they concentrated on the challenges and opportunities. The author emphasized that insect viruses as biological control agent represent an important component in the integrated pest management (IPM) programs, mainly because it is specific and safe for the environment and compatible with the other integrated pest management components. Baculoviruses are active in controlling early larval stage of lepidopteran insects. These viruses were isolated from insect from different insect orders such as Lepidoptera, Diptera, Orthoptera, and Coleoptera [35–37].

This chapter will cover the history of insect viruses, followed by a discussion on the large-scale production and commercialization of insect viruses with some examples. Thereafter, the emergence of resistance development and strain composition as challenges for insect viruses will be discussed with highlights on the recent reports in this field. Finally, the future and perspectives for the use of insect viruses as biocontrol agent will be summarized.

Third Part: Biocontrol Products for Suppressing Plant Diseases

The third part of this book contains four chapters and dealt with using biological control products for managing some plant diseases like phyto-pathogenic bacteria, fungal plant diseases, plant parasitic nematodes, and plant viral diseases.

Tenth Chapter: Biological Control of Phyto-pathogenic Bacteria

The authors confirmed that the pathogenic bacteria can attack many plants causing different symptoms, including necrosis, tissue maceration, wilting, and hyperplasia and resulting diseases and damage to crops. Phyto-pathogenic bacteria can be controlled by different methods such as cultural practices, resistant varieties, soil sterilization, seed disinfection, and hot water treatment [38, 39].

The biological control of pathogenic bacteria occurs in nature, where biocontrol agents can reduce the inoculum density of many plant pathogens, where the biological agents have different mechanisms included; competition, antibiosis, parasitism, plant growth stimulation and induce systemic resistance as well as it considered as environmentally friendly measure [40–42]. In this chapter, the authors indicated that the common phyto-pathogenic bacterial species able to be controlled by various biocontrol agents; the beneficial effects of common biocontrol agents and field application of biological control agents will be reviewed.

Eleventh Chapter: Fungal Plant Diseases Management

In this chapter, the authors discussed using different biocontrol agents for controlling the fungal plant diseases. Plant fungal diseases are the most destructive diseases, where the fungal pathogens attack many economical crops causing yield losses [43]. On the other hand, the fungal species can attack only one plant species or many plant species.

This chapter discussed using various biological control agents instead of chemical control for the fungal plant pathogens [44, 45]. The author mentioned that the biological control has many advantages in relation to soil fertility, plant, animal, and human health, where the use of chemicals such as fungicides that accumulate in the plant crop tissue then transferred to human causing several diseases such as liver and kidney failures.

Twelfth Chapter: Biological Control of Plant Parasitic Nematodes

The author mentioned that plant parasitic nematodes spend at least some part of their lives in soil—one of the most complex environments. Their activities are not only influenced by variation in soil physical factors such as weather conditions and aeration but also by a vast array of living organisms, including other nematodes, bacteria, fungi, algae, protozoa, insects, mites, and other soil animals [46, 47].

For biological control to be successful, it must be supported by a backbone of basic ecological research [47, 48]. In this chapter, the author aimed to have a sound knowledge of the population dynamics of nematodes, of the threshold levels needed to cause economic damage, of the role that parasites, predators, and other soil organisms play in regulating nematode populations and other components of the soil ecosystem. Noweer [48] investigated the effect of the nematode-trapping fungus *Dactylaria brochopaga* and the nematode egg parasitic fungus *Verticillium chlamydosporium* in controlling citrus nematode infesting mandarin, and interrelationship with the co-inhabitant fungi; similar results were recorded by Noweer [47].

Thirteenth Chapter: Plant Viral Diseases

The authors tried in this chapter to discuss the potential management tools to control plant viral diseases. The authors mentioned that viruses are essentially obligate parasites with fairly simple and unicellular organized only by one type of nucleic acid, either DNA or RNA. Mostly, all types of living organisms including animals, insects, plants, fungi, and bacteria are able to be infected by different viruses [50].

The main goal of this chapter is to shed light on the plant viral diseases that affect plants and crops. The losses resulted from viral infection affect the production and the crop yields which lead to hunger complained in the third world as a whole. The authors reported that they identified several ways that may be used to eliminate most of these viruses, which are effective as a protection and not a cure [51, 52]. Several microbes, plant extracts, nanomaterials and etc. that are used to reduce viral infection. The authors ended their chapter by using a local Egyptian trail bacillus spp. filtrates mixed with nano-clay to protect potato from virus infection.

Fourth Part: Bio-products Against Abiotic Factors

This part of the book consists of one chapter dealing with the "Biochemical indicators and biofertilizer application for diagnosis and alleviation micronutrient deficiency in plant".

Fourteenth Chapter: Biochemical Indicators of Micronutrient Deficiency

Micronutrients like iron (Fe), copper (Cu), zinc (Zn), manganese (Mn), and boron (B) are important for an abundance of physiological functions in plant growth, development, and oxidative stress response. The authors stated that at low concentrations in electron transport and antioxidant systems, micronutrients are desired for cellular structures and protein stabilization [53].

The authors mentioned that metabolic processes in root and shoot levels are affected in such micronutrient deficiencies [54–56]. In this chapter, the authors epitomize sources and factors that influence the behavior and availability of micronutrients in soils, the possible approaches to control the availability of soil micronutrients, and the present physiological and biochemical changes in plants. They explained some strategies to increase the availability of micronutrients in soil for plant uptake.

Final Part: Conclusions

The Conclusions chapter, which summarizes the covered topics in this book with an update of the recently published research, highlights the extracted conclusions from the chapters and presents a set of recommendations for further future research on biological, organic, and supportable agriculture environment.

Consequently, biological control has certainly been highly succeeded over a huge area over the world, even though the same pest continues to be a problem in another place. Suppression level of any insect may show role of natural enemies, which it is highly likely that any contribution, would give to the insect suppression a useful contribution to the integrated management of that insect pest.

Nabil El-Wakeil
Mahmoud Saleh
Mohamed Abu-hashim

References

1. Waterhouse DF, Sands DPA (2001) Classical biological control of arthropods in Australia. ACIAR Monograph No. 77, 560 p
2. Li LY (1994) Worldwide use of *Trichogramma* for biological control on different crops: a survey. In: Wajnberg E, Hassan SA (eds) Biological control with egg parasitoids. CAB International, Oxon, U.K., pp 37–53
3. Van Lenteren JC (2000) Success in biological control of arthropods by augmentation of natural enemies. In: Gurr G, Wratten S (eds) Biological control: measures of success. Kluwer Academic Publishers, Dordrecht, pp 77–103
4. Smith SM (1996) Biological control with *Trichogramma*: advances, successes, and potential of their use. Ann Rev Entomol 41:375–406
5. Van Lenteren JC, Bueno VHP (2003) Augmentative biological control of arthropods in Latin America. Biocontrol 48:123–139
6. Hegazi E, Herz A, Hassan SA, Khafagi WE, Agamy E, Zaitun A, Abd El-Aziz G, Showeil S, El-Said S, Khamis N (2007) Field efficiency of indigenous egg parasitoids to control the olive moth *Prays oleae*, and the jasmine moth *Palpitaunionalis*, in an olive plantation in Egypt. Biol Control 43:171–187
7. El-Heneidy AH, Sekamatte BM (1998) Survey of larval parasitoids of the cotton bollworms in Uganda. Bull Soc Ent Egypt 76:125–134
8. Briggs CJ, Latto J (2001) Interactions between the egg and larval parasitoids of a gall-forming midge and their impact on the host. Ecol Entomol 26:109–116
9. Ferracini C, Ingegno BL, Navone P, Ferrari E, Mosti M, Tavella L, Alberto A (2012) Adaptation of indigenous larval parasitoids to *Tuta absoluta* (Lepidoptera: Gelechiidae) in Italy. J Econ Entomol 105:1311–1319
10. Hagvar EB, Hofsvang T (1991) Aphid parasitoids (Hymenoptera: Aphidiidae) biology, host selection and use in biological control. Biocontrol News Inform 12:13–41
11. Ibrahim AMA (1987) Studies on aphidophagous parasitoids with special reference to *Aphidiusuzbekistanicus* (Luz), Ph.D. thesis, Faculty of Agriculture Cairo University, p 202

12. Jonsson M, Wratten SD, Landis DA, Gurr GM (2008) Recent advances in conservation biological control of arthropods by arthropods. Biol Control 45:172–175
13. van Lenteren JC, Manzaroli G (1999) Evaluation and use of predators and parasitoids for biological control of pests in greenhouses. In: Albajes R, Gullino ML, van Lenteren JC, Elad Y (eds) Integrated pest and disease management in greenhouse crops. Kluwer, Dordrecht, The Netherlands, pp 183–201
14. van Lenteren JC, Bueno VHBP (2003) Augmentative biological control of arthropods in Latin America. Bio Control 48:123–139
15. El-Wakeil NE, Vidal S (2005) Using of *Chrysoperlacarnea* in combination with *Trichogramma* species for controlling *Helicoverpaarmigera*. Egypt J Agric Res 83:891–905
16. Gerson U (2008) The Tenuipalpidae: an under-explored family of plant feeding mites. Syst Appl Acarol 2:83–101
17. Halliday RB (2006) New taxa of mites associated with Australian termites (Acari: Mesostigmata). Internat J Acarol 32:27–38
18. Moreira GF, de Morais MR, Busoli AC, de Mraes GJ (2015) Life cycle of *Cosmo laelaps jaboticaba lens* on *Frankliniella occidentalis* and two factitious food sources. Exp ppl Acarol 65:219–226
19. El-Sawi SA, Momen FM (2006) *Agistemusexsertus* Gonzalez (Acari: Stigmaeidae) as a predator of two scale insects of the family Diaspididae (Homoptera: Diaspididae). Archiv Phytopathol Plant Prot 39:421–427
20. Momen FM (2012) Inffiuence of life diet on the biology and demographic parameters of *Agistemus olive* Romeih, a specific predator of eriophyid mites (Acari: Stigmaeidae and Eriophyidae). Trop Life Sci Res 23:25–34
21. Steinhaus EA (1951) Possible use of *Bacillus thuringiensis* as an aid in the biological control of the alfalfa caterpillar. Hilgardia 20:359–381
22. Salama HS, Zaki F (1986) Effects of *Bacillus thuringiensis* Berliner on prepupal and pupal stages of *Spodoptera littoralis*. Insect Sci Appl 7:747–749
23. Salama HS, Salem S, Zaki F, Matter M (1990) Control of *Agrotis ypsilon* on some vegetable crops in Egypt using the microbial agent *Bacillus thuringiensis*. Anz Schadlingsk Pffianz Umwelt 63:147–151
24. Salama HS, Salem S, Matter M (1991) Field evaluation of the potency of *Bacillus thuringiensis* on lepidopterous insects infesting some field crops in Egypt. Anz Schadlingsk Pffianz Umwelt 64:150–154
25. Saker M, Salama HS, Salama M, El-Banna A, Abdel-Ghany N (2011) Production of transgenic tomato plants expressing Cry 2 Ab gene for the control of some lepidopterous insects endemic in Egypt. J Genet Eng Biotechnol 9:149–155
26. Soper et al (1988) Isolation and characterization of *Entomophaga maimaiga* sp. nov., a fungal pathogen of gypsy moth *Lymantria dispar*, from Japan. J Invertebr Pathol 51:229–241
27. Abdel-Raheem MA (2013) Susceptibility of the red palm weevil, *Rhynchophorus ferrugineus* Olivier to some Entomopathogenic fungi. Bull NRC 38:69–82
28. Abdel-Raheem MA, Al-Keridis LA (2017) Virulence of three Entomopathogenic fungi against white ffiy, *Bemisiatabaci* (Genn.) (Hemiptera: Aleyrodidae) in Tomato Crop. J Entomol 14:155–159
29. Ainsworth GC (1956) Agostino Bassi, 1773–1856. Nature 177:255–257
30. Gaugler R, Boush GM (1979) Nonsusceptibility of rats to the entomogenous nematode *Neoaplectana carpocapsae*. Environ Entomol 8:658–660
31. Lacey LA, Georgis R (2012) Entomopathogenic nematodes for control of insect pests above and below ground with comments on commercial production. J Nematol 44:218–225
32. Lacey LA, Grzywacz D, Shapiro-Ilan DI, Frutos R, Brownbridge M, Goettel MS (2015) Insect pathogens as biological control agents: back to the future. J Invert Pathol 132:1–41
33. Saleh MME, Hussein MA, Hafez GA, Hussein MA, Salem HA, Metwally HM (2015) Foliar application of entomopathogenic nematodes for controlling *Spodoptera littoralis* and *Agrotis ipsilon*(Lepidoptera: Noctuidae) on corn plants. Adv Appl Agric Sci 3:51–61

34. Saleh MME, Kassab AS, Abdelwahed MS, Alkhazal MH (2010) Semi-field and field evaluation of the role of entomopathogenic nematodes in the biological control of the red palm weevil *Rhynchophorus ferrugineus*. Acta Hort 882:407–412

35. Beas-Catena A, Sánchez-Mirón A, García-Camacho F, Contreras-Gómez A, Molina-Grima E (2014) Baculovirus biopesticides: an overview. J Anim Plant Sci 24:362–373

36. Hunter-Fujita FR, Entwistle PF, Evans HF, Crook NE (1998) Insect viruses and pest management. In: Insect viruses and pest management [Internet]. cited 2019 Apr 12. Available from: https://www.cabdirect.org/cabdirect/abstract1105344

37. Lacey LA, Grzywacz D, Shapiro-Ilan DI, Frutos R, Brownbridge M, Goettel MS (2015) Insect pathogens as biological control agents: back to the future. J Invertebr Pathol 132:1–41

38. Gadoury DM, McHardy WE, Rosenberger DA (1989) Integration of pesticide application schedules for disease and insect control in apple orchards of the northern United States. Plant Dis 73:98–105

39. Agrios G (1997) Plant pathology, 4th edn. Academic Press, pp 1–635

40. Andrews JH (1992) Biological control in the phyllosphere. Annu Rev Phytopathol 30:603–635

41. Raaijmakers JM, Weller DM (2001) Exploiting genotypic diversity of 2,4-diacetylphloroglucinol-producing *Pseudomonas* spp.: characterization of superior root-colonizing *P. ffluorescens* strain Q8r1-96. Appl Environ Microbiol 67:2545–2554

42. Abd El-Kahir H (2004) Efficacy of starner in controlling the bacterial soft rot pathogen in onion. Ann Agric Sci 49:721–731

43. Crous PW, Hawksworth DL, Wingfield MJ (2015) Identifying and naming plant-pathogenic fungi: past, present, and future. Annu Rev Phytopathol 53:247–267

44. Moussa TAA, Ali DMI (2008) Isolation and identification of novel disaccharide of a-L-Rhamnose from *Penicillium chrysogenum*. World Appl Sci J 3:476–486

45. Rashad YM, Al-Askar AA, Ghoneem KM et al (2017) Chitinolytic *Streptomyces griseorubens* E44G enhances the biocontrol efficacy against *Fusarium* wilt disease of tomato. Phytoparasitica 45(2):227–237

46. Aboul-Eid HZ, Noweer EMA, Ashour NE (2014) Impact of the nematode-trapping fungus, *Dactylaria brochopaga* as a biocontrol agent against *Meloidogyne incognita* infesting superior grapevine. Egypt J Biol Pest Cont 24:477–482

47. Noweer EMA (2017) A field trial to use the nematode-trapping fungus *Arthrobotrys dactyloides*to control the root-knot nematode *Meloidogyne incognita* infesting bean plants. Comm Appl Bio Sci 82:275–280

48. Noweer EMA, Al-Shalaby MEM (2009) Effect of *Verticilium chlamydosporium* combined with some organic manure on *Meloidogyne incognita* and other soil micro-organisms on Tomato. Int J Nematol 19:215–220

49. Noweer EMA (2018) Effect of the nematode-trapping fungus *Dactylaria brochopaga* and the nematode egg parasitic fungus *Verticilium chlamydosporium* in controlling citrus nematode infesting mandarin, and interrelationship with the co inhabitant fungi. Int J Eng Technol 7:19–23

50. Fenner F, Maurin J (1976) The classification and nomenclature of viruses. Arch Virol 51:141–149

51. Abdelkhalek A, Eldessoky D, Hafez E (2018) Polyphenolic genes expression pattern and their role in viral resistance in tomato plant infected with *Tobacco mosaic virus*. Biosci Res 15:3349–3356

52. Abdelkhalek A, ElMorsi A, AlShehaby O, Sanan-Mishra N, Hafez E (2018) Identification of genes differentially expressed in *Iris yellow spot virus* infected onion. Phytopathologia Mediterr 57:334–340

53. O'Neil MA, Ishiim T, Albersheimm P, Darvill AG (2004) Rhamnogalacturonan II: structure and function of a borate cross-linked cell wall pectic polysaccharide. Ann Rev Plant Biol 55:109–139

54. El-Fouly MM, El-Baz FK, Youssef AM, Salama ZA (1998) Carbonic anhydrase, aldolase, and catalase activities as affected by spraying different concentrations and forms of zinc and iron on faba bean and wheat. Egypt J Sci 22:1–11

55. Salama ZA (2001) Diagnosis of copper deficiency through growth, nutrient uptake and some biochemical reactions in Pisum sativum L. Pak J Biol Sci 4:1299–1302
56. Reda F, Abdelhamid MT, El-Lethy SR (2014) The role of Zn and B for improving *Vicia faba* L. tolerance to salinity stress. Middle East J Agric Res 3:707–714

Contents

Parasitoids, Predacious Insects and Mites for Managing Agricultural Pests

Egg Parasitoid Production and Their Role in Controlling Insect Pests

Saad H. D. Masry and Nabil El-Wakeil

Abstract Egg parasitoids have been used successfully as inundative or augmentative biological control agents against a wide range of agricultural pests; *Trichogramma* and other egg parasitoids are currently the most widely produced and applied natural enemies over the world. *Trichogramma evanescens* as well as many species have also been deeply researched in different crops; maize, cotton, rice and sugarcane and vegetables like tomato, cabbage, pepper and potato as well as fruit orchards; apple, olive, pomegranate and grape as well as for controlling the forest and stored product insect pests. *Trichogramma* species have a short generation time and can be easily mass-produced and they could kill the Lepidopteran pests during the egg stage. The endogenous species is selected for release on the environmental basis that it is better adapted to the proposed climate and habitat than those are exotic parasitoids. Efficiency of the parasitoids is depended on the factitious hosts which reared on; *Sitotroga cerealella*, *Ephestia kuehniella* and *Corcyra cephalonica*. There are biological components in a mass rearing facility: the host and the parasitoid. To ensure high product quality and to avoid contamination with other parasitoid species, facilities usually rear only a few parasitoid strains or species at any given time. Storage of both host eggs as well as parasitized ones is a requirement to manage commercial or otherwise supply of parasitoid material in times of need. There is a need for regulatory evaluating the quality of parasitoid wasps. There are many factors affect the field release, like weather, crop, host, predation, pesticide uses, and parasitoid quality that influence the release and disappearance rate. In field applications over world, releases of several million female wasps/ha through the season proved to be very effective in suppressing the key lepidopteran pests of many crops accomplishing

S. H. D. Masry
Department of Plant Protection and Molecular Diagnosis, Arid Land Research Institute (ALRI), City of Scientific Research and Technological Applications, Alexandria, Egypt

Abu Dahbi Agriculture and Food Safety Authority, Research and Development Division, Al Ain, UAE

N. El-Wakeil (✉)
Pests and Plant Protection Department, National Research Centre, Dokki, Cairo, Egypt
e-mail: nabil.elwakeil@yahoo.com; nabil.el-wakeil@landw.uni-halle.de

Institute of Agriculture and Nutritional Sciences, Martin-Luther-University Halle-Wittenberg, Halle, Germany

© Springer Nature Switzerland AG 2020
N. El-Wakeil et al. (eds.), *Cottage Industry of Biocontrol Agents and Their Applications*, https://doi.org/10.1007/978-3-030-33161-0_1

parasitism up to 91%. Therefore, egg parasitoids are a promising biocontrol agent for mainly Lepidopteran insects.

Keywords Egg parasitoids · Mass rearing · Host selection · Quality control · Field application · Lepidopteran pests

1   Introduction

Biological control of insect pests is promising as an important component of Integrated Pest Management (IPM) globally, with the realization of environmental and human hazards associated with the use of chemical pesticides. Currently, world agriculture is transforming into a more professional approach, support to eco-friendly pest control techniques as key thrusts in promoting export horticulture and organic farming [1–4].

Among other different biocontrol techniques, a procedure that has shown efficient results in controlling pest outbreaks, mainly from the order Lepidoptera, is the use of egg parasitoids from the genus *Trichogramma* (Hymenoptera: Trichogrammatidae) [5]. Egg parasitoids are one of the effective natural enemies of important agricultural and forestry pests. Research on egg parasitoids receives maximum attention because they are amenable to mass rearing, not only on the target hosts but, also on alternate or factitious hosts. Besides, utilization of the appropriate egg parasitoids at the right time can prevent hatching of the pest's eggs to larvae, thus preventing damage at the initial stage. Trichogrammatid egg parasitoids are the most widely mass-produced biocontrol agents, both globally and in Egypt [3, 6–8]. *Trichogramma* is a dominant genus of egg parasitoids and among the earliest and most widely exploited natural enemies [9].

Egg parasitoids have been used successfully for many decades as inundative biological control agents against a wide range of economically important agricultural and forest pests; they are currently the most widely produced and released natural enemies in biological control throughout the world [6, 10]. *Trichogramma* species have been carried out in more than 50 countries and commercial releases occur in approximately 32 million ha every year [11]. Egg parasitoids has been used to control many of insect pests which infest crops such as cotton, sorghum, soybean, sugar cane, tomato, and grape [12, 13]. Some of the advantages in using *Trichogramma* spp. in biological control programs are: theses parasitoids are easily reared on alternative hosts and are highly aggressive in parasitizing eggs of different pest species as well [8, 14]. A development strategy has been conducted to optimize the technology of field releases as well as to increase the efficacy of mass production of *Trichogramma* and storage at cold temperature. Establishing a new industrial process for *Trichogramma* production has been formed for controlling *Ostrinia nubilalis* biologically in France [15]. Who mentioned that egg parasitoids are promising an eco-friendly technology and *Chilo sacchariphagus* could be managed by *Trichogramma chilonis* as applied in their program for 10 years.

Egg parasitoids other than Trichogrammatidae have an economic significance and play a precise role in IPM programs, these parasitoid wasps are found in the following families: Mymaridae, Scelionidae and Platygastridae. The biological control programs with those wasp groups is occurred through the collaborative governmental, specialist institutions and international cooperators working on quarantine, mass production and release [16]. The scelionid egg parasitoids have been recorded from eggs of hemipterans or lepidopterans. In the subfamily Telenominae, *Eumicrosoma cumeus* (Nixon) and *Eumicrosoma phaex* (Nixon) were originally recorded as egg parasitoids of the lygaeid, *Cavelerius* sp. [17]. Anwar and Zeya [18] recorded 14 mymarid species in 5 genera from different states in India. In the following paragraphs the characteristics, biology, mass rearing and importance of these species were presented.

1.1 Eulophidae

The Eulophidae comprise 297 genera found in 4 subfamilies as follows: Euderinae, Eulophinae, Entedoninae and Tetrastichinae. The Tetratischinae are usually primary endoparasitoids of eggs of Lepidoptera and Coleoptera [19].

1.2 Platygastridae

Wasp species of family Platygastridae are commonly endoparasitoids of many genera in orders Coleoptera, Diptera and Homoptera [20]. Some species of *Fidiobia* were collected as parasitoids of Coleoptera (curculionid and chrysomelid) eggs [20]. Six species of the genus *Fidiobia* are recorded. Of these, 5 species are known to parasitize eggs of genera *Diaprepes*, *Hypera* and *Entypotrachelum* and one species is known to parasitize eggs of *Fidia viticida* [21, 22]. In Brazil, genus *Fidiobia* parasitized of weevil eggs which infested citrus orchards [23].

1.3 Mymaridae

There are many genera and species recorded as egg parasitoids belonging this family.

1.3.1 *Anagrus epos* Girault

A collection of this wasp species had been established in UCR quarantine from the parasitized egg masses of *Cuerna fenestella* Hamilton [24]. Fresh eggs of *H.*

vitripennis were offered to the mated females of *A. epos*. After mass production of *A. epos*, it had been released in California against the GWSS [24, 25].

1.3.2   *Gonatocerus ashmeadi* **Girault**

This species was inserted and released in California, which obtained from southeastern USA and northeastern Mexico. It is by far the most abundant natural enemy of GWSS in California [26]. This species likely established itself in California long before the arrival of GWSS on eggs to California [24]. *G. ashmeadi* could successfully suppressed the GWSS infestation in Hawaii and Chile [24].

1.3.3   *Gonatocerus morrilli* **(Howard)**

Gonatocerus morrilli were introduced into California fields from Texas and Tamaulipas (Mexico) [25, 27], who indicated that *G. walkerjonesi* Triapitsyn, was a species native to California, which accurately had been released in California.

1.3.4   *Gonatocerus fasciatus* **Girault**

Gonatocerus species are egg parasitoids of Proconiini [28] which produce up to 7 wasps per each GWSS egg. Its emergence holes could be recognized easily by host egg. In northern California, *G. fasciatus* was discovered by Triapitsyn [24], who suggests that this species is native there. This wasp species had been first introduced into California from Louisiana as mentioned by Triapitsyn et al. [28] and later mass released against *H. vitripennis* [25].

Using of Egg parasitoids is favored by simple mass rearing systems, persistent economic efficiency of the parasitoids and the suitable technologies for their commercial use [6, 7, 11, 14, 29].

2   Historical Perspective

In the early stages of development, biological control of necessity followed mainly a 'trial and error' approach [30]. The release of *Trichogramma* for biological control of lepidopterous pests has been considered for more than 120 years, although the mass rearing of these hymenopterous parasitoids was not proposed in USA until the 1920s. Flanders' work inspired activity during the 1930s, but this activity quickly dissipated in the West with the rise of chemical insecticides. It was left primarily to scientists in the former USSR (from 1937) and China (from 1949) to develop *Trichogramma* as biological-control agents [31]. In the 1960s, the Europeans and

Americans revitalized research on *Trichogramma;* in the 1970s, they began mass-rearing and release [32–34].

It has been almost 40 years since an exhaustive review of inundative releases has been conducted [35]. Although the genus *Trichogramma* is not the only group to be used with this approach, much of our understanding of inundative release comes from studies with these minute egg parasitoids. At the same time as Stinner's review [35]; a compendium of international research was published that marked the beginning of an exchange of information among scientists in North America, Europe, the former USSR, and China [36]. This exchange resulted in an explosion of research that continues today. International symposia with published proceedings have been conducted every four years since 1982 [37], and informal sessions have been held at the last three International Congresses of Entomology. In addition, there have been eight international symposia on quality control [38].

Trichogramma has also been deeply researched in different crops started by Farghaly [39]; Zaki [40] and El-Wakeil [41] in maize fields. There were many wide biocontrol projects in 1980s until now in Egypt using *Trichogramma* in controlling insect pests i.e. sugar cane [42–45], cotton [29, 46, 47], grape [7], olive [13, 48] and tomato [49, 50].

3 Life Cycle of Egg Parasitoids

3.1 Life Cycle of Trichogramma spp.

Trichogramma spp. (Hymenoptera: Trichogrammatidae) have a short generation time and can be easily mass-produced and they could kill the Lepidopteran pests during the egg stage before caterpillars can emerge and damage the crop [33, 36]. *Trichogramma* is solitary endoparasitoid, it seeks out and parasitizes host eggs, more than one egg may be inserted into each host egg and this is based, at least in part, on the egg size. After hatching, the parasitoid larvae feed on the contents of the host egg. The wasps pupate within the egg and adults chew an emergence hole to escape (Fig. 1). It takes about 10 days from the time of parasitism to emergence of wasps at a constant 27 °C [51].

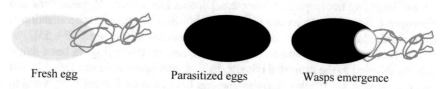

Fresh egg   Parasitized eggs   Wasps emergence

Fig. 1 Life cycle of *Trichogramma* species. Drawn by El-Wakeil

(a) **(b)** **(c)**

Fig. 2 Adult *Trissolcus* sp. (**a**), *Trissolcus* sp. investigating its host egg mass (**b**), Adult *Trissolcus* sp. that failed to emerge from a host egg (**c**) (https://biocontrol.entomology.cornell.edu/parasitoids/ trissolcuseuschisti.php, March 2015)

3.2 Life Cycle of Scelionid Wasp

Adult female *Trissolcus euschisti* (Ashmead) (Hymenoptera: Scelionidae) (Fig. 2a) locate an egg host through chemical cues that are found by drumming the antennae over the host eggs (Fig. 2b). One wasp egg is deposited into each host egg. The adult female will then remain on or near the egg mass to defend it against other parasitoids. As parasitoid egg hatches out of host egg, it starts eating the developing embryo of the host, and pupates inside the host egg. *T. euschisti* takes 25 days to develop to male and 30 days to female inside the host egg at a mean temperature of 27 °C. If temperature falls to 15 °C, or exceeds 33 °C, *T. euschisti* may develop but fail to emerge from the host egg (Fig. 2c). Males will emerge approximately 1 day before females and will wait for their female siblings to hatch out so they can mate with them before dispersing [52]. In temperate climate zones, *T. euschisti* can have up to 5 generations per year.

4 Selection of Parasitoids to Release

The selection of the most appropriate parasitoid for release starts with the best species. This process is difficult, because there is considerable interspecific variation in the more than 145 species known, and the taxonomy of the genus is poorly understood [53]. Members of the genus are polyphagous egg parasitoids on ten orders of insects, including Lepidoptera, Coleoptera, Diptera, Heteroptera, Hymenoptera, and Neuroptera. As more species are discovered, however, increasing specialization is recognized. Molecular studies help clarify the taxonomy of this genus [54, 55].

The local species is generally selected for release on the ecological basis that it is better adapted to the proposed climate, habitat, and host conditions [56, 57]. For example, at least six species of *Trichogramma* have been used around the world to control *Ostrinia* spp. In their native regions, the most common are *T. nubilale* Ertle and Davis and *T. pretiosum* Riley in the United States, *T. ostriniae* Pang et Chen and *T. dendrolimi* Matsumura in China, and *T. evanescens* (Westwood) and *T. brassicae* Bezdenko in Europe. Use of the local species is the basis of inundative theory and is only contraindicated when there is no native species or when pre-introductory screening suggests otherwise [56, 58]. In diverse habitats, this competition could lead to the elimination of local species or strains when non-native species are released [59]. A survey of local species should be conducted before parasitoids are released, because natural levels of parasitism, although sometimes negligible, can be as high as 40–100% [60, 61]. The recent move by national agencies in some countries to restrict the importation of organisms for biological control makes it important that effective native species be identified.

Once the species has been selected, the population (=strain) to release must be determined. Parasitoid is play both interspecific and intraspecific variations in biology and behavior that are strongly influenced by environmental factors. These populations have been compared in terms of development, fecundity, egg absorption, sex ratio, longevity, host age selection, oviposition, host preference, and activity [14, 62], as well as in terms of their response to environmental conditions [63]. The ultimate choice of strain will depend on how it ranks in terms of those biological attributes considered advantageous for the environment into which it will be released and the type of releases.

The final aspect of selection is that of founding populations, i.e. where and how many collections (both individuals and populations) are needed to initiate a vigorous colony. This field is most important, because it is based on the population genetics. A few studies have examined the genetic variability of traits, including fecundity [64], reactive distance, walking behavior [60], and sex allocation [64]. *Trichogramma* are haplodiploid organisms (most females arise from fertilized diploid eggs, and most males from unfertilized haploid eggs), however, and are characterized by high rates of sib-mating an estimated 55–64% in the field [65] and naturally low heterozygosity. This feature suggests that the degree of heterozygosity normally required to maintain a vigorous colony of a *Trichogramma* species might be less than expected and that healthy colonies may be founded ca. 500 individuals. Considerable work is needed to develop an understanding of how *Trichogramma* maintain sufficient levels of variation under conditions of small population size. The similarity in climate between collection and release sites is a further important consideration that can have a major impact on successful establishment in new environments [66].

5 Mass Production of Trichogrammatid Parasitoids

The members of the family Trichogrammatidae, who are amongst the tiniest insects, measuring less than a millimeter, are all parasitic on the eggs of other insects which mostly are lepidopterans. The tiniest amongst them are the species of *Megaphragma*, measuring 0.18 mm [67]. In other words, these are next only to mymarids. The species of *Trichogramma* (and also Trichogrammatoidea) have been selected for laboratory mass production and releases against noxious lepidopterous crop pests in the field. Amongst the natural lepidopterous hosts of Trichogramma spp., Noctuidae has maximum number of host species, followed by Tortricoidea, Pyraloidea, etc.

[68]. Their small size did not prevent entomologists all over the world to exploit them for biological control of harmful moths and butterflies [69–71]. This is mainly because:

(1) Their short life cycle, most of them in tropical climate having 8–10 days.
(2) Their high breeding potential, a female can produce 25 and 80 offsprings.
(3) Their high percent of female progenies (60–90%).
(4) Most of them adapt to laboratory factitious host lepidopterans like rice moth *Corcyra cephalonica*, grain moth *Sitotroga cerealella* and flour moth *Ephestia kuehniella*. Their laboratory production and field releases were commenced during the 20th century.

Mass rearing and releasing of natural enemies are considered one of the vital tactics in IPM programs [72]. Much research has led to the technical standardization and regulation of commercial production processes [73]. This has resulted in a scientific foundation for the commercial production and application of augmentative agents. Most Chinese *Trichogramma* manufacturers are located in Northeast China. Among them, several manufacturers have reached an annual production capacity of 20 billion insects. They may be released every 50–60 days at a density of 0.18–0.21 million wasps/ha per release [74]. At these release rates *Trichogramma* can parasitize 37–40% of the cotton bollworm eggs and reduce bollworm populations up to 60% [75–77]. Mass rearing is the production of insects competent to achieve program goals with an acceptable cost/benefit ratio and in numbers per generation exceeding ten thousand to one million times the mean productivity of the population of native females [78].

5.1   *Production System for* Trichogramma *Mass Rearing*

Once a parasitoid colony has been selected, the next step is to rear large numbers for release. This process has been accomplished in various ways during the past 100 years and has been the focus of considerable attention [32, 79, 80]. Mass production of *Trichogramma* is a growing field, and many facilities have been established or expanded in 80s. Two types of rearing systems have evolved those with short-term high daily output and those with long-term low daily output [34]. A range in production from 4 to 1000 million parasitoids/day has been found, depending on the mode of output; short-term output usually has the higher values [31]. Consistent levels of output of 100 million female parasitoid/week, although rarely specified, are not uncommon for the larger facilities [34]. Major commercial facilities are currently found in Europe, America, China and the former USSR [81]. Numerous other smaller facilities can be found over the world with government, private, and cooperative support [82–84]. Most of the larger facilities produce parasitoids on a year-long basis, whereas the smaller facilities produce parasitoids for local periodic releases [34].

El-Wakeil [14] compared the efficiency of the parasitoid *Trichogramma evanescens* (Westwood) reared on eggs of three different factitious hosts; *Sitotroga*

Fig. 3 Host eggs used for *Trichogramma* rearing facility **a** *Galleria mellonella*, **b** *Ephestia kuehniella*, **c** *Sitotroga cerealella*, **d** *Helicoverpa armigera*. Photograph by El-Wakeil

cerealella, Ephestia kuehniella and *Galleria mellonella* (Fig. 3). Rates of parasitism on *Helicoverpa armigera* eggs, emergence rates of parasitoids and their longevity were the highest for wasps reared on *H. armigera*. Wasps reared on *S. cerealella* gave comparable rates. However, wasps from *E. kuehniella* and *G. mellonella* gave the lowest ones. Parasitized eggs of *H. armigera* and *S. cerealella* produced more parasitoid females than eggs of *E. kuehniella* and *G. mellonella*.

There are generally worldwide biological components in a mass rearing facility: the rearing host and the parasitoid. Larger facilities have these compartmentalized into smaller units for host production and parasitization; these units are then replicated in the facility according to the desired level of output. A common factor in almost all facilities is that they allocate at least two thirds their space and energy for the production of the rearing host(s) and the remainder for the parasitoid [85]. Development of rearing and application of *Trichogramma dendrolimi* for controlling the citrus insect pests are discussed by Niu et al. [86]. Their results led to a significant reduction in needing of classical pesticides for the control some citrus insects.

5.2 Mass Production of Other Egg Parasitoids

Eggs of *Diaprepes* root weevils could be obtained from its adults which may be kept in 30 × 30 × 30 cm Plexiglas cages with water and foliage of *Conocarpus erectus* as a food source and an oviposition substrate [87]. Strips with eggs of the *Diaprepes* root weevil are removed daily from the cage and either hung inside a similar cage containing a colony of *Q. haitiensis*, water and smears of honey [87]. Eggs are removed from the ovipositing cage or chamber after approximately 3 days, the strips are opened and the parasitized eggs exposed to facilitate parasitoid emergence.

Conocarpus erectus leaves containing *D. abbreviatus* eggs are positioned into a 30 × 30 × 30 cage in a room held at 27 °C, 12:12 L:D. Adults of *A. vaquitarum* are inserted inside the cage and providing honey and water. After 4–5 days, parasitized eggs are removed from the cage and placed in emergence containers for almost 2 weeks [88, 89].

Fidiobia citri was reared successfully on *Pachnaeus* eggs, producing ca 11.6 parasitoids/egg mass, versus 0.34 parasitoid wasps got from *D. abbreviatus* eggs. Accordingly, it does not act using *D. abbreviatus* effectively as a host for rearing *F. citri* [90]. Moreover, trials exposing eggs of *Diaprepes* in citrus groves across the state of Florida to determine possible native egg parasitoids provided no parasitism [90].

5.3 Rearing Host

Two major biological aspects of host rearing are the species to use (including artificial host eggs) and whether the eggs can be stored to extend the production period of the facility. To date, the consequential choice in host rearing has been limited to species that produce either small or large eggs. Flanders [69] originally proposed a small host egg of *Sitotroga cerealella*, and producers in slightly more than half the countries today use this species. Several countries, including France and Canada, have switched to *Ephestia kuehniella*, because of better production from the rearing medium, ease in mechanization, and improved sanitation conditions. The third small egg species, the rice meal moth, *Corcyra cephalonica* (Stn.) is used in various Asian countries because of its local availability. However, *Trichogramma* emerging from either *Ephestia* or *Sitotroga* are equivalent in the field. Somewhat better performance has been noted in the laboratory by parasitoids from *Ephestia*, possibly because of their slightly larger size [14, 91, 92].

A relatively recent development in *Trichogramma* rearing is in vitro production on artificial host media [93]. This area has been researched in since 1975 [31, 94]. Two approaches have been taken. In the first approach, the natural insect egg hemolymph is partially replaced with egg yolk and milk solids. In the second approach, a completely artificial diet is created from biochemical analysis of the insect and its egg [95]. Eighteen species of *Trichogramma* have now been reared from egg to adult in various forms of artificial media [93]. The closest system to commercial production is that developed for *T. dendrolimi* on the basis of insect hemolymph [60, 93]. This diet has been packaged in plastic host egg-cards (produced at 1200 egg-cards per hour), and the resultant parasitoids have been used on more than 1300 ha with parasitism equal to parasitoids from natural host eggs [60, 96]. The development of completely artificial hosts is an important goal and, when realized, will lead to major reductions in the size of facilities, the cost of the product, and changes in the strategy for implementation in the field.

An essential part of producing the rearing host is some means of storing the eggs to ensure a continuous supply; sterilization and cool temperatures are the most common features. Sterilization increases the storage and flexibility of unparasitized small host eggs and is achieved by either cold storage or freezing for short periods of time or by irradiation using ultraviolet or gamma sources. Bigler [32] reported a maximum storage of four weeks for irradiated *Ephestia* eggs held at 2 °C and 90% RH; Vieira and Tavares [97] suggested that eggs still can produce high-quality parasitoids after storage at 0.7 °C and 60% RH in the dark for up to 3.5 months. A more promising approach to long-term storage is liquid nitrogen for all host egg types.

High-quality parasitoids have been produced both in China, where eggs of both silkworm and *Corcyra* have been stored from 8 to 30 months in liquid nitrogen and have produced high-quality parasitoids, and in USA, where eggs of *Sitotroga* have been stored for 21 days [98].

5.4   Factitious Hosts Selection

5.4.1   *Corcyra cephalonica* (Stainton)

Amongst the parasitoids, as mentioned, *Trichogramma* species are the best available ones which could be mass produced with comparative ease. Factitious host election is the main factor to be considered first. In Indian conditions, the rice moth *C. cephalonica* (Fig. 4), is the best host which is accepted by most *Trichogramma* species except a couple of new species discovered recently, described by Nagaraja and Prashanth [99]. But a problem encountered in this is the caterpillars' facultative cannibalistic as well as predaceous habits. In *Trichogramma* rearing using egg cards the unparasitized eggs left on the card hatch and commence feeding on other eggs, both fresh as well as parasitized ones, hence becoming facultative cannibalistic as well as predaceous; by this way, the entire parasitoid culture might be lost. In order to avoid this, experiments were conducted earlier for killing the developing embryos of

(a)   **(b)**

Fig. 4 Rearing of Rice meal moth, *Corcyra cephalonica* (Stainton): *C. cephalonica* reared on rice grains (**a**), eggs collected (**b**) (cited from Perveen and Sultan [100])

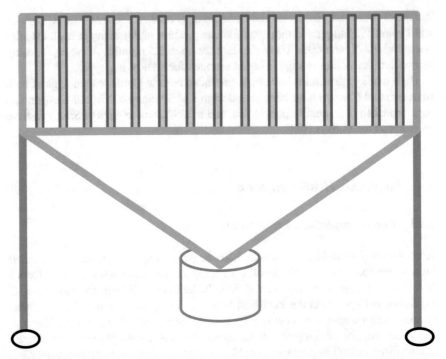

Fig. 5 Rearing unit of grain moth, *Sitotroga cereallela* (Olivier). Designed by El-Wakeil

the host egg without affecting the shape and nutritious contents by exposing them to (1) high temperatures, (2) freezing by keeping them in the freezer, and (3) irradiating under UV light. The UV irradiation for 10–45 min was found to be the safest and the eggs remained as fresh as ever and are readily parasitized [100].

5.4.2 *Sitotroga cerealella* (Olivier)

In fact, this is a safer factitious host, unparasitized larvae of which would hatch but they are not cannibalistic/facultatively predaceous like *Corcyra*. This is the host insect which is bred on a massive scale producing several kilograms of eggs every day in Russia and India as well as in Egypt (Fig. 5); it is used for mass production. Automation in breeding of *S. cerealella* is rather more practical than *Corcyra* [14, 100].

5.4.3 *Ephestia kuehniella* Zeller

This host is also mass bred and used for *Trichogramma* mass breeding and releasing against various horticultural and agricultural pests in many European countries, the

Fig. 6 Rearing facilities of *Ephestia kuehniella*. Photograph by El-Wakeil

USA and middle east [14] (Fig. 6). St-Onge et al. [101] used also *Ephestia kuehniella* as hosts for *Trichogramma* parasitoids mass breeding; *E. kuehniella* eggs must be sterilized to prevent larvae from emerging and eating the unhatched parasitized eggs. Three sterilization methods were examined: UV irradiation, freezing at −15 °C and vitrification (liquid nitrogen submersion). Vitrification resulted in significantly lower parasitoids production with a global emergence rate of 28.7%, compared to UV irradiation (75.1%), freezing at −15 °C (77.4%) and control (80.9%). Host eggs sterilization method did not affect sex-ratio, occurrence of malformation in adults, and female walking speed. Fecundity was significantly reduced in the females emerging from UV irradiated (37.2 offsprings) and vitrified (36.9 offsprings) eggs, compared to control (43.1 offsprings [101].

Mass production of natural enemies mainly egg parasitoids is obligatory to ensure the obtainability of biocontrol agents for suppressing the key insect pests. Many factitious hosts need to be sterilized to avoid development their embryos as confirmed by St-Onge et al. [101] in additional it allows their storage for a longer period. *Ephestia* or *Sitotroga* eggs are frequently used as hosts for *Trichogramma* wasps but they must be sterilized for avoiding larvae emerging. Sterilization or cold methods of *Ephestia*eggs didn't affect wasp efficiency, and didn't occurrence of malformation in adults [101, 102].

5.5 Conditions Required for Mass Production

The mass-production laboratory should have well-ventilated premises, which should also have facility for fixing air conditioner, heater and humidifier, if required, especially in the regions having extremes of weather conditions. The air conditioning becomes useful to protect the breeding stock from extreme cold as well as extreme heat. Otherwise, the productions should be carried out under normal conditions without air conditioning [70].

If *Corcyra* is being mass produced, it should be protected from larval parasitoids *Bracon hebetor* (Say) (Hymenoptera: Braconidae) and pupal parasitoids like *Antrocephalus*. For this, the doors must be provided with antechamber with fine 0.5 mm steel wire-mesh doors and the windows with additional mesh of the above size, which should always be kept closed. However, the conditions explained for *Corcyra* are good enough to protect this culture also. The common intruders in both cultures are the predatory mites. These should be tackled using some acaricide, safe to the host as well as the parasitoid cultures [55].

For manual collection of moths, vacuum cleaners fitted with hose for sucking moths are used. The hose used for moth suction is connected to the wire-mesh moth collecting and egg-laying cylindrical cage within airtight plastic container. This container is connected to the suction of vacuum cleaner. The egg-laying moth cylinders are kept on trays of suitable dimensions for collecting eggs. This facilitates egg laying as well as egg collection by brushing the exposed paper/cloth strips. The moth collection apparatus holds good for both the above hosts. When *Corcyra* is used as the laboratory host, the eggs have to be UV treated for 10–30 min to arrest the development so that no larva hatches and at the same time the eggs remain fresh enough to attract parasitism. This sterilization is not necessary for *Sitotroga* since the larvae hatching from unparasitized eggs are not cannibalistic as well as facultative predators as in *Corcyra* [14, 100].

5.6 Parasitoid Rearing

To ensure high product quality and to avoid contamination, facilities usually rear only a few parasitoid strains or species at any given time. Improved techniques for identifying different populations rapidly by using DNA markers are being developed, and their integration into facilities may help screen for such rearing problems [103]. The ratio of the number of parasitoids to host eggs in the parasitization units is also important. High ratios may lead to superparasitism, high numbers of male progeny, and poor product quality, whereas low ratios may result in poor parasitization and inefficient use of host eggs [104]. The acceptance and allocation of offspring in host eggs by *Trichogramma* are influenced by the density of the host [105], and parasitoid fecundity or clutch size is adjusted according to host availability relative to abundance [106, 107], host egg size [108], and spacing between eggs [109]. Ratios of females to small host eggs of 1:10 are often used to maintain parasitism of 70–80% and sex ratios of 50–80% females in rearing facilities [70].

Once uniform parasitism of the host eggs has been achieved through manipulation of lighting and temperature, their emergence must be programmed. Programming can be as simple as allowing the parasitoids to complete development at a specified temperature and photoperiod [110] or it may involve more complex manipulation of environmental conditions to achieve synchronization, long-term storage and delayed emergence. In general, storage at low temperatures (6–12 °C) during the pupal stage is considered best for *Trichogramma* [111, 112], although such storage has never extended much longer than two weeks without losses in parasitoid quality [113]. Those species that are more cold hardy; *T. brussicue, ostriniae, evanescens* and *dendrolimi* and/or undergo diapause (initiated by temperature and photoperiod effects on the maternal generation and developing larvae) can tolerate longer storage [104]. The specific conditions that promote parasitoid storage and diapause are being pursued actively to allow rearing facilities to economize and maintain the genetic quality of their stock better. Other factors that may affect the spread of emergence include superparasitism and intrinsic competition [114].

The maintenance of parasitoid quality is critical to the reputation of a production facility, and the quality may be compromised after rearing the parasitoid for many generations under uniform conditions and on an atypical host. Two important changes can occur: loss of tolerance to natural physical extremes and loss of preference for the target host. The first change has been rarely studied, although rearing the parasitoid under fluctuating temperatures has been recommended to maintain tolerance. Unfortunately, this recommendation is difficult to implement in a commercial rearing facility [115].

Brassicas are major vegetable crops which it needs increasing steadily natural enemy production. Therefore, efforts to develop biological pest management strategies for brassica vegetables have been conducted. The successful examples of *Trichogramma* application with other IPM management practices in brassicas show the great potential of enhancement biological control to reduce chemical pesticide input and improve vegetable production [77, 116].

The loss of preference for the target host is a controversial area, because this effect has been demonstrated for some *Trichogramma* species [46, 117], but not for others [118]. Approaches taken to counter this effect include the setting of maximum limits for the number of generations that can be reared in the facility, periodically switching the parasitoids to different hosts, or both. The first approach is used in France, where 100 female *T. brassicae* are collected annually and maintained in isofemale lines for three generations and then mixed together for a maximum of 20–25 generations [119]. In Switzerland, *T. brassicae* is reared for a maximum of six generations on *Ephestia;* if kept longer, it is switched to the target host [32]. This switching to the target or any factitious host is also recommended in Germany and Australia [120], although parasitization problems can occur in the initial generation [118].

Standardized biological and behavioral tests to monitor quality in rearing facilities had been studied. Molecular markers may help identify genetic shifts in populations

[121]. Whether quality changes occur in continuous rearing must be determined to support recommendations for rejuvenating commercial cultures routinely [122].

5.7 Adult Food

In case of parasitoids, the adult food is usually honey. The honey solution may be mixed with 5–10% of vitamin E, since it helps maintaining the insect vigor. It may also be necessary to use 10% yeast solution as additional food for lengthening adult life [123].

5.8 Body Size of the Progeny

Body size of an individual species plays an important role in fecundity, longevity, and searching ability. Normal size of progeny is an important factor which is obtained with optimum availability of food and space during development in a host egg [117]. The female tends to super parasitise if hosts are not adequate, resulting in either no emergence or emergence of progeny of small size. It is therefore a must to provide enough host eggs. Body size of offspring depends on the number of individuals developing in a single host egg. The number of developing individuals in a host egg naturally depends on the size of the host egg. The 'optimum offsprings' meant the ones developing to normal size adults with normal longevity and parasitising capacity [71, 124].

5.9 Storage

Storage of both host eggs as well as parasitized ones is a requirement to manage commercial or otherwise supply of parasitoid material in times of need. The supplies on demand are not likely to be uniform throughout the year. Therefore, an effective storage system which ensures no deterioration of the quality of the product material has to be evolved [125]. The methods used in some countries are explained hereunder.

Storage of host eggs as well as parasitized eggs without deterioration of the quality for a long time is equally important in quality control measures. Liquid nitrogen is an important medium in which the live and parasitised eggs are stored for long periods. Cocoons of silk worms were stored under −5 °C for up to 2 months after they were held for 24 h at 15 °C for up to 2 months. Thus, it was reported that the whole pupae with the cocoons could be stored at −3 to −5 °C for 3–5 months. Fresh eggs of silkworms could be stored in ice below −10 °C for 6 months. The *Trichogramma* reared by adopting storage systems were *T. dendrolimi* Mats., *T. chilonis* Ishii, and *T. closterae* Pang and Chen [124, 125]. Induced diapause during off seasons of market activities is also a method used for prolonging the life of *Trichogramma*. Garcia et al. [126] used a technology under very low temperatures to induce diapause in *Trichogramma cordubensis*.

In Egypt, Abd El-Gawad et al. [127] evaluated the impact of cold stored pupae of *T. evanescenes* which mass reared in the eggs of *S. cereallela* and its pupae stored at 4, 7 and 10 °C for 1–8 weeks. The results showed that these two factors of storage temperatures and time substantially were negatively influenced produced adults and F1. However, Chen and Leopold [71] assessed the effects of refrigerated storage on the suitability of eggs of the glassy-winged sharpshooter, *Homalodisca coagulata* (Say), as hosts for propagation of the parasitoid *Gonatocerus ashmeadi* Girault (Fig. 7a). Moreover, Bayram et al. [128] investigated the effect of cold storage of both pupae and adults of *Telenomus busseolae* Gahan (Fig. 7b). Storage had a significant adverse effect on mean adult emergence. *T. busseolae* could be stored for 4 weeks at 4 and 8 ± 1 °C but only 2 weeks storage was possible at 12 ± 1 °C. Percent survival was reduced with longer storage times.

In their studies Zhang et al. [129] investigated the biological parameters of diapaused, non-diapaused, and cold-stored *T. dendrolimi* for two generations on host eggs of *Antheraea pernyi* under laboratory conditions. They mentioned that the non-diapaused *T. dendrolimi* had a higher emergence rate, longer longevity, and a lower proportion of deformed individuals than diapaused *T. Dendrolimi* in F1 [129]. In a 3-year augmentative field release from 2014 to 2016, diapaused *T. dendrolimi* showed effective parasitism on eggs of the *Ostrinia furnacalis*. These results indicated that diapaused *T. dendrolimi* can be an efficient alternative method for mass rearing of *T. dendrolimi* for long-term cold storage.

(a)   **(b)**

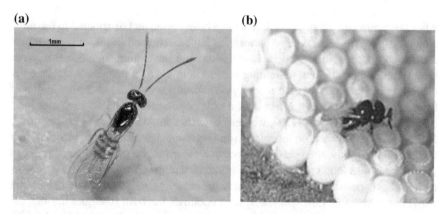

Fig. 7 a Adult of *Gonatocerus ashmeadi* (Girault), http://www.nhm.ac.uk/resources/research-curation/projects/chalcidoids/images/chalc532.jpg. **b** Adult of *Telenomus busseolae* (Gahan), https://www.futura-sciences.com/planete/dossiers/zoologie-insectes-secours-environnement-801/page/4/

5.10   Evaluation of Quality

Singh et al. [130] have listed out the quality parameters for trichogrammatids, which include clear species identification, per cent parasitism and percent emergence ≥ 90%, percent females 50% or more, and one card to contain 16,000–18,000 eggs.

There is a need for regulatory agencies to periodically evaluate the quality of trichogrammatids produced by insectaries based on simple and reliable methods. *Trichogramma* producers should reach a common consensus on the method of producing egg cards and also the recommendations to be given to farmers. This would ensure the proper utility of trichogrammatids in biocontrol programmers. Ballal et al. [78] assessed the post-shipment quality (measured in terms of some biological parameters) of *T. chilonis* supplied by nine production units in India. Considering the recommendation of 150,000 parasitized eggs per ha of cotton, only three units could reach the quality standards, while the parasitizing efficiency of *T. chilonis* supplied by two units was less than 50%. Based on their evaluation of egg cards supplied by different units in South India, Romeis et al. [131] observed that due to the low product quality, up to 85% fewer parasitoids were being actually released than claimed by the producer.

6   Strategies in the Field

6.1   Approaches for Release

After large numbers of egg parasitoids have been reared and distributed to the field, they are ready for release. Egg parasitoids have been used in all three biological control approaches: introduction, inoculation, and inundation. Inundative releases achieve an immediate, non-sustaining reduction in the host population. In inoculative releases, however, it is the progeny of parasitoids released at the beginning of the season that have a later effect on the host population. Inundative releases have predominated in the West [132], whereas countries in Asia and parts of the former USSR have put emphasis on inoculation and occasional introductions [115]. Warmer climates favor multivoltine pests and inoculative releases, because the parasitoids can multiply during a long growing season. Inundative releases, which are timed specifically to the ovipositional period of the pest, are more appropriate in northern climates with uni- or bivoltine host species. Several countries use both strategies through the repeated annual applications of *Trichogramma* and other egg parasites [44, 115, 133, 134].

The different approaches to the use of egg parasites have resulted in two different perspectives: the inundative approach, which tends to view the parasitoid as a fast-acting replacement for chemical insecticides, and the inoculative/introduction approach, which sees the parasitoid as one aspect of integrated pest management. Considerably less experimental research has gone into inoculative releases than into inundative releases because of the ecological complexity involved and lack of funding. This situation is unfortunate, because the few integrated studies with *Trichogramma* and microbials, such as *Bacillus thuringiensis* (Bt), *Nosema,* and *Beauveria,* have been positive [135]. In Egypt, El-Mandarawy et al. [136] evaluated the efficiency of biological control agents against the *O. nubilalis* Hübner, using inundative releases of the egg parasitoid, *T. evanescens* alone or integrated with *Bacillus thuringiensis* (Bt). They found that the mean rates of parasitization reached 74.72, 76.83 and 77.23% in the plots treated with the parasitoid alone and 72.90, 74.21 and 75.56% in the plots treated with *Trichogramma* plus Bt., in 2001, and 2003 respectively.

Several authors suggest that releases of more than one species of nature enemies will improve efficacy [120, 137], and some countries have combined *Trichogramma* releases with those of other parasitoids (e.g. *Hubrobracon*) to provide acceptable levels of control for pests in cotton, tomato, and pine [115, 138]. In Colombia, releases that integrate *Trichogramma* with both Bt and *Apunteles* or *Telenomus* have reduced the use of insecticide by 50% on tomato and cassava [139].

Trichogramma releases may have an immediate effect not only on the target pest, but possibly on other insect populations as well [107]. Direct effects may be seen on non-target Lepidoptera in the crop and surrounding area, and indirect effects may be seen on the natural enemy complexes associated with them [59, 140]. Lopez and Morrison [141] showed that predators of *Helicoverpa* in cotton where *Trichogramma* wasps were released were unaffected compared with insecticide plots. The effects on non-targets will become increasingly important, not only because of the world-wide interest in biodiversity but because producers are now being asked to address this aspect before *Trichogramma* can be registered. Releasing *Trichogramma* in vegetable crops have led Chinese farmers and producers to reduce quantity of chemical pesticides which it is practiced since 1950s in greenhouses [73, 142, 143].

Mass releases of *T. dendrolimi* in apple orchards provided effective control of *Adoxophyes orana* and *Grapholitha molesta* [73]. The parasitism rate of *T. dendrolimi* stabilized in the range of 81–99% in the experimental orchards. The parasitism rates of *A. orana* eggs in the field reached to 93–94%. Chinese scientists in apple orchards have achieved great results for saving and increasing the quality fruit without chemical residue achievements. Overall, augmentative of *Trichogramma* have been highly developed for Lepidopteran, Coleopteran and Dipteran insect pests [9, 144, 145].

A control program of litchi stink bug, *Tessaratoma papillosa* by the egg endoparasitoid *Anastatus japonicus* is an example of successful classical biological control as conducted by Li et al. [146], who focused on research activities and experiences drawn from many years of experimentation and field work in an attempt to promote biological control and reduce insecticide use to produce healthier food and a safer environment. The straightforwardness of mass-rearing and easy access to high quality factitious host eggs have made it possible to control *T. papillosa* with this parasitoid [147–149].

6.2 Timing, Frequency and Rates of Release

The timing, frequency, and rates of release all depend on the approach taken. With inoculative releases, relatively few parasitoids are required very early in the season, possibly independent of the ovipositional period of the pest. In contrast, inundative releases require large numbers to be synchronized closely with the start of oviposition of a uni- or bivoltine pest [44, 115]. The earlier oviposition can be predicted, the better for the rearing facility and the field program. If large numbers of parasitoids are needed in a short time, then some facilities may require several weeks' or months' notice.

Different methods, including calendar date, plant development, pheromone or light traps, and egg-laying, have been used to synchronize inundative releases with the start of host oviposition. Plant development is the least accurate method, unless it is linked to pest phenology [150]. Light traps [151, 152] and, where available, pheromone traps [153] appear to be the best predictors, because they collect adult moths before oviposition starts. Studies that compare trap catch, oviposition, and efficacy have shown consistently that the best results are achieved when the *Trichogramma* are released a few days before rather than at the start of oviposition [153].

Synchronization with the host also means that programming of parasitoid emergence must be considered. Although most facilities ship parasitoid material ready to emerge and the majority of releases have used material emerging within hours of release, this is not always the case. Some strategies mix different stages of *Trichogramma* development, thereby staggering emergence, particularly if only one or a few releases can be made [153, 154]. This approach ensures that there are always some females actively searching throughout host oviposition [155]. In practice, this approach is limited if the released eggs are exposed to predation or extreme temperatures.

The first release is usually aimed at just before host oviposition, not only to achieve high levels of parasitism but also to enable released material to multiply in the natural host eggs. Such an approach ensures that a continuous supply of superior-quality parasitoids is produced from the target pest in the field [32]. This field multiplication reduces the need for many releases and is the basis for some of the more successful programs [156]. The goal of release is to maintain a level of more than 80% parasitism on freshly laid host eggs [157]. When single releases and field multiplication do not achieve this goal, then multiple releases at various frequencies are used. Few studies have actually compared different frequencies of similar final release rates [132, 153].

Similarly, the actual rates of release vary considerably, even for the same pest, crop, and country. For example, the total rates of release for *T. brassicae* alone, which is reared from small host eggs against Os*trinia* in Europe, range from 150,000 to 2.8 million wasps/ha. Rates in the several millions of wasps/ha are generally cited in arboreal situations, such as forestry [153], and in fruit or nut orchards [48, 156], whereas those in agricultural crops, such as corn, cotton, and tomato, range from 500,000 to 1,000,000 wasps/ha, with averages of 200,000–600,000 wasps/ha

(a)   (b)

Fig. 8 Rearing of *Trichogramma*, card with 20,000 eggs (**a**) and parasitized eggs in card ready for release (**b**) (Biological control laboratory, NRC, Egypt). Photograph by El-Wakeil

(Fig. 8). This range is probably related to the range in dimensional volume of the crop [158].

Much of the confusion in application rates results from the inconsistency with which they are reported. The rearing host often is not identified; and numbers may be cited per release rather than the total number per hectare. In addition, some individuals refer to numbers of host eggs, whereas others refer to numbers of parasitized (black) eggs (Fig. 8b), wasps/parasitoids, or females per hectare. To compare these values, assumptions must be made regarding parasitoid size and quality, parasitism rates (usually 60–70%), emergence rates (usually 80–90%), and sex ratio (usually 50–70%). This lack of uniformity makes it difficult to compare studies and provide specific information on which rate should be used [70, 159].

The reason why different rates of application are selected is not usually given, although an expected ratio of female wasps: hosts on the order of 1:2 or 1:10 for egg masses and 1:20 for eggs and the volume of the crop are often starting points [70, 160]. More emphasis has been placed on establishing the correct spacing of points for given release rates or their vertical location within the crop. Higher rates generally result in better parasitism. This is not always the case, however, as many factors influence the outcome of the release [123, 161].

6.3   Factors that Affect Release

There are many factors affect the field release: weather, the crop, host, predation, use of pesticides and parasitoid quality all influence the release and disappearance rate of *Trichogramma* [162].

6.3.1   Weather

Weather is probably the most pervasive, in that it is a complex of metrologic variables that affect the development, emergence, survival, activity, and fecundity of parasitoid. The most influential components are temperature and humidity; in the extreme, both these components have been linked to poor field results [162, 163]. Parasitoid development is directly related to temperature; thus, extremes in the field disrupt not only survival and performance but also emergence [164]. Rare field studies suggest that *Trichogramma* avoid dew extreme temperature [60, 70, 165], areas of bright light intensity, heavy rain, and winds greater than 1.1 km/h [60, 166]. Thus, if inclement weather cannot be avoided, the rate and frequency of release must be adjusted upward, with specific regard to changes in the pattern and extent of emergence.

6.3.2 Crop

As the bottom component in the tritrophic interaction and the principal factor in habitat location, the crop is another important variable for egg parasitoids as it has both physical and chemical effects (Tables 1, 2 and 3). Different levels of parasitism can be found on the same host in different crops, and *Trichogramma* and *Telenomus* are much more habitat-specific than host-specific [165, 167]. Because *Trichogramma* are thought to search for host eggs randomly, parasitism is directly proportional to the size of the plant [168], its surface area, and the complexity (number of planes

Table 1 Mean of gland and hair density of leaves (1st, 3rd and 5th) on 3 cotton cultivars[a]

Cotton cultivar	Gland density of leaves	Hair density of leaves
	No. of gland/Ø cm leaf	No. of hair/Ø cm of leaf
Giza86	<40 (less)	>200 (high)
Giza89	40–80 (middle)	50–200 (middle)
Alex4	>80 (high)	<50 (less)

[a]Cited from El-Wakeil [47]

Table 2 Mean (±SE) parasitism rates of *T. pretiosum* and *T. minutum* on *Helicoverpa* eggs on different leaf positions of three different cotton cultivars in the greenhouse[a]

Cultivar	Leaf position	Parasitism rate[b]	
		T. pretiosum	*T. minutum*
Giza 89	1	84.3 ± 3.1 ab	82.2 ± 2.3 ab
	3	81.0 ± 4.6 abc	79.5 ± 6.4 bc
	5	90.6 ± 2.1 a	86.7 ± 1.9 a
Giza 86	1	80.1 ± 2.0 bc	78.1 ± 2.3 bc
	3	77.4 ± 3.9 c	75.7 ± 5.6 c
	5	87.2 ± 2.3 a	81.7 ± 1.3 abc
Alex 4	1	86.7 ± 3.2 a	84.8 ± 2.8 ab
	3	83.1 ± 5.3 ab	81.3 ± 6.5 abc
	5	92.8 ± 2.2 a	89.8 ± 2.1 a

[a]Cited from El-Wakeil [47]
[a]Different letters indicate significant differences

Table 3 Correlation values (r) between parasitism rates and trichome densities and number of glands on different cotton cultivars[a]

Cultivar	Leaf position	Trichome		Glands	
		T. pretiosum	T. minutum	T. pretiosum	T. minutum
Giza 89	1	−0.41	−0.44	0.39	0.35
	3	−0.35	−0.28	0.28	0.24
	5	−0.51	−0.48	0.52	0.51
Giza 86	1	−0.63	−0.60	0.36	0.31
	3	−0.53	−0.51	0.24	0.19
	5	−0.59	−0.62	0.49	0.46
Alex 4	1	−0.43	−0.40	0.46	0.42
	3	−0.38	−0.32	0.37	0.35
	5	−0.48	−0.43	0.69	0.62

[a]Cited from El-Wakeil [47]

and angles) of the plant and its leaf surfaces [169]. From the chemical perspective, the plant provides volatile cues that, although not specific or longrange, arrest and stimulate searching and parasitism in parasitoid [170]. These factors all influence the rate of release necessary and the resulting expected level of parasitism [160, 171]. Qasim et al. [172] evaluated *Trichogramma* effectiveness, showing 43% in okra. While, infestation in the controlled field reached to 63%.

6.3.3   Host Abundance and Location

The abundance and location of the host also influence parasitoid releases. Parasitism tends to be higher in areas that have more hosts [13, 161], although some releases have not shown this response [173]. *Trichogramma* generally demonstrate either independent on tow functional responses to host density [105], with better parasitism seen in hosts that lay eggs in clusters rather than singly [174]. Most species use kairomones to locate hosts from varying distances, sizes and egg shapes on closer examination. In addition to plant cues, sex pheromones [167], chemicals on the wing scales of the hosts [11, 163], and chemicals on the surface of the host eggs [124] all delay flight initiation, suppress positive phototaxis, intensify searching, increase retention time, and decrease speed of movement. Releasing kairomones with parasitoid has given positive in cotton [175].

6.3.4   Predation

If emergence of *Trichogramma* is delayed in the field, then high losses may occur through predation. Depending on the diversity and location of the crop system, major predators include *Geocoris, Nabis, Orius, Hippodamia, Coleomegilla, Chrysopa,* ants, spiders, and small vertebrates. Predation of released material can be significant, with losses up to 50% in corn [123, 151] and 91–98% in cotton [175]. Studies suggest that anthocorid predators are more likely to accept unparasitized host eggs than those that contain pupae of *Trichogramma*, although younger stages of *Trichogramma* are equally susceptible. It is difficult to predict which predators will be the most significant in a system, as this aspect depends extensively on the climate, structure of the plant community, and cultural practices. Attempts to reduce predation have been made by using specialized wax capsules for release points. The effect of predation on *Trichogramma* efficacy releases merits increased focus, as it significantly influences cost and success of releases.

6.3.5 Pesticides

Pesticides also have been shown to reduce the effect of *Trichogramma* significantly [176]. Many studies have compared the relative toxicity of pesticides, including insecticides, fungicides, and herbicides, to *Trichogramma* in screening trials [177–179]. In general, parasitoids are affected more by insecticides than by the other two groups, with the greatest mortality of adult *Trichogramma* seen 5–10 days after the use of selective pesticides, 15 days after moderately toxic pesticides, and 20–30 days after toxic pesticides [177]. *Bacillus thuringiensis* does not appear to affect parasitism when fed to adult wasps but can reduce parasitism when applied to the surface of host eggs [122, 180, 181].

6.3.6 Parasitoid Quality

Parasitoid quality is the final component that affects releases. Quality (longevity, fecundity, and searching capacity) can be increased two to ten times by providing a food source to adult wasps [115, 173, 182]. In the field, this food source may be obtained from host feeding, nectar and plant fluids of damaged leaves [165]. Although we know little about plants as nurseries or refugee for *Trichogrummu,* several countries use plantings of nectariferous plants successfully, either in the fields or in adjacent areas [31, 115]. If these plants are not widely available, one way of improving parasitoid quality is to supply a sugar source (e.g. honey or molasses) with the parasitoids on release [123]. This approach has also been proposed in rearing facilities, although neither approach has been undertaken commercially.

Since the right food source is very important for the reproduction of egg parasitoids, in order to reach a high parasitism efficacy. Nonetheless, adult *T. laeviceps* need a sugar-rich food source to increase their parasitation performance; or the farmers should grow flowers as nectar resources for them in fields. Barloggio et al. [183] found that flowers attracted more parasitoids than in fields without flowers. Also fecundity and the number of female offspring produced were higher for females kept on these flowers. It is found that the existence of food resources such as flowering plants, can has encouraging effect on search ability and parasitism rate. Diaz et al. [184] mentioned that presence of flowering plants play a vital role on the longevity and parasitism rate of *Trichogramma*; i.e. red clover (*Trifolium pratense* L.) increased the longevity and parasitism rate of *T. atopovirilia* in Colombia, suggesting that so long as food resources to parasitoids should be part of a habitat diversification strategy to control noctuid pests [184]. The flowering plants enhance *Trichogramma* efficiency. Géneau et al. [185] conducted experiments to find selective plants which improve the efficacy and longevity of wasps that attacking the lepidopteran pests in Switzerland as well as in USA [186].

The selection for increased parasitoid quality, such as fecundity and tolerance to environmental extremes and pesticide residues, is important to the success of releases. Lopez and Morrison [141] found that continuous rearing under variable temperatures and light regimes did not produce more heat-resistant parasitoids; however, selection studies to improve tolerance to heat of *T. pretiosum* [111], fecundity in *T. brassicae*, and parasitism at 15 °C in *T. minutum* have all been positive [187].

6.3.7 Disperse of Parasitoid

The dispersal or movement of parasitoid is important to releases from two aspects: uniformity of parasitism within a crop and reduction caused by dispersal outside the crop. These tiny wasps actually have two modes of dispersing: either on their own or through phoresy on adult moths. Greenhouse studies suggest that there may be an early migratory phase for *T. evanescens;* on emergence, it responds to light by flying upward instead of searching for hosts [188].

The ability of *Trichogramma* to disperse on its own appears to be high within single plants but low between plants [173], which suggests that the parasitoids avoid open areas where they lose their ability for directed flight. Most studies show that parasitism decreases with distance from the release point with distances of 4–50 m [38, 173]. Upwind dispersal is usually impossible [65]; thus, most studies find significantly greater parasitism downwind [173]. Although significant losses of 60–75% by parasitoids dispersing out of fields have been cited [189], this occurrence generally is not considered a major factor. Vertical movement within the plant usually is related to the location of host eggs and the release point [60].

7   Role of Egg Parasitoids in Controlling Different Insects

The farmers suffered severe attacks by both native and invasive insect pests as reported by Zhang et al. [190]. Yang et al. [143] conducted many research projects on biological control of some insect pests. Mass rearing and release techniques have been applied and developed of *Dastarcus helophoroides*. By releasing this parasitoid, 92.6% of *Monochamus alternatus* were parasitized.

7.1 Egg Parasitoids in Cotton

In recent times, cultivating cotton (*Gossypium barbadense* L.) has become a costly affair due to pest menace. The crop is known to suffer from 162 species of insects and mites [191]. It is attacked by various pests during the different stages of its development. The Cotton leaf worm, *Spodoptera littoralis* and the cotton bollworms, pink bollworm, *Pectinophora gossypiella* and spiny bollworm, *Earias insulana* and American bollworm *Heliothis armigera* cause the greatest part of yield losses resulted from nearly one million feddans cultivated annually [192, 193].

Augmentative release of laboratory-reared *Trichogramma* sp. egg parasitoids (Fig. 9) of PBW has shown some promise for early-season control. In large scale cotton fields, biweekly release of this parasitoid significantly reduced boll infestations during July in comparison with control plot. Parasitoid released also increased the yield by 10–13% and reduced seed damage by 22–50%. The parasitoid only attacked 7–15% of the eggs laid under the calyx later in the season, a level insufficient for pest control [194]. Egg parasitoid; however, were almost exclusively used through inundative releases to increase the parasitization rate sufficiently to reduce crop damage. *T. evanescens* was mass-reared and released from 0 to 3 times in different treatments of cotton fields. It was very successful in finding and parasitizing the eggs of host. The overall parasitism was about 24.5% on *P. gossypiella* eggs, 19.6.6% on *E. insulana* and 6.2% on *S. littoralis* [193].

In greenhouse experiments, the vertical and horizontal distribution of *H. armigera* eggs was manipulated. *T. minutum* and *T. pretiosum* differed significantly in searching behavior as measured by parasitization rates on three cotton cultivars.

Furthermore, parasitization rates were negatively correlated with distance between the releasing site and egg batches on the cotton plants. Morphological traits, i.e. black glands or trichome densities of the cotton cultivars played a significant role. The parasitization rates on cultivars with glands and lower trichome density were higher than with no glands and high trichome density [29, 47].

In India, the key pests of cotton are *Earias vittella, Earias insulana, Pectinophora gossypiella* and *Helicoverpa armigera*. Also, there are a few sporadic pests like *Spodoptera litura* and *Phycita infusella*, which sometime cause damage [195]. The exotic parasitoid, *Trichogramma pretiosum*, was introduced in Punjab, Maharashtra,

Fig. 9 Cards used to release *Trichogramma* wasps in cotton fields (El-Wakeil [29])

Haryana and Karnataka against bollworms, and their recovery was reported [196]. Out of the six species of native egg parasitoids recorded from the different parts of the country, only *Trichogramma chilonis* and *T. achaeae* were tried in fields. However, Prasad et al. [197] released *T. achaeae* and *T. pretiosum* in cotton fields against *E. insulana*, *E. vittella* and *P. gossypiella* and reported significant reduction in the damage to cotton.

The adult parasitoids of *Chelonus blackburni* Cameron (Braconidae) at the rate of 3000 adults/ha and *T. chilonis* at the rate of 200,000 adults/ha were released immediately after appearance of the eggs of *H. armigera*. In October, the parasitoids were recovered from *H. armigera* with 32.35 and 6.25%, respectively [198].

In Brazil, Bueno et al. [199] released *Trichogramma pretiosum* Riley, used to control *Anticarsia gemmatalis* Hübner, (Lepidoptera, Noctuidae) and *Spodoptera frugiperda* in cotton crops. However, another egg parasitoid, *Telenomus remus* Nixon, (Hymenoptera: Scelionidae), has been observed parasitizing eggs of five different species of *Spodoptera*, even on overlapping layer egg masses [63]. Each female of *T. remus* produces 270 eggs during its reproductive lifespan [200].

7.2 Egg Parasitoids in Sugarcane and Maize Crops

Sugarcane *Saccharum officinarum* is considered an economic crop in the tropical and subtropical areas. Successful biological control trials have been practiced in Egypt since 1984 to control the lesser sugarcane borer *Chilo agamemnon* by releasing mass reared *T. evanescens* early in the season, during May and June. The sugarcane area treated with *Trichogramma* increased gradually from 5 to 500 feddans (200 ha) in the 1989 season. The obtained results indicated that there were subsequent increases in the rate of parasitism after the release all over the growing season and up to 60% reduction in the rate of infestation by the pest at harvest time [45, 58].

The egg parasitoid *Telenomus remus* Nixon has been released in corn field areas, as part of IPM programs, in Venezuela, where the parasitism rates reached up to 90% [201], demonstrating the high potential of this control agent of several species of *Spodoptera*.

El-Wakeil and Hussein [133] evaluated egg parasitoid for controlling two corn borers (*C. agamemnon* and *O. nubilalis*) in maize fields. The parasitism percentages by *Trichogramma* were higher when using 30 cards/acre level compared to 20 cards/acre level. At season's end, the numbers of *C. agamemnon* and *O. nubilalis* larvae were significantly reduced on the *Trichogramma* release compared to on the control (Fig. 10).

Early studies were conducted using egg parasitoid *Platytelenomus hylas* (Nixon) (Hym.: Scelionidae) for controlling *Sesamia cretica* (Led.) eggs in maize fields as mentioned by Hafez et al. [202]. Ragab et al. [203] studied the parasitism rates by *P. hylas* on *S. cretica* eggs. The percentage of parasitism in eggs increased markedly and steadily as the season advances in the studied localities in *S. cretica* eggs.

Fig. 10 *Trichogramma* egg
cards are placed on maize
plants. Photograph by
El-Wakeil

7.3 Egg Parasitoids in Rice Crop

The major targets of egg parasitoids occurring in cereal crops include mainly stem
borers (*Chilo, Scirpophaga*), leaf folders (*Cnaphalocrocis*), the gall midge and
leaf/plant hopper. Among trichogrammatids commonly occurring in cereal ecosystem are the '*minutum*' group (*T. chilotraeae, T. pretiosum, T. chilonis*) and the 'japonicum' group (*T. japonicum, T. pallidiventris*). The other families of egg parasitoids
include Scelionidae, Mymaridae, Eulophidae and Platygasteridae. In rice ecosystem,
especially on *Scirpophaga incertulas, Trichogramma, Telenomus* and *Tetrastichus*
species appear to complement in natural biocontrol. Trials on inundative release of
T. japonicum and *T. chilonis* have shown positive impact on yellow stem borer and
leaf folder in rice [204].

Sherif et al. [205] used *T. evanescens* for controlling rice stem borer,
C.agamemnon. They concluded that the reductions in dead hearts and white heads
averaged 75–80% indicating that *Trichogramma* release could efficiently control in
rice fields.

7.4 Egg Parasitoids in Vegetable Crops

In vegetable ecosystem, the egg parasitoids are mainly (*Trichogramma, Trichogrammatoidea*) on Lepidoptera and scelionids (*Telenomus, Trissolcus*) on Lepidoptera
and Heteroptera, besides mymarids on leafhopper and thrips. *T. chilonis* is found
promising as natural and augmentation control agent for several lepidopteran pests.
Inundative releases of *T. chilonis, T. brasiliensis* and *T. pretosium* for *Helicoverpa*
control in tomato and okra have been demonstrated as effective. The scope of mass
rearing of *Telenomus remus* on *Corcyra cephalonica* has been demonstrated, while

(a)   **(b)**   **(c)**   **(d)**

Fig. 11 Tomato tunnelsin Egypt (**a**), Eggs of *Tuta absoluta* on tomato leaves (**b**), *Trichogramma* card (parasitized eggs) (**c**), *Trichogramma* egg cards attached into tomato tunnels (**d**). Photograph by Masry

further improvements in their efficiency are required. Utilization of semiochemicals is helping for improving the field performance of mass-released parasitoids [206].

The tomato leafminer *Tuta absoluta* (Meyrick) (Lepidoptera: Gelechiidae) has become an economically important pest in the major tomato-producing countries in the Mediterranean Basin countries of Europe and North Africa [207, 208]. Egg parasitoids of *T. absoluta* have been belonged to the order Hymenoptera. The most important *T. absoluta* egg parasitoids are found in the families Trichogrammatidae, Encyrtidae and Eupelmidae. *T. pretiosum*, *T. exiguum*, *T. evanescens*, *T. minutum* and *T. cacoeciae* are more general parasitoids, by which it's likely to parasitize a range of different species [209]. These egg parasitoids wasps have been widely used to control *T. absoluta* [207, 210–214]. Moreover, the parasitic wasp *Anastatus* sp. was reported on various lepidopteran and hemipterans in different parts of the world [215, 216]. Also, it was recorded as an egg parasitoid on or used for controlling *T. absoluta* [211].

El-Arnaouty et al. [50] compared the efficiency of two *Trichogramma* species, the indigenous *T. euproctidis* and *T. achaeae*, for controlling *T. absoluta* in Egypt. The results show that both *Trichogramma* species were significantly efficient keeping down *T. absoluta* mines. *Trichogramma* was strongly effective on *Tuta* as well as to be used in an IPM program as mentioned by Abdel Razek et al. personal communication (2019) (Fig. 11).

7.5 Egg Parasitoids of Fruit Orchrads

Among the temperate fruit crops, there is a good scope for biocontrol of the codling moth with repeated release of *Trichogramma embryophagum* while the Indian gypsy moth is subjected to varying levels of natural parasitisation by *Anastatus kashmirensis* Mathur. In tropical fruits, against lemon butterfly, both *Trichogramma chilonis* and *Telenomus incommodus* provide varying levels of natural parasitism around the year, with *T. chilonis* offering scope for augmentative biocontrol. Eggs of fruit sucking moths are found to be naturally parasitized by *Trichogramma*, *Telenomus* and

Fig. 12 *Trichogramma* card
attached to grape tree.
Photograph by El-Wakeil

Ooencyrtus. Fruit borer, *Meridarchis scyrodes* appears to be effectively controlled
by *T. chilonis* releases [217].

In Egypt, the olive moth, *Prays oleae*, and the jasmine moth, *Palpita unionalis*
are serious pests in modern olive plantations, causing significant yield loss by fruit
fall as well as by damage on leaves, flowers and fruits. The egg parasitoid species
Trichogramma bourarachae, *T. cordubensis*, *T. euproctidis* and *T. evanescens* were
released in several applications in an intensively managed olive plantation for biological control of these pests. The results indicated that larval densities of target pests
were significantly reduced up to 83% on *Trichogramma* release trees in comparison
to control trees [13, 48, 218].

El-Wakeil et al. [7] evaluated the efficacy of *T. evanescens* in controlling the
European grape berry moth *Lobesia botrana* in two vineyards in northern Egypt and
found that the release cards should be distributed in every three grape rows and on
height 130–170 cm to obtain good parasitism rates. *T. evanescens* could be a potential
candidate for controlling *L. botrana* (Fig. 12).

7.6 Using Other Egg Parasitoids for Controlling Insect Pests of Orchard Fruits

In the 1990s, there were lot of efforts for establishing egg parasitoids *Quadrastichus
haitiensis* [16]. This species is reared on *Exopthalmus quadrivitattus*, *Diaprepes
abbreviatus* and *Pachnaeus litus* eggs (Coleoptera: Curculionidae) from Port-au-
Prince, Haiti [219, 220]. It was also the most abundant egg parasitoid of citrus
weevils found in Guadeloupe [221]. Thus, releases of *Q. haitiensis* were done into
ornamental and citrus fields in Florida beginning in 2000 [16].

The most important natural enemies of *D. abbreviatus* are *Aprostocetus
vaquitarum,* which is recorded in the Caribbean Region. The tetrastichinae *A.*

vaquitarum were collected in the Dominican Republic on *Diaprepes* spp. eggs in citrus during 2000 and introduced into Florida. Succeeding, Florida University had mass-reared as well as release this parasitoid and then evaluated it in the citrus orchards.

Smith et al. [222] recorded *Fidiobia citri* (Nixon) a parasitoid of the fuller rose weevil *Asynonychus cervinus* (Boheman) in Australia. *Fidiobia citri* parasitized up to 50% of the fuller rose weevil in California [223] and this parasitoid had been collected in Florida from eggs of the blue green weevil, *Pachnaeus* spp. during 1999. *Diaprepes abbreviatus* eggs and *Pachnaeus* eggs were exposed to *F. citri* [222]. The genus *Fidiobia* was released and released against some curculionid pests [224].

Homalodisca vitripennis glassy-winged sharpshooter (GWSS) is known vector of the plant diseases [225] on grapes. The establishment of *H. coagulata* in California in the 1990s, later in Hawaii (USA) and even more lately in Chile Pilkington et al. [25], who encouraged interest to conduct studies on proconiine sharpshooter, including survey of their egg parasitoids in USA [25, 27]. *Gonatocerus* Nees is one of the mymarid egg parasitoids of *H. vitripennis* [24, 226, 227]. Morgan et al. [27] surveyed the complex of Mymaridae wasps; they mentioned that *G. ashmeadi* Girault is a dominate species in their studies. Egg parasitism rate of GWSS eggs in California was high during the summer months, sometimes reaching almost 100% [228].

8   Integration with Other Biocontrol Agents

Egg parasitoids wasps were integrated worldwide use with other biocontrol agents such as: other parasitoids, predators, pathogens and nematodes etc.

8.1   Integration with Other Parasitoids

Telenomus podisi, Trissolcus basalis, Trissolcus urichi (Hymenoptera: Scelionidae) were tested in soybean fields to evaluate the parasitism behavior on eggs of *Nezara viridula, Euschistus heros, Piezodorus guildinii* and *Acrosternum aseadum* (Heteroptera: Pentatomidae) in Brazil. For all parasitoid species, the results demonstrated the existence of a main host species that maximizes the reproductive success. Exploitative competition was observed for egg batches at the genus level (*Telenomus* vs. *Trissolcus*) and interference competition at the species level (*T. basalis* vs. *T. urichi*). *Trissolcus urichi* was the most aggressive species, interfering with the parasitism of *T. basalis*. The selection of parasitoid species for use in augmentative biological control programs should take into account the diversity of pentatomids present in soybean in addition to the interactions among the different species of parasitoids [229].

8.2 Integration with Predators

El-Wakeil and Vidal [140], in organic cotton farms, studied the combination between parasitism of *Trichogramma* species and predation of *Chrysoperla carnea* (Stephen) on *H. armigera*. The results referred that the Predation rates of *Chrysoperla* on parasitized *Helicoverpa* or *Sitotroga* eggs decreased with aging of parasitized eggs. The efficiency of *T. chilonis* combining with *C. carnea* and neem extract against *H. armigera* was investigated in tomato field. The lowest number of *H. armigera* larvae (0.68/plant) recorded in combined treatments, which was significantly lower than those of other plots [230].

8.3 Integration with Insect Pathogens

In Egypt, El-Mandarawy et al. [136] evaluated the efficiency of biological control agents against one of the main pests, infesting maize plants, the European corn borer, *Ostrinia nubilalis*, using inundative releases of the egg parasitoid, *T. evanescens*, alone or integrated with *Bacillus thuringiensis* (*Bt*) application. After two releases, with a total of 96,000 parasitoids/feddan, the mean rates of parasitism reached 77.23% in the plots treated with the parasitoid alone and 75.56% in the plots treated with *T. evanescens* plus *Bt*, respectively.

Also, Khidr et al. [231] in tomato fields, tested control methods during spring and summer plantations, 2012 for controlling *T. absoluta*. Based on reduction percentages in the number of larvae, the efficacy of the tested treatments could be descending arranged as follows *Bacillus thuringiensis* + Neem, *B. thuringiensis* + *Trichogramma evanescens* + mass trapping, *B. thuringiensis* + *Trichoderma harzianum*, *T. harzianum* + Neem, *T. harzianum* + mass trapping and *T. evanescens* + Neem.

Furthermore, El-Wakeil and Hussein [133] evaluated the combination between the egg parasitoids and entomopathogenic nematodes for controlling three corn borers (*Sesamia cretica*, *Chilo agamemnon* and *Ostrinia nubilalis*) in corn fields. Two entomopathogenic nematodes (EPNs) *Heterorhabditis bacteriophora* and *Steinernema carpocapsae* and an egg parasitoid *Trichogramma*. At season's end, the numbers of *C. agamemnon* and *O. nubilalis* larvae were significantly reduced on the *Trichogramma* release plots compared to on the control plots. The overall reduction in corn borer larvae on the treated plots using EPNs and later *Trichogramma* resulted in an increased yield compared to on the control plots. The results suggest that EPN and *Trichogramma* together can play a crucial role to control the corn borers.

9   Future Prospects

Egg parasitoids are prevalent everywhere, but the species collected from one cropping system may not work very well in another cropping system. Therefore, selection of the strain/ecotype plays a crucial role. For example, In India, six ecotypes of *Trichogramma chilonis* Ishii collected from six different parts of were compared with the laboratory strain of the same species. All the ecotypes were significantly superior to the laboratory strain in parasitizing the eggs of *Corcyra cephalonica* Stainton, *Helicoverpa armigera* Hübner and *Spodoptera litura* Fabricius and showed a distinct preference for *H. armigera* eggs [232].

Tritrophic interactions should be studied before using the egg parasitoids in a target cropping system. Parasitoids are attracted to volatile compounds released by plants in response to herbivore feeding. The release of volatile signals by plants occurs not only in response to tissue damage but is also specifically initiated by exposure to herbivore salivary secretions. Although some volatile compounds are stored in plant tissues and immediately released when damage occurs, others are induced by herbivore feeding and released not only from damaged tissue but also from undamaged leaves. Additionally being highly detectable and reliable indicators of herbivore presence, herbivore-induced plant volatiles may convey herbivore-specific information that allows parasitoids to discriminate closely-related herbivore species [233].

Egg parasitoids stored for about one week at 8–10 °C in the refrigerator was successfully and could reach 23 days without adversely affecting their mergence and parasitization efficiency [234]. Storing of excess of egg parasitoids for some period will help in regulating the inundative releases when required. *T. chilonis* could not survive above 38 °C under field condition, and therefore, it was suggested to release in tomato for the control of *T. absoluta* after the onset of monsoons when the temperature drops below 35 °C [235]. *T. chilonis*, *T. brasiliensis* and *T. pretiosum* were released for the control of *H. armigera* and *Plutella xylostella* on several crops, and the greatest reduction in the larval population (92.4%) was observed on tomato [236]. Therefore, egg parasitoids are a promising biocontrol agent for agriculture pests mainly from the order Lepidoptera insects.

Biological control using egg parasitoids is one of the most important strategies which will be the development of extension support to deliver the product to the user and allow them to get into the field in a form that can have an effect. Information regarding where, when, and how to release in different grower situations have been included with these egg parasitoids. This package, which will provide a service rather than a product alone, could come from the producers, government extension, or consulting.

Acknowledgements The authors are grateful to Prof. Mahmoud Saleh for reviewing this manuscript in the early stage. Special thanks go to the team of National Research centre library, where we found most of the requisite literature resources.

References

1. Onstad DW, McManus ML (1996) Risks of host range expansion by parasites of insects-population ecology can help us estimate the risks that biological control agents pose to non-target species. Bioscience 46:430–435
2. Jervis MA (1997) Parasitoids as limiting and selective factors: can biological control be evolutionarily stable? Trends Ecol Evol 12:378–380
3. Cônsoli FL, Parra JRP, Zucchi RA (eds) (2010) Egg parasitoids in agroecosystems with emphasis on Trichogrammatid. Springer, New York, 479 p. ISBN 978-1-4020-9109-4
4. El-Wakeil NE, Abdalla AMM, El Sebai TN, Gaafar NMF (2015) Effect of Organic sources of insect pest management strategies and nutrients on cotton insect pests. In: Gorawala P, Mandhatri S (eds) Agricultural research updates. Nova Science Publisher, New York, USA, pp 51–84
5. Parra JRP, Zucchi RA, Silveira Neto S (1987) Biological control of pests through egg parasitoids of the genera *Trichogramma* and/or *Trichogrammatoidea*. Mem Inst Oswaldo Cruz 82:153–160
6. Li LY (1994) Worldwide use of *Trichogramma* for biological control on different crops: a survey. In: Wajnberg E, Hassan SA (eds) Biological control with egg parasitoids. CAB International, Oxon, U.K., pp 37–53
7. El-Wakeil NE, Farghaly HT, Ragab ZA (2009) Efficacy of inundative releases of *Trichogramma evanescens* in controlling *Lobesia botrana* in vineyards in Egypt. Arch Phytopathol Plant Prot 81:705–714
8. Sithanantham S, Ballal CR, Jalali SK, Bakthavatsalam N (2013) Biological control of insect pests using egg parasitoids. Springer, India, p 424. ISBN 978-81-322-1181-5
9. Zhou H, Yu Y, Tan X, Chen A, Feng J (2014) Biological control of insect pests in apple orchards in China. Biol Cont 68:47–56
10. Van Lenteren JC (2000) Success in biological control of arthropods by augmentation of natural enemies. In: Gurr G, Wratten S (eds) Biological control: measures of success. Kluwer Academic Publishers, Dordrecht, pp 77–103
11. Smith SM (1996) Biological control with *Trichogramma*: advances, successes, and potential of their use. Ann Rev Entomol 41:375–406
12. Van Lenteren JC, Bueno VHP (2003) Augmentative biological control of arthropods in Latin America. Biocontrol 48:123–139
13. Hegazi E, Herz A, Hassan SA, Khafagi WE, Agamy E, Zaitun A, Abd El-Aziz G, Showeil S, El-Said S, Khamis N (2007) Field efficiency of indigenous egg parasitoids to control the olive moth *Prays oleae*, and the jasmine moth *Palpita unionalis*, in an olive plantation in Egypt. Biol Cont 43:171–187
14. El-Wakeil NE (2007) Evaluation of efficiency of *Trichogramma evanescens* reared on different factitious hosts to control *Helicoverpa armigera*. J Pest Sci 80:29–34
15. Goebel FR, Roux E, Marquier M, Frandon J, Khanh HDT, Tabone E (2010) Biocontrol of *Chilo sacchariphagus* a key pest of sugarcane: lessons from the past and future prospects. Proc Int Soc Sugar Cane Technol 27:1–8
16. Peña JE, Hall DG, Nguyen R, McCoy CW, Amalin D, Stansly P, Adair R, Lapointe S, Duncan R, Hoyte A (2004) Recovery of parasitoids (Hymenoptera: Eulophidae and Trichogrammatidae) released for biological control of *Diaprepes abbreviatus* in Florida. Proc Int Citrus Congr 3:879–884
17. Mahalakshmi K, Manickavasagam S, Rajmohana K (2012) Distributional records of genera of scelionid egg parasitoids from south India. Madras Agric J 99:576–579
18. Anwar PT, Zeya SB (2012) Record of some species of Mymaridae from different states of India (Hymenoptera: Chalcidoidea). Bionotes 14:52–53
19. Noyes JS (2008) Universal chalcidoidea database. http://www.nhmac.uk/entomology/chalcidoids/index.html. Accessed 23 Apr 2008

20. Notton D (1998) Platygastroidea. In: Taxonomy and biology of parasitic hymenoptera. Department of Entomology, The Natural History Museum, London and the Department of Biology, Imperial College, University of London, 18–25 Apr 1998

21. Buhl PN (1998) On some new or little known NW European species of Platygastridae (Hymenoptera, Proctotrupoidea). Fragmenta Entomol 30:295–334

22. Evans GA, Peña JE (2005) A new Fidiobia species (Platygastridae) reared from eggs of *Diaprepes double*irii (Curculionidae) from Dominica. Florida Entomol 88:61–66

23. Guedes JVC, Parra J, Loiacono M (2001) Parasitismo natural de posturas de curculionideos da fraiz dos citros por Fidiobia spp. (Hym: Platygastrioidea). VII Simposio de Control Biologico 2001. MG, Livro Resumos, Poços de Caldas, p 340

24. Triapitsyn SV (2006) A key to the Mymaridae (Hymenoptera) egg parasitoids of proconiine sharpshooters (Hemiptera: Cicadellidae) in the Nearctic region, with description of two new species of Gonatocerus. Zootaxa 1203:1–38

25. Pilkington LJ, Irvin NA, Boyd EA, Hoddle MS, Triapitsyn SV, Carey BG, Jones WA, Morgan DJW (2005) Introduced parasitic wasps could control glassy-winged sharpshooter. Calif Agric 59:223–228

26. Vickerman DB, Hoddle MS, Triapitsyn S, Stouthamer R (2004) Species identity of geographically distinct populations of the glassy-winged sharpshooter parasitoid *Gonatocerus ashmeadi*: morphology, DNA sequences, and reproductive compatibility. Biol Cont 31:338–345

27. Morgan DJW, Simmons GS, Higgins LM, Shea K (2002) Glassy-winged sharpshooter biological control in California: building framework for active adaptive management, pp 140–143. In: Hoddle MS (ed) Proceedings of 3rd California conference on biological control, 162 p

28. Triapitsyn SV, Morgan DJW, Hoddle MS, Berezovskiy VV (2003) Observations on the biology of *Gonatocerus fasciatus* Girault (Hymenoptera: Mymaridae), egg parasitoid of *Homalodisca coagulata* (Say) and *Oncometopia orbona* (Fabricius) (Hemiptera: Clypeorrhyncha: Cicadellidae). Pan-Pac Entomol 79:75–76

29. El-Wakeil NE (2003) New aspects of biological control of *Helicoverpa armigera* in organic cotton production. Ph.D. dissertation, Goettingen University, Germany, 140 pp

30. DeBach P (1964) Biological control of insect pests and weeds. Chapman and Hall, London

31. Li LY (1982) Integrated rice insect pest control in the Guangdong Province of China. Enromophaga 27:81–88

32. Bigler F (1986) Mass production of *Trichogmmma maidis* Pint. et Voeg. and its field application against *Ostrinia nubilalis* Hbn. in Switzerland. J Appl Entomol 102:23–29

33. Hassan SA (1993) The mass rearing and utilization of *Trichogramma* to control lepidopterous pests: achievements and outlook. Pest Manag Sci 37:387–391

34. Newton PJ (1993) Increasing the use of trichogrammatids in insect pest management: a case study from the forests of Canada. Pestic Sci 37:381–386

35. Stinner RE (1977) Efficacy of inundative releases. Ann Rev Entomol 225:15–31

36. King EG, Bouse LF, Bull DL, Coleman RJ, Dickerson WA, Lewis WJ, Lopez JD, Morrison RK, Phillips JR (1986) Management of *Heliothis* spp. in cotton by augmentative release of *Trichogramma pretiosum*. J Appl Entom 101:2–10

37. Wajnberg E, Vinson SB (1990) *Trichogramma* and other egg parasitoidr. In: Proceedings of international symposium 3rd Sun Antonio, Texas, no 56. Les Colloques de l'INRA, Paris, 246 pp

38. Bigler F (1991) Quality control of mass reared Arthropods, Wageningen, The Netherlands. In: Proceedings of 5th workshop of the IOBC global working group. International Organisation for Biological Control, p 205

39. Farghaly HT (1974) Parasitization of corn borers eggs *O. nubilalis* and *C. agamemnon* by *T. evanescens* in Kaliubia and Alexandria Governorates of Egypt. In: Proceedings of 2nd pest control conference, Alexandria, Egypt, pp 395–406

40. Zaki FN (1985) Reactions of the egg parasitoid *Trichogramma evanescens* Westw. to certain insect sex pheromones. Z Angew Entomol 99:448–453

41. El-Wakeil NE (1997) Ecological studies on certain natural enemies of maize and sorghum pests. M.Sc. thesis, Faculty of Agriculture, Cairo University, Egypt, p 212
42. Negm AA, Temerak SA (1979) Studies on certain behavioral attributes of the oophagous wasp, Trichogramma evanescens. 1: under field conditions through new interpretations of data on parasitism of Chilo agamemnon. 2: in sugar cane fields. Assiut J Agric Sci 10:3–13
43. Abbas MST, EL-Heneidy AH, El-Sherif SI, Embaby MM (1987) On utilization of Trichogramma evanescens to control the lesser sugarcane borer Chilo agamemnon in sugarcane fields in Egypt. Bull Soc Ent Egypt 17:57–62
44. El-Heneidy AH, Abbas MS, Embaby MM (1989) On utilization of Trichogramma evanescens west to control the lesser sugar-cane borer, Chilo agamemnon Bles in sugarcane fields in Egypt. 2. Proper technique and numbers of release. In: Proceedings of 1st international conference on economic entomology, vol 2, pp 87–92
45. El-Heneidy AH, Abbas MS, Embaby MM, Ewiese MA (1990) Utilization of Trichogramma evanescens to control the lesser sugarcane borer, Chilo agamemnon in sugarcane fields in Egypt. 5—An approach towards large scale release. Trichogramma and other egg parasitoids San Antonio (TX, USA). 23-27.09.1990, INRA, Paris (les Colloques no.56), pp 187–189
46. Abd El-Hafez AA, El-Khayat EF, Shalaby FF, El-Sharkawy MAA (2001) Acceptance and preference of pink bollworm and some lepidopterous eggs for parasitism by Trichogramma. Egypt J Agric Res 79:123–132
47. El-Wakeil NE (2011) Impacts of cotton traits on the parasitization of Heliocoverpa armigera eggs by Trichogramma species. Gesunde Pflanzen 63:83–93
48. Agamy E (2010) Field evaluation of the egg parasitoid, Trichogramma evanescens West. against the olive moth Prays oleae (Bern.) in Egypt. J Pest Sci 83:53–58
49. El-Heneidy AH, El-Awady SM, El-Dawwi HN (2010) Control of the tomato fruit worm, Helicoverpa armigera by releasing the egg parasitoid, Trichogramma evanescens in tomato fields in southern Egypt. Egypt J Pest Cont 20:21–26
50. El-Arnaouty SA, Pizzol J, Galal HH, Kortam MN, Afifi AI, Beyssat V, Desneux N, Biondi A, Heikal IH (2014) Assessment of two Trichogramma species for the control of Tuta absoluta in North African tomato greenhouses. Afr Entomol 22:801–809
51. Hoffmann MP, Walker DL, Shelton AM (1995) Biology of Trichogramma ostriniae reared on Ostriniae nubilalis and survey for additional hosts. Entomophaga 40:387–402
52. Saber M, Hejazi MJ, Kamali K, Moharramipour S (2005) Lethal and sublethal effects of fenitrothian and deltamethrin residues on egg parasitoid Trissolcus grandis (Hymenoptera: Scelionidae). J Econ Entomol 98:35–40
53. Xu H-Y, Yang N-W, Wan F-H (2013) Competitive interactions between parasitoids provide new insight into host suppression. PLoS ONE 8:e82003
54. Goebel R, Tabone E, Rochat J, Fernandez E (2001) Biological control of the sugarcane stem borer Chilo sacchariphagus (lep: Pyralidae) in reunion island: current and future studies on the use of Trichogramma spp. In: Proceedings of South Africa sugar technology association, vol 75, pp 171–174
55. Harvey JA, Wagenaar R, Bezemer TM (2009) Life-history traits in closely related secondary parasitoids sharing the same primary parasitoid host: evolutionary opportunities and constraints. Entomol Exp Appl 132:155–164
56. Hassan SA (1994) Strategies to select Trichogramma species for use in biological control. In: Wajnberg E, Hassan SA (eds) Biological control with egg parasitoids. CAB International, Oxon, UK, pp 55–73
57. Harvey JA, Poelman EH, Tanaka T (2013) Intrinsic inter- and intraspecific competition in parasitoid wasps. Ann Rev Entomol 58:333–351
58. Tohamy TH (2008) Better conditions for releases of the egg parasitoid, Trichogramma evanescens for controlling the lesser sugarcane borer, Chilo agamemnon Bles, in sugarcane fields in Minia region. Egypt J Biol Pest Cont 18:17–26
59. Howarth FG (1991) Environmental impacts of classical biological control. Ann Rev Enromol 36:485–509

60. Wajnber E (1995) *Trichogramma* and other egg parasitoids. In: International symposium of 4th Cairo, Egypt, no 73. Les Colloques de l'INRA, Paris, 226 pp

61. Khafagi WE, Hegazi EM (2008) Does superparasitism improve host suitability for parasitoid development? A case study in the *Microplitis rufiventris—Spodoptera littoralis* system. Biol Cont 53:427–438

62. Pavlik J (1993) Variablity in the host acceptance of European corn borer, *Ostriniu nubilalis* Hbn. in strains of the egg parasitoid *Trichogrumma* spp. J Appl Entomol 115:77–84

63. Bueno RCOF, Parra JRP, Bueno AF (2009) Biological characteristics and thermal requirements of a Brazilian strain of the parasitoid *Trichogramma pretiosum* reared on eggs *Pseudoplusia includens* and *Anticarsia gemmatalis*. Biol Cont 51:355–361

64. Brotodjojo RRR, Walter GH (2006) Oviposition and reproductive performance of a generalist parasitoid (*Trichogramma pretiosum*) exposed to host species that differ in their physical characteristics. Biol Cont 39:300–312

65. Kazmer DJ, Luck RF (1995) Field tests of the size-fitness hypothesis in the egg parasitoid *Trichogmmma pretiosum*. Ecology 76:412–425

66. Bale JS, van Lenteren JC, Bigler F (2008) Biological control and sustainable food production. Phil Trans R Soc B 363:761–776

67. Borror DJ, Triplehorn CA, Johnson NF (1989) An introduction to the study of insects, 6th edn. Saunders College, Holt, Rinehart and Winston, Toronto/London/Sydney/Tokyo, p 877

68. El-Wakeil NE, Abdalla A (2012) Cotton pests and the actual strategies for their management control. In: Giuliano B, Vinci EJ (eds) Cotton: Cultivation, varieties and uses. Nova Science Publishers, New York, USA, pp 1–56

69. Flanders SE (1930) Mass production of egg parasites of the genus *Trichogramma*. Hilgardia 4:465–501

70. Bari MN, Rabbi MF, Choudhury DAM, Ameen M, Howlader MMA (2005) *Trichogramma zahiri* Polaszek: its development period, sex ratio and parasitism pattern on rice hispa eggs. Bangladesh J Entomol 15:45–55

71. Chen WE, Leopold RA (2007) Progeny quality of *Gonatocerus ashmeadi* reared on stored eggs of *Homalodisca coagulata*. J Econ Entomol 100:685–694

72. Wu KM, Lu YH, Wang ZY (2009) Advance in integrated pest management of crop in China. Chin Bull Entomol 46:831–836

73. Wang LX (2007) Study on the amount of *Trichogramma* released in orchards. Heilongjiang Agric Sci 6:56–57

74. Lu YH, Wu KM, Jiang YY, Guo YY, Desneux N (2012) Widespread adoption of Bt cotton and insecticide decrease promotes biocontrol services. Nature 487:362–365

75. Ba HGL, Ai HMT, Wu PE (2008) Preliminary study on biological control of cotton bollworm, *Helicoverpa armigera* with *Trichogramma* in Tulufan in Xinjiang autonomous region. China Cotton 35:17–18

76. Luo S, Naranjo SE, Wu K (2014) Biological control of cotton pests in China. Biol Control 68:6–14

77. Liu SS, Rao A, Vinson SB (2014) Biological control in China: past, present and future—an introduction to this special issue. Biol Cont 68:1–5

78. Ballal CR, Srinivasan R, Chandrashekhar B (2005) Evaluation of quality of *Trichogramma chilonis* Ishii from different production units in India. Biol Cont 19:1–8

79. Greenberg SM, Nordlund DA, Wu Z (1998) Influence of rearing host on adult size and ovipositional behavior of mass produced female *Trichogramma minutum* Riley and *Trichogramma pretiosum* Riley. Biol Cont 11:43–48

80. Fatima B, Ashraf M, Ahmad N, Suleman N (2002) Mass production of *Trichogramma chilonis*: an economical and advanced technique. In: Proceedings of BCPC conference: pests and diseases, Brighton, UK, 18–21 Nov, pp 311–316

81. Wang ZY, He K, Yan S (2005) Large-scale augmentative biological control of Asian corn borer using *Trichogramma* in China: a successful story. In: 2nd international symposium on biological control of arthropods, pp 487–494

82. Bentur JS, Kalcde MB, Rajendran B, Patel VS (1994) Field evaluation of the egg parasitoid, *Trichogramma japonicum* Ash. (Hym., Trichogrammatidae) against the rice leaf folder, *Cnuphalocrocis medinalis* (Lep., Pyralidae) in India. J Appl Enromol 117:257–261
83. Migiro L, Sithanantham S, Gitonga L, Matoka CM (2003) Evaluation of some larval feeding materials for rearing *Corcyra cephalonica* as host for Trichogrammatids in Kenya. In: Proceedings of AAIS symposium, ICIPE, Nairobi, African Journals Online, South Africa, p 90
84. Sithanantham S, Ranjith C (2010) *Trichogramma* mass production system improvement. 1. Study of moth emergence and egg production pattern in host (*Corcyra*) culture. Hexapoda 17:37–40
85. Bouse LF, Morrison RK (1985) Transport, storage, and release of *Trichogramma pretiosum*. Southwest Entomol 8:36–48
86. Niu JZ, Hull-Sanders H, Zhang YX, Lin JZ, Dou W, Wang JJ (2014) Biological control of arthropod pests in citrus orchards in China. Biol Cont 68:15–22
87. Castillo J, Jacas JA, Peña JE, Ulmer BJ, Hall DG (2006) Effect of temperature on life history of *Quadrastichus haitiensis* (Hymenoptera: Eulophidae), an Endoparasitoid of *Diaprepes abbreviatus* (Coleoptera: Curculionidae). Biol Cont 36:189–196
88. Ulmer B, Jacas J, Peña JE, Duncan RE, Castillo J (2006) Effect of temperature on life history of *Aprostocetus vaquitarum* (Hymenoptera: Eulophidae), an egg parasitoid of *Diaprepes abbreviatus* (Coleoptera: Curculionidae). Biol Cont 39:19–25
89. Jacas JA, Peña JE, Duncan RE (2005) Successful oviposition and reproductive biology of *Aprostocetus vaquitarum* (Hymenoptera: Eulophidae) a predator of Diaprepes abbreviates (Coleoptera: Curculionidae). Biol Cont 33:352–359
90. Hall D, Peña J, Franqui R, Nguyen R, Stansly P, Mccoy C, Lapointe S, Adair R, Bullock R (2001) Status of biological control by egg parasitoids of *Diaprepes abbreviatus* Coleoptera: Curculionidae) in citrus in Florida and Puerto Rico. Biol Cont 46:61–70
91. Bigler F, Meyer A, Bosshart S (1987) Quality assessment in *Trichogramma maidis* Pintureau et Voegelté reared from eggs of the factitious hosts *Ephestia kuehniella* Zell. and *Sitotroga cerealella* (Olivier). J Appl Entomol 104:340–353
92. Corrigan JE, Laing JE (1994) Effects of the rearing host species and the host species attacked on performance by *Trichgrammma minutum*. Environ Entomol 23:755–760
93. Grenier S (1994) Rearing of *Trichogramma* and other egg parasitoids on artificial diets. In: Wajnberg E, Hassan SA (eds) Biological control with egg parasitoids. CAB International, Wallingford, U.K., pp 73–92
94. Xie ZN, Xie YQ, Li LY, Li YH (1991) A study of the oviposition stimulants of *Trichogramma neustadt*. Acta Entomol Sinica 34:54–59
95. Cônsoli FL, Parra JRP (1997) Development of an oligidic diet for in vitro rearing of *Trichogramma galloi* and *Trichogramma pretiosum* Riley. Biol Cont 8:172–176
96. Liu ZC, Liu JF, Wang CX, Yang WH, Li DS (1995) Mechanized production of artificial host egg for the mass rearing of parasitic wasps. In: Wajnberg E (ed) *Trichogramma* and other egg parasitoids. Proceedings of 4th international symposium Cairo, Egypt, 4–5 Oct. INRA, Paris, France, pp 163–164
97. Vieira V, Tavares J (1995) Rearing of *Trichogramma cordubensis* on Mediterranean flour moth cold-stored eggs. In: Wajnberg E (ed) *Trichogramma* and other egg parasitoids. International symposium, 4th Cairo, Egypt. Les Colloques de l'INRA No. 73, Paris, 226 pp
98. Goulart MMP, Bueno A-F, Bueno RCOF, Diniz AF (2011) Host preference of the egg parasitoids *Telenomus remus* and *Trichogramma pretiosum* in laboratory. Rev Bras Entomol 55:129–133
99. Nagaraja H, Prashanth M (2010) Three new species of *Trichogramma* (Hymenoptera: Trichogrammatidae) from southern India. Biol Cont 24:203–209
100. Perveen F, Sultan R (2012) Effects of the host and parasitoid densities on the quality production of *Trichogramma chilonis* on lepidopterous (*Sitotroga cereallela* and *Corcyra cephalonica*) eggs. Arthropods 1:63–72

101. St-Onge M, Cormier D, Todorova S, Lucas É (2014) Comparison of *Ephestia kuehniella* eggs sterilization methods for *Trichogramma* rearing. Biol Cont 70:73–77
102. Ayvaz A, Eyüp K, Karabörklü S, Tunçbilek AS (2008) Effects of cold storage, rearing temperature, parasitoid age and irradiation on the performance of *Trichogrammaevanescens*. J Stored Prod Res 44:232–240
103. Zouba A, Chermiti B, Kadri K, Fattouch S (2013) Molecular characterization of *Trichogramma bourarachae* strains (Hymenoptera: Trichogrammatidae) from open field tomato crops in the South West of Tunisia. Biomirror 4:5–11
104. Corrigan JE, Laing JE, Zubricky JS (1995) Effects of parasitoid: host ratio and time of day of parasitism on development and emergence of *Trichogramma minutum*, parasitizing eggs of *Ephestia kuehniella*. Ann Entomol Soc Am 88:773–780
105. Reznik SY, Umarova TY (1991) Host population density influence on host acceptance in *Trichogramma*. Entomol Exp Appl 58:49–54
106. Bai B, Smith SM (1993) Effect of host availability on reproduction and survival of the parasitoid wasp *Trichogramma minutum*. Ecol Entomol 18:279–286
107. Stouthamer R, Luck RF (1993) Influence of microbe-associated parthenogenesis on the fecundity of *Trichogramma deion* and *T. pretiosum*. Entomol Exp Appl 67:183–192
108. Schmidt JM, Smith JJB (1987) Measurement of host curvature by the parasitoid wasp *Trichogramma minutum*, and its effect on host examination and progeny allocation. J Exp Biol 129:151–164
109. Schmidt JM, Smith JJB (1985) The mechanism by which the parasitoid wasp *Trichogramma minutum* responds to host clusters. Entoml Exp Appl 39:287–294
110. McLaren IW, Rye WJ (1983) The rearing, storage, and release of *Trichogramma ivelae* Pang and Chen (Hym.: Trichogrammatidae) for control of *Heliothis punctiger* Wall. (Lep.: Noctuidae) on tomatoes. J Aust Entomol Soc 22:119–124
111. Maceda A, Hohmann CL, Santos HR (2003) Temperature effects on *Trichogramma pretiosum* Riley and *Trichogramma toideaannulata*. Braz Arch Biol Technol 46:27–32
112. Gharbi N (2014) Influences of cold storage period and rearing temperature on the biological traits of *Trichogramma oleae*. Tunisian J Plant Prot 9:143–153
113. Tezze AA, Botto EN (2004) Effect of cold storage on the quality of *Trichogramma nerudai* (Hymenoptera: Trichogrammatidae). Biol Cont 30:11–16
114. Moreira MD, Dos Santos MC, Beserra EB, Torres JB, De Almeida RP (2009) Parasitism and superparasitism of *Trichogramma pretiosum* Riley (Trichogrammatidae) on *Sitotroga cerealella* (Oliver) (Gelechiidae) eggs. Neotrop Entomol 38:237–342
115. Voronin KN (1982) Biocenotic aspects of *Trichogramma* utilization in integrated plant protection. Les Trichogrammes. Les Colloques de l'INRA, INRA, Paris, pp 269–274
116. He YR, Lv LH, Chen KW (2005) Parasitizing ability and interspecific competition of *Trichogramma confusum* and *T. pretiosum* Riley on the eggs of *Plutella xylostella* (L.) in the laboratory. Acta Ecol Sinica 25:837–841
117. Hohmann CL, Luck RF (2004) Effect of host availability and egg load in *Trichogramma platneri* Nagarkatti (Hymenoptera: Trichogrammatidae) and its consequences on progeny quality. Braz Arch Biol Technol 47:413–422
118. Bourchier RS, Smith SM, Corrigan JE, Laing JE (1994) Effect of host switching on performance of mass-reared *Trichogramma minutum*. Biocontrol Sci Technol 4:353–362
119. Frandon J, Kabiri F, Pizzol J, Daumal J (1991) Mass rearing of *Trichograrnma brassicae* used against the European c m borer *Ostrinia nubilalis*. In: Bigler F (ed) Proceedings of 5th workshop of the IOBC global working group. Quality control of mass reared Arthropods Wageningen, The Netherlands. International Organisation for Biological Control, pp 146–151
120. Seymour J, Foster J, Brough E (1994) Workshop report: use of *Trichogramma* as a biocontrol agent in Australia. Coop Res Ctr Trop Pest Mgmt, Brisbane, Australia, p 54
121. Santos NR, Almeida RP, Padilha IQM, Araújo DAM, Creão-Duarte AJ (2015) Molecular identification of *Trichogramma* species from regions in Brazil using the sequencing of the ITS2 region of ribosomal DNA. Braz J Biol 75:391–395

122. El-Wakeil NE, Gaafar N, Sallam A, Volkmar C (2013) Side effects of insecticides on natural enemies and possibility of their integration in plant protection strategies. In: Trdan S (ed) Insecticides-development of safer and more effective technologies. InTech Open Science, pp 1–55

123. Yu DSK, Byers JR (1994) Inundative release of *Trichogramma brarsicae* Bezdenko for control of European corn borer in sweet corn. Can Entomol 126:291–301

124. Feng XQ, Hirai K, Kegasawa K (1989) Recent advances of *Trichogramma* utilization for controlling agricultural insect pests. Misc Publ Tohoku Nat Agric Exp Stn 9:133–148

125. Lalitha Y, Jalali SK, Venkatesan T, Sriram S (2010) Production attributes of *Trichogramma* reared on Eri silkworm eggs *vis-à-vis Corcyra* eggs and economics of rearing system. In: Blueprint for the future of arthropod rearing and quality assurance–abstracts of 12th workshop of the IOBC global working group on arthropod mass rearing and quality control (IOBC–AMRQC) held at IAEA, Vienna, 18–22 Oct, IOBC-AMQRC, pp 47

126. Garcia PV, Wajnberg E, Pizzol J, Oliviera LM (2002) Diapause in the egg parasitoid *Trichogramma cordubensis*: role of temperature. J Insect Physiol 48:349–355

127. Abd El-Gawad HAS, Sayed AMM, Ahmed SA (2010) Impact of cold storage temperature and period on performance of *Trichogramma evanescens* Westwood (Hymenoptera: Trichogrammatidae). Aust J Basic Appl Sci 4:2188–2195

128. Bayram A, Ozcan H, Kornosor S (2005) Effect of cold storage on the performance of *Telenomus busseolae* Gahan (Hymenoptera: Scelionidae), an egg parasitoid of *Sesamia nonagrioides* (Lefebvre) (Lepidoptera: Noctuidae). Biol Cont 35:68–77

129. Zhang JJ, Zhang X, Zang LS, DuWM HouYY, RuanCC DesneuxN (2018) Advantages of diapause in *Trichogramma dendrolimi* mass production on eggs of the Chinese silkworm, *Antheraea pernyi*. Pest Manag Sci 74:959–965

130. Singh SP, Murphy ST, Ballal CR (eds) (2001) Augmentative biocontrol. In: Proceedings of ICAR-CABI workshop, 29th June to 1st July 2001, Project Directorate of Biological Control, Bangalore. CABI, Wallingford, 250 pp

131. Romeis J, Shanower TG, Jyothirmayi KNS (1998) Constraints on the use of *Trichogramma* egg parasitoids in biological control programs in India. Biocontrol Sci Technol 8:289–299

132. Hoffmann MP, Wright MG, Pitcher SA, Gardner J (2002) Inoculative releases of *Trichogramma ostriniae* for suppression of *Ostrinia nubilalis* (European corn borer) in sweet corn: field biology and population dynamics. Biol Cont 25:249–258

133. El-Wakeil NE, Hussein MA (2009) Field performance of entomopathogenic nematodes and an egg parasitoid for suppression of corn borers in Egypt. Arch Phytopathol Plant Prot 42:228–237

134. Samin N, Shojai M, Koçak E, Ghahari H (2011) Distribution of scelionid wasps (Hymenoptera: Scelionidae) in Western Iran. Klapalekiana 47:75–82

135. Michele P, Alves LFA, Lozano E, Roman JC, Pietrowski V, Neves PMOJ (2015) Interactions between *Beauveria bassiana* and *Trichogramma pretiosum* under laboratory conditions. Entomol Exp Appl 154:213–221

136. El-Mandarawy MBR, Samea SAA, El-Naggar MAZ (2004) Applications of *Trichogramma evanescens* Westwood (Hymenoptera: Trichogrammatidae) and *Bacillus thuringiensis* for controlling *Ostrinia nubilalis*. Egypt J Biol Pest Cont 14:21–29

137. Zouba A, Chermiti B, Chraiet R, Mahjoubi K (2013) Effect of two indigenous *Trichogramma* species on the infestation level by tomato miner *Tuta absoluta* in tomato greenhouses in the south-west of Tunisia. Tunisian J Plant Prot 8:87–106

138. Grieshop MJ, Flinn PW, Nechols JR (2006) Biological control of Indian meal Moth (Lepidoptera: Pyralidae) on finished stored products using egg and larval parasitoids. J Econ Entomol 99:1080–1084

139. Navarro MA (1988) Biological control of *Scrobipalpula absoluta* (Meyrick) by *Trichogramma* sp. in the tomato *Lycopersicon esculentum* Mill.). Colloques de l'INRA 43:453–458

140. El-Wakeil N, Vidal S (2005) Using of *Chrysoperla carnea* in combination with *Trichogramma* species for controlling *Helicoverpa armigera*. Egypt J Agric Res 83:891–905

141. Lopez JD, Morrison RK (1985) Parasitization of *Heliothis* spp. eggs after augmentative releases of *Trichogramma pretiosum* Riley. Southwest Entomol 8:110–137
142. Zhang ZH, Chen QY, Gao LH, Wang JJ, Shen J, Zhang XY (2010) The development strategy of protected vegetable industry in China. Vegetables 6:1–3
143. Yang NW, Zang LS, Wang S, Guo JY, Xu HX, Zhang F, Wan FH (2014) Biological pest management by predators and parasitoids in the greenhouse vegetables in China. Biol Cont 68:92–102
144. Liu YS, Guo JY, Wan FH, Zheng FQ, Ye BH (2011) Biological control of fruit pests. Jindun Press, Beijing
145. Xie J, Liu X, Zhang ZK, Xu WJ, Li QY (2012) Current status and prospects of researches on mycoinsecticides. J Jilin Agric Sci 37:49–53
146. Li DS, Liao C, Zhang BX, Song ZY (2014) Biological control of insect pests in litchi orchards in China. Biol Cont 68:23–36
147. Zou HJ (2008) The effect of releasing *Anastatus* sp. to control *Tessaratoma papillosa* in the field. China Plant Prot 28:26–27
148. Qin LF, Yang ZQ (2010) Population dynamics and interspecific competition between *Trissolcus halyomorphae* and *Anastatus* sp. in parasitizing *Halyomorphahalys* (Stal) eggs. Chin Agric Sci Bull 26:211–225
149. Chen JN, Zeng XN, Chen BX, Dong YZ, Lu H (2010) Occurrence and control of geometrid moth in litchi orchard. Chin J Trop Crops 31:1564–1570
150. Thomson L, Bennett D, Glenn D, Hofmann A (2003) Developing *Trichogramma* as a pest management tool. In: Koul O, Dhaliwal GS (eds) Predators and parasitoids. Taylor and Francis, London, UK, pp 65–85
151. Bigler F, Brunetti R (1986) Biological control of *Ostrinia nubilalis* Hbn. By *Trichogramma maidis*. On corn for seed production in southern Switzerland. J Appl Entomol 102:303–308
152. Neil KA, Specht HB (1990) Field releases of *Trichogramma pretiosum* (Trichogrammatidae) for suppression of corn earworm, *Heliothis zea* (Lep.: Noctuidae), egg populations on sweet corn in Nova Scotia. Can Entomol 122:1259–1266
153. Smith SM, Carrow JR, Laing JE (1990) Inundative release of the egg parasitoid *Trichogramma minutum* against forest insect pests such as *Choristoneum fumiferana*: The Ontario Project 1982-1986. Mem Entomol Soc Can 153:1–87
154. Prokrym DR, Andow DA, Ciborowski JA, Sreenivasam DD (1992) Suppression of *Ostrinia nubilalis* by *Trichogramma nubilalis* in sweet corn. Entomol Ex App 6:485–493
155. Smith SM, You M (1990) A life system simulation model for improving inundative releases of the egg parasite *Trichogramma minutum* against the spruce budworm. Ecol Model 51:123–142
156. Hassan SA, Kohler E, Rost WM (1988) Mass production and utilization of *Trichogramma*: 10. Control of the codling moth *Cydia pomonella* and the summer fruit tortrix moth *Adoxophyes orana* (Lep.: Tortricidae). Entomophaga 33:413–420
157. Suverkropp BP, Dutton A, Bigler F, Van Lenteren JC (2008) Oviposition behaviour and egg distribution of the European corn borer, *Ostrinia nubilalis*, on maize, and its effect on host finding by *Trichogramma* egg parasitoids. B Insectol 61:303–312
158. Wang CL (1988) Biological control of *Ostrinia furnacalis* with *Trichogramma* sp. in China. In: Voegelté J, Waage J, van Lenteren JC (eds) *Trichogramma* and other egg parasites. International symposium on Trichogramma 2nd Guangzhou, PR China. LesColloques de l'INRA, INRA, No. 43, Paris, 644 pp
159. Oliveira HN, Zanuncio JC, Pratissoli D, Picanço MC (2003) Biological characteristics of *Trichogramma maxacalii* on eggs of *Anagasta kuehniella*. Braz J Biol 63:647–653
160. Andow DA, Prokrym DR (1990) Plant structural complexity and host-finding by a parasitoid. Oecologia 82:162–165
161. Smith SM (1988) Pattern of attack on spruce budworm egg masses by *Trichogramma minutum* released in forest stands. Environ Entomol 17:1009–1015
162. Firake DM, Khan MA (2014) Alternating temperatures affect the performance of *Trichogramma* species. J Insect Sci 41:1–14

163. Bari MN, Jahan MK Islam KS (2015) Effects of temperature on the life table parameters of *Trichogramma zahiri* (Hymenoptera: Trichogrammatidae), an egg parasitoid of *Dicladispa armigera* (Chrysomelidae: Coleoptera). Environ Entomol 8:1–11

164. Botto EN, Horny C, Klasmer P, Gerding M (2004) Biological studies on two neotropical egg parasitoid species: *Trichogramma nerudai* and *Trichogramma* sp. (Hymenoptyera: Trichogrammatidae). Biol Sci Technol 14:449–457

165. Keller MA, Lewis WJ, Stinner RE (1985) Biological and practical significance of movement by *Trichogramma* species: a review. Sowthwest Entomol 8:138–155

166. Bueno RCOF, Carneiro TR, Pratissolli D, Bueno AR, Fernandes OA (2008) Biology and thermal requirements of *Telenomus remus* reared on fall armyworm *Spodoptera frugiperda* eggs. Ciênc Rural 38:16

167. Miguel B, Stefano C, Pamela R-L, Kamlesh CR, Moraes B, Carolina M, Richard AJ (2003) Kairomonal effect of walking traces from *Euschistus heros* (Pentatomidae) on two strains of *Telenomus podisi* (Scelionidae). Physiol Entomol 28:349–355

168. Boo KS, Yang JP (2000) Kairomones used by *Trichogramma chilonis* to find *Helicoverpa assulta* eggs. J Chem Ecol 26:359–375

169. Romeis J, Babendreier D, Wäckers FL, Shanower TG (2005) Habitat and plant specificity of *Trichogramma* egg parasitoids-underlying mechanisms and implications. Basic Appl Ecol 6:215–236

170. Poelman EH, Zheng S-J, Zhang Z, Heemskerk NM, Cortesero A-M Dicke M (2011) Parasitoid-specific induction of plant responses to parasitized herbivores affects colonization by subsequent herbivores. PNAS 18:647–652

171. İslamoğlu M, Koçak E (2014) Behavioral responses of egg parasitoid *Trissolcus semistriatus* (Nees) (Scelionidae) to odors of five plant species. Acta Zool Bulg 66:59–64

172. Qasim M, Husain D, Ul Islam S, Ali H, Islam W, Hussain M, Wang F, Wang L (2018) Effectiveness of *Trichogramma chilonis* Ishii against spiny bollworm in Okra and susceptibility to insecticides. J Entomol Zool Stud 6:1576–1581

173. Yu DSK, Laing JE, Hagley EAC (1984) Dispersal of *Trichogramma* spp. in an apple orchard after inundative releases. Environ Entomol 13:371–374

174. Parra JRP, Zucchi RA (2004) *Trichogramma* in Brazil: feasibility of use after twenty years of research. Neotrop Entomol 33:271–281

175. Davies AP, Carr CM, Scholz BCG, Zalucki MP (2011) Using *Trichogramma* Westwood (Hymenoptera: Trichogrammatidae) for insect pest biological control in cotton crops: an Australian perspective. Aust J Entomol 50:424–440

176. Suh CPC, Orr DB, Van Duyn JW (2000) Effect of insecticides on *Trichogramma exiguum* (Trichogrammatidae: Hymenoptera) preimaginal development and adult survival. J Econ Entomol 93:577–583

177. Cônsoli FL, Parra JRP, Hassan SA (1998) Side effects of insecticides used in tomato fields on the egg parasitoid *Trichogramma pretiosum* Riley (Hym.: Trichogrammatidae), a natural enemy of *Tuta absoluta* (Lep., Gelechiidae). J Appl Entomol 122:43–47

178. Shoeb MA (2010) Effect of some insecticides on the immature stages of the egg parasitoid *Trichogramma evanescens* (Trichogrammatidae). Egypt Acad J Biol Sci 3:31–38

179. Saad ASA, Tayeb EH, Awad HA, Abdel Rehiem ASA (2015) *Trichogramma evanescens* release in correlation with certain pesticides against the spiny bollworm, *Earias insulana* infestation in early and late cotton cultivation. Middle East J Appl Sci 5:290–296

180. Salama HS, Zaki FN (1985) Biological effects of *Bacillus thuringiensis* on the egg parasitoid *Trichogramma evanescens*. Insect Sci Appl 6:145–148

181. Alsaedi G, Ashouri A, Talaei-Hassanloui R (2017) Assessment of two *Trichogramma* species with *Bacillus thuringiensis* var. krustaki for the control of the tomato leafminer *Tuta absoluta* in Iran. Open J Ecol 7:112–124

182. Bai B, Luck RF, Forster L, Stephens B, Janssen JAM (1992) The effect of host size on quality attributes of the egg parasitoid, *Trichogramma pretiosum*. Ent Exp Appl 64:37–48

183. Barloggio G, Tamm L, Nagel P, Luka H (2018) Selective flowers to attract and enhance *Telenomus laeviceps*: a released biocontrol agent of *Mamestra brassicae* (Lepidoptera: Noctuidae), pp 1–9. https://doi.org/10.1017/s0007485318000287

184. Díaz MF, RamírezA PovedaK (2012) Efficiency of different egg parasitoids and increased floral diversity for the biological control of noctuid pests. Biol Cont 60:182–191
185. Géneau CE, Wäckers FL, Luka H, Daniel C, Balmer O (2012) Selective flowers to enhance biological control of cabbage pests by parasitoids. Basic Appl Ecol 13:85–93
186. Witting-Bissinger BE, Orr DB, Linker HM (2008) Effects of floral resources on fitness of the parasitoids *Trichogramma exiguum* (Hymenoptera: Trichogrammatidae) and *Cotesiacongregata* (Hymenoptera: Braconidae). Biol Cont 47:180–186
187. Haitao Q, Bin C, Zaolin Z, Qiuhui D (2013) Effect of some environmental and biological factors on reproductive characters of *Trichogramma* spp. Afr J Agric Res 8:2195–2203
188. Suverkropp BP, Bigler F, Van Lenteren JC (2009) Dispersal behaviour of *Trichogramma brassicae* in maize fields. B Insectol 62:113–120
189. Bigler F, Bieri M, Fritschy A, Seidel K (1988) Variation in locomotion between laboratory strains of *Trichogramma maidis* and its impact on parasitism of eggs of *Ostrinia nubilalis* in the field. Entomol Exp Appl 49:283–290
190. Zhang YZ, Huang DW, Zhao TH, Liu HP, Bauer LS (2005) Two new species of egg parasitoids (Hymenoptera: Encyrtidae) of wood-boring beetle pests from China. Phytoparasitica 33:253–260
191. Sundaramurthy VT, Basu AK (1983) The impact of integrated insects management system on the productivity of cotton. Cotton Dev 13:35–38
192. Salama HS (1983) Cotton-pest management in Egypt. Crop Protect 2:183–191
193. Gergis MF, Hamid AA, Mostafa SA, Fouda ME (2001) Biologically based new approach for management of cotton key pests in middle Egypt. In: Proceedings of Beltwide cotton conference, National cotton council, Memphis TN, vol 2, pp 876–882
194. Naranjo SE, Peter C, Hagler JR (2004) Conservation of natural enemies in cotton: role of insect growth regulators in management of *Bemisia tabaci*. Biol Cont 30:52–72
195. Yadav DN (2013) Egg parasitoids in cotton ecosystem In: Sithanantham S, Ballal CR, Jalali SK, Bakthavatsalam N (eds) Biological control of insect pests using egg parasitoids. Springer, London, pp 301–316
196. Thontadarya TS, Rao JK (1977) Field recovery of *Chelonus blackburni* Cameron (Braconidae) from the cotton spotted bollworm, *Earias vittella*. Curr Sci 46:687
197. Prasad RP, Roitberg BD, Henderson DE (1999) The effect of rearing temperature on flight initiation of *Trichogramma sibiricum* Sorokina at ambient temperatures. Biol Cont 16:291–298
198. Singh SP (1994) Fifteen years of AICRP on biological control. Technical Bulletin No. 8. Project Directorate of Biological Control, Indian Council of Agricultural Research, Bangalore
199. Bueno RCOF, Parra JRP, Bueno AF, Moscardi F, Oliveira JRG, Camillo MF (2007) Sem barreira. Rev Cultivar 12–15
200. Morales J, Gallardo JS, Vásquez C, Rios Y (2000) Patrón de emergência, longevidad, parasitismo y proporción sexual de *Telenomus remus* (Hymenoptera: Scelionidae) com relación al cogollero del maíz. Inst Biotecnol Appl Agropecuária 12:47–54
201. Ferrer F (2001) Biological control of agricultural insect pests in Venezuela, advances, achievements, and future perspectives. Biocontrol News Inf 22:67–74
202. Hafez M, Fayad YH, El-Kifl AH (1978–79) Impact of the egg parasite *Platytelenomus hylas* (Nixon) on the population of the sugar-cane borer *Sesamia cretica* (Led.) in Egypt. Bull Soc Ent Egypt Econ Ser 11:49–55
203. Ragab ZA, Awadallah KT, Farghaly HTh, Ibrahim AMA, El-Wakeil NE (1999) Parasitism rates by *platytelenomus hylas* (Nixon) on *sesamia cretica* (Led.) eggs in certain governorates in Egypt. Egypt J Appl Sci 14:339–350
204. Katti G, Padmakumari AP, Pasalu IC (2013) Egg parasitoids in cereal crops ecosystem. In: Sithanantham S, Ballal CR, Jalali SK, Bakthavatsalam N (eds) Biological control of insect pests using egg parasitoids. Springer, Heidelberg, pp 331–371
205. Sherif MR, Hendawy AS, El-Habashy MM (2008) Utilization of *Trichogramma evanescens* (Ashmead) for controlling rice stem borer, *Chilo agamemnon* in rice fields in Egypt. Egypt J Biol Pest Cont 18:11–16

206. Krishnamoorthy A, Mani M, Visalakshy PNG (2013) Egg parasitoids in vegetable crops ecosystem: research status and scope for utilization. In: Sithanantham S, Ballal CR, Jalali SK, Bakthavatsalam N (eds) Biological control of insect pests using egg parasitoids. Springer, London, pp 301–316

207. Desneux N, Wajnberg E, Wyckhuys KAG, Burgio G, Arpaia S, Narváez-Vasquez CA, González-Cabrera J, Ruescas DC, Tabone E, Frandon J (2010) Biological invasion of European tomato crops by *Tuta absoluta*: ecology, geographic expansion and prospects for biological control. J Pest Sci 83:197–215

208. Hanafy HEM, El-Sayed W (2013) Efficacy of bio-and chemical insecticides in the control of *Tuta absoluta* (Meyrick) and *Helicoverpa armigera* (Hübner) infesting tomato plants. Aust J Basic App Sci 7:943–948

209. Knutson A (2005) A guide to the use of *Trichogramma* for biological control with special reference to augmentative releases for control of bollworm and budworm in cotton. Texas Agricultural Extension Service. B-6071, 5-98, 42 pp

210. Zouba A, Mahjoubi K (2010) Biological control of *Tuta absoluta* with release of *Trichogramma cacoeciae* in tomato greenhouses in Tunisia. Afr J Plant Sci Biotech 4:85–87

211. Öztemiz S (2012) The tomato leafminer [(*Tuta absoluta* Meyrick (Lepidoptera: Gelechiidae)] and its biological control. KSU J Nat Sci 15:47–57

212. Öztemiz S (2013) Population of *Tuta absoluta* and natural enemies after releasing on tomato grown greenhouse in Turkey. Afr J Biotechnol 12:1882–1887

213. Vasconcelos GR (2013) Strain selection and host effect on *Trichogramma pretiosum* Riley, 1879 (Trichogrammatidae) quality for *Tuta absoluta* (Meyrick, 1917) (Lepidoptera: Gelechiidae) control in tomato crops. M.Sc. thesis, University Moura Lacerda, Brazil, 98 pp

214. Ghoneim K (2014) Parasitic insects and mites as potential biocontrol agents for a devastative pest of tomato, *Tuta absoluta* in the world: a review. Int J Adv Res 2:81–115

215. Kim JH, Broadbent AB, Lee SG (2001) Quality control of the mass-reared predatory mite, *Amblyseius cucumeris* (Acarina: Phytoseiidae). J Asia-Pacific Entomol 4:175–179

216. Marchiori CH (2003) Occurrence of the parasitoid *Anastatus* sp. in eggs of *Leptoglossus zonatus* under the maize in Brazil. Ciênc Rural 33:767–768

217. Mani M, Krishnamoorthy A, Gupta PR (2013) Egg parasitoids of fruit crop pests. In: Sithanantham S, Ballal CR, Jalali SK, Bakthavatsalam N (eds), Biological control of insect pests using egg parasitoids. Springer, London, pp 389–396

218. Hegazi E, Herz A, Hassan SA, Agamy E, Khafagi WE, Showeil S, Zaitun A, Mostafa S, El-Hafez AM, EL-Shazly A, El-Said S, Abo Abdala L, Khamis N, EL-Kemny S (2005) Naturally occurring *Trichogramma* species in olive farms in Egypt. Insect Sci 12:185–192

219. Armstrong A (1987) Parasitism of *Tetrastichus haitiensis* Gahan on egg masses of *Diaprepes abbreviatus* in Puerto Rico. J Agric Univ Puerto Rico 71:407–409

220. Schauff ME (1987) Taxonomy and identification of the egg parasites (Hymenoptera: Platygastridae, Trichogrammatidae, Mymaridae, and Eulophidae) of citrus weevils (Coleoptera: Curculionidae). Proc Entomol Soc Wash 89:31–42

221. Ulmer B, Duncan RE, Pavis C, Pena JE (2008) Parasitoids attacking citrus weevil eggs in Guadeloupe. Florida Entomol 91:311–314

222. Smith D, Beattie GAC, Broadley R (1997) Citrus pests and their enemies: integrated pest management in Australia. Queensland Department of Primary Industries Series QI97030, 272 p

223. Anonymous (2004) http://www.ars-grin.gov/cgi-bin/nirgp/probl./taxon.P73657. Accessed 8 Apr 2004

224. Coulson JR, Vail PV, Dix ME, Norlund DA, Kauffman W (2000) 110 years of biological control research and development in the USDA 1883–1983 USDA, ARS, 645 p

225. Blua MJ, Phillips PA, Redak RA (1999) A new sharpshooter threatens both crops and ornamentals. Calif Agric 53:22–25

226. Triapitsyn SV, Hoddle MS, Morgan DJW (2002) A new distribution and host record for *Gonatocerus triguttatus* in Florida, with notes on *Acmopolynema sema* (Hymenoptera: Mymaridae). Florida Entomol 85:654–655

227. Triapitsyn SV, Bezark LG, Morgan DJW (2002) Redescription of *Gonatocerus atriclavus* Girault (Hymenoptera: Mymaridae), with notes on other egg parasitoids of sharpshooters (Homoptera: Cicadellidae) in northeastern Mexico. Pan-Pac Entomol 78:34–42
228. Triapitsyn SV, Mizell RF III, Bossart JL, Carlton CE (1998) Egg parasitoids of *Homalodisca coagulata* (Homoptera: Cicadellidae). Florida Entomol 81:241–243
229. Sujii ER, Costa Maria LM, Pires CSS, Colazza S, Borges M (2002) Inter and intra-guild interactions in egg parasitoid species of the soybean stink bug complex. Pesq Agropec Bras 37:1541–1549
230. Usman M, Inayatullah M, Usman A, Sohail K, Shah SF (2012) Effect of egg parasitoid, *Trichogramma chilonis*, in combination with *Chrysoperla carnea* and neem seed extract against tomato fruitworm, *Helicoverpa armigera*. Sarhad J Agric 28:253–257
231. Khidr AA, Gaffar SA, Nada MS, Taman AA, Salem FA (2013) New approach for controlling tomato leafminer, *Tuta absoluta* in tomato fields in Egypt. Egypt J Agric Res 91:335–348
232. Kumar P, Shenhmar M, Brar KS (2004) Field evaluation of trichogrammatids for the control of *Helicoverpa armigera* (Hübner) on tomato. Biol Cont 18:45–50
233. De Moraes CM, Lewis WJ, Tumlinson JH (2000) Examining plant-parasitoid interactions in tritrophic systems. An Soc Entomol Bras 29:189–203
234. Khosa SS, Brar KS (2000) Effect of storage on the emergence and parasitization efficiency of laboratory reared and field collected population of *Trichogramma chilonis* Ishii. Biol Cont 14:71–74
235. Yadav DN, Anand J, Devi PK (2001) *Trichogramma chilonis* on *Plutella xylostella* (Lepidoptera: Plutellidae) in Gujarat. Indian J Agric Sci 71:69–70
236. Shirazi J (2007) Comparative biology of *Trichogramma chilonis* Ishii on eggs of *Corcyra cephalonica* (Stainton) and *Helicoverpa armigera* (Hübner). Biol Cont 21:37–42

Propagation and Application of Larval Parasitoids

Huda Elbehery, Mahmoud Saleh and Nabil El-Wakeil

Abstract Potential of larval parasitoids through monitoring the abundance, mass production and the field application were aimed and reviewed. There are two larval parasitoid kinds; Endoparasitoids: it develops within the body of their larvae and Ectoparasitoids grows out of larvae bodies. Chemical communication between hosts and larval parasitoids is one of most interested aspects of reciprocal communication between insects. The host scent plays an important role in the attraction for specific larval parasitoids. In general, there are biocontrol agents in nature (field crops, vegetables, fruit orchards and forests), and they have been preventing some species from insect outbreaks. Mass rearing of biocontrol agents contains the production of thousand/millions of insects (host and bio-agent), objectives to control some insect pests. One goal in pest control is to develop managing strategies that kill insect pests without harming the environment or other organisms. An ideal method would be to enhance larval parasitoids to control some insect pest species. This can be accomplished by mass releasing the biological control agents for controlling many of agricultural insect pests. The mass-production and release of beneficial insects are affected by several factors like as supplemental food, environmental conditions and the host. There are many practices which could conserve the larval parasitoids under different ecological conditions such as plant-provided food, food sprays, semiochemicals, and induced plant responses. There are landscape factors affect insect populations and associated natural enemies such as farm scale, crop diversity, pest density, field size and shape and field margins. Combination among larval parasitoids with other biocontrol agents seems to be well suited to protecting many crops, vegetables or fruit orchards from insect infestations. The different biocontrol strategies would use in the appropriate routines against certain insects in large areas. This could result in significant and important synergistic effects on pest population suppression.

Keywords Larval parasitods · Abundance · Mass production · Field application · Interaction

H. Elbehery (✉) · M. Saleh · N. El-Wakeil
Pests and Plant Protection Department, National Research Centre (NRC), Dokki, Cairo, Egypt
e-mail: helbehery@gmail.com

© Springer Nature Switzerland AG 2020
N. El-Wakeil et al. (eds.), *Cottage Industry of Biocontrol Agents and Their Applications*, https://doi.org/10.1007/978-3-030-33161-0_2

1 Introduction

Biological control history may be considered from Egyptian records of 4000 years ago where domestic cats were depicted as useful in rodent control [1]. Insect parasitism was documented in 1602 by the Italian Aldrovandi, who detected the cocoons of *Apanteles glomeratus* being attached to larvae of *Pieris rapae*, which it is known later larval parasitoids. Most biological control agents, namely parasitoids at work in the agricultural environments are naturally occurring ones, which provide excellent regulation of many insect pests with little or no assistance from humans [2–7]. The natural abundance of some biocontrol agents is one cause that many plant-feeding insects do not become economic pests. There is a great potential for increasing the benefits from natural enemies, through the elimination or reduction in the use of chemical pesticides [8].

The use of insect parasitoids and predators to control insect pests has many advantages over traditional chemical controls. These natural enemies leave no harmful chemical residues. Natural enemies released in a storage facility continue to reproduce as long as hosts are available and environmental conditions are suitable [9]. Unlike chemicals that need to be applied to a wide area, natural enemies can be released at a single location and they will find and attack pests located deep inside crevices or within a grain mass [8]. Parasitoids and predators that attack insect pests are typically very small, and have a short life cycle and a high reproductive capacity. They can easily be removed from bulk grain using normal cleaning procedures before milling. In many ways the stored-product environment is favorable for biological control strategies in of frame of IPM [10]. Environmental conditions are generally favorable for natural enemies, and storage structures prevent these beneficial insects from leaving. Several reviews have been published on the use of insect parasitoids to control many insect pest species [11, 12].

Parasitoid is an organism that lives and feed on/ in an organism called host and finally causes death of the host [13]. In many species female parasitoids lay their eggs on/inside the host that they are attacking. It can be classified according to the stage used. There are parasitoids of eggs, larvae/nymphs, pupae, and adults. The eggs hatch and larva begins feeding on its host stage in order to complete its development. When fully fed the parasitoid, larva spins a cocoon and pupates either externally or internally and then emerges as an adult to feed, mate and continue the life cycle. Most of adult parasitoids feed on flower nectar, pollen and honeydew. While there are some adult parasitoids also feeding on fluids that exude from a host's puncture wound, caused by the wasp's ovipositor [13].

We aimed in this book chapter to discuss potential of larval parasitoids in sustainable agriculture through monitoring the abundance of natural enemies, the mass production and the field application of these natural enemies as well as combining larval parasitoids with other biocontrol agents.

2 Types of Larval Parasitoids

2.1 Ectoparasitoids

The parasitoids which develop externally on its larval host called ectoparasitoids.

A. *Bracon* spp.: The braconid wasp is an important potential biological control agent of several species of Lepidoptera, mainly pyralid moths and different hosts such as *Sesamia cretica*, *Ostrinia nubilais* Hbn., *Eublemma amabilis* Sharma, *Tuta absoluta*, *Pectinophora gossypiella*, *Corcyra cephalonica* Stainton, *Sitotroga cerealella* Oliv., *Galleria mellonella* Linn., *Maruca testulalis* Geyer, *Helicoverpa armigera* (Hübner) Hardwick, *Spodoptera litura* Fab and *Earias vittella* [14] (Fig. 1a).

B. (*Netelia producta*): Is Ichneumon medium-sized orange wasps that attacks both helicoverpa and armyworm caterpillars. The hatched larva of parasitoid develops externally, hanging on behind the head of the host. The parasitoid doesn't arrest the development of their host where they do not complete its larval development until the host caterpillar has tunnelled into the soil and formed its pupation chamber. After the host forms its chamber, the parasitoid kills the host and the *Netelia* larva spins a black furry cocoon within this chamber (Fig. 1b).

A. female *Bracon* sp. parasitizing host larvae   B. *Netelia producta* (Orange ectoparasitoid)

C. Adult of *Eulophus pennicornis*   D. Adult *Goniozus nephantidis*

Fig. 1 Different ecto-larval parasitoids [14–16]

C. *Eulophus pennicornis* (Nees) (Hymenoptera: Eulophidae): Is a gregarious ectoparasitoid. It attacks a range of macrolepidopteran species, including the tomato moth, *Lacanobia oleracea* L. (Lepidoptera: Noctuidae) (Fig. 1c) [15].

D. *Goniozus nephantidis* Muesebeck (Hymenoptera: Bethylidae): It is a gregarious larval ectoparasitoid. It dominant and responsible for the reduction in the population density of Black-headed caterpillar, *Opisina arenosella* Walker (Fig. 1d) [16].

2.2 Endoparasitoid

The endoparasitoids develop within the body of their larvae; in the following paragraphs, some endo-larval parasitoids will be presented.

A. *Microplitis demolitor* Wilkinson (Hymenoptera: Braconidae): Small waspes, with black wings and orange upper abdomen and legs. It is The most an important solitary parasitoid that parasitizes several species of noctuid larvae as *Helicoverpa* and *Heliothis* spp. The female deposits a single egg in early-stage host larvae. The parasitoid larva feeds internally and chews a hole in the side of its host to emerge and pupate externally. The whole lifecycle takes about 10–12 days (Fig. 2a) [17].

B. *Cotesia flavipes* Cameron (Hymenoptera: Braconidae): is a gregarious endoparasitoid, the main biological control agent to manage stem borers in sugarcane (*Diatraea* spp.) [29]. Females of parasitoid oviposit directly into the host haemocoel, and the number of deposited eggs depends on the host age [18]. This parasitoid remains within the host during its whole embryonic and larval development, being both dependent on the temperature conditions and development status of the host. Parasitoids exit from the body cavity of the host to pupate after producing a characteristic silk cocoon (Fig. 2b) [19].

C. *Apanteles taragamae* (Hymenoptera: Braconidae): It is a common braconid endoparasitoid associated with various lepidopterous crop pests. This species was recorded on *Spilosoma obliqua* Wlk. (Lepidoptera: Arctiidae), *Eucosoma critica* Meyrick (Lepidoptera: Eucosomidae), *Mythimna unipunctata* Haworth (Lepidoptera: Noctuidae) and *Diaphania indica* (Saunders) [20]. *A. taragamae* develops in the larvae of the coconut black-headed caterpillar, *Opisina arenosella* Wlk. (Lepidoptera: Xylorictidae) as a solitary parasitoid [21]. It is also major parasitoid of *Diaphania indica* and cowpea pod borer *Maruca vitrata* on *Sesbania cannabina* (Fig. 2c) [22].

D. *Campoletis chlorideae* Uchida (Hymenoptera: Ichneumonidae): Is an important larval endoparasitoid. It parasitizes different insect species of Lepidoptera [23]. *Helicoverpa armigera* (Hübner) is the most preferred host of *C. chlorideae* on a number of crops, such as cotton, groundnut, chickpea, pigeonpea, sorghum and pearl millet (Fig. 2d) [24].

A. female *Microplitis demolitor*

B. Adult *Cotesia flavipes*

C. *Apanteles* larvae emerging from
   Pieris larva

D. *Campoletis chlorideae*

E. Male *Meteorus autographae* wasp

F. *Sturmiopsis inferens*

Fig. 2   Different endo-larval parasitoids [17–28]

E. *Meteorus autographae* Muesebeck (Hymenoptera: Braconidae): It develop as
    endoparasitoids of Coleoptera and Lepidoptera larvae [25], the adult wasp is
    tiny does not exceed 6 mm orange, with black eyes and antennae. This parasitoid
    attacks larval stage of noctuid such as the eastern blackheaded budworm; *Agrotis
    ipsilon* (Hufn.), the armyworm; *Pseudoplusia includens* (Wlkr.), the soybean
    looper; *Spodoptera eridania* (Cram.), the southern armyworm; *S. exigua* (Hbn.),
    the beet armyworm; *S. frugiperda* (Sm.), the fall armyworm; *S. ornithogalli*
    (Guen) (Fig. 2e).

F. *Sturmiopsis inferens* (*Diptera:* tachinid): Is an important solitary endoparasitoid
    of sugarcane shoot borer *Chilo infuscatellus* Snell. [26]. It also parasitises the

stalk borer, *C. auricilius* Ddgn. 1977) [27], and Gurdaspur borer, *Acigona steniellus* Hampson (Fig. 2f) [28].

3 Impact of Kairomone on the Larval Parasitoid Behaviour

Chemical communication is widespread among many groups of organisms, including insects [30]. Pheromones are interspecific semiochemicals which communicate between individuals of the same species while allelochemicals (i.e., kairomone, allomone and synomone) communicate between different species and may be classified according to the benefits they provide to the producer and receiver. Where those that benefit the receiver but disadvantage the producer are kairomones. Allomones benefit the producer by modifying the behavior of the receiver although having a neutral effect on the receiver. Synomones benefit both the producer and the receiver. Semiochemicals are known to serve a major role as cues to help parasitoids in locating and recognizing their hosts [31]. So the use of kairomones for biocontrol of insect pests has been of interest for several decades due to the fundamental importance of host-plant selection by phytophagous insects, as well as the potential of natural enemies to co-opt those processes in ensuring their own survival.

Chemical communication between hosts and ectoparasitoids is one of most interested aspects of reciprocal communication between insects. The host scent plays an important role in the attraction for specific ectoparasitoids. The study of the impact of kairomones on the behaviour of the parasitoid is so rare [32]. The female parasitoid is able to detect the host location even when surrounded by a very complex odor background [33].

For example, fifth instar larvae of *Ephestia kuehniella* produce secretion by Mandibular glands acts as kairomone for the larval ectoparasitoid *Bracton hebetor* [34]. They found that the kairomone from *E. cautella* is responsible for the stinging behavior of its parasitoid *B. hebetor* by decreasing the location time, and they found the hosts early. The behavioral response of the parasitoids to the kairomone varies with kairomone concentration and distribution. The application of host-extract (kairomone) to the host larvae (*E. kuehniella)* led to reduce the time of host location by the ectoparasitoid *B. hebetor.* Also, the application of host kairomone boosts the parasitism by increase the oviposition response of the parasitoid. The cause of these increase in the oviposition response might due to the stimulation and continuous enhance of an intensified searching behaviour of the parasitoid [32].

Also some host plants emits or release volatiles scent that have been demonstrated in many studies, it serve as kairomone, for attracting beneficial entomophagous as parasitoid and predators to the host plant [35]. *Microplitis croceipes* and *Cardiochiles nigriceps* larval parasitoids of *Heliothis* sp. were substantially more efficient at locating their hosts on some plants than on others.

4 Pillars of Biological Control Industry

4.1 Abundance of Natural Enemies

In general, there are biocontrol agents in nature (field crops, vegetables, fruit orchards and forests), and they have been preventing some species from insect outbreaks. For example, larval parasitoids, by definition, are those that attack and complete their life cycle Endo/ Ecto a host larva. They may be either solitary or gregarious, but in all cases, they prevent the host larvae to complete their life cycles (Fig. 3). These species being abundant in some inhabiting continents were chosen for commercial production and field application [1].

The following larval parasitoids *Apanteles ruficrus, Bracon hebetor, Tachina larvarum,* and *Microplitis* sp. were recorded on corn borers *S. cretica, C. agamemnon* and *O. nubilalis* in maize fields [36]. El-Heneidy and Sekamatte [37] surveyed the larval parasitoids of the cotton bollworms and recorded twenty one parasitoid species belonging 7 families. Of the 21 species, 10, 7 and 4 were found on *Helicoverpa armigera, Earias insulana* and *Pectinophora gossypiella,* respectively. The larval parasitoid *Aganaspis daci* was retrieved from naturally infested *Bactrocera zonata*

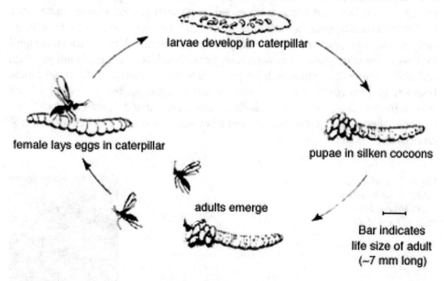

larvae develop in caterpillar

female lays eggs in caterpillar

pupae in silken cocoons

adults emerge

Bar indicates
life size of adult
(~7 mm long)

Fig. 3 Life cycle of Braconid larval parasitoid *Cotesia glomerata* in cabbage worm https://www.google.com.eg/imgres?imgurl=https%3A%2F%2Fi.pinimg.com%2Foriginals%2Fbb%2F76%2F47%2Fbb7647c7c47bc14dbdde44896e92a6fb.jpg&imgrefurl=https%3A%2F%2Fwww.pinterest.com%2Fpin%2F322288917061264502%2F&docid=KYsMsFLXrbnbVM&tbnid=oiOgCHkjpSzfJM%3A&vet=10ahUKEwi73cPr0ufkAhVIxoUKHQqZA0EQMwgxKAEwAQ..i&w=425&h=254&bih=623&biw=1366&q=Life%20cycle%20of%20Braconid%20larval%20parasitoid%20Cotesia%20glomerata%20in%20cabbageworm&ved=0ahUKEwi73cPr0ufkAhVIxoUKHQqZA0EQMwgxKAEwAQ&iact=mrc&uact=8

or *Ceratitis capitata* in guava or citrus orchards in Giza, Egypt [5]. Elbehery [38] found two larval parasitoids *Bracon* spp. and *Apanteles* spp. on the potato tuber moth *Phthorimaea operculella,* the cotton leafworm *Spodoptera littoralis* and the black cut worm *Agrotis ipsilon.*

4.2 Mass Production of Natural Enemies

Mass rearing of biocontrol agents contains the production of thousand/millions of insects (host and bio-agent), objectives to control some insect pests. Mass production started from smaller-scale or research rearing, or intermediate-sized rearing, upon which basic research about the target (often an agricultural pest) and the larval parasitoids are conducted. Therefore, to produce natural enemies, two species have to be reared, the pest and the natural enemy. The potential for rearing a large number of natural enemies increased as artificial diets began to be developed since the 1960s for Lepidoptera, Coleoptera, and Diptera [1].

The full grown larvae *Galleria mellonella* were used for rearing the Braconid wasps (Fig. 4). Pairs of female and male *Bracon* spp. adults—1 to 2 days old were released in the jar covered with muslin cloth then fitted with rubber lids for parasitism and egg lying. Drops of honey were added on jar wall as food source which made vital effects on efficiency of parasitoid wasps [39]. The parasitoids were transferred daily, to a new jar prepared with a corrugated paper sheet with fresh host larvae until the death. The corrugated paper sheet with parasitized larvae was kept until pupation and adult emergence, which will be prepared to mass release in the target fields. Kares et al. [40] reared the ecto-larval parasitoid species, *Bracon brevicornis* on three different hosts; *Ostrinia nubilalis*, *Sesamia cretica* and *Agrolis ipsilon*. Mean numbers of parasitoid's progeny per host larva were 9.6 (on *O. nubilalis*), 9.3 (on *S. cretica*) and 7.3 (on *A. ipsilon*).

Fig. 4 The full-grown larvae *G. mellonella* with adults *Bracon hebetor* (cited from Saleh et al. [1])

4.3 Field Application of Natural Enemies

One goal in pest control is to develop managing strategies that kill insect pests without harming the environment or other organisms. An ideal method would be to enhance larval parasitoids to control some insect pest species. This can be accomplished by mass releasing the biological control agents for controlling many of agricultural insect pests [1].

A pilot trial to mass release parasitoid *Aganaspis daci* against *Bactrocera zonata* under field conditions in Egypt were conducted by El-Heneidy et al. [5] and obtained a significant reduction of insect population. Zaki et al. [41] reported that the highly significant effect of kairomones was recorded by increasing parasitism rates by *Bracon brevicornis* on *S. cretica* and *O. nubilalis*. Zaki et al. [42] released *Diaeretiella rapae* in cabbage field to control *Brevicoryne brassicae* at the rate of 1:50 resulted in 29% parasitism. Releasing *Eretmocerus mundus* for controlling *B. tabaci* in cabbage at the rate of 5 adults/m^2 resulted in 32% parasitism.

5 Factors Affecting the Larval Parasitoid Production

The biopesticide industry is undertaking rapid change, reproducing enlarged global trade in agricultural supplies [43]. Currently biopesticides include $\approx$15% of the Egyptian insecticide market. Biopesticide research on the subcontinent is at a relatively early stage, but evolving rapidly, and focusing on indigenous biocontrol agents. The mass-production and release of beneficial insects are the fundamentals of augmentative biological control [44]. However, production of these biological agents requires the additional step of rearing prey/hosts. The costs of mass-production are balanced by the economic and environmental benefits of the use of the beneficial biological agents [45]. The appropriate use, augmentative releases of parasitoids may consider alternative for the suppression of different pest populations. The percent emergence of adults is the main indicator of success, which has been achieved due to the following advances.

5.1 Supplemental Food

Adult parasitoids are free-living, and hence they must forage for food resources. Numerous of these adults are feed on different plant derived food such as pollen and nectar [46]. That feeding of adult parasitoids with honey or sugar solution has a positive effect on the longevity of various *Diadegma* species [47–50]. Therefore, artificial diets can play important roles as floral nectar resource substitutes in laboratory rearing of parasitoids for experimental studies and/or for field release for the control of pest species. *Diadegma mollipla* parasitoid survived for much longer than males on both honey and sugar solutions larvae. So artificial diets in the form of honey and sugar solutions sustain and significantly extend adult lifespan in *D. mollipla*, and more in females than in males [51].

Improving artificial diets is considered an important way to make more cost-effective the mass production of natural enemies. However, when artificial diets are less nutritious than the natural prey or hosts [52], then their quality as biological control agents decreases [53]. Recently, many studies were targeted at improving an artificial diet to the development requirements of entomophagous species [54–56].

5.2 Environmental Conditions

Temperature and moisture are the main factors that affect the optimum development of mass-production of larval parasitoid. Temperature adjustment aimed to attainment of average useful lifespan for parasitoids, with over 50% of the adult parasitoids remaining alive. When immature stages are developed at suitable environmental conditions, however, the adults emerge in the required time, facilitating the coordination of releasing process [57]. For example braconid wasp (*Cotesia melitaearum*) pupal cocon develops slowly in a cold climatic condition, whereas its host *Melitaea cinxia* (Lepidoptera; Nymphalidae) larvae searched for open sunny microclimate resulting in increased body temperature and rapid development [58].

5.3 The Host

Host quality is, therefore, a crucial determinant of parasitoid fitness as it influences the developmental rate and duration, survival, sex ratio, fecundity, body size and progeny longevity [59]. In order to high progeny fitness, a female parasitoid will select the highest quality hosts to lay its eggs [60]. Similarly, female parasitoids will deposit female eggs in high-quality hosts and male eggs in low-quality ones, thus ensuring higher immature survival rates and production of more female progeny. *Diadegma mollipla* deposited more female eggs in larger than in smaller host [51], also the type of the host is a crucial factor for achieving powerful development of the parasitoid. For example, the fecundity of *B. hebetor* was affected with the host, where daily and total deposited eggs were significantly higher when reared on *G. mellonella*, on the other hand it significantly less on *Ephestia kuehniella* and *C. cephalonica*. The hosts affect the biology of the *B. hebetor*. *G. mellonella* is considered more suitable hosts for the parasitoid than the other hosts [61].

6   Practices of Larval Parasitoids Preservation

There are many practices which could conserve the larval parasitoids under different ecological conditions [7].

6.1   Plant-Provided Food

Wäckers et al. [62] stated that adults of parasitoids and gall midges can increase their longevity, flight activity and oviposition by feeding on nectar. Another approach can be to select crop varieties with increased levels of plant-provide food resources [63]. Thus, the availability of plant-provided food can be a driving force in biocontrol success program [64].

6.2   Food Sprays

Artificial or natural food supplements can be sprayed or dusted onto the crop to support parasitoid wasps in crops, vegetables and fruit orchards, where nectar and pollen are absent or only present at low densities [65]. The development of inexpensive alternative food sources is one of the major opportunities and challenges for enhancing biological control in different crop [66].

6.3   Applying Semiochemicals

Behaviour of natural enemies is directed by semiochemicals. Attraction of natural enemies with synthetic compounds, similar to plant volatiles, is being tested in crops [67, 68]. Natural enemies may also respond to odours that are produced by their host species, such as sex pheromones (kairomones). Kairomones may be used to attract released parasitoids in order to help them establish. Applying attractants in combination with food sprays may promote oviposition of released parasitoids into the target crop. Hexane extract of corn borer larvae was applied on corn plants to enhance performance of larval parasitoid *Bracon brevicornis* adults against the corn borers *Ostrinia nubilalis* and *Sesamia cretica* [41, 69].

6.4   Planting Suitable Non-crop Plants Near Fields

Many studies recommended that preservation biological control agents may be improved by planting suitable non-crop plants near fields either to support migration into the crop or to provide a shelter when field crops are harvested and plants removed. Field surroundings may also contribute to the migration of parasitoids into fields [70]. These methods may contribute to early establishment of natural enemies in new season in the spring [7].

6.5   Induced Plant Responses

Induced plant resistance against insects includes direct traits, such as the production of toxins and feeding deterrents that reduce survival, host preference, fecundity or developmental rate of pests, and indirect traits, that attract and/or retain natural enemies [71]. The latter contains traits such as the plant producing volatiles and floral nectar [72]. Insect induced plant volatiles help parasitoid wasps to detect their hosts in a crop [71, 73], whereas floral nectar production is increased in response to insect attack, guiding these wasps to find their hosts [74]. Preservation such natural enemies might be enhanced in different crops by breeding varieties that produce more volatiles and nectar [7, 75].

7   Does Landscape Affect Larval Parasitoid Activity?

There are numerous landscape factors affect insect populations and associated natural enemies such as farm scale, crop diversity, crop types, pest density, field size and shape, field margins as well as other direct factors like shelter, alternative hosts and habitat and semi-natural habitat cover (Fig. 5).

To maximize parasitoid survival in agroecosystems, resources such as alternate hosts, food for adults (e.g., pollen and nectar), accessibility of overwintering habitats, constant food supply, and appropriate microclimates must be present [2, 76, 78, 79]. These resources may occur within fields through intercropping [80], or by promoting the existence of a weedy background [81, 82]. Resources for natural enemies may also occur at a larger, between-field scale, through the presence of hedgerows, woodlots, and old fields [76, 77] (Fig. 5).

The practice explanation may be due to bottom-up influences of the vegetation structure of complex landscapes. Extra field-vegetation differed considerably among the three complex landscapes, which may translate into variations in the abundance of alternative hosts of *M. communis,* the parasitoid species that accounts for 72.5% of total parasitism of *P. unipuncta* in this study. All alternative hosts of *M. communis,*

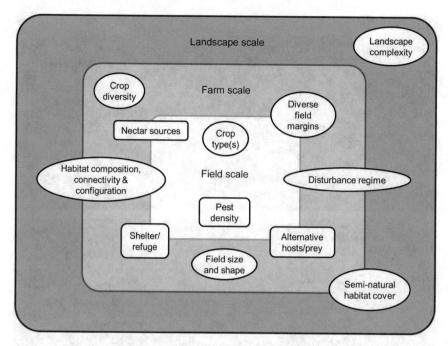

Fig. 5 Landscape scale factors affect larval parasitoid activities [76, 77]

except *Dargida procincta* Grote, which does not occur in Michigan, are exposed to larvae that feed primarily on trees and shrubs [83–86].

Menalled et al. [87] mentioned that landscape complexity may increase rates of parasitism and parasitoid diversity. To find differences in rates of parasitism between a complex and a simple landscape are complicated and not generalizable for a short term [86]. Menalled et al. [87] identified three possible reasons for the observed departures from the forecasted model of higher rates of parasitism in complex landscapes versus simple ones (Fig. 6). Other causes may affect the parasitoid activity such as host range, overwinter requirements and their alternate hosts were investigated by Dyer and Landis [78, 79]; Shaw [88]; Corbett and Rosenheim [89].

8 Interaction Between Larval Parasitoids and Other Bioagents

The interactions between natural enemies can affect the abundance and population of the host species; i.e. combining egg parasitoids with larval parasitoids could be a practical integrated biological control program for mainly lepidopteran insects [90]. These effects are of both academic interest to ecology and of applied interest in the field of biological control. There are much arguing has occurred over whether the

Fig. 6 Effect of landscape factors on parasitoid wasp activities [86, 87] https://www.thesolutionsjournal.com/article/landscape-features-improve-pest-control-agriculture/

best strategy for the reduction of pest density is to release a single or multiple natural enemies. Theoretical models on this topic have given conflicting advice, supporting both the release of multiple species. Academic models on this point have given different advice, supporting both the release of multiple species [91, 92] and the single species release strategy [93, 94]. In practice, biological control workers have often applied any available natural enemy species, and in some cases have achieved success with more than one species [95–98].

The ability of larval parasitoid and egg parasitoids to control the new invasive insect *Tuta absoluta* infestation, makes them potential candidates for mass production and biological control, adding to the list another example of the adaptation of an indigenous parasitoid to an exotic pest and highlighting the importance of a rich and variegated biodiversity [99–102].

Using egg parasitoid *Trichogramma deion* in combination with larval parasitoid *Habrobracon hebetor* for preventing infestations of *Plodia interpunctella* were conducted by Grieshop et al. [90]. They concluded that *H. hebetor* had a significant impact on the number of live *P. interpunctella*, suppressing populations by ≈71%. By releasing *T. deion* significantly increased the level of pest control (87%). By adding *H. hebetor*, there was a significant reduction of *P. interpunctella* reached to 96.7%. Grogshop's findings [90] suggest that, in most conditions, a combined release of both egg parasitoids such as *T. deion* and larval parasitoid like *H. hebetor* would have the best insect management impact.

Another combination strategy is using hymenopteran larval parasitoids with entomopathogenic nematodes, which expect that it may finally benefit the compatibility of these two groups. Steinernematidae and Heterorhabditidae have a good safety record especially regarding their effects on predators and parasitoids [103]. Generally, parasitoids are better at exploiting uninfected hosts because of their search capabilities, whereas nematodes have limited search capacity and require appropriate moisture. Hochberg et al. [104] mentioned that complementarity between parasitoid and pathogens in terms of their extrinsic and intrinsic qualities helped to increase their efficacy as stated by Lacey et al. [105].

Synchronizing use of parasitoids and entomopathogenic nematodes (EPNs) for codling moth (CM) control may produce an antagonistic interaction between the two groups resulting in death of the parasitoid larvae. Lacey et al. [105] evaluated two ectoparasitic ichneumonid species, *Mastrus ridibundus* and *Liotryphon caudatus* for biological control codling moth (CM) larvae combining with *Steinernema carpocapsae*. Exposure of *M. ridibundus* and *L. caudatus* developing larvae to infective juveniles (IJs) of *S. carpocapsae* (10 IJs/cm^2; LC_{80-90} for CM larvae) within CM cocoons resulted in 70.7 and 85.2% mortality, respectively. On the other hand, diapausing full grown parasitoid larvae were mainly completely protected from nematode penetration within their own tightly woven cocoons. *M. ridibundus* and *L. caudatus* females were able to detect and avoid ovipositing on nematode-infected cocooned CM moth larvae as early as 12 h after treatment of the host with IJs. This ability of these parasitoids to avoid nematode-treated larvae and to seek out and kill cocooned CM larvae that survive nematode treatments enhances the complementarily of EPNs and *M. ridibundus* and *L. caudatus* [105].

Other results clearly demonstrate the vulnerability of developing larvae of both *M. ridibundus and L. caudatus* to infection by *S. carpocapsae*. The susceptibility of parasitic Hymenoptera to entomopathogenic nematodes has been reported by other researchers for both internal and external hymenopterous parasitoids [106–108].

Premature death of the host with subsequent death of internal parasitoids is the most frequently reported consequence of host-parasitoid-pathogen interaction rather than direct infection of the parasitoid [109, 110]. The outcome of antagonism among parasitoids and pathogens is hardly recognized [104].

9 Case Study

As an alternative approach, biological control has been actively pursued. Several natural enemies occurring in its native area have been reported as fully documented by Desneux et al. [111], and the efficacy of different biocontrol agents has been evaluated for the implementation of biological control strategies [1, 6, 7, 112]. Natural enemies have been investigated in many countries with the aim of using them in biological control programs; in particular, research has been conducted on the larval parasitoids, such as *Apanteles gelechiidivoris* Marsh, *Pseudapanteles dignus* (Muesebeck), *Bracon* spp. (Hymenoptera: Braconidae), *Dineulophus phthorimaeae* (de Santis) (Hymenoptera: Eulophidae), and *Diadegma* spp. (Hymenoptera: Ichneumonidae) [113–118]. However, none of these beneficial organisms seem to have so far been decisive in controlling *T. absoluta,* and the research is ongoing [100].

Honey bee products were used as natural diets for rearing ectoparasitoid *Bracon hebetor* under laboratory conditions. *B. hebetor* was allowed to fed on the tested diets and lying eggs on host larva *Galleria Mellonella*. The results showed that, females of larval parasitoid *Bracon hebetor* produced a significantly higher number of eggs than the control (*G. mellonella* larvae only) when feed on honey bee products Table 1 [39].

Saleh [119] mentioned that the kairomonal effect of hexane extract of *Ostrinia nubilalis* larvae and *Sesamia cretica* on *Bracon brevicornis* adults was evaluated through olfactometer tests. Females were attracted to the kairomone of *S. cretica* more than to that of *O. nubilalis*. The kairomone of *S. cretica* increased the parasitisation from 7.74 to 17.05%, while the kairomones of *O. nubilalis* and of *Spodoptera littoralis* were not significantly effective. Spraying molasses solution (10%) on the

Table 1 Effect of different tested diets for *Bracon hebetor* females on some biological aspects of the parasitoid [39]

Diets	Total no. of laid eggs	Ovipositional period (days)	Incubation period (days)	Hatchability (%)	Pupation (%)	Emergence (%)
Larvae only	142.7 ± 29.2d	12.7 ± 2.3cd	1.5 ± 0.1a	83.90 ± 3.2b	90.39 ± 1.6a	91.40 ± 2.1a
Sugar solution	147.1 ± 23.5d	8.3 ± 0.7d	1.36 ± 0.1ab	84.94 ± 1.7b	91.60 ± 5.4a	92.27 ± 3.5a
Bee honey	292 ± 69.3 cd	15.4 ± 1.3bcd	1.18 ± 0.1bc	91.25 ± 4.3ab	91.99 ± 4.2a	95.04 ± 2.1a
Royal jelly	527.1 ± 65.5b	22.1 ± 3.2ab	1.22 ± 0.1bc	92.16 ± 4.1ab	98.26 ± 1.7a	93.64 ± 4.1a
Pollen grains + bee honey	439.3 ± 439.3bc	17 ± 2.6bc	1.10 ± 0.1c	96.65 ± 1.5a	91.89 ± 4.6a	97.26 ± 1.7a
Royall jelly + bee honey	738.2 ± 107.2a	27.7 ± 4.3a	1.20 ± 0.1bc	97.37 ± 1.3a	94.58 ± 1.9a	98.70 ± 1.3a
F value	12.412**	8.079*	3.307*	3.431*	0.543 NS	1.372 NS

Means followed by the same letter (s) in the column aren't significantly different at 5% probability
*significant, **highly significant, *NS* non significant

corn stalks before releasing *B. brevicornis* parasitoids increased the rate of parasitism from 7.74 to 28.21%. The concentration of 5% gave insignificant increase in the parasitism rate [32, 41, 119].

10 Conclusion and Future Prospects

Agriculture production has to increase largely by 7% in the near future in order to face the population increase. This production increase should be achieved through the sustainable agriculture system as well as using biocontrol agents, i.e., larval parasitoids, which preserves the environmental recourses and limits the use of chemical pesticides. Therefore, field application of biological control agents should be a main part of sustainable agriculture. Because of the increasing consciousness about the hazards of traditional chemical pesticides to both human and environment, many countries have converted to biocontrol as alternatives to chemical control.

In Egypt, combination larval parasitoids with other biocontrol agents seems to be well suited to protecting many crops, vegetables or fruit orchard trees from insect infestations; however, sometimes few species will probably be ineffective in some areas or crop species, then the producers would use the other biocontrol agents. *H. hebetor* seems to be well suitable to some stored product insects, for example, it could use to find *P. interpunctella* larvae in damaged packages or storages. Therefore, the combination of releasing both egg parasitoids with larval parasitoids (i.e., *Trichogramma evanescens* and *B. hebetor* should provide the best control over both the short and long term. However, it is important to realize that unlike pest mortality resulting from egg parasitism by *T. evanescens* and larval mortality caused by *H. hebetor* will only prevent future infestation. For adequate suppression of corn borers to occur, releases both parasitoid wasps should probably be applied, as early as possible, so that the parasitoids have a better chance of managing pest populations before they reach economic levels.

In the light of presented satisfactory results, additional studies on other biocontrol species are required to clarify their potential parasitoid species for combining them with other active, detect their efficiency on the primary hosts, and evaluate them in biological control and IPM programs in commercial tomato plantations. The compatibility of the two biocontrol groups for insect pest control could be easily managed if mass releases of parasitoids have been applied 48 h or more after for example EPNs applications.

The different biological control strategies would use in the appropriate routines and against certain insect pests in large areas, the proposed approach might be highly effective and environmentally acceptable. Furthermore, if this kind of biological control could be used in conjunction with other control strategies, such as the cultural control methods or sterile insect technique and botanical insecticides; this could result in significant and important synergistic effects on pest population suppression.

References

1. Saleh MME, El-Wakeil NE, Elbehery H, Gaafar N, Fahim S (2019) Biological pest control for sustainable agriculture in Egypt, Springer Publisher. In: Negm AM, Abu-Hashim M (eds) Sustainability of agricultural environment in Egypt: Part II ISBN 978-3-319-95356-4, © Springer Nature Switzerland AG 2019—Soil-water-plant Nexus, Hdb Environ Chem 77:145–188. https://doi.org/10.1007/978-3-319-95357-1
2. DeBach P, Rosen D (1991) Biological control by natural enemies, 2nd edn. Cambridge University Press, Cambridge, pp 440. ISBN 0-521-39191-1
3. Tawfik MFS (1997) Biological control for insect pests (in Arabic), 2nd edn. Academic Bookshop, Cairo, 757 p
4. Gurr G, Wratten S (2000) Biological control: measures of success. Springer Science+Business Media, Dordrecht, 429 pp
5. El-Heneidy AH, Hosny ME, Ramadan MM (2016) Adaptation and first field release of *Aganaspis daci*, a larval parasitoid of the peach fruit fly, *Bactrocera zonata*, in Egypt. In: Proceedings of 9th ISFFEI, pp 395–400. ISBN: 978-616-358-207-2
6. El-Wakeil NE, Abd-Alla AM, El Sebai TN, Gaafar NM (2015) Effect of organic sources of insect pest management strategies and nutrients on cotton insect pests. In: Gorawala P, Mandhatri S (eds) In frame of book Agricultural research updates, chap 2, vol 10, pp 49–81. ISBN: 978-1-63482-745-4
7. El-Wakeil NE, Saleh MME, Gaafar N, Elbehery H (2017) Conservation biological control practices. In: Frame of book biological control of pest and vector insects, chap 3. Published by Intech Open Access. ISBN 978-953-51-5041-1
8. El-Wakeil N, Gaafar N, Sallam A, Volkmar C (2013) Side effects of insecticides on natural enemies and possibility of their integration in plant protection strategies. In: Trdan S (ed) Agricultural and biological sciences "insecticides—development of safer and more effective technologies". Intech, Rijeka, Croatia, pp 1–54
9. Brower JH, Smith L, Vail PV, Flinn PW (1996) Biological control. In: Subramanyam BH, Hagstrum DW (eds) Integrated management of insects in stored products. Marcel Dekker, New York, pp 223–286
10. Schöller M, Prozell S, Al-Kirshi AG, Reichmuth CH (1997) Towards biological control as a major component of integrated pest management in stored product protection. J Stored Prod Res 33:81–97
11. Adler C, Schöller M (1998) Integrated protection of stored products. In: IOBC wprs bulletin 21 international organization for biological and integrated control of noxious animals and plants, Dijon, France, 173 pp
12. Schöller M, Flinn PW (2000) Parasitoids and predators. In: Subramanyam B, Hagstrum DW (eds) Alternatives to pesticides in stored-product IPM. Springer, Boston, MA
13. Araújo A, Jansen AM, Bouchet F, Reinhard K, Ferreira LF (2003) Parasitism, the diversity of life, and paleoparasitology. Mem Inst Oswaldo Cruz 98:5–11
14. Dabhi (2011) Comparative biology of *Bracon hebetor* say on seven lepidopteran hosts. Karnataka J Agric Sci 24:549–550
15. Shaw MR (1987) Host associations of species of Eulophus in Britain (Hymenoptera: Eulophidae). Entomol Gaz 38:59–63
16. Venkatesan T, Jalali SK, Srinivasa Murthy K, Rabindra RJ, Rao NS (2006) Field evaluation of different doses of *Goniozus nephantidis* (Muesebeck) for the suppression of *Opisina arenosella* Walker on coconut. Inter J Cocon R&D (CORD). 22:78–84
17. Burke GR (2016) Analysis of genetic variation across the encapsidated genome of *Microplitis demolitor* bracovirus in parasitoid wasps. PLoS ONE 11:e0158846. https://doi.org/10.1371/journal.pone.0158846
18. Brewer FD, King EG (1981) Food consumption and utilization by sugarcane borers parasitized by *Apanteles flavipes*. J Georgia Entomol Soc 16:181–185

19. Botelho PS, Macedo N (2002) *Cotesia flavipes* para o controle de *Diatraea saccharalis*, pp 409–421. In: Parra JRP, Botelho PSM, Corrêa-Ferreira BS, JMS Bento (eds) Controle Biológico no Brasil: Parasitóides e Predadores, São Paulo, Manole, 646 p

20. Nixon GEJ (1965) A reclassification of the tribe Migrogasterini {Hymenoptera: Braconidae). Bull Br Mus (Nat. Hist) Entamol Suppl 2:6–7

21. Rao RY, Cherian MC, Ananthanarayanan KP (1948) Infestation of *Nephantis serinopa* Meyr. in South India and their control by biological method. Indian J Entomol 10:205–247

22. Huang C, Peng W-K, Talekar NS (2003) Parasitoids and other natural enemies of *Maruca vitrata* feeding on *Sesbania cannabina* in Taiwan. Biocontrol 48:407–416

23. Yan ZG, Wang CZ (2006) Similar attractiveness of maize volatiles induced by *Helicoverpa armigera* and *Pseudaletia separata* to the generalist parasitoid *Campoletis chlorideae*. Entomol Exp Appl 118:87–96

24. Kumar N, Kumar A, Tripathi CPM (1994) Functional response of *Campoletis chlorideae* Uchida (Ichneumonidae), a parasitoid of *Heliothis armigera* (Hubner) (Noctuidae) in an enclosed experimental system. Biol Agric Hort 10:287–295

25. Shaw MR, Huddleston T (1991) Classification and biology of braconid wasps (Hymenoptera: Braconidae), vol 7. Royal Entomological Society, London, 126 pp. http://www.royensoc.co.uk/sites/default/files/Vol07_Part11.pdf

26. David H, Easwaramoorthy S (1986) Biological control. In: David H, Easwaramoorthy S, Jayanthi RO (eds) Sugarcane entomology in India. Sugarcane Breeding Institute, Coimbatore, pp 383–421

27. Singh OP (1977) Integrated control of sugarcane stalk borer, *Chilo auricilius* Ddgn. Sug News 9:36–43

28. Chaudhary JP, Yadav SR, Singh SP (1980) Some observations on the biology and host preference of *Sturmiopsis inferens* Townsend (Tachinidae: Diptera). Indian J Agric Res 14:147–154

29. Postali JR, Machado PS, De Sene A (2010) Biological control of pests and a key component for sustainable sugarcane production. In: Barbosa LA (ed) Sugarcane bioethanol R&D for productivity and sustainability. Blucher Brazilian science and technology, Sao Paulo, Brazil, pp 441–450

30. Rajchard J (2013) Kairomones important substances in interspecific communication in vertebrates: a review. Vet Med 58:561–566

31. Lewis WJ, Martin WRJR (1990) Semiochemicals for use with parasitoids: status and future. J Chem Ecol 16:3067–3089

32. Shonouda ML, Nasr FN (1998) Impact of larval-extract (kairomone) of *Ephesiia kuehniella* Zell. on the behaviour of the parasitoid *Bracon hebetor*. J Appl Ent 122:33–35

33. Kigathi RN, Unsicker SB, Reichelt M, Kesselmeier J, Gershenzon J, Weisser WW (2009) Emission of volatile organic compounds after herbivory from *Trifolium pratense* (L.) under laboratory and field conditions. J Chem Ecol 35:1335–1348

34. Strand MR, Williams HJ, Vinson SB, Mudd A (1989) Kairomonal activities of 2-acylcyclohexane-1, 3- diones produced by *Ephestia kuehniella* Zeller in eliciting searching behaviour by the parasitoid *Bracon hebetor* Say. J Chem Ecol 15:1491–1500

35. Murali Baskaran RK, Sharma KC, Kumar J (2017) Seasonal and relative abundance of stem-borer and leaf-folder in wet land rice eco-system. J Entomol Zool Stud 5:879–884

36. Fayad YH, Hafez M, El-Kifl AH (1979) Survey of the natural enemies of the three corn borers *Sesamia cretica*, *Chilo agamemnon* and *Ostrinia nubilalis* in Egypt. Agric Res Rev 57:29–33

37. EL-Heneidy AH, Sekamatte BM (1998) Survey of larval parasitoids of the cotton bollworm s in Uganda. Bull Soc Entomol Egypt 76:125–134

38. Elbehery H (2013) Biological, ecological and genetical studies on the parasitoid, *Bracon* spp. (Braconidae). Faculty of Science, Ain Shams University, Egypt 142 pp

39. Abd El-Wahab TE, El-Behery HHA, Farag NA (2016) Evaluation of some honey bee products as artificial diets for rearing the parasitoid *Bracon hebetor* Say (Hymenoptera: Braconid). Egypt J Biol Pest Cont 26:309–312

40. Kares EA, Ebaid GH, El-Sappagh IA (2009) Biological studies on the larval parasitoid species *Bracon brevicornis* reared on different insect hosts. Egypt J Biolog Pest Control 19:165–168

41. Zaki FN, El-Saadany G, Gomaa A, Saleh MME (1998) Increasing rates of parasitism of the larval parasitoid *Bracon brevicornis* (Hym., Braconidae) by using kairomones, pheromones and a supplementary food. J Appl Ent 122:565–567

42. Zaki FN, El-Shaarawy MF, Farag NA (1999) Release of two predators and two parasitoids to control aphids and whiteflies. J Pest Sci 72:19–20

43. Kumar KK, Sridhar J, Murali-Baskaran RK, Senthil-Nathan S, Kaushal P, Dara SK, Arthurs S (2018) Microbial biopesticides for insect pest management in India: current status and future prospects. J Invertebr Pathol. https://doi.org/10.1016/j.jip.2018.10.008

44. Morales-Ramos JA, Rojas MG (2003) Natural enemies and pest control: an integrated pest management concept. In: Koul O, Dhaliwal GS (eds) Predators and parasitoids. Taylor and Francis, London, UK, pp 17–39

45. Nation JL (2002) Insect physiology and biochemistry. CRC Press, Boca Raton, USA

46. Gillespie MAK, Gurr GM, Wratten SD (2016) Beyond nectar provision: the other resource requirements of parasitoid biological control agents. Ent Exp Appl 159:207–221

47. Broodryk SW (1971) The biology of *Diadegma stellenboschense*, a parasitoid of potato tuber moth. J Entomol Soc S Afr 34:413–423

48. Idris AB, Grafius E (1995) Wildflowers as nectar sources for *Diadegma insulare* (Hymenoptera: Ichneumonidae), a parasitoid of the diamondback moth (Lepidoptera: Yponomeutidae). Environ Entomol 24:1726–1735

49. Ooi PAC (1980) Laboratory studies of *Diadegma cerophagus* (Hym.: Ichneumonidae), a parasite introduced to control *Plutella xylostella* (Lep.: Yponomeutidae) in Malaysia. Entomophaga 25:249–259

50. Yang JC, Chu YI, Talekar NS (1993) Biological Studies of *Diadegma semiclausum* (Hym. Ichneumonidae), a parasite of Diamondback Moth. Entomophaga 38:579–586

51. Sithole R, Chinwada P, Lohr BL (2018) Effects of host larval stage preferences and diet on life history traits of *Diadegma mollipla*, an African parasitoid of the Diamondback Moth. Biocont Sci Technol 28:172–184

52. Grenier S (2009) In vitro rearing of entomophagous insects—past and future trends: a minireview. Bull Insectol 62:1–6

53. Riddick EW (2009) Benefit and limitations of factitious prey and artificial diets on life parameters of predatory beetles, bugs, and lacewings: a mini-review. Biocontrol 54:325–339

54. Riddick EW, Chen H (2014) Production of coleopteran predators. In: Morales-Ramos JA, Rojas MG, Shapiro-Ilan DI (eds) Mass production of beneficial organisms: invertebrates and entomopathogens. Elsevier, Amsterdam, The Netherlands, pp 17–55

55. De Clercq P, Coudron TA, Riddick EW (2014) Production of heteropteran predators. In: Morales-Ramos JA, Rojas MG, Shapiro-Ilan DI (eds) Mass production of beneficial organisms: invertebrates and entomopathogens. Elsevier, Amsterdam, The Netherlands, pp 57–100

56. Dindo ML, Grenier S (2014) Production of dipteran parasitoids. In: Moralesramos JA, Rojas MG, Shapiro-Ilan DI (eds) Mass production of beneficial organisms: invertebrates and entomopathogens. Elsevier, Amsterdam, The Netherlands, pp 101–143

57. Cancino J, Montoya P (2006) Advances and perspectives in the mass rearing of fruit fly parasitoids in Mexico. N, Brazil (Web)

58. Salim M, Gökçe A, Naqqash MN, Bakhsh A (2016) An overview of biological control of economically important lepidopteron pests with parasitoids. J Entomol Zoology Stud 4:354–362

59. Godfray HCJ (1994) Parasitoids, behavioral and evolutionary ecology. Princeton University Press, Princeton, New Jersey

60. Charnov EL (1976) Optimal foraging: the marginal value theorem. Theor Popul Biol 9:129–136

61. Farag NA, Ismail IA, Elbehery HHA, Abdel-Rahman RS, Abdel-Raheem MA (2015) Life table of *Bracon hebetor*. (Braconidae) reared on different hosts. Internat J Chem Tech Res 9:123–130

62. Wäckers FL, van Rijn PCJ, Bruin J (eds) (2005) Plant-provided food for carnivorous insects: a protective mutualism and its applications. Cambridge University Press, Cambridge, UK

63. Koptur S (2005) Nectar as fuel for plant protectors. In: Wäckers FL, van Rijn PCJ, Bruin J (eds) Plant-provided food for carnivorous insects: a protective mutualism and its applications. Cambridge University Press, Cambridge, UK, pp 75–108

64. Gurr GM, Wratten SD, Barbosa P (2000) Success in conservation biological control of arthropods. In: Gurr GM, Wratten SD (eds) Biological control: measures of success. Kluwer, Dordrecht, pp 105–132

65. Wade MR, Zalucki MP, Wratten SD, Robinson KA (2008) Conservation biological control of arthropods using artificial food sprays: current status and future challenges. Biol Cont 45:185–199

66. Messelink GJ, Bennison J, Alomar O, Ingegdno BL, Tavella L, Shipp L, Palevsky E, Wackers FL (2014) Approaches to conserving natural enemy populations in greenhouse crops: current methods and future prospects. Biocontrol 59:377–393

67. Simpson M, Gurr GM, Simmons AT, Wratten SD, James DG, Leeson G, Nicol HI, Orre-Gordon GUS (2011) Attract and reward: combining chemical ecology and habitat manipulation to enhance biological control in field crops. J Appl Ecol 48:580–590

68. Kaplan I (2012) Attracting carnivorous arthropods with plant volatiles: the future of biocontrol or playing with fire? Biol Cont 60:77–89

69. El-Wakeil NE (1997) Ecological studies on certain natural enemies of maize and sorghum pests, M.sc. Faculty of Agriculture Cairo University, Egypt, p 212

70. Gerling D, Alomarb O, Arnó J (2001) Biological control of *Bemisia tabaci* using predators and parasitoids. Crop Prot 20:779–799

71. Paré PW, Tumlinson JH (1999) Plant volatiles as a defense against insect herbivores. Plant Physiol 121:325–331

72. Araj SE, Wratten S, Lister A, Buckley H (2009) Adding floral nectar resources to improve biological control: potential pitfalls of the fourth trophic level. Basic Appl Ecol 10:554–562

73. Lavandero IB, Wratten SD, Didham RK, Gurr G (2006) Increasing floral diversity for selective enhancement of biological control agents: a double-edged sward? Basic Appl Ecol 7:236–243

74. Wäckers FL, Bonifay C (2004) How to be sweet? Extrafloral nectar allocation by *Gossypium hirsutum* fits optimal defense theory predictions. Ecology 85:1512–1518

75. Kappers IF, Aharoni A, van Herpen T, Luckerhoff LLP, Dicke M, Bouwmeester HJ (2005) Genetic engineering of terpenoid metabolism attracts bodyguards to Arabidopsis. Science 309:2070–2072

76. van Emden HF (1990) Plant diversity and natural enemy efficiency in agroecosystems. In: Mackauer M, Ehler L, Roland J, West KJ, Miller JC (eds) Critical issues in biocontrol. Intercept, Andover, UK, pp 63–80

77. Coombes DS, Sotherton NW (1986) The dispersal and distribution of polyphagous predatory coleoptera in cereals. Ann Appl Biol 108:461–474

78. Dyer LE, Landis DA (1996) Effects of habitat, temperature, and sugar availability on longevity of *Eriborus terebrans*. Environ Entomol 25:1192–1201

79. Dyer LE, Landis DA (1997) Influence of noncrop habitats on the distribution of *Eriborus terebrans* (Hymenoptera: Ichneumonidae) in cornfields. Environ Entomol 26:924–932

80. Perrin RM (1977) Pest management in multiple cropping systems. Agro-ecosystems 3:93–118

81. Dempster JP (1969) Some effects of weed control on the numbers of the small cabbage white (*Pieris rapae*) on Brussels sprouts. J Appl Ecol 6:339–345

82. Foster MS, Ruesink WG (1984) Influence of flowering weeds associated with reduced tillage in corn on a black cutworm (Lepidoptera: Noctuidae), parasitoid *Meteorus rubens* (Nees von Esenbeck). Environ Entomol 13:664–668

83. Krombein KV, Hurd PD, Smith DR Jr, Burks BD (1979) Catalog of hymenoptera in America north of Mexico. Smithsonian Institution, Washington, DC

84. Covell CV (1984) A field guide to the moth of eastern North America. Houghton Mifflin, Boston, Massachusetts, USA

85. West KJ, Miller JC (1989) Patterns of host exploitation by *Meteorus cummunis* (Hymenoptera: Braconidae). Enviro Entomol 18:537–540

86. Marino PC, Landis DA (1996) Effect of landscape structure on parasitoid diversity in agroecosystems. Ecol Appl 6:276–284
87. Menalled FD, Marino PC, Gage SH, Landis DA (1999) Does agricultural landscape structure affect parasitism and parasitoid diversity? Ecol Appl 9:634–641
88. Shaw MR (1994) Parasitoid host ranges. In: Hawkins BA, Sheehan W (eds) Parasitoid community ecology. Oxford University Press, Oxford, UK, pp 111–114
89. Corbett A, Rosenheim JA (1996) Impact of a natural enemy overwintering refuge and its interaction with the surrounding landscape. Ecol Entomol 21:155–164
90. Grieshop JG, Flinn PW, Nechols JR (2006) Biological control of Indian meal moth on finished stored products using egg and larval parasitoids. J Econ Entomol 99:1080–1084
91. May RM, Hassell MP (1981) The dynamics of multiparasitoid host interactions. Am Nat 117:234–261
92. Hogarth WL, Diamond P (1984) Interspecific competition in larvae between entomophagous parasitoids. Am Nat 124:552–560
93. Kakehashi M, Suzuki Y, Iwasa Y (1984) Niche overlap of parasitoids in host parasitoid systems: its consequences to single versus multiple introduction controversy in biological control. J Appl Ecol 21:115–131
94. Briggs CJ (1993) The effect of multiple parasitoid species on the gallforming midge *Rhopalomyia californica*, Ph.D. thesis, California University, Santa Barbara
95. Huffaker CB, Kennett CE (1966) Studies of two parasites of olive scale, *Parlatoria oleae* (Colvee). IV. Biological control of *Parlatoria oleae* (Colvee) through the compensatory action of two introduced parasites. Hilgardia 37:283–355
96. DeBach P, Rosen D, Kennett CE (1971) Biological control of coccids by introduced natural enemies. In: Huffaker CB (ed) Biological control. Plenum, New York, pp 165–194
97. Force DC (1974) Ecology of insect host parasitoid communities. Science 184:624–632
98. Takagi M, Hirose Y (1994) Building parasitoid communities: the complementary role of two introduced parasitoid species in a case of successful biological control. In: Hawkins BA, Sheehan W (eds) Parasitoid community ecology. Oxford University Press, New York, pp 437–448
99. Briggs CJ, Latto J (2001) Interactions between the egg and larval parasitoids of a gall-forming midge and their impact on the host. Ecol Entomol 26:109–116
100. Ferracini C, Ingegno BL, Navone P, Ferrari E, Mosti M, Tavella L, Alberto A (2012) Adaptation of indigenous larval parasitoids to *Tuta absoluta* (Lepidoptera: Gelechiidae) in Italy. J Econ Entomol 105:1311–1319
101. Urbaneja A, Gonzalez-Cabrera J, Arn J, Gabarra R (2012) Prospects for the biological control of *Tuta absoluta* in tomatoes of the Mediterranean basin. Pest Manag Sci 68:1215–1222
102. Biondi A, Desneux N, Amiens-Desneux E, Siscaro G, Zappalà L (2013) Biology and developmental strategies of the Palaearctic parasitoid *Bracon nigricans* (Braconidae) on the Neotropical moth *Tuta absoluta* (Gelechiidae). J Econ Entomol 106:1638–1647
103. Bathon H (1996) Impact of entomopathogenic nematodes on nontarget hosts. Biocontrol Sci Technol 6:421–434
104. Hochberg ME, Hassell MP, May RM (1990) The dynamics of host–parasitoid–pathogen interactions. Am Nat 135:74–94
105. Lacey LA, Unruh TR, Headrick HL (2003) Interactions of two idiobiont parasitoids (Ichneumonidae) of codling moth (Tortricidae) with the entomopathogenic nematode *Steinernema carpocapsae* (Rhabditida: Steinernematidae). J Inverteb Pathol 83:230–239
106. Kaya HK (1978) Interaction between *Neoaplectana carpocapsae* (Nematoda: Steinernematidae) and *Apanteles militaris* (Hymenoptera: Bracondiae), a parasitoid of the armyworm, *Pseudaletia unipuncta*. J Invertebr Pathol 31:358–364
107. Zaki FN, Awadallah KT, Gersraha MA (1997) Competitive interaction between the braconid parasitoid, *Meteorus rubens* Nees and the entomogenous nematode, *Steinernema carpocapsae* (Weiser) on larvae of *Agrotis ipsilon*. J Appl Entomol 121:151–153
108. Sher RB, Parrella MP, Kaya HK (2000) Biological control of the leafminer, Liriomyza trifolii (Burgess): implications for intraguild predation between *Diglyphus begini* Ashmead and *Steinernema carpocapsae* (Weiser). Biol Cont 17:155–163

109. Begon M, Sait SM, Thompson DJ (1997) Two's company, three's a crowd: host–pathogen–parasitoid dynamics. In: Gange AC, Brown VK (eds) Multitrophic interactions in terrestrial systems. Blackwell, Oxford, pp 307–332

110. Brooks WM (1993) Host–parasitoid–pathogen interactions. In: Beckage NE, Thompson SN, Federici BA (eds) Parasites and pathogens of insects, vol 2. Pathogens. Academic Press, San Diego, pp 231–272

111. Desneux N, Wajnberg E, Wyckhuys KAG, Burgio G, Arpaia S, Narváez-Vasquez CA, González-Cabrera J, Ruescas DC, Tabone E, Frandon J (2010) Biological invasion of European tomato crops by *Tuta absoluta*: ecology, geographic expansion and prospects for biological control. J Pest Sci 83:197–215

112. Batalla-Carrera L, Morton A, García-del-Pino F (2010) Efficacy of entomopathogenic nematodes against the tomato leafminer *Tuta absoluta* in laboratory and greenhouse conditions. Biocontrol 55:523–530

113. Colomo MV, Berta DC, Chocobar MJ (2002) El complejo de himenópteros parasitoides que atacan a la "polilla del tomate" *Tuta absoluta* en la Argentina. Acta Zool Lilloana 46:81–92

114. Marchiori CH, Silva CG, Lobo AP (2004) Parasitoids of *Tuta absoluta* collected on tomato plants in Lavras, state of Minas Gerais, Brazil. Braz J Biol 64:551–552

115. Miranda MMM, Picanc MC, Zanuncio JC, Bacci L, da Silva EM (2005) Impact of integrated pest management on the population of leafminers, fruit borers, and natural enemies in tomato. Cienc Rural 35:204–208

116. Bajonero J, Córdoba N, Cantor F, Rodríguez D, Cure YJR (2008) Biology and life circle of *Apanteles gelechiidivoris* (Hymenoptera: Braconidae) parasitoid of *Tuta absoluta* (Lepidoptera: Gelechiidae). Agron Colomb 26:417–426

117. Sánchez NE, Pereyra PC, Luna MG (2009) Spatial patterns of parasitism of the solitary parasitoid *Pseudapanteles dignus* on the tomato leafminer *Tuta absoluta* (Meyrick) (Lepidoptera: Gelechiidae). Environ Entomol 38:365–374

118. Luna MAG, Wada VI, Sánchez NE (2010) Biology of *Dineulophus phtorimaeae* and field interaction with *Pseudapanteles dignus* (Hymenoptera: Braconidae), larval parasitoids of *Tuta absoluta* in tomato. Ann Entomol Soc Am 103:936–942

119. Saleh MME (1992) Ecological studies on certain parasitoids of corn borers, Ph.D. Faculty of Agriculture, Ain Shams University

Propagation, Manipulation, Releasing and Evaluation of Aphid Parasitoids in Egypt

Ahmed Amin Ahmed Saleh

Abstract Studies were carried out in the Arab Republic of Egypt to study certain aphid species infestation, their associated parasitoids on Cabbage, Cauliflower, faba bean, oleander plant (Dafla), cucumber, corn and cowpea plants. Survey and seasonal abundance of the aphid species *Brevicoryne brassicae* L., *Aphis craccivora* (Koch), *Aphis nerii* Fonsecolombe., *Hyalopterus pruni* (Geoffroy), *Aphis gossypii* Glover, *Rhopalosiphum maidis* Fitch *Rhopalosiphum padi* L. and *Hypermoyzus lactucae* L. and its parasitoids were studied. Obtained results of the study showed that *Diaeretiella rapae* (M'Intosh) was the dominant parasitoid on aphid species. Longevity was affected by temperature and food but the sex ratio was not affect by host species. The behavior of this parasitoid at varying host densities was also studied. Result showed a decrease of host-searching and first sting times with increasing host density but number of sting and number of mummies increased as host density increases. The obtained results evaluated the role of the parasitoid *D. rapae* in controlling *B. brassicae* and *A. craccivora* under greenhouses and field conditions. Results emphasized that the efficiency of this aphid parasitoid had been decreased with the highly population density of aphids. It may be concluded that when this aphid parasitoid used as biological control agents against *B. brassicae* and *A. craccivora* should be used with low host density.

Keywords Aphid · Parasitoids · Biology · Mass production · Release · Egypt

1 Introduction

Aphids are one of insect groups whose economic importance increases with agriculture development [1]. Natural enemies include predatory arthropods (insects, true spiders and mites), parasitoids, predators, microbial pathogens, antagonistic fungi. Yet, most attention has focused on the regulation of arthropod pests by their arthropod natural enemies, and on habitat management [1–6]. Biological control is a satisfactory tactic in an integrated pest management strategy. Control of insect pests by

A. A. A. Saleh (✉)
Plant Protection Research Institute, Agricultural Research Center, Dokki, Giza, Egypt
e-mail: amin_ahmed4u@yahoo.com

© Springer Nature Switzerland AG 2020   73
N. El-Wakeil et al. (eds.), *Cottage Industry of Biocontrol Agents and Their Applications*, https://doi.org/10.1007/978-3-030-33161-0_3

parasitoids is defined as the action of parasitoids that maintains a pest population at a low population size. Parasitism of aphid is density dependent [3, 7, 8]. The parasitoid *Diaeretiella rapae* (Hymenoptera: Aphidiidae) was considered by several authors to be important in the control of the cabbage aphid, *Brevicoryne brassicae* [2, 3]. The aphid species are major insect pests of many plants in several parts of the world. It is not advisable to use chemical pesticides to control such pests especially on the edible crops which are consumed as human food [4–7]. The routine application of these pesticides to control aphids has two adverse effects, firstly they are mostly non selective and kill aphid natural enemies, which causes further disturbance of the ecosystem. Secondly, as in other species, resistant strains of aphids may be expected to develop soon [3, 8, 9]. In order to reduce the environmental pollution by pesticides, biological control is one of the most important bases of integrated control programs and pest management. [10–20]. Some aphid parasitoids (Aphidiidae, Hymenoptera) are well known as active bio-control agents for many aphid species in the world and Egypt [21–27].

There were many species of indigenous hymenopteran parasitoids effectively attacking aphids. These parasitoids might have considerable potential in integrated pest management programmes for aphids infesting vegetables [25]. *D. rapae* is an important primary parasitoid of a wide range of aphid species world wide and Egypt as well, including major aphid pests such as cabbage aphid, *Brevicoryne brassicae* (L.), green peach aphid *Myzus persicae* Sulz, Russian wheat aphid *Diuraphis noxia* Mord, wheat aphids *Rhopalosiphum padi* (L.) and *Schizaphis graminum* Sulz., cotton aphid *Aphis gossypii* Glov., faba bean aphid *Aphis craccivora* Koch, corn leaf aphid *Rhopalosiphum maidis* F., reed plants aphid *Hyalopterus pruni* Geoffroy and oleander aphid, *Aphis nerii* Boyer [13–18].

Diaeretiella rapae is well known as a potential bio-agent for many aphid species in different countries [19, 20, 26–35]. Several hypotheses concerning the apparent adaptive significance of such effects of parasitoids can be proposed. For example, a paralyzed host may exhibit reduced defensive capabilities and also reduced tissue uptake of haemolymph nutrients, thereby providing a greater supply of nutrients for parasitoid [20, 23].

D. rapae played a significant role in suppressing populations of *B. brassicae* and should be taken into consideration in control programs aim at protecting cruciferous vegetable crops against aphid pests [33–36]. Large-scale production system was established to rear parasitoids. This consisted of rearing laboratory containing cages, of potted plants. These plants are infested with aphids and then used to infest larger cages or houses [15]. The present study aims to focus the light on the parasitoids of seven aphid species; *Brevicoryne brassicae* (L.), *Aphis craccivora* Koch, *Aphis nerii* Boyer, *Aphis gossypii*, *Rhopalosiphum maidis*, *R. padi* and *Hyalopterus pruni*, Geoffroy. These species are considered the main pests infesting: cabbage, cauliflower, faba bean, Cowpea, Dafla (oleander plants), cucumber, maize and Hagana (reed plants) [14–19].

The study is concerned with the following aspects

1. Survey and Seasonal abundance of certain aphids (*B. brassicae, A. craccivora, A. nerii, R. padi, R. maids* and *H. pruni*) and estimation of the percentages of parasitism on the population sizes of the aphid species and the parasitoid *D. rapae* during four seasons.
2. Estimating the effect of certain weather factors (Temperature and relative humidity) on the population size of the aphid parasitoid population, *D. rapae* and its aphid hosts.
3. Host suitability and host stage preference of certain aphid species for *Diaeretiella rapae*.
4. Studying the biological aspects, host suitability, physiology of the most aphid parasitoid, and *D. rapae* on certain aphid species.
5. Evaluating the role of the parasitoid, *D. rapae* in controlling the cabbage and cauliflower aphid under field conditions.
6. Performance of parasitoid *D. rapae* and the biology on certain aphid species on both laboratory and field.
7. Mass production and estimating the role of the parasitoid, *D. rapae* in controlling the cabbage and cauliflower aphid species, *Brevicoryne brassicae* (L.) and cowpea aphid *Aphis craccivora* (Koch) in greenhouses and in the field.

2   Seasonal Abundance of Aphid Parasitoids

2.1   Survey and Abundance of Aphid Parasitoids

Brevicoryne brassicae was the dominant aphid species infesting cabbage and cauliflower crops. Only one parasitoid species namely: *Diaeretiella rapae* was found to emerge from the mummified aphids of *B. brassicae*. In this respect, *B. brassicae* was the aphid species infesting cabbage and cauliflower crops in Egypt [28, 32–37]. The main parasitoid emerged from the mummified aphid was *D. rapae*. *Aphis craccivora* was the dominant aphid species infesting faba bean crop [14]. Three parasitoids were emerged from the mummified aphids. They were *D. rapae*, *Ephedrus persica* and *Trioxys sp.* (Table 1).

Results agree with [14, 17, 26, 38] in Egypt, they stated that the hymenopterous parasitoids, *D. rapae*, *L. fabarum*, *Ephedrus* sp. and a hyperparasitoid, *Aphidencyrtus sp.* emerged from mummified aphid *A. craccivora*.

Hyalopterus pruni was the major aphid species infesting Hagna plants (common weedy graminouses plants). The three parasitoids *D. rapae*, *Aphidius colemani* and *Aphelinus* sp. emerged from the mummified aphids (Table 1). *Aphidius Colemani* and *D. rapae* are considered among important parasitoid species on the aphid *H. pruni*. *Aphis nerii* was the main aphid species infesting Dafla plants.

Table 1 Aphid parasitoids surveyed from aphid species on different host plants in Egypt

Host plant	Aphid species	Parasitoids	
		Scientific name	Status
Cabbage and cauliflower	*Brevicoryne brassicae* (L.)	*Diaeretiella rapae* (M'Intosh)	Primary
		Pachyneuron sp. *Alloxysta sp.* (Cynipidae)	Secondary
Faba bean	*Aphis craccivora* (Koch)	*Diaeretiella rapae* (M'Intosh) *Ephedrus persicae* Froggatt	Primary
		Pteromalidae	Hyper
Dafla (Oleander plants)	*Aphis nerii* (Boyer)	*Diaeretiella rapae* (M'Intosh) *Aphidius matricariae* Haliday	Primary
		Pteromalidae	Hyper
Hagna (Reed plants)	*Hyalopterus pruni* (Geoffroy)	*Diaeretiella rapae* (M'Intosh) *Aphidius colemani* viereck *Aphelinus sp.* (Nees)	Primary
Cucumber plants	*Aphis gossypii* Glover	*Lysiphlebus fabarum* Marshall *Diaeretiella rapae* (M'Intosh)	Primary
		Pachyneuron sp.	Hyper
Cowpea	*Aphis craccivora* (Koch)	*Lysiphlebus fabarum* Marshall *Diaeretiella rapae* (M'Intosh)	Primary
		Aphidencyrtus sp.	Hyper
Corn	*Rhopalosiphum maidis*	*Diaeretiella rapae* (M'Intosh)	Primary
	Rhopalosiphum padi Linnaeus	*Paron sp.*	Primary

Nerium oleander [39–42]. The three parasitoids *D. rapae*, *Aphidius matricariae* and *Aphelinus sp.* emerged from the mummified aphids. On the same host plant, [43] in Greece showed that the most common parasitoid species attacking the oleander aphid *A. nerii* in Greece were *A. Colemani*, *Binodoxys angelicae*, *D. rapae* and *P. volucre*. Five primary parasitoids, *D. rapae* and *Aphidius sp.*; while *Pachyneuron sp.*, *Alloxysta sp.* and *Aphidencyrtus sp.* were Hyperparasitoids on the aphid *A. nerii* [19].

Aphis gossypii was the aphid species infesting cucumber crop. The primary parasitoids emerged from the mummified aphid were: *Lysiphlebus fabarum*, *D. rapae*,

Binodoxys angelica. Also, one secondary parasitoids *Pachyneuron* sp. emerged from the mummified aphid. The parasitoid, *Ephedrus cerasicola, Lysiphlebus testaceipes* and *A. colemani* emerged from mummified aphid *A. gossypii* on cucumber. Meanwhile, [44, 45] mentioned that *A. matricariae* parasitoid on aphid *A. gossypii* on cucumber plant in Germany. However, in Egypt, studies on the seasonal fluctuations of *A. gossypii* and associated predators and parasitoids proved that Chrysopid species the most abundant predators and *Trioxy auctus* was first record on *A. gossypii* in Egypt [10]. On the other hand, host acceptance and host suitability of the cotton aphid *A. gossypii* for *Lysiphlebus japonicas* and *A. colemani* [43–46].

The following is a list of hymenopterous primary and hyper parasitoid species that emerged from cowpea aphid *A. craccivora* during the period of study:

Primary parasitoids: *Lysiphlebus fabarum Diaeretiella rapae* and *Trioxys sp.* Hyperparasitoids: *Aphidencyrtus* sp. [47–51].

However [47, 52, 53] reported that *A. craccivora* was parasitized by *Aphidius colemani* and *L. fabarum* in Cowpea fields, *L. fabarum* recorded as a parasitoid on *A. craccivora* in Egypt [38]. Moreover, *Trioxys angelicae* (Hal.) recorded as parasitoid of *A. craccivora* [19, 38, 48]. The present results agree with those of [19, 26, 29] who found that *D. rapae*, *L. fabarum* and *Ephedrus* sp. as parasitoids on *A. craccivora* in Egypt. *Lysiphlebus fabarum, A. matricariae* and *Trioxys* sp. were found attacking *A. craccivora* on faba bean cultivar [18, 54–76].

2.2 Seasonal Abundance of Aphid Parasitoids and Percentages of Parasitism

2.2.1 Seasonal Abundance of *B. brassicae* on Cabbage Plants and Its Parasitoid Species

The parasitoid, *D. rapae*, showed different numerical responses to increased densities of aphids [46]. *Pachyneuron aphidis* was reported as a hyperparasitoid on *D. rapae* by [41, 42]. The highest number of *B. brassicae* was (590 individuals/20 inch^2) while, in the 2nd season the highest number of *B. brassicae* was (656 individuals). Two peaks of *B. brassicae* were (530 individuals), and (590 individuals) during 2010 season [12] were recorded. Several authors attribute lack of control of *B. brassicae* by *D. rapae* to a delay in parasitoid response to aphid population increase [54, 77], while in the 2nd season, three peaks for *B. brassicae* were recorded on cabbage plants during the 2nd and 4th weeks of December, and 4th week of January where (577, 656 and 497) individuals, respectively were recorded [31].

However, the aphid *B. brassicae* is a major pest on cruciferous plants, in several parts of the world especially cabbage and cauliflower in Egypt [10, 14, 51, 52]. The aphid parasitoid *D. rapae* is a primary parasitoid on *B. brassicae* infesting cabbage had three peaks (257), (306) and (224 individuals) in the first season, while four peaks were recorded in the 2nd season (372), (225), (290) and (226 individuals) [29]. The

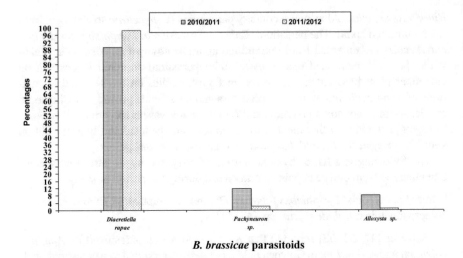

Fig. 1 Occurrence percentages of primary and hyper parasitoid species on cabbage crop with *Brevicoryne brassicae* during 2010/2011 and 2011/2012 seasons

maximum number of mummified aphids was (306 individuals) in the first season and (372 individuals) in the second one [29]. The percentages of parasitism ranged between 0.49 and 13.37% during the period from the 3rd week of January and the 1st week of April. At the end of the season, it fluctuated between 55 and 100%; in the last week of April and 1st week of May, 1990 respectively. In the 1991 season it was between 0.99 and 9.71% in the 1st week of January and the 3rd week of March. At the end of the season the percentages were 54.67, 97.29 and 100% in the 3rd, 4th week of April and 1st week of May.

The rate of parasitism of *B. brassicae* was about four times as that of *M. persicae*. There were no significant differences in the longevity, development time and percentage of successful emergence between *D. rapae* individuals reared on the two hosts [18, 54–56, 77, 78].

Figure 1 shows the intensity percentage of *D. rapae* to the total catch of this parasitoid during two years of study, it was 80.07 and 97.02%. The highest total parasitism percentage was 83.84% which was recorded in the 4th week of February in the first season while it was 79.45% in the 3rd week of February in the 2nd one [27]. *Diaeretiella rapae* played the major role towards suppressing *B. brassicae* populations [29].

2.2.2 Seasonal Abundance of *B. brassicae* on Cauliflower and Its Parasitoid Species

The rates of parasitism were usually 0–3%, and rarely exceeded 10%. Rates of parasitism generally decreased during the phase of rapid increase of aphid numbers and showed a marked increase only when aphid populations were in their final sharp decline phase. Results indicated that the parasitoids did not substantially influence the aphid populations. Rates of parasitism of the primary parasitoids averaged 50% (with a maximum of 100%) [57]. Highest population abundance of *B. brassicae* was (643 individuals) in the first season of 2011, but in the second season, it reached 676 individuals [27]. Meanwhile, the lowest number was 376 individuals in 2011 season [10], while it was (396 individuals) during 2012 [3, 27]. As for parasitoid *D. rapae*, the main primary parasitoid of *B. brassicae* on cauliflower, it had three peaks 79, 230, and 282 individuals in the first season, while four peaks were recorded in the second one (127, 264, 326 and 278 individuals) [10, 15]. The maximum number of mummified aphids was (282 individuals) in the first season, and 326 individuals in 2011/2012 season. Parasitoids were made available for the surrounding crops [56].

Later this system became also a source of its specificity for the cabbage aphid; *D. rapae* clearly dominated the parasitoids (more than 90% of individuals). Parasitism was high enough to free cauliflower heads from cabbage aphids [31]. Figure 2 shows the intensity percentage of the primary parasitoid *D. rapae* to the total catch of these parasitoids during two years of study, it occupied 88.26 and 87.13% in the two study seasons, respectively. The highest total parasitism rate was 56.85% it was recorded in the 2nd week of February in the first season while it was 65.85% in the 1st week of March in the 2nd season [10, 29]. However, [33] in Egypt, studied the seasonal abundance of *B. brassicae* and its associated parasitoids on cauliflower plants in Egypt. Also, he found that four to six peaks of abundance were recorded for the aphids on the leaves during the two seasons of 1990 and 1991 [33, 51–53, 79, 80]. The percentages of parasitism on *B. brassicae* by *D. rapae* in Egypt reached the peak

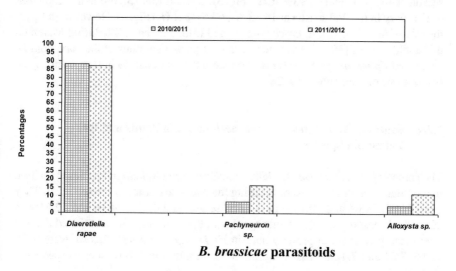

B. brassicae parasitoids

Fig. 2 Occurrence percentages of primary and hyper parasitoid species on cauliflower crop infested with *Brevicoryne brassicae* during 2010/2011 and 2011/2012 seasons

of 56.85% in the 2nd week of February in 2011 season and 65.8% in the 1st week of March in 2012 season on cauliflower plants [10, 23].

2.2.3 Seasonal Abundance of *A. craccivora* on Faba Plants and Its Parasitoid Species

The highest population abundance of *Aphis craccivora* was (892 individuals) during the 1st week of November, 2010 in the 1st season, but in the 2nd season it reached 866 individuals during the 3rd week of October. Meanwhile, the lowest number (330 individuals) occurred in the 4th week of January in the first season, while in 2011/2012 season, it reached 397 individuals during the 1st week of February. Faba bean is attacked by several insect pests including the cowpea aphid, *A. craccivora* which is considered a key pest of faba bean cultivation in Egypt according to Saleh [10].

From Fig. 3, it could be noted that the maximum number of mummified aphids was recorded in the 2nd week of December (70 individuals) [19], when the temperature and relative humidity were 15.78 °C and 66.79% R.H. in the first season and (69 individuals) in the 2nd season during the 4th week of January (Fig. 4).

Figure 5 shows the intensity percentages of *D. rapae*, *Ephedrus persica* and *Trioxys sp.* in relation to the total catch of these parasitoids during the two seasons of study, it was 48.75, 17.35 and 20.34% in the first season and 60.2, 20.05 and 5.18% in the 2nd season of study [19]. The parasitism percentage increased sharply to reach its maximum 13.94% which was recorded in the 4th week of January in the first season at 14.99 °C and 63.93% R.H. (Fig. 3), while it was 13.10% in the 2nd week of February in the 2nd season at 14.83 °C, 60.57% R.H. (Fig. 4). However, in Egypt the rate of parasitism on *A. craccivora* ranged from 15.4 and 22% during March on this aphid species [38]. The seasonal means of parasitism rates of the parasitoid *D. rapae* and *Ephedrus sp.* on *A. craccivora* were 8.17 and 6.45% in Egypt during the two seasons of their study [19, 28].

2.2.4 Seasonal Abundance of *Aphis nerii* on Dafla Plants and Its Parasitoid Species

The percentages of *D. rapae*, *Aphidius matricariae* and *Aphelinus* species in relation to the total catch of all parasitoids during the two years of study are recorded. They were 54.07, 24.65 and 10.74% in the first season and 49.12, 24.16 and 12.41% in 2011/2012 season (Fig. 6). Results agree with those of [54] who mentioned that the total mean parasitism of primary parasitoids *D. rapae* and *Aphidius sp.* were 8.74, 12.66, 7.01 and 7.41% in Zagazig and Mansoura during two seasons, respectively (Fig. 6).

The percentage of parasitism gradually increased to reach the maximum 35.14% in 2010/2011 season recorded in the 4th week of February [46], while it was 40.78% in the 2nd week of March in the 2nd season at 18.88 °C, 58.86% R.H. [77, 78, 81].

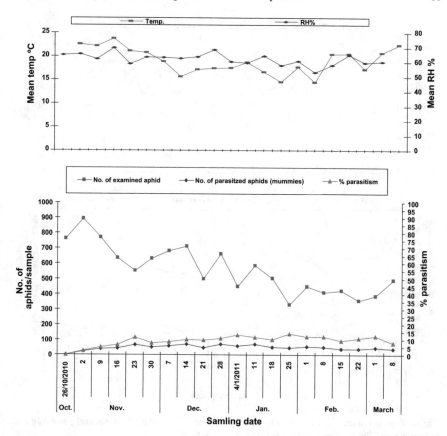

Fig. 3 Population density of *A. craccivora*, number of mummified aphids and percentages of parasitism on faba bean plants during 2010/2011 season

However, mentioned that the parasitism rates on *A. nerii* tend to range between 1 and 10% [46]. Although the percentages of parasitism ranged from 30 to 45.69% on some host plants; *Asclepias* species and Dafla plants [47, 54].

In Zagazig district:

1. *Diaeretiella rapae*

In the first season (Fig. 7), the parasitoid disappeared completely by the end of September and continued to the 1st week of December, *D. rapae* was found in very high density resulting in (100% parasitism) during the period extended from the 2nd week of December to the 1st week of January. During the next period until the 1st week of March the parasitoid, *D. rapae* remained in high relative density (85.71–92) [18]. The mean density of the parasitoid in that season was 88.45% during the 2nd season (Fig. 7). A similar trend was noticed, where *D. rapae* was the most dominant species representing 93.33–100% of the total parasitoids during the period from the

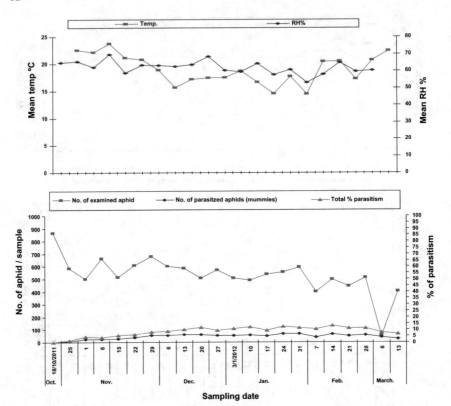

Fig. 4 Population density of *Aphis craccivora*, number of mummified aphids and percentages of parasitism on faba bean plants during 2011/2012 season

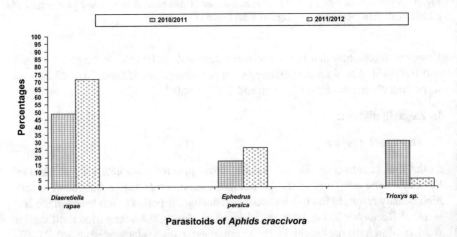

Fig. 5 Occurrence percentages of primary parasitoids species on faba bean crop infested with *A. craccivora* during 2010/2011 and 2011/2012 seasons

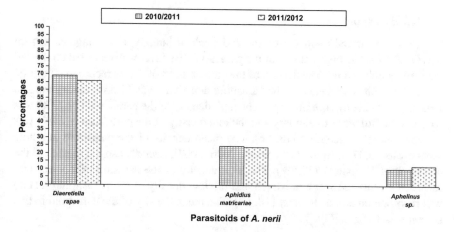

Fig. 6 Occurrence percentages of primary parasitoid species on Dafla plant infested with *Aphis nerii* during 2010/2011 and 2011/2012 seasons at Sharkia governorate

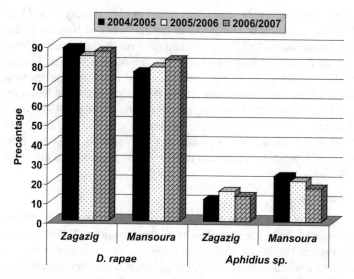

Fig. 7 Occurrence percentage of different aphid parasitoid species on Dafla plants infested with *A. nerii* during 2004–2005, 2005–2006 and 2006–2007 seasons

2nd week of December to late December. *D. rapae* remains in high relative density (90.67–91.67). The means of the parasitoid in whole season was 84.63%. In the last season (Fig. 7), *D. rapae* was found in very high density (100%) during late November and continued to the 3rd week of December. The mean season percentage of parasitism was 86.95% [18, 28].

2. *Aphidius* spp.

Aphidius spp. started to appear on the 2nd week of January, Percentage parasitism was (14.92%). The high density of the parasitoid (16.42%) was recorded on the 3rd week of March and the low density of the parasitoid (8.0%) was recorded on the 1st week of March. The means of the parasitoid density in was 11.55% during the 2nd season (2005–2006), *Aphidius* spp. The high density of the parasitoid (22.73%) was recorded on 3rd week of January and the low density of the parasitoid (6.67%) was recorded on late December [18, 56]. The mean density of the parasitoid in whole season was 15.37%. In the last season (2006–2007), *Aphidius* spp. appeared at the 4th week of December (12.5%). The high density of the parasitoid (16.94%) was recorded on the 2nd week of March and the low density of the parasitoid (9.41%) was recorded on late of January [18, 53]. The mean density of the parasitoid in this season was 13.05% (Fig. 7).

In Mansoura district:

1. *Diaeretiella rapae*

In the 1st season (Fig. 7), *D. rapae* was found in high density (100%) during the period from the 1st week of November to the 2nd week of November, The lowest density of the parasitoid (57.89%) was recorded on the 3rd week of January [18, 56]. The mean density of the parasitoid in that season was 76.68%. During the 2nd season (Table 6 and Fig. 7), *D. rapae* was the most dominant species representing 86.67–100% of the total parasitoids during the period from late November to the 2nd week of December. The lowest density of the parasitoid (69.49%) was recorded on late January [18]. The mean parasitoid density was 79.11%. In the last season (Table 7), *D. rapae* was found in high density (100%) during the period from the 3rd week to end of November. The mean density of the parasitoid in the entire season was 82.68% [46].

2. *Aphidius* spp.

Aphidius spp. started to appear on the 3rd week of November, the percentage of parasitism was 18.75%. The high density of the parasitoid (28.57%) was recorded on the 3rd week of December. The means density of the parasitoid in this season was 23.32% during the 2nd season (2005–2006), *Aphidius* spp., disappeared completely at 2nd week of September to 4th of November [18, 56]. The high density of the parasitoid (30.51%) was recorded on 4th week of January and the low density of the parasitoid (11.11%) was recorded on 1st week of December. The mean density of the parasitoid in whole season was 20.89% [18]. In the last season (2006–2007), *Aphidius* sp. appeared at 2nd week of December; it was (16.28%). The high density of the parasitoid (24.44%) was recorded on 2nd week of December [33]. The mean density of the parasitoid in this season was 17.32% (Fig. 7).

Percentage of parasitism ranged between 1.16 and 34.01% during the period from the 2nd of December to the 4th of February in the first season (2004–2005). The mean total percentage of parasitism in that season was 9.77%. During the 2nd season (2005–2006), percentage of parasitism ranged between 2.35 and 28.03% during the period from 2nd of December and 4th of March. The mean total percentage of parasitism in whole season was 8.69%. Percentage of parasitism ranged between 0.58 and 45.69% during the period from 4th week of November and 3rd week of March [3, 15, 28, 77]. The mean total percentage of parasitism in last season (2006–2007) was 12.66%. In Zagazig district. In Mansoura district, percentage of parasitism ranged between 1.8 and 10.28% during the period from 1st week of November and 3rd week of February during (2004–2005). The mean total of parasitism was 3.39% in this season. In the season of (2005–2006) it was between 3.72 and 22.11% in the 4th of November and 2nd week of February. The mean total of parasitism was 7.32% in 2nd season. In the last season (2006–2007) it was between 2.39 and 19.18% in the 3rd week of November and 1st week of February. The mean total of parasitism was 7.34% in last season.

As in many aphid-parasitoid systems [3, 77, 78], parasitoid did not play a significant role in regulating population growth of *A. nerii* on *Asclepias* species. Aphid densities continued to grow exponentially up to the end of the growing season [81]. Aphid parasitoids often exploit a small number of available hosts, and parasitism rates tend to range between 1 and 10% [53]. Although, parasitism rates reached over 30% on some host plants, i.e. *Asclepias* species [81]; *Pachyneuron* species achieved 100% and established on the 1st week of December. Then, it declined to the lowest activity (40%). During late season of 2007, the highest activity with (100%) on the 1st and 2nd week of January [28, 55].

Pachyneuron sp. (100%) was established on 1st week of December. Then, it declined to the lowest activity (40%). During late season of 2007, the highest activity with value was (100%) on the 1st and 2nd week of January [28, 55]. The lowest activity was 36.36% on 1st week of March in Zagazig (Fig. 8).

The highest activity (100%) was established on the last week of January in the first season 2005. in the 2nd season 2006, the 1st record of *Pachyneuron* sp. (100%) was established on 2nd week of January. Then, it was fluctuated to record the lowest activity with value (40%) in the 4th week of February. During the last season 2007, the 1st appearance of hyper-parasitoid, P. sp. was (100%) in 1st week of January [18]. Then, it was the highest activity with value was (33.33%) on the last week of January in Mansoura district (Fig. 8). The recorded hyperparasitoid on *D. rapae* were *Pachyneuron* sp., *Atrichoptilusnea avenues* Masi (Pteromalidae), *Alloxysta minuta* Horting (Cynepidae) and *Pachyneuron aphidis* Boch and *Alloxysta* sp. in Egypt [12, 18, 29–31, 51, 79]. Hyperparasitism has traditionally been viewed in the context of applied ecology as being harmful and so it is believed to have usually a negative impact on beneficial primary parasitoids. To obtain hyperparasitoids, mummies (one or two days after mummification) were exposed to hyperparasitoids for two days and individually isolated in micro-tubes [18]. There is a contrary speculation as to hyper parasitoids possible positive role in maintaining a proper balance between the primary parasitoids and their hosts by preventing an excessive build up of parasitoid numbers [39, 55].

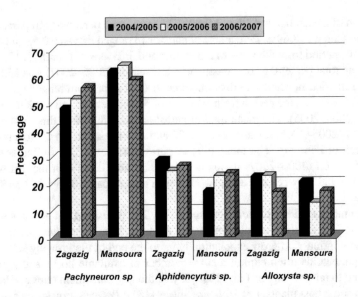

Fig. 8 Occurrence percentages of hyperparasitoids on primary parasitoid infested *A. nerii* during 2004–2005, 2005–2006 and 2006–2007 seasons

2.2.5 Seasonal Abundance of *Aphis gossypii* and Its Parasitoids on Cucumber Plants

Lysiphlebus fabarum adults began to appear on 2nd week of April (8 individuals) then the number of parasitoid in increased gradually to reach the maximum (39 individuals) in the last week of April (Fig. 9), while it had one peak in the 2nd season 2016. This peaks occurred in the 4th week of April (85 individuals) [58]. The data also cleared that *D. rapae* a parasitoid for *A. gossypii* on cucumber had two peaks in the 1st season 2015. These peaks were 17 and 5 individuals, while the 3rd peak occurred in the 4th week of April (39 individuals). The data also cleared that *Binodoxys angelica* a parasitoid for *A. gossypii* on cucumber had two peaks in the first season 2015, while it had one peak in the 2nd season 2016 [58]. This peaks occurred in the 4th week of April (19 individuals) (Fig. 10). The data also cleared that *Pachyneuron* sp. a hyper parasitoid had few number of *Pachyneuron* sp. Adult was recorded during the two seasons. *Pachyneuron* sp was recorded during the period of April–June in the two seasons [58]. The percentages were 53.7, 22.6, 15.1% and 8.6% in the first season, 2015, and 52.2, 27.7, 12.1 and 7.89% in the 2nd season 2016, Fig. 11.

The percentages of parasitism ranged from 3.14 to 21.0% in the first season 2015; while the percentages of parasitism starting by 2.66% in the 2nd week of April and it increased until reached the peak of 42.66% in the 2nd season 2016 the development and evaluation of an open rearing system for the control of *A. gossypii* by *L. testaceipes* in greenhouse on sweet pepper plant in Brazil [61]. The parasitism percentage of *A. house A. gossypii* ranged from 5 to 13%. The highest total parasitism

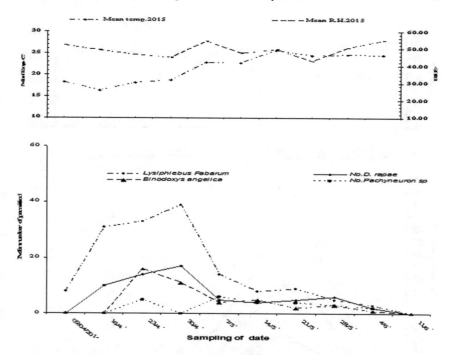

Fig. 9 Population density of *Aphis gossypii* aphid and number of parasitoid aphids in cucumber field during 2015 season

ratio was 21.0% during the 4th week of April in the first season 2015 as shown in Fig. 9, while it was 42.66% during the 4th week of April in the 2nd season 2016 as shown in Fig. 10. The parasitism and development of *L. testaceipes on A. gossypii* was studied by [57–59] on sweet papper. Percentage of parasitism was higher, 44.2% adult emergence 92.6%. However, the development and evaluation of an open rearing system for the control of *A. gossypii* by *L. testaceipes* in greenhouse on sweet pepper plants was recorded by [53] in Brazil. The parasitism percentage of *A. gossypii* ranged from 5 to 13%. On the other hand, [44] recorded that *L. testaceipes* and *Ephedrus cerasicola* parasitized 26–23% of *A. gossypii*, *A. colemani* parasitized 72.80% of aphid while *Aphidius matricariae parasitized* less than 6% of *A. gossypii*.

2.2.6 Seasonal Abundance and Estimation of Parasitism in the Cowpea Field

The primary parasitoid *Lysiphlebus fabarum* was the most dominant species with mean relative densities 69.10 and 63.26% during 2016 and 2017 seasons, respectively [19]. While *Trioxys* sp. recorded 19.22 and 20.83%, respectively, the Pteromalids was represented by 11.68 and 15.91%, respectively (Tables 2 and 3).

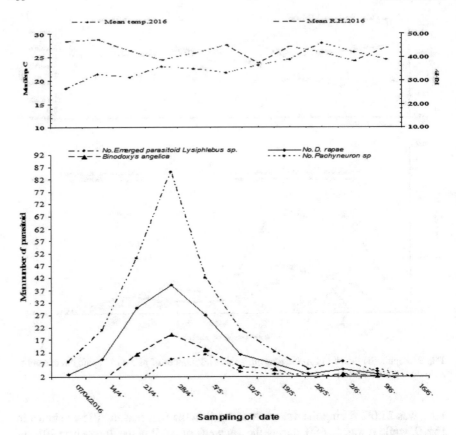

Fig. 10 Population density of *Aphis gossypii* aphid and number of parasitoid aphids in cucumber field during 2015/2016 season at Diarb Nagem district

Table 3 showed that the percentage of parasitism ranged from 1.69 to 19.11%, on the 3rd week of July and second week of September respectively, during the first season [19, 56]. In the 2nd season (Table 2), percentage of parasitism ranged between 0.90 and 13.80%, on the 4th week of July and 1st week of October at means 34 °C and 59% R.H., and 37 °C and 60% R.H. respectively [28]. Total means of parasitism rates by *L. fabarum*, *Trioxys* sp. and *Aphidencertus* sp, recorded 4.56 and 6.63% during 2016 and 2017 seasons respectively. The percentage were 69.10, 19.22, and 11.68% in the first seasons 2016 and 63.26, 20.83, and 15.91 in the seasons [39].

2.2.7 Seasonal Abundance of *R. maidis* on Maize Plants and Its Parasitoid Species

Infestation started to appear from the beginning of the 1st week of July with (601 individuals/sample) *R. maidis* and (280 individuals/sample) *R. padi* (Table 4) while

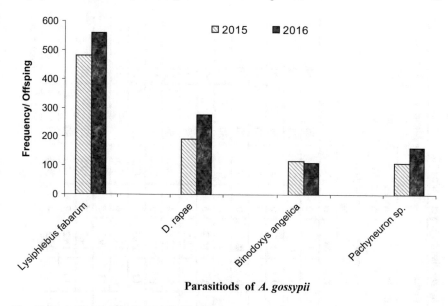

Parasitiods of *A. gossypii*

Fig. 11 Percentages of different parasitoid species on cucumber plant infested with *A. gossypii* during 2015 and 2016 seasons at Diarb Nigem district

it appeared during the 4th June with (592 individual/sample) *R. maidis* and (256 individual/sample) *R. padi*, in 2010 season (Table 5). *D. rapae* was the primary parasitoid emerged from *R. maidis* mummies and *Praon* sp. was the primary parasitoid emerged from *R. padi* mummies. Percentage of parasitism ranged between 0.75 to 3.00% and 0.84 to 3.49% for *D. rapae* on *R. maidis* in both seasons 2009 and 2010. Mean while the percentage parasitism of *Praon* sp. on *R. padi* ranged between 0.80 to 3.45% and 0.79 to 2.44% in both seasons respectively. The mean percentage of parasitism were 1.21 ± 0.47 and 1.71 ± 0.46 (*D. rapae*) and it were 1.38 ± 0.53 and 1.11 ± 0.35 (*Praon* sp.) in both seasons, respectively [28, 56]. However, AL Hag et al. (1996) recorded that *D. rapae* as an important parasitoid of *R. maidis* on wheat and barley fields in Saudi Arabia. Meanwhile [63] concluded that the main parasitoids of *R. padi* on maize plant were *Aphidius* sp., *Praon* sp., and *Aphelius* sp. On the other hand, *D. rapae* is an important primary parasitoid of a wide range of aphid species in the world and Egypt, *B. brassicae*, *Myzus persicae*, *Diuraphis noxia*, *A. gossypii*, *A. craccivora*, *R. padi*, *R. maidis* and *A. nerii* [15, 54].

Table 2 Parasitism percentage of *A. craccivora* (Koch.) infested cowpea plants, cultivated in Zagazig district, Egypt, during 2016

Sampling date (weeks)	*A. craccivora*	No. of parasitized aphid (mummies)			Parasitism %	Emerged parasitoids						Mean °C	Mean R.H.
						Primary parasitoids				Hyper parasitoids			
						Lysiphlebus fabarum		*Trioxys* sp.		*Aphidencertus* sp.			
		A	B	Total		No	No	No	No	No	No		
June 3rd	201	0	0	0	0	0	0	0	0	0	0	36	61
4th	304	0	0	0	0	0	0	0	0	0	0	36	59
July 1st	497	0	0	0	0	0	0	0	0	0	0	24	61
2nd	331	0	0	0	0	0	0	0	0	0	0	30	60
3rd	530	0	9	9	1.69	5	100	0	0	0	0	30	60
4th	590	0	18	18	3.05	9	100	0	0	0	0	30	55
5th	771	0	29	29	3.76	15	83.3	3	16.7	0	0	30	57
August 1st	509	0	37	37	7.27	21	77.8	6	22.2	0	0	32	56
2nd	463	7	43	50	10.8	30	76.9	5	12.8	4	10.3	32	51
3rd	690	29	35	64	9.28	29	69.0	10	23.8	3	3.14	30	59
4th	501	43	29	63	12.6	25	55.6	12	26.7	8	17.8	31	56
September 1st	492	30	18	48	9.76	20	58.8	8	23.5	6	17.6	29	57
2nd	450	47	39	86	19.1	46	69.7	13	19.7	7	10.6	31	64
3rd	685	51	28	79	11.5	35	74.5	7	14.3	5	10.6	29	58
4th	607	47	33	80	13.5	38	69.1	10	18.2	7	12.7	30	56
October 1st	372	14	10	24	6.45	7	50.0	3	21.3	4	28.6	27	61

(continued)

Table 2 (continued)

Sampling date (weeks)	A. craccivora	No. of parasitized aphid (mummies)			Parasitism %	Emerged parasitoids						Mean °C	Mean R.H.
						Primary parasitoids				Hyper parasitoids			
		A	B	Total		Lysiphlebus fabarum		Trioxys sp.		Aphidencertus sp.			
						No	No	No	No	No	No		
2nd	297	8	5	13	4.38	4	40.0	2	20.0	4	40.0	28	64
Total	8290				113	248	924	79	219	48	151		
Mean	487				6.63	14.6	54.3	4.6	12.9	2.8	8.90		

RD = Relative density

A = No. of mummies counted at the date of inspection

B = No. of mummified host appeared during laboratory rearing

Table 3 Parasitism percentage of A. craccivora (Koch.) infested cowpea plants, cutivated in Zagazig, Egypt, during 2017

Sampling date (weeks)	A. craccivora	No. of parasitized aphid (mummies)			Parasitism %	Emereged parasitoids						Total	Mean °C	Mean R.H.
						Primary parasitoids				Hyper parasitoids				
		A	B	Total		Lysiphlebus sp. fabarum		Trioxys sp.		Aphidencertus sp.				
						No	No	No	No	No	No			
4th	370	0	0	0	0	0	0	0	0	0	0	0	29	59
July 1st	410	0	0	0	0	0	0	0	0	0	0	0	40	56
2nd	394	0	0	0	0	0	0	0	0	0	0	0	40	67
3rd	500	0	0	0	0	0	0	0	0	0	0	0	40	63
4th	650	0	0	0	0	0	0	0	0	0	0	0	38	59
5th	442	0	4	4	0.90	2	100	0	0	0	0	12	34	59
August 1st	581	0	22	22	3.79	12	100	0	0	0	0	12	37	59
2nd	554	0	25	25	4.51	13	86.67	2	13.33	0	0	15	37	58
3rd	767	0	43	43	5.61	19	67.90	9	32.14	0	0	28	36	61
4th	507	0	38	38	7.50	17	73.91	5	21.74	0	0	22	36	55
September 1st	595	6	46	46	7.73	20	64.52	9	29.03	2	6.45	31	32	62
2nd	840	15	41	56	6.67	22	62.86	7	20	6	17.14	35	34	63
3rd	651	33	20	53	8.14	20	64.52	6	19.35	5	16.13	31	34	60
4th	570	19	27	46	8.07	15	55.56	4	14.81	8	29.63	27	35	55
October 1st	500	39	30	69	13.8	20	51.28	9	23.07	10	25.64	39	37	60

(continued)

Table 3 (continued)

Sampling date (weeks)	A. craccivora	No. of parasitized aphid (mummies)			Parasitism %	Emereged parasitoids						Total	Mean °C	Mean R.H.
		A	B	Total		Primary parasitoids				Hyper parasitoids				
						Lysiphlebus sp. fabarum		Trioxys sp.		Aphidencertus sp.				
						No	No	No	No	No	No			
2nd	394	11	15	26	6.6	5	33.33	3	20	7	46.67	15	33	57
3rd	315	5	8	13	4.13	2	28.57	1	14.29	4	57.14	7	30	62
Total	9040				77.45	167	63.26	55	20.83	42	15.91	264		
Mean	531.76				4.56	9.82	3.72	3.24	1.22	2.47	0.93	15.53		

RD = Relative density

A = No. of mummies counted at the date of inspection

B = No. of mummified host appeared during laboratory rearing

Table 4 Monthly mean percentages of parasitism *R. maidis* by *D. rapae* on maize plants during two successive seasons of 2009 and 2010

Seasons	Sampling dates	No. of examined aphid	No. of parasitized			C.	Parasitism %
			A	B	Total		
2009	July	558	0	0	0	558	0
	August	375	2	3	5	370	1.33
	September	391	3	6	9	382	2.30
Mean		441.33 ± 41.37	1.66	3	4.66	436.67	1.21 ± 0.47
2010	June	521	0	0	0	521	0
	July	482	3	5	8	474	1.66
	August	498	5	8	13	485	2.61
	September	429	4	7	11	418	2.56
Mean		482.5 ± 14.75	3	5	8	474.5	1.71 ± 0.46

RD = Relative density
A = No. of mummies counted at the date of inspection
B = No. of mummified host appeared during laboratory rearing

Table 5 Monthly mean percentages of parasitism *R. padi* by *Praon* sp. on maize plants during two successive seasons of 2009 and 2010

Seasons	Sampling dates	No. of examined aphid	No. of parasitized			C.	Parasitism %
			A	B	Total		
2009	July	274	0	0	0	274	0
	August	257	1	3	4	253	1.56
	September	231	2	4	6	225	2.59
Mean		254 ± 9.67	1	2.33	3.33	250.66	1.38 ± 0.53
2010	June	252	0	0	0	252	0
	July	263	0	2	2	261	0.76
	August	283	2	4	6	277	2.12
	September	255	2	2	4	251	1.57
Mean		263.25 ± 5.27	1	2.0	3	260.25	1.11 ± 0.35

RD = Relative density
A = No. of mummies counted at the date of inspection
B = No. of mummified host appeared during laboratory rearing

3 Biological Studies

3.1 Life Cycle of D. rapae Reared on B. brassicae, A. craccivora and A. nerii

D. rapae is a solitary koinobiont endoparasitoid, i.e. one parasitoid larvae grows within the aphid body, consuming host internal tissues without stopping host growth (aphids can still produce some offspring in the first days after parasitism, although much less than un-parasitized aphids) [64–69]. parasitoids can be classified into four types of species with differing lifetime reproductive strategies and life history traits. Type 1 includes pro-ovigenic species, which possess their total egg load at emergence and do not mature additional eggs during adult life, and Type 2 includes weakly-synovigenic species which also emerge with an initial load of mature eggs but can mature additional one [27]. s. Type 3 and Type 4 include species which do not possess mature eggs at emergence, and mature a substantial part, or all, of their eggs during adult life. Typically, Type 1 and 2 have high fecundities, short lives and concentrate ovipositions at the beginning of their life, whereas Type 3 and 4 have low fecundities, long lives, and oviposit all along their life. *D. rapae* is weakly synovigenic and lays the majority of its eggs in the first four days of life (Jamont et al. 2013), and it lives a short time compared to other parasitoid species [69–73].

The period of egg stage in host body differed, significantly in different aphid species. The incubation period of egg lasted an average 4.29 ± 0.50 (2–5) days on *B. brassicae*. This period averaged 4.04 ± 0.49 (2–5) days on *A. craccivora* and 4.49 ± 0.41 (2–6) days on *A. nerii*. Therefore, *D. rapae* oviposition period was extended from 3 to 4.5 days, which is a period of particularly high fecundity. Indeed, *D. rapae* is a synovigenic parasitoid, i.e. it can mature eggs during its adult stage [74] of mature eggs in ovaries of *D. rapae* is at its highest between the 2nd and 4th day after emergence [82–85].

The larval stage duration lasted an average of 6.11 ± 0.64 with a range of 4–8 days on *B. brassicae*, while it was 5.89 ± 0.50 (3–8) days on *A. craccivora*, opposed to 6.62 ± 0.39 (4–8) days on *A. nerii*. The prepupal and pupal periods altogether ranged from 5 to 7 days, with an average of 6.31 ± 0.30 days on *B. brassicae* [56]. While, this period was 4–6 days with an average of 5.18 ± 0.19 days on *A. craccivora* and 7.27 ± 0.41 (5–9) days on *A. nerii* (Table 6). When larval growth is near to completion and most host tissue has been consumed, the *D. rapae* larva pupates inside its host, which turns brown and round-shaped, forming an aphid 'mummy' from which the adult eventually emerges. As an adult, *D. rapae* does not feed on host tissue (unlike some parasitoid species), but it can feed on carbohydrate-rich fluids, such as floral and extrafloral nectar, honeydew, or synthetic sucrose solutions. After emergence, it can live 6–14 days when fed on nectar, which is about 2–5 times longer than if it has access to water only and no food [74, 82, 83].

The mean of total developmental period from deposition of egg to adult's emergence ranged from 11 to 20 days with an average 16.97 ± 0.14 days on *B. brassicae*, opposed to 14.65 ± 0.15 (9–19) days on *A. craccivora* and 18.38 ± 0.91 (11–23)

Table 6 Effect of aphid host species on the developmental periods of the parasitoid *D. rapae* immature stages at 19.5 °C and 63.63% R.H.

Host plant	Host aphid	Egg stage	Larval stage	Pupal stage	Total developmental period/days
Cabbage	*B. brassicae*	4.29 ± 0.50[a] (2–5)	6.11 ± 0.64[a] (4–8)	6.31 ± 0.30[a] (5–7)	16.97 ± 0.14[a] (11–20)
Faba bean	*A. craccivora*	4.04 ± 0.49[a] (2–5)	5.89 ± 0.50[a] (3–8)	5.18 ± 0.19[b] (4–6)	14.65 ± 0.15[b] (9–19)
Dafla	*A. nerii*	4.49 ± 0.41[a] (2–6)	6.62 ± 0.39[a] (4–8)	7.27 ± 0.41[a] (5–9)	18.38 ± 0.91[a] (11–23)
LSD$_{0.05}$		0.49947	0.79915	0.9989	1.9978

Mean under each variety having different letters in the same raw denote a significant different ($p \leq 0.05$)
Data expressed as Mean ± S. D
*$p \leq 0.05$; **$p \leq 0.01$

days on *A. nerii* (Table 6). The parasitoid *D. rapae* completed its life-cycle in a period ranged from 8 to 18 days at 26 ± 1 °C [64]. The biology of *T. agnelicae* on *A. craccivora* and found that the parasitoid completed its developmental cycle within 15.4 days at 21.7 °C and R.H. 61.34% [48].

The parasitoid *D. rapae* completed its life-cycle in *B. brassicae* throughout mean periods of 17.2, 12.6 and 10.3 days at 20, 25 and 30 °C, respectively [65]. *D. rapae* completed its life-cycle in a period of 12–18 days at 19.5 °C on *B. brassicae* and 11–15 days on *A. craccivora* [26]. On the other hand, the total development period of the parasitoid *D. rapae* lasted 16–24 days, with an average of 19.87, 24.39, 16.34 and 18.55 days in *B. brassicae*, *A. nerii*, *A. craccivora* and *H. purni*, respectively [27].

3.2 Effect of Temperature and Food Supply on the Longevity of Parasitoid D. rapae

Considering that parasitoids are known to visit and feed on flowers, feeding on nectar benefits parasitoid longevity, fecundity and searching performance, and monocultures are generally devoid of sugar resources, floral subsidies are intuitively expected to augment parasitism rates and pest control [66]. The presence of honeydew might mitigate potential exploitative competition for nectar, by providing a sugar source when nectar is intensely consumed by other insects such as pollinators [67]. From data presented in Tables 7, 8 and 9, it could be, generally, observed that adult females lived for a longer period than males, irrespective of host aphid species from which the adults emerged, the nutritive solution on which the adults were supplied, and also the temperature at which the adults were kept. It could be, also observed that keeping

Table 7 Effect of temperature and food supply on longevity of *D. rapae* adult emerged from *B. brassicae* mummies

Group	Treatment	Temp. (°C)	Adult longevity (days)			
			Female		Male	
			Range	Mean ± SE	Range	Mean ± SE
A	–	16.9	4–8	6.05 ± 0.4^d	3–5	3.9 ± 0.2^d
B	+	16.9	6–9	8.22 ± 0.36^c	5–7	6.89 ± 0.29^c
C	–	9	16–23	19.35 ± 0.80^b	7–10	9.20 ± 0.30^b
D	+	9	36–46	39.58 ± 1.03^a	23–37	28.78 ± 1.12^a
$LSD_{0.05}$				1.8828		1.8828

– Starved; + supplied with droplets of bee honey
Mean under each variety having different letters in the same raw denote a significant different ($p \leq 0.05$)
Data expressed as Mean ± S. D
*$p \leq 0.05$; **$p \leq 0.01$

Table 8 Effect of temperature and food supply on longevity of *D. rapae* adult emerged from *A. craccivora* mummies

Group	Treatment	Temp. (°C)	Adult longevity (days)			
			Female		Male	
			Range	Mean ± SE	Range	Mean ± SE
A	–	17.5	3–5	4.90 ± 0.18^d	2–4	3.64 ± 0.13^d
B	+	17.5	5–8	7.54 ± 0.29^c	5–7	6.32 ± 0.26^c
C	–	9	12–16	14.94 ± 0.40^b	6–9	7.74 ± 0.30^b
D	+	9	29–36	33.84 ± 0.78^a	19–27	24.36 ± 0.89^a
$LSD_{0.05}$				0.94142		0.94142

– Starved; + supplied with droplets of bee honey
Mean under each variety having different letters in the same raw denote a significant different ($p \leq 0.05$)
Data expressed as Mean ± S. D
*$p \leq 0.05$; **$p \leq 0.01$

the emerged adults at low temperature (9 °C) led the *D. rapae* adults to survive for 3–4 times or more longer than longevities recorded for adults kept at 16.9 °C. From data in the same Tables 7, 8 and 9, it could be also observed that longevities of *D. rapae* adults emerged from *B. brassicae* mummies were the longest (3.9–28.78 days for males and 6.05–39.58 days for females) compared to those emerged from *A. craccivora* mummies (3.64–24.36 and 4.9–33.48 days, respectively). Two recent studies provide such diet comparison in aphid parasitoids, however their results are contradictory; *Lysiphlebus testaceipes* lived as long on buckwheat nectar as on host (*Aphis gossypii*) honeydew diets [76], but the longevity of *D. rapae* was 3.5 times shorter when fed *B. brassicae* L. honeydew than *Vicia faba* L. extra-floral nectar

Table 9 Effect of temperature and food supply on longevity of *D. rapae* adult emerged from *A. nerii* mummies

Group	Treatment	Temp. (°C)	Adult longevity (days)			
			Female		Male	
			Range	Mean ± SE	Range	Mean ± SE
A	−	18.2	3–5	4.23 ± 0.24^d	2–3	3.10 ± 0.11^d
B	+	18.2	5–8	6.73 ± 0.34^c	4–6	5.49 ± 0.34^c
C	−	9	10–16	13.38 ± 0.63^b	5–8	6.85 ± 0.31^b
D	+	9	27.36	31.74 ± 0.93^a	17–25	21.6 ± 0.91^a
$LSD_{0.05}$				0.4707		0.4707

− Starved; + supplied with droplets of bee honey
Means followed by the same letters in a column are not significantly different at 1% level of probability (Duncan's Multiple Range test)

[74]. In that case, *D. rapae* longevity on honeydew was not significantly different from control wasps given access to water only, suggesting that *B. brassicae* honeydew has no nutritional value to this parasitoid. While those emerged from *A. nerii* mummies showed the shortest longevities (3.1–21.6 and 4.23–31.74 days for males and females, respectively) (Table 9).

Parasitoids may exploit sugar present in honeydew; e.g., in citrus orchards, the parasitoid *Aphytis melinus*, whose host does not produce honeydew, feeds on honeydew of various phloem-feeding insects [68]. It is necessary to know and list these sources, because if they provide enough, accessible and constantly-available carbohydrates, floral subsidies may not be visited by parasitoids.

Increased reproductive output of *D. rapae* implies that more eggs are matured, but also that more hosts are found, i.e. nutrition also affects the behaviour of female parasitoids. Few studies documented how nectar feeding can affect the activities performed by some parasitoids, and the available knowledge on *D. rapae* behaviour does not cover the effect of metabolic state. To fully understand how nectar affects *D. rapae* fitness, behavioural observations of wasps fed with buckwheat nectar or water only, and then exposed to hosts [69].

As for the effect of supplying honey droplets for feeding *D. rapae* compared to the starved adults, data in Table 7 show that among adults emerged from *B. brassicae* mummies fed males lived at 16.9 °C for 6.89 days and at 9 °C for 28.78 days, opposed to 3.9 and 9.2 days, respectively for the starved adults. Correspondent longevities for females were 8.22 and 39.58 days for fed adults and 6.05 and 19.35 for starved adults [15]. Those emerged from *A. craccivora* lived for 6.32 days at 17.5 °C and 24.36 at 9 °C in case of fed males and 3.64 and 7.47 days for starved males. These values were 7.54 and 33.84 (fed) and 4.9 and 14.94 days for starved females (Table 8). Feeding also altered the suite of actions performed by parasitoids. For example, the aphid parasitoid *A. ervi* allocated more time to walking (explorative behaviour) and attacking hosts when fed nectar than when starved, in which case it stayed stationary [70].

Values recorded for those emerged from *A. nerii* mummies were 5.49 and 21.6 (fed males); 3.1 and 6.85 days (starved males), opposed to 6.73 and 31.74 (fed females) and 4.23 and 13.38 days for starved females at 18.2 and 9 °C, respectively (Table 9). However, the adult life span of adult parasitoids is affected by many factors such as temperature, humidity, food, presence or absence of hosts, etc. [39]. On the other hand, the longevity was affected by temperature and food supply of the parasitoid *D. rapae* [26, 27, 48].

3.3 *Effect of Host Aphid Species on* D. rapae *and Adults' Emergence*

The sex-ratio and percentage of adults' emergence of *D. rapae* in the field and of two laboratory generations on three aphid species; *B. brassicae*, *A. craccivora* and *A. nerii* are assessed. On *B. brassicae*, in the field the percentage of parasitoid emergence was 84.88% with the sex-ratio 2.25 females: 1 male. While, in the laboratory, the percentage of adults' emergence in the first generation was 76.97% with the sex-ratio 1.29 female: 1 male, while in the 2nd generation, 69.04% emergence occurred from host mummies with the sex-ratio 1.02 female: 1 male (Table 10).

Meanwhile, on *A. craccivora* in the field, the percentage of *D. rapae* emergence from host mummies was 78.96% with the sex-ratio 1.81 female: 1 male. While, in the laboratory the percentage of adults' emergence and sex-ratio were 71.67% and 1.19:1, respectively opposed to 67.93% and 1.04 female: 1 male, respectively in the 2nd generation [71]. As for the third species of aphids (*A. nerii*), the percentages of parasitoid emergence from host mummies were 71.31, 65.46 and 57.63%, respectively with sex-ratios (female: male) 1.42:1, 1.1:1 and 1.06:1 for adults emerged from mummies collected in the field and the two laboratory reared generations, respectively (Table 10).

Statistical analysis confirmed significant differences between percentages of *D. rapae* emergence from *B. brassicae*, *A. craccivora* and *A. nerii*, and also between those from mummies from the field and each of the two successive generations. Highest percentage of emergence was that from field collected *B. brassicae* mummies. Among the emerged adults, the sex-ratio was almost 1:1 except for those emerged from field collected *B. brassicae* mummies among which the sex-ratio was in favor to females, being 2.25:1. In this respect, [71] reported that sex-ratio of the parasitoid *D. rapae* (females: males) was 1.7:1 by rearing the parasitoid for five successive generations, sex-ratio was almost 1:1 in the first three generations, but males dominated in the 4th and 5th generations.

Table 10 Effect of host aphid species on *D. rapae* and adults' emergence

Host aphid	Source parasitoid	Total no of mummies	Total no. of adults emerged	Percentage of adults' emergence	Females	Males	Sex ratio male: female
B. brassicae	In the field	979	831	84.88 a	575	256	1:2.25
	1st generation	534	419	76.97 c	236	183	1:1.29
	2nd generation	407	281	69.04 e	142	139	1:1.02
A. cracivora	In the field	385	304	78.96 b	196	108	1:1.81
	1st generation	293	210	71.67 d	114	96	1:1.19
	2nd generation	237	161	67.93 e	82	79	1:1.04
A. nerii	In the field	251	179	71.31 d	105	74	1:1.42
	1st generation	194	127	65.46 f	65	62	1:1.105
	2nd generation	118	68	57.63 g	33	35	1:1.06
LSD$_{0.05}$				1.7454**			

Mean under each variety having different letters in the same raw denote a significant different ($p \leq 0.05$)

Data expressed as Mean ± S. D

$*p \leq 0.05$; $**p \leq 0.01$

3.4 Behaviour of the Parasitoid D. rapae on Different Host Densities

Feeding increased the time allocated to searching, and greatly decreased time spent immobile. Searching, defined as rapid walking, was almost absent in unfed parasitoids [27]. Results may suggest that, in 30 min, fed parasitoids could explore an area that is 40 times larger than unfed parasitoids, and according to model studies, a 40-fold increase in search rate over a parasitoid's lifetime would divide pest equilibrium population by the similar rate of 40, which would represent a very significant improvement of biocontrol [72]. However, as ingested sugars are consumed via metabolic activity [73]. *D. rapae* may maintain an intense searching behaviour if they feed frequently on nectar, which would also increase their longevity [74, 83]. After alighting on a Brussels sprout plant, female *D. rapae* tended to walk up the stem, stopping at leaf-nodes and walking up petioles to explore leaves, and eventually reached the top of the plant which was checked intensively for hosts, before the parasitoid flew off [75]. Because nectar-fed *D. rapae* spent more time in rapid walk than did starved individuals, they might be faster in their journey from leaf to leaf [69, 75]. As reported by [18] the leaf-reaching is a measure of the attractiveness potency of the semiochemicals emitted by the food plants and the host insects. Data in Table 11 indicate that the leaf-arrival time decreased with increasing the host population density. While, the number of oviposition (No. of stings) and number of resultant aphid mummies increased with increasing the host density. The maximum leaf-arrival time was 8.41 ± 0.25 min when the host density was 20 and it started to decrease as the host density increased which became the minimum 0.86 ± 0.17 min at host density of 80 individuals (Fig. 12). The time of the first sting increased as the host population increased. It was minimum 8.92 ± 0.53 min at host density of 20 individuals, while its maximum value was 21.13 ± 0.30 min at host density of 80 individuals (Fig. 13). The number of stings (oviposition) increased as the host population increase. The lowest value was 6.6 ± 0.5 at population of 20 individuals and reached the maximum value 65.8 ± 2.86 when the host population became

Table 11 Behavior of the parasitoid *D. rapae* on cabbage at different *B. brassicae* densities

Host density	Leaf-arrival time (min)	First sting time (min)	No. of stings (oviposition)	No. of mummies
20	8.41 ± 0.25 a	8.92 ± 0.53 c	6.6 ± 0.5 d	3.4 ± 0.39 d
40	5.88 ± 0.23 b	14.57 ± 0.37 b	28.2 ± 2.07 c	8.2 ± 0.73 c
60	1.83 ± 0.23 c	15.88 ± 0.51 b	49.5 ± 0.81 b	13.2 ± 1.01 b
80	0.86 ± 0.17 d	21.13 ± 0.30 a	65.8 ± 2.86 a	19.00 ± 0.54 a
$LSD_{0.05}$	0.6765**	1.3286**	5.5076**	2.1409**

Means followed by the same letter in a column are not significantly different at 0.05% level
*$p \leq 0.05$; **$p \leq 0.01$

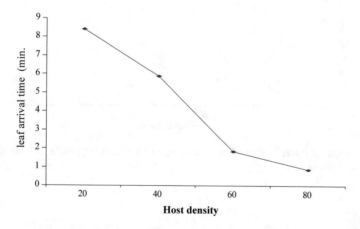

Fig. 12 Graphical representation of leaf arrival time (min) (mean $\pm$ SD) by *D. rapae* against *B. brassicae*

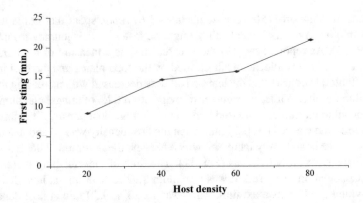

Fig. 13 Graphical representation of first sting (min) (mean ± SD) by *D. rapae* against various densities of *B. brassicae*

80 individuals (Fig. 14). Also, the number of the formed mummies increased by increasing the population density. Its minimum value was 3.4 ± 0.39 at host population of 20 individuals and increased gradually to become maximum (19.00 ± 0.54) when the host population became 80 individuals (Fig. 15). The increased number of antennal encounters, oviposition and number of mummies with increase of host density might be due to increased concentration of the kairomones excreted by the host aphids. These kairomones enhance the activity of the parasitoid, thus increasing its potentiality to locate more host individuals [27, 71, 86].

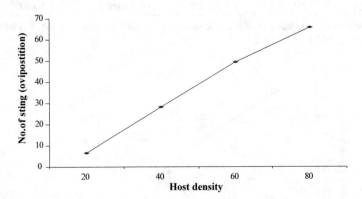

Fig. 14 Graphical representation of number of stings (oviposition) (mean ± SD) by *D. rapae* against *B. brassicae*

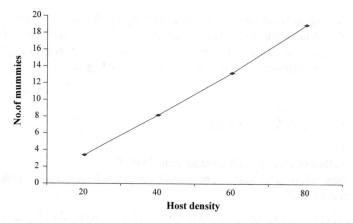

Fig. 15 Graphical representation of No. of mummies (mean ± SD) by *D. rapae* against *B. brassicae*

3.5 *Conservation of the Parasitoid* **D. rapae**

Storage and release of the adult parasitoids is also important when males are lacking. In this case, it is possible to hold virgin females after they have deposited male eggs and subsequently, mate with their own male progeny [7, 87]. The maximum percentages of emergence (91.43 and 57.14%) were obtained from mummies of *B. brassicae* and *A. nerii*, respectively, when the mummies were kept at (9 ± 0.4 °C) for 5 h before storage for one week at (5 ± 0.4 °C). While, the maxima percentages of mortality (65.71 and 88.67%) occurred when the mummies were kept in the refrigerator at (9 ± 0.4 °C) for 15 h before storage for 4 weeks at (5 ± 0.4 °C) in *B. brassicae* and *A. nerii* mummies, respectively. The total number of mummies were kept at different periods before storage (5, 10 and 15 h) and then stored for one week recorded the mean percentages of 84.76, and 51.43% emergence from the 120 mummies of *B. brassicae* and *A. nerii*, respectively, while the correspondent mean mortality percentages were 15.24 and 48.57% [15]. Mummies kept for the same pre-storage periods and then stored for two weeks recorded 71.43 and 39.05% emergence from the total mummies 28.57 and 60.95% mortality in *B. brassicae* and *A. nerii*, respectively. The highest total percentage of emergence after three and four weeks storage periods were (54.29 and 39.05%) in case of *B. brassicae* mummies, and (28.57 and 15.24%) in the mummies of *A. nerii*. Meanwhile, the mean mortality percentages after storage for the same periods were (45.71 and 60.95%) and (71.43 and 84.76%) in the mummies of *B. brassicae* and *A. nerii*, respectively.

The mummies of aphidiids had a higher rate of emergence if kept when newly formed at low temperature, therefore the age in which the parasitoids had been stored seemed to be the best [88, 89]. Similar conclusion was reported by [5] in Egypt who stored freshly formed mummies of *A. uzbekistanicus*, for one month at 3 °C after being kept in the refrigerator at 8 ± 0.5 °C for periods of 6, 12 and 15 h, different results about *Aphidius matricariae* Hal. storage for two weeks in southern France and Also for two and four weeks had been tested [90, 91]. These results are in general agreement with the work of [92] in France and [9, 15] in Egypt who stored *D. rapae* in freshly formed mummies of *B. brassicae*, for more than two months at 5 °C after being kept in the refrigerator at 8 ± 0.4 °C for 9, 12 and 15 h.

4 Biochemical Assessment

Determination of total soluble protein colorimetric
Determination of total soluble protein in homogenated aphids was carried out at as described by [92].

– Determination of total lipids: Total lipids were estimated by the method of [93].
– Free amino acids determination: Total amino acids were calorimetrically assayed by ninhydrin reagent according to the methods described by Lee [94].
– Total carbohydrates: Total carbohydrates were estimated in acid extract of aphids by the phenol-sulphuric acid reaction of [95]. Total carbohydrates were extracted and prepared for assay according to [96].

Several hypotheses concerning the apparent adaptive significance of such effects of parasitoids can be proposed. For example, a paralyzed host may exhibit reduced defensive capabilities and also reduced tissue uptake of haemolymph nutrients, thereby providing a greater supply of nutrients for parasitoid [97].

Data presented in Table 12 show the level of total protein in the supernatant of the homogenated aphid species *Hypermoyzus lactucae* and *H. pruni* recorded the highest significant level; (34.733 ± 1.22 and 34.533 ± 1.23 mg/g. t. wt., respectively). While *B. brassicae* gave the lowest significant one (12.89 ± 0.560 mg/g. t. wt.), P = 0.0000.

Table 12 Nutrient compounds of different aphid species

Species	Total protein (mg/g. b. wt.)	Total carbohydrates (mg/g. b. wt.)	Total lipids (mg/g. b. wt.)	Free amino acids (mg/g. b. wt.)
B. brassicae	12.896 ± 0.56 c	10.49 ± 0.33 c	4.616 ± 0.27 a	1.65 ± 0.09 b
H. pruni	34.533 ± 1.23 a	23.10 ± 0.95 a	1.573 ± 0.09 b	0.596 ± 0.06 c
A. craccivora	30.00 ± 1.04 b	10.42 ± 0.40 c	1.21 ± 0.006 bc	4.266 ± 0.15 a
A. nerii	32.033 ± 0.98	10.73 ± 0.45 c	0.88 ± 0.04 c	0.726 ± 0.01 c
H. lactucae	34.733 ± 1.2 a	12.903 ± 0.64 b	1.143 ± 0.04 c	0.556 ± 0.04 c
LSD$_{0.05}$	3.260	1.884	0.419	0.298
P	0.0000***	0.0000***	0.0000***	0.0000***

Mean under each variety having different letters in the same raw denote a significant different (p≤0.05)
Data expressed as Mean ± S.D.
*p≤0.05; **-***p≤0.01

Results also indicate significant increase in the total carbohydrate was regarded in the case of *H. pruni* (23.10 ± 0.95 mg/g. t. wt). Reversely, *A. nerii*, *B. brassicae* and *A. craccivora* recorded the lowest significant reduction (10.73 ± 0.45, 10.49 ± 0.33 and 10.42 ± 0.40 mg/g. t. wt.), respectively P = 0.0000, Table 12.

As for total lipids, *B. brassicae* manifested the highest significant level of total lipids (4.616 ± 0.27 mg. g. t. wt.) followed by *H. pruni* 1.573 ± 0.09, *A. craccivora* 1.21 ± 0.006, *Hypermoyzus lactucae* 1.143 ± 0.04 and *A. nerii* 0.88 ± 0.04 mg/g. t. wt., P = 0.0000. *A. craccivora* and *B. brassicae* produced the highest significant increase in free amino acids (4.266 ± 0.15 and 1.65 ± 0.09 mg/g. b. wt.), respectively, Table 12. On the other hand, *Hypermoyzus lactucae* recorded the least significant decrease (0.556 ± 0.04 mg/g. t. wt., P = 0.0000) [56]. Available informations indicated that, there are few researches about the relation between parasitism percentages and nutrition components of different species of aphids. The parasitoid aphid continues to feed, grow and develop. The host represents an open resource system in the future, as opposed to current resources [56]. Parasitoid larvae grew at different rates in different aphids of similar size, which suggest that quality is a specific attribute to each host species [98]. The host species may influence the rate of development and the survival of a parasitoid. A host may be unsuitable due to the lack of some necessary nutritional or hormonal resource [99, 100]. Our data on *D. rapae* showed that host aphid species, *B. brassicae*, *A. nerii* and *A. craccivora* were nutritionally and physiologically suitable for parasitoid development but *B. brassicae* was considered the best host among the species tested, adult emergence and higher parasitization. That is probably attributed to its higher nutrient composition of total lipids and free amino acids. These amino acids rapidly incorporate to produce large amounts of proteins that necessary for the developing parasitoid. Lipids can convert to proteins to substitute the reduction in protein content or produce supplementary energy used for growth and development. In addition, they include important hormones and pheromones [101, 102].

5 Mass Production and Field Application of Aphid Parasitoids

5.1 Cabbage Insect Pests

5.1.1 Laboratory Experiments

Biological control is satisfactory program in an integrated pest management. Control of insect pests by parasitoids is defined as the action of parasitoids that maintains a pest population at a low level. Parasitism of aphid has been shown to be density dependent [3]. Most cases of biological control release have concentrated only on the agent and its prey/host and have ignored the fact that the released agent may become part of a food web comprising species native to the region of introduction

Table 13 Effect of parasitoid density on percentage of parasitism in Petri dishes under laboratory conditions (16.0 ± 1 °C and 75.0 ± 2% R.H.)

Parasitoid density	Mean ± SD	Mean ± SD	Mean ± SD	Mean ± SD	Mean ± SD
	No. of aphid parasitised	No. of adults emerged	No. of adults non emerged	Percentage of adult emergence	Percentage of parasitism
1♀	25.2 E ± 0.72	20.6 E ± 4.16	4.6 D ± 2.07	82.20 A ± 4.57	12.7 E ± 2.71
2♀	41.6 D ± 4.98	33.6 D ± 4.28	8.0 D ± 1.58	80.74 A ± 3.27	20.8 D ± 2.49
4♀	59.4 C ± 9.13	45.8 C ± 5.12	13.6 C ± 4.51	77.53 B ± 4.26	29.7 C ± 4.56
8♀	72.6 B ± 8.41	55.8 B ± 7.33	16.8 C ± 1.92	76.77 B ± 2.41	36.1 B ± 3.81
12♀	79.0 B ± 4.18	59.8 B ± 3.96	18.8 B ± 4.55	75.66 C ± 1.69	40.5 B ± 2.85
16♀	95.6 A ± 6.91	70.8 A ± 4.82	24.8 A ± 4.54	74.16 C ± 3.77	47.8 A ± 3.46
F value	71.66	65.38	29.45	3.89	72.96

Means followed by the same letter in a column are not significantly different at 5% level

and others which had been previously introduced [69]. The maximum percentage of adult emergence of the parasitoids from mummies was 82.2 for *D. rapae* at one parasitoid per cage, while the minimum of 74.16 was recorded at 16 parasitoids per cage. With the increase of parasitoid density, the percentage of parasitism was 47.8 for *D. rapae* at 16 parasitoids per cage with a minimum 12.7 at one parasitoid per cage. There were significant differences in the percentage of parasitism and number of parasitized aphid among one, two, four and eight parasitoid per cage and 16 parasitoids/cage (Table 13). The maximum number of aphid parasitized was 95.6 for *D. rapae* at 16 parasitoids per cage with a minimum of 25.2 was at one parasitoid per cage.

Bionomics of *Diaeretiella rapae* at Varying Densities in the Pots

D. rapae played a significant role in suppressing populations of *B. brassicae* and should be taken into consideration in any control programs aimed at protecting Brassica crop against aphid pests Bahana and Karuhize [36].

The maximum percentage of parasitism was 82.32 for *D. rapae* at one parasitoid per cage and a minimum of 73.48 at 16 parasitoids per cage. There were significant differences between the percentage of adult emergence among one parasitoid per cage and two, four, eight and 16 parasitoids per cage but no such difference occurred between two parasitoids per cage and four, eight and 12 and 16 parasitoids per cage. With the increase of parasitoid density the rate of parasitism and the number of

Table 14 Effect of parasitoid number on percentage of parasitism in the pots under the laboratory conditions (16.0 ± 1 °C and 75.0 ± 2% R.H.)

Parasitoid density	Mean ± SD	Mean ± SD	Mean ± SD	Mean ± SD	Mean ± SD
	No. of aphid parasitised	No. of adults emerged	No. of adults non emerged	Adult emergence %	Parasitism %
1♀	29.0 F ± 6.32	24.0 E ± 5.87	4.8 E ± 0.84	82.32 A ± 3.5	14.5 F ± 3.16
2♀	54.0 E ± 7.11	42.6 D ± 3.29	11.4 D ± 4.34	79.4 A ± 5.65	27.0 E ± 3.55
4♀	71.2 D ± 8.61	55.2 C ± 9.31	16.0 C ± 2.45	77.19 A ± 4.99	35.6 D ± 4.31
8♀	86.8 C ± 6.72	65.8 B ± 8.14	21.0 B ± 1.87	75.59 B ± 3.86	43.4 C ± 3.36
12♀	104.2 B ± 4.82	77.8 A ± 6.26	26.4 A ± 3.97	74.33 B ± 3.97	52.3 B ± 2.39
16♀	114.8 A ± 7.92	84.2 A ± 3.19	30.6 A ± 4.83	73.48 B ± 2.52	57.4 A ± 3.96
F value	104.07	61.55	40.47	3.14	104.66

Mean under each variety having different letters in the same raw denote a significant different ($p \leq 0.05$)

Data expressed as Mean ± S. D

$*p \leq 0.05$; $**p \leq 0.01$

parasitized aphids increased [15]. The maximum percentage of parasitism was 57.4 for *D. rapae* at 16 parasitoids per cage and a minimum of 14.5 was recorded at one parasitoid per cage. The maximum number of aphid parasitized for *D. rapae* (114.8) was recorded at 16 parasitoids per cage and a minimum of 29.0 was recorded at one parasitoid per cage (Table 14).

Bionomics of *Diaeretiella rapae* at Varying Densities in the Field

The maximum percentage emergence was 83.71 for *D. rapae* at one parasitoid per cage and the minimum was 72.45 at 16 parasitoids per cage. There were significant differences in the percentages of adult emergence among parasitoid density of one parasitoid per cage, eight parasitoids per cage and 16 parasitoids per cage but no such differences occurred between one, two and eight parasitoids per cage and between 12 parasitoids and 16 parasitoids per cage. The maximum percentage of parasitism was 93.4 for *D. rapae* at 16 parasitoids per cage and the minimum was 43.9 at one parasitoid per cage [15]. The maximum number of parasitized aphid was 186.8 for *D. rapae* at 16 parasitoids per cage and the minimum was 87.8 at one parasitoid per cage. There were significant differences in the total numbers of parasitized aphid and the total percentage of parasitism at all densities (Table 15). The total percentage of

Table 15 Effect of parasitoid densities on percentage of parasitism in the field under 18.0 ± 1 °C and $73.0 \pm 2\%$ R.H.

Parasitoid density	Mean ± SD	Mean ± SD	Mean ± SD	Mean ± SD	Mean ± SD
	No. of aphid parasitised	No. of adults emerged	No. of adults non emerged	Percentage of adult emergence	Percentage of parasitism
1♀	87.8 F ± 11.82	73.6 C ± 10.78	14.2 D ± 1.30	83.71 A ± 1.39	43.9 F ± 5.91
2♀	101.6 E ± 9.84	81.6 C ± 7.70	20.0 C ± 2.55	80.33 A ± 1.15	50.8 E ± 4.92
4♀	115.4 D ± 8.32	82.6 C ± 20.37	25.6 C ± 1.82	81.39 A ± 8.94	57.7 D ± 4.16
8♀	114.0 C ± 12.86	108.0 B ± 6.04	38.2 B ± 5.54	75.19 B ± 2.97	72.4 C ± 6.01
12♀	169.0 B ± 7.04	125.0 A ± 5.96	44.0 B ± 3.53	73.96 C ± 1.76	84.5 B ± 3.52
16♀	186.8 A ± 5.40	135.4 A ± 11.59	51.4 A ± 9.50	72.45 C ± 5.15	93.4 A ± 2.70
F value	76.79	35.40	26.25	3.61	44.78

Means followed by the same letter in a column are not significantly different at 5% level of probability

parasitism increase in the field and decreased in the laboratory, which was probably due to some weather factors [103–108].

5.1.2 Evaluation of *D. rapae* in Controlling *B. brassicae* Under Field Conditions

The maximum total percentage of adult emergence was 80.85 for *D. rapae* at 1:10 ratio/cage and the minimum (75.53) was recorded at 1:50 ratio/cage. There were no significant differences in percentage of adult emergence at all tested ratios [109]. The maximum total of percentage parasitism was 54.4 for *D. rapae* at 1:10 ratio/cage and the minimum 22.31 was recorded at 1:50 ratio/cage. There were significant differences in total percentage parasitism at all ratios. The maximum total of number of aphid parasitized was 330.8 for *D. rapae* at 1:50 ratio/cage and the minimum 163.20 was recorded at 1:10 ratio/cage [15]. There were significant differences in the total number of aphid parasitized among 1:10 ratio, 1:20 ratio, 1:30 ratio and 1:40 ratio but no significant between 1:40 and 1:50 ratios (Table 16).

Table 16 Effect of different parasitoid densities (*D. rapae*) as mummies/host (*B. brassicae*) ratios on percentage of parasitism under field conditions

Parasitoid density	Mean ± SD	Mean ± SD	Mean ± SD	Mean ± SD	Mean ± SD
	No. of aphid parasitized	No. of adults emerged	No. of adults non emerged	Percentage of adult emergence	Percentage of parasitism
1:10	163.2 D ± 1.73	123.2 D ± 4.31	40.0 B ± 13.04	80.85 A ± 1.93	54.40 A ± 3.91
1:20	215.0 C ± 14.65	168.0 C ± 6.32	47.0 B ± 17.76	79.72 A ± 3.67	35.83 B ± 2.44
1:30	264.6 B ± 15.04	209.0 B ± 3.27	55.6 A ± 12.18	78.43 A ± 3.64	29.62 C ± 1.64
1:40	316.4 A ± 16.92	252.4 A ± 6.92	64.0 A ± 12.45	78.47 A ± 6.66	26.37 D ± 1.41
1:50	330.8 A ± 13.95	267.4 A ± 2.09	63.4 A ± 7.44	75.53 A ± 7.24	22.31 E ± 1.15

Means followed by the same letter in a column are not significantly different at 5% level

5.2 Inundative Release for D. rapae in Green Houses and in the Field

5.2.1 In the Green House

The parasitoids were released in the field and glass houses using cabbage leaves infested with colonies of parasitized aphids. Four density levels of 1:5, 1:10, 1:15 and 1:20 were planned in four green houses for the parasitoid and its host (*B. brassicae*) combination [15]. The initial numbers of the parasitoid mummies and aphid nymphs used for each treatment were 5880:29,400, 4015:40,150, 2881:43,215 and 3762:75,240 individuals, respectively, in 2003 season and 4046:20,230, 3245:32,450, 2640:39,600 and 3488:69,760 individuals, in 2003–2004 respectively [109].

Inundative Release for *D. rapae* in Green Houses and in the Field (1:5)

Two peaks of *B. brassicae* were recorded on cabbage plants. These peaks occurred in the 3rd week of January (625 individuals) and 4th week of February (855 individuals), respectively [109]. Three peaks for *B. brassicae* in the 2nd season were recorded on cabbage plants [42]. These peaks occurred in the 4th week of December (875 individuals), 1st week of January (860 individuals) and the 4th week of January (860 individuals) (Fig. 16). The aphid parasitoid *D. rapae, as* parasitoid of *B. brassicae* on cabbage plants had two peaks in the 1st season, while three peaks in the 2nd season were recorded [15]. The maximum number of mummified aphids was recorded in the 4th week of January (705 individuals) [109].

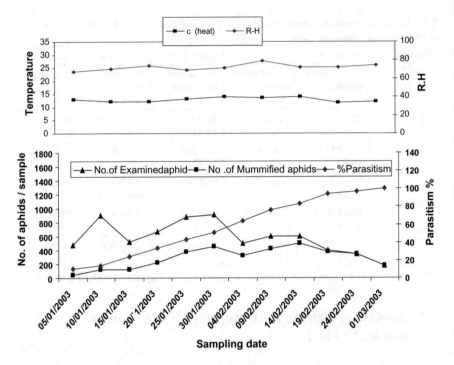

Fig. 16 Population density of *B. brassicae* and percentage of parasitism in cabbage field under green house at the parasitoid/host ratio 1:5 during 2003–2004 season

Population Size of *Brevicoryne brassicae* and Its Parasitoid

Three peaks of *B. brassicae* on cabbage plants occurred in the 2nd week of January (600 individuals), the 3rd week of January (1230 individuals) and the 2nd week of February (800 individuals) (Fig. 17). Three peaks also occurred in the 2nd season under almost similar conditions of temperature and relative humidity [109]; the parasitoid *D. rapae* occurred also one peak in the 1st season; while other two peaks occurred in the 2nd season. The highest total parasitism rate (100%) was recorded in the 3rd week of February in the 1st season [42], while it reached the same percent (100%) in the 1st week of February in the 2nd season (Fig. 18).

Population Size of *B. brassicae* and Its Parasitoid *D. rapae* (Ratio 1:15)

There were many species of indigenous hymenopteran parasitoids effectively attacking aphids. These parasitoids might have considerable potential in integrated pest management programmes for aphids infesting vegetables [25]. Four peaks of *B. brassicae* on cabbage plants, these peaks occurred in the 2nd week of January (600 individuals), 3rd week of January, 1st week of February (650 individuals) and 3rd

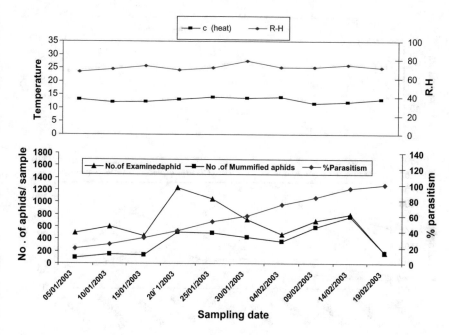

Fig. 17 Population density of *B. brassicae* and percentage of parasitism in cabbage field under green house conditions at the parasitoid/host ratio 1:10 during 2003 season

week of February (460 individuals) [15]. Three peaks also occurred in the 2nd season under almost similar conditions of temperature and relative humidity [109]. The parasitoid *D. rapae* also had three peaks in the first season, while four peaks occurred in the second one [15] The highest total parasitism ratios of 100% was recorded in the 4th week of February in the first season at 14.62 °C and 69.8% R.H. (Fig. 19), while it reached to 100% in the 2nd week of January in the 2nd season (Fig. 20).

Population Size of *B. brassicae* and Its Parasitoid *D. rapae* (Ratio 1:20)

Three peaks of *B. brassicae* were recorded in the 2nd week of January (900 individuals), 4th week of January (910 individuals), and 2nd week of February (600 individuals), respectively in the first season (Fig. 21). Five peaks were recorded for *B. brassicae* in the second season in the 3rd week of December (850 individuals), 4th week of December (1000 individuals), 2nd week of January (800 individuals), 3rd week of January (1055 individuals) and 2nd week of February (650 individuals) [15]. The aphid parasitoid *D. rapae* showed three peaks in the first season, while five peaks occurred in the 2nd season [109]. The highest parasitism ratio of 100% was obtained in the 4th week of February in the 1st season (Fig. 21), while it reached the same ratio in the 2nd week in the 2nd season (Fig. 22).

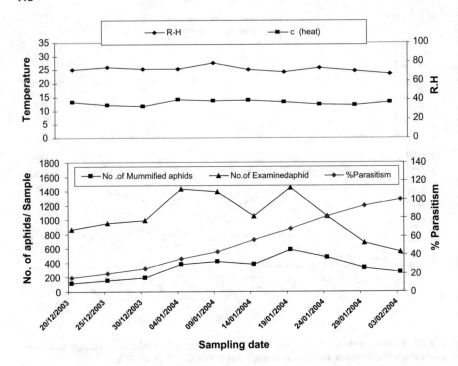

Fig. 18 Population density of *B. brassicae* and percentage of parasitism in cabbage field under green house at the parasitoid/host ratio 1:10 during 2003–2004 season

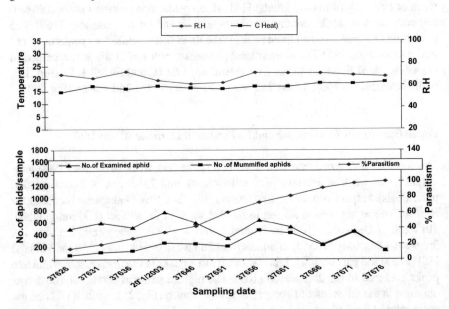

Fig. 19 Population size of *B. brassicae* on percentage of parasitism in cabbage of field under green house at the parasitoid/host ratio 1:15 during 2003 season

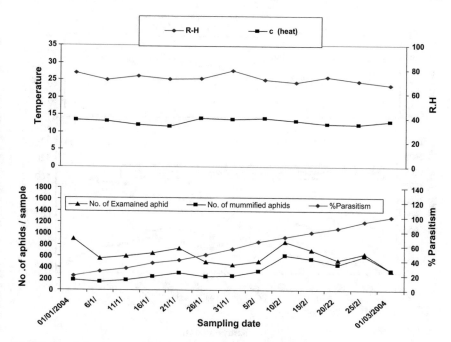

Fig. 20 Population density of *B. brassicae* on percentage of parasitism in cabbage field under green house at the parasitoid/host ratio 1:15 during 2003–2004 season

5.2.2 In the Field

However, natural enemies do not always drive the size of their hosts and preys down, because in some cases hosts can reproduce faster than, or ahead of, their parasitoids [110]. However, in the field, the lack of synchrony between *D. rapae* and *B. brassicae* *lessen* or reduces the impact of the parasitoid on its host [111].

The initial average numbers of the mummies and aphid nymphs was found 465:4215 in m^2 (1:9.28) to reach (1:5) it must add 398 mummies per m^2. The total number was 9950 mummies' [109]. In the second plot the initial numbers of the mummies and aphid nymph was found 378:5720 in m^2 must add in 25 m^2 (1:15.13) to reached (1:10) it must add 194 mummies per m [109, 112]. The same sampling techniques were taken in second cabbage field to compare the percentage of parasitism with release parasitoid and without release in the same time and in the same date [15].

Population Size of *B. brassicae* and Its Common Parasitoid *D. rapae* at Ratio 1:5

Three peaks of *B. brassicae* were recorded in the 1st week of February (650 individuals), 1st week of March (978 individuals) and 3rd week of March (575 individuals).

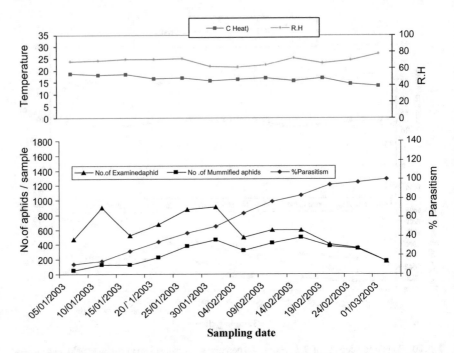

Fig. 21 Population density of *B. brassicae* and percentage of parasitism in cabbage field under green house at the parasitoid/host ratio 1:20 during 2003 season

The parasitoid *D. rapae* also showed three peaks [15]. It can be noted in Fig. 23 that the maximum number of mummified aphids was recorded in the 1st week of March (666 individuals) [109]. The percentage of parasitism increase until it reached 100% in the 3rd week of March 55 days after releasing (Fig. 24).

Population Size of *B. brassicae* and Its Parasitoid *D. rapae* at (Ratio 1:10)

Four peaks of *B. brassicae* were recorded on cabbage plants. These peaks occurred in the 1st week of January (900 individuals), 3rd week of January (735 individuals), 2nd week of February (850 individuals) and 4th week of February (640 individuals). The parasitoid *D. rapae* had four peaks occurred [15]. The maximum number of mummified aphids was recorded in the 2nd week of February (615 individuals) [109]. The percentage of parasitism increased until it reached 100% in the 1st week of March 65 days after from releasing in this time (Fig. 25).

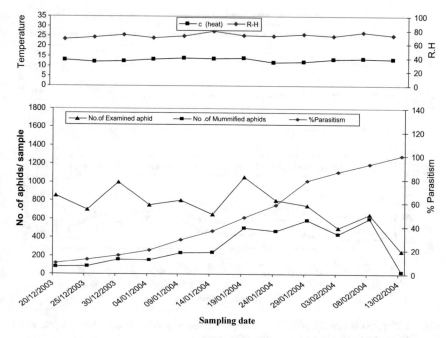

Fig. 22 Population density of *B. brassicae* on percentage of parasitism in cabbage field under green house at the parasitoid/host ratio 1:20 during 2003–2004 season

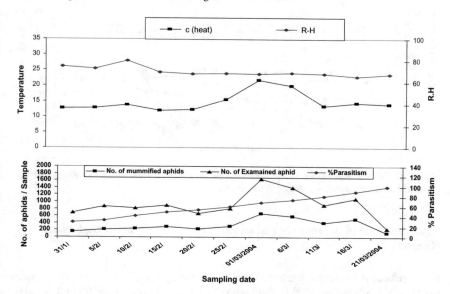

Fig. 23 Population density of *B. brassicae* percentage of parasitism on cabbage plants at the parasitoid/host ratio1:5 in the field during 2004 season

A. A. A. Saleh

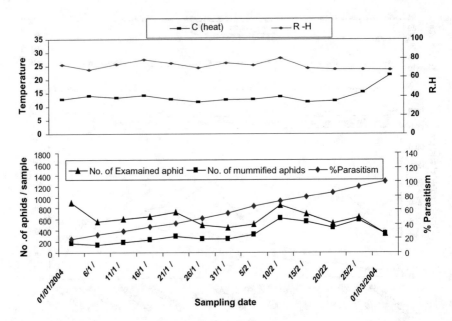

Fig. 24 Population density of *B. brassicae* percentage of parasitism on cabbage plants at the parasitoid/host ratio 1:10 in the field during 2004 season

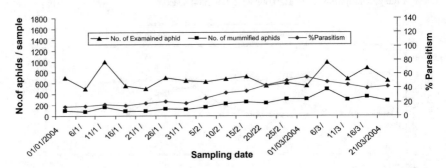

Fig. 25 Population density of *Brevicoryne brassicae*, number of mummified aphids and percentage of parasitism in cabbage fields during 2004 season

Population Size of *B. brassicae* and Its Parasitoid *D. rapae* in Cabbage Field (Control)

Four peaks of *B. brassicae* were recorded on cabbage plants. These peaks occurred in the 1st week of January (700 individuals), 2nd week of January (1000 individuals), 4th week of January (700 individuals) and 1st week of March (1000 individuals) [48]. The aphid parasitoid *D. rapae* also had seven peaks. The maximum number of mummified aphids in the 1st week of March was (488 individuals) [109]. The highest total parasitism ratio was 55.45% and recorded in the 1st week of March (Fig. 26).

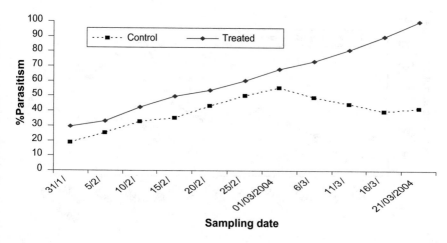

Fig. 26 Number of mummified aphids and percent of parasitism of the aphid *B. brassicae* by *Diaeretiella rapae* at the parasitoid aphid ratio of 1:5 in the field in the year 2004

Release of Cabbage Leaves with Mummies in the Field

In the field, aphids are distributed in patches, or colonies, which parasitoids search for and exploit. The optimal foraging theory predicts that parasitoids optimise the time spent in each patch to maximise their reproductive output [103]. This theory is supported by studies showing that patch residence time can be affected by various parameters such as the perception of other patches, host density, number of hosts attacked, and various environmental conditions [103, 104], and also parasitoid's energy reserves, however this last parameter has only rarely been included in empirical studies [103].

The initial average numbers of the mummies and aphid nymphs was 465:4215 in 1 m^2 (1:9.28). In order ratio to reach (1:5), we must add 398 mummies per m^2. The total number was 9950 mummies'. In the 2nd plot the initial numbers of the mummies and aphid nymph was 378:5720 in 1 m^2 must add in 25 m^2 (1:15.13) to reach (1:10) it must add 194 mummies per m^2.

The mummified aphids increased until they reached 666 individuals on March 1st in the treated plot. The number of mummified aphids was significantly higher than that in the control plot (309 individuals) (F test, p = 0.0381). On March 16, the number of mummified aphid rise again [15]. The percentage of parasitism reached 100% in the treated plot, while it was about 55.45% in the control plot and decreased after that (Fig. 26). The number of mummified aphids and percentage of parasitism at the parasitoid aphid ratio of 1:10. It may be seen that the mummified aphid reached 299 individuals on January 21. The number did not exceed 443 individuals in the treated plots [109]. The number of mummified aphid in treated plot was significantly higher than that in the control plot (Fig. 27). The percentage of parasitism reached 100% in the treated plot while, it was only 55.45% in the control one [109, 113].

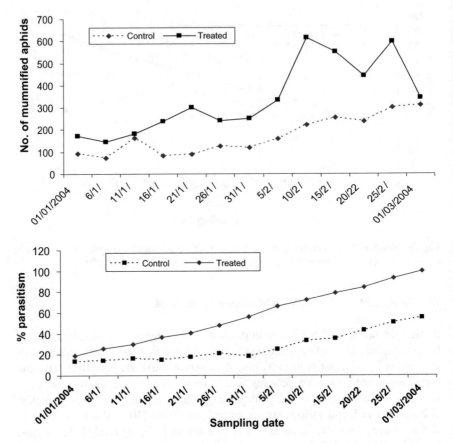

Fig. 27 Percent of parasitism of the aphid *B. brassicae* by *D. rapae* at the parasitoid aphid ratio of 1:10 in the field in the year 2004

5.3 Cauliflower

5.3.1 Bionomics of *D. rapae* at Varying Densities Under Cages in the Laboratory

The parasitoid density influenced the percentage of parasitism; giving maximum percentage was 87.75c/o for *D. rapae* kept at rate of 15 parasitoid females per cage and the minimum was 48.60c/o at three parasitoids/cage [109]. There were significant differences in the total numbers of parasitized aphid and the total percentage of parasitism at all densities [15]. The maximum number of parasitized aphids by *D. rapae* (175.5) was recorded at 15 parasitoids per cage and a minimum of 97.2 was recorded at three parasitoids/cage (Table 17). The highest percentage of adult's emergence was 82.41% for *D. rapae* at three parasitoids/cage. These findings agree with that of [102, 105, 106, 109].

Table 17 Effect of parasitoid density on parasitism rates and adult emergence percentages in the field under 19 ± 10 °C and $74 \pm 3\%$ R.H.

Parasitoid density	Mean ± SD				
	No. of emerged adults	No. of non emerged adults	No. of parasitized aphid (mummies)	Percentage of parasitism	Percentages of adult emergence
3♀	80.1 E ± 4.89	17.1 D ± 2.12	97.2 E ± 6.25	48.6 E ± 2.96	82.41 A ± 3.11
6♀	95 D ± 3.96	28.5 C ± 1.27	123.5 D ± 5.21	61.75 D ± 2.53	80.13 B ± 2.39
9♀	112 C ± 2.71	32.5 C ± 1.41	144.5 C ± 4.03	72.25 C ± 2.25	76.18 C ± 2.02
12♀	120 B ± 5.21	39.0 B ± 3.22	159 B ± 7.01	79.5 B ± 3.03	73.35 E ± 1.68
15♀	132.1 A ± 8.71	43.4 A ± 6.31	175.5 A ± 5.09	87.75 A ± 2.60	75.27 D ± 4.09
LSD$_{0.05}$	1.68407	4.125106	5.179249	0.729223	0.0197474

Means followed by the same letter in a column are not significantly different at 5% level of probability

5.3.2 In the Green House

Aphid parasitoids have considerable potential as biological control agents but their efficiency is dependent upon their presence in the right place at the right time and right host: parasitoid ratio [114]. Understanding parasitoid behavior, together with identification of physical and chemical cues regulating the behavior, is providing exciting opportunities for manipulation of parasitoids in the field, as populations introduced through inundative releases [15]. The initial numbers of the parasitoid mummies and aphid nymphs used for each treatment were 5730:22,920, 3894:31,152, 3436:41,232 and 2872:45,952 individuals, respectively [109]. Four density levels of 1:4, 1:8, 1:12 and 1:16 were planned in four green houses. The percentages of parasitism were calculated according to [30]. The lowest total parasitism ratios reached 32.37, 19.91, 11.80 and 9.18 with the parasitoid host ratios 1:4, 1:8, 1:12, and 1:16 respectively after eight days from releasing [112].

After that the total parasitism ratios increased until reached 100, 78.70, 73.57 and 36.62% at the last same parasitoid ratios respectively after 28 days from released. The average parasitism ratios in green houses were 73.94, 60.74, 54.29 and 39.98 at the parasitoid host ratios 1:4, 1:8, 1:12 and 1:16 respectively (Table 18).

Table 18 Percentages of parasitism on *B. brassicae* after release of *D. rapae* at parasitoid: host ratios under green houses on cauliflower plants during 2010/2011

Sampling dates	Dates after release	Parasitoid: host ratio			
		1:4	1:8	1:12	1:16
16-12-2010	4	0	0	0	0
20-12	8	32.37	19.91	11.89	9.18
24-12	12	49.67	25.71	18.19	12.29
28-12	16	68.09	36.62	23.79	15.90
1-1-2011	20	73.48	48.2	41.58	20.13
5-1	24	89.74	65.74	54.49	28.75
9-1	28	100	78.70	73.57	36.62
13-1	32	100	93.28	81.75	58.25
17-1	36	100	100	92.03	78.93
21-1	40	100	100	100	86.6
25-1	44	100	100	100	93.08
Average		73.94 ± 10.19	60.74 ± 11.01	54.29 ± 11.17	39.98 ± 10.11

5.3.3 In the Field

Two density levels of 1:8, 1:12 were planned for the parasitoid and its host (*B. brassicae*) combination on cauliflower. The initial average numbers of the mummies and aphid nymphs was found 498:5984 in m^2 (1:12.02) to reached (1:8) it must add 250 mummies per m^2. The total number was 8750 mummies. In the second plot the initial numbers of the mummies and aphid nymph was found 397:5986 in m^2 must add in 35 m^2 (1:15.08) to reached (1:12) it must add 101 mummies per m^2. The total number 35 m^2 was 3535 mummies [112].

The lowest total parasitism ratios reached 13.32 and 9.69 with the parasitoid host ratios 1:8 and 1:12 while in the farmer was 8.44% after four days from released. On the other hand the total parasitism ratios increased until reached 98.57 and 84.73 with the same parasitoid host ratios 1:8 and 1:12 while in the farmer was 52.74% after 40 days [109]. Meanwhile, the average parasitism ratios in the field were 58.39, 50.01 and 32.49% at the parasitoid host ratios 1:8, 1:12 and in the farmer during the period of release [109]. Finally the parasitoid *D. rapae* can be used as biological control agent's *B. brassicae* in cauliflower plantation under green houses and under field conditions at the parasitoid host ratios of 1:8 and 1:12 on cauliflower (Table 19).

Table 19 Percentages of parasitism on *B. brassicae* after release of *D. rapae* at different parasitoid: host ratios in the field on cauliflower plants during 2011/2012

Sampling dates	Dates after release	Parasitoid: host ratio		In the farm
		1:8	1:12	
15-12-2011	4	13.32	9.69	8.44
19-12	8	17.67	14.22	10.81
31-12	12	20.62	16.67	14.13
27-12	16	33.88	21.68	16.44
31-12	20	41.89	32.47	22.00
4-1-2012	24	51.47	40.61	27.7
8-1	28	62.68	53.07	39.22
12-1	32	74.25	62.75	48.67
16-1	36	86.28	70.98	45.12
20-1	40	98.57	84.73	52.47
24-1	44	100	93.92	48.17
28-1	48	100	100	56.46
Average		58.39	50.01	32.49

5.4 Evaluation of Parasitoid D. rapae in Controlling A. craccivora Under Field Conditions

It is concluded that the parasitoid *D. rapae* could have the potential to be suitable biological control agent against *B. brassicae*, *A. craccivora* and *A. nerii*. [112]. The percentages of parasitism were gradually increased as the time increased from 5 to 40 days after treatment. The highest rate of parasitism (75%) which a significant effect was recorded after 40 days at a parasitoid/host ratio 1:2 followed by 56.18, 48.32 and 32.48% at the parasitoid/host ratios of 1:4, 1:6 and 1:8, respectively. The effect of time on percentages of parasitism was significant only after 40 day, while it was non significant at all periods tested with the ratio of 1:2 [112]. The mean percentages of parasitism during the different periods after treatment revealed that the average total effect of the ratio 1:2 was superior to other ratios tested. The percentages of parasitism were increased at lower host densities, as the parasitoid was unable to attack a greater proportion of its host at increased densities [113]. The effect of different population sizes of *T. indicus*, *T. angelicae* and its host *A. craccivora* on searching efficiency and rates of parasitism are investigated [102, 105, 113, 115]. They found that the percentages of parasitism increased at lower host densities. The mummified aphids increased until they reached 666 individuals on March 1st in the treated plot. The number of mummified aphids was significantly higher than that in the control plot (309 individuals) (F test, p= 0.0381). On March 16, the number of mummified aphid rise again [15]. The percentage of parasitism reached 100% in the treated plot, while it was about 55.45 % in the control plot and decreased after that (Fig. 26). The number of mummified aphids and percentage of parasitism at the parasitoid aphid ratio of 1:10. It may be seen that the mummified aphid reached 299 individuals on January 21. The number did not exceed 443 individuals in the treated plots [109]. The number of mummified aphid in treated plot was significantly higher than that in the control plot (Fig. 27). The percentage of parasitism reached 100% in the treated plot while, it was only 55.45 % in the control one [109, 113].

5.5 Relation of Parasitoid Densities on Parasitization Rate

5.5.1 In the Laboratory

The use of *D. rapae* in the laboratory resulted in 58% reduction in the level of aphid infestation, while the combination of both biological control agents led to a 66% reduction [107]. *D. rapae* different densities had influence on the percentage of emergence of adult parasitoids, giving maximum percentage i.e. 80.93, 75.54 and 63.85% at one parasitoid per jar and a minimum ones i.e. 69.33, 49.03 and 43.99% at 11 parasitoids per jar on *B. brassicae*, *A. craccivora* and *A. nerii*, respectively. There were significant differences between the percentage of parasitism among one parasitoid per jar and each of 3, 5, 7 and 11 parasitoids per jar. The increase of parasitoid density increased the rate of parasitism and the number of parasitized aphids. The maximum percentage of parasitism were 35.7, 29.7 and 25.2% for *D. rapae* at 11 parasitoid per jar and a minimum of 11.30, 7.30 and 5.9% were recorded at one parasitoid per jar on *B. brassicae*, *A. craccivora* and *A. nerii*, respectively [116]. The increase of the number of parasitoid increased the percentages of the parasitism increased, while the percentage emergence of adult parasitoids decreased on the same aphid species, respectively (Figs. 28, 29 and 30).

The maximum numbers of mummies in laboratory were 71.4, 59.4 and 50.40 for *D. rapae* at 11 parasitoids per jar and minimum of 22.6, 14.6 and 11.8 mummies

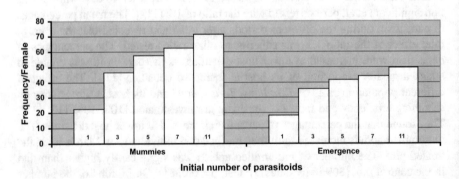

Fig. 28 Fecundity of the parasitoid of *D. rapae* as number of mummies formed from *B. brassicae* under laboratory conditions

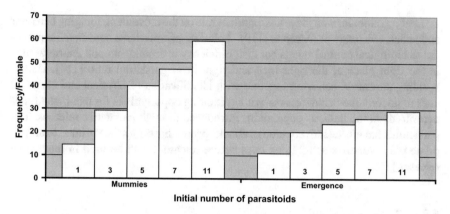

Fig. 29 Fecundity of the parasitoid of *D. rapae* as number of mummies formed from *A. craccivora* under laboratory conditions

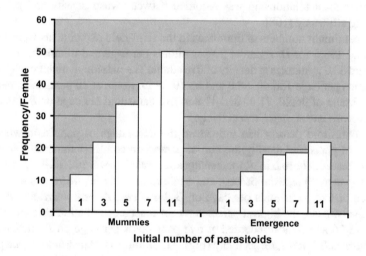

Fig. 30 Fecundity of the parasitoid of *D. rapae* as number of mummies formed from *A. nerii* under laboratory condition

at one parasitoid per jar. Also, the maximum numbers of emerged adults were 49.4, 29.2 and 22.2 for *D. rapae* at 11 parasitoids per jar and minimum of 18.2, 11.0, 7.6 at one parasitoid per jar on *B. brassicae*, *A. craccivora* and *A. nerii*, respectively (Figs. 28, 29 and 30).

In general, differences in the density of host aphids and the respective parasitoids influence the behavioral characteristics. A higher density of parasitoid may increase [114]. The proportion of male progeny, because male need less food resources than females [115]. Preference of certain host species has been demonstrated in laboratory studies where parasitoids more often oviposit in some species than in other, when both the host species are offered separately or simultaneously [117, 118].

Leaf epicuticular wax plays an important role on the movement, foraging behavior and attack efficiency of *D. rapae* [119]. Aphid parasitoids have considerable potential as biological control agents but their efficiency is dependent upon their presence in the right place at the right time and right host: parasitoid ratio [114]. Understanding parasitoid behavior, together with identification of physical and chemical cues regulating the behavior, is providing exciting opportunities for manipulation of parasitoids in the field, as populations introduced through inundative releases. The parasitoids having selectively bred to attack specific hosts and then primed to appropriate plant volatiles as foraging cues before release, could be used in inundative releases [109].

5.5.2   In the Semi-field

The functional response of the parasitoid to different host densities was that of type II. A significant relationship was recorded between wasp density and per capita searching efficiency [108].

The maximum numbers of mummies in the semi-field experiments were 185.60, 166.4 and 158.6 for *D. rapae* at 20 parasitoids per cage and minimum of 124.60, 97.40 and 83.0 mummies at density of five adults. The maximum number of emerged parasitoids per cage 136.40, 88.8 and 63.40 for *D. rapae* at 20 parasitoids per cage and minimum of 96.20, 71.60 and 47.8 at five parasitoid per cage on *B. brassicae*, *A. craccivora* and *A. nerii*, respectively.

The parasitoid density had influenced the percentage of parasitism where the increase of parasitoid density the rate of parasitism and the number of parasitized aphid increased. The maximum percentages of parasitism were 92.20, 83.2 and 79.3% for *D. rapae* at 20 parasitoids/cage and a minimum of 61.80, 48.70 and 41.5% was recorded at five parasitoids per cage on *B. brassicae*, *A. craccivora* and *A. nerii*, respectively. Also, the highest percentages for emergence of adult parasitoids were 77.80, 73.56 and 57.9% recorded at five parasitoids per cage on *B. brassicae*, *A. craccivora* and *A. nerii*, respectively (Figs. 31, 32 and 33). *Diaeretiella rapae* played the major role towards suppressing *B. brassicae* population. With the increase of parasitoid density the fecundity of the parasitoids *D. rapae* as number of mummies and emerged adults increased [51]. The previous results indicated that the percentage of parasitism were increased at lower host densities as the parasitoid was enable to attack high number of its host at increased densities [113, 118].

Parasitoid density in relation to host density had influenced percentage of parasitism. Highest percentage reached 91.40% at 16 *D. rapae* female parasitoids per cage while the minimum was 55.6% at one female per cage. The percentage of parasitism increased with increase of numbers of parasitoid [106]. The functional response and rate of parasitism by *D. rapae* on different densities of *Diuraphis noxia* (Mordvilho). The increases in host density lead to a decrease in the proportion of hosts parasitized by the parasitoid, there was intra-specific competition among individuals of the female parasitoids in high density [117, 119].

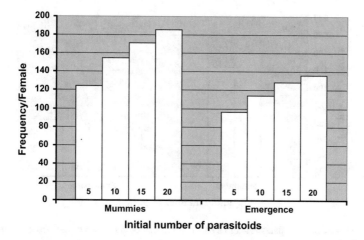

Fig. 31 Fecundity of the parasitoid of *D. rapae* as number of mummies formed from *B. brassicae* under semi-field conditions

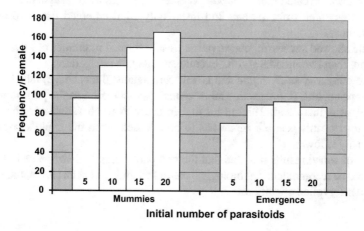

Fig. 32 Fecundity of the parasitoid of *D. rapae* as number of mummies formed from *A. craccivora* under semi-field conditions

6 Conclusions

From results explained in this study, the following may be concluded:

1. *Diaeretiella rapae* is an important primary parasitoid of a wide range of aphid species and is considered a promising biological control agent against aphid species especially cabbage aphid in cabbage and cauliflower fields.
2. It recommended to be an item of Integrated Pest Management Programs in Egyptians fields designed to control *B. brassicae*.

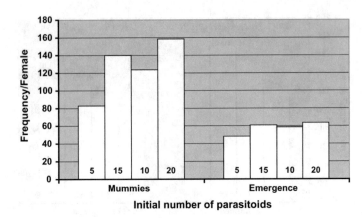

Fig. 33 Fecundity of the parasitoid of *D. rapae* as number of mummies formed from *A. nerii* under semi-field conditions

3. The Data revealed that *D. rapae* accepted all stages of *B. brassicae*, no adults were emerged from 1st and 2nd instar only adult emerged from 3rd and 4th instars.
4. Data showed successful conservation freshly formed mummies of *D. rapae*, for more than two months at 5 °C in colld storage in a refrigerator.
5. The efficiency of *D. rapae* and L. fabarum against the cabbage, faba bean and cowpea aphid, *B. brassicae* and *A. craccivora* increased as the parasitoid: host ratio was increased. This meant that in order to reach satisfactory biological control of this pest, *D. rapae* has to be released when the aphid's population density is low.
6. The obtained results revealed that the two parasitoids *L. fabarum* and *D. rapae* could be recommended as biological control candidates against this aphid species under Egyptian conditions.

7   Recommendations

This chapter highlights the following recommendations for future considerations by researchers, farmers and interested stakeholders:

1. The obtained results revealed that successful storage of freshly formed mummies of *D. rapae*, at 5 °C in the refrigerator for more than two months.
2. *Diaeretiella rapae* was the most common on all host aphids concerned in this study (*Brevicoryne brassicae*, *Aphis craccivora*, *A. nerii*, *Aphis gossypii*, *Rhopalosiphum maidis* and *Hyalopterus pruni*).
3. The obtained results revealed that the parasitoids *D. rapae* could be recommended as biological control agents against aphid species under Egyptian conditions.

4. It could be generally concluded that *D. rapae* is a promising biological control agent, against the cabbage aphid in cabbage and cauliflower fields.
5. Finally, it is to be recommended to consider the releasing time of this parasitoid under Egyptian field conditions in IPM program designed for the control of *B. brassicae* to decrease the environmental pollution by using the traditional insecticides.

References

1. Stary P (1976) Aphid parasite (Hymenoptera: Aphidiidae) of the Mediterranean area. Trans Czechoslovak Acad Sci Ser Math Nat Sci 86:1–95
2. Jonsson M, Wratten SD, Landis DA, Gurr GM (2008) Recent advances in conservation biological control of arthropods by arthropods. Biol Control 45:172–175
3. Walker GP, Nault LR, Simonet DE (1984) Natural mortality factors acting on potato aphid (*Macrosiphum euphorbiae*) populations in processing tomato field in Ohio. Environ Entomol 13:724–732
4. Hagen KS, Van den Bosch R (1986) Impact of pathogens parasites and predators on aphids. Annu Rev Entomol 13:325–384
5. Hagvar EB, Hofsvang T (1991) Aphid parasitoids (Hymenoptera: Aphidiidae) biology, host selection and use in biological control. Biocontrol News Inform 12:13–41
6. Doutt RL, Smith CF (1971) The pesticide syndrome-diagnosis and suggested prophylaxis. Biol Control 8:3–15
7. Ibrahim AMA (1987) Studies on Aphidophagous parasitoids with special reference to *Aphidius uzbekistanicus* (Luz). PhD thesis, Faculty of Agriculture, Cairo University, 202 pp
8. Sarhan AA (1976) Studies on the biological control of cotton white fly *Bemisia tabaci* (Genn.) in Egypt. MSc thesis, Faculty of Agriculture, Cairo University, 164 pp
9. Saleh AAA (2008) Storage of *Diaeretiella rapae* (M'Intosh) (Hym., Aphidiidae) mummies in three aphid species; *Brevicoryne brassicae* (L.), *Aphis nerii* Boyer De Fonscolombe and *Aphis gossypii* (Glov.). Egypt J Biol Pest Control 18:39–42
10. Saleh AAA (2012) Evaluation of release of *Diaeretiella rapae* (M'Intosh) for controlling the cruciferous aphid *Brevicoryne brassicae* L. on cauliflower plants at Sharkia governorate. Egypt J Plant Prot Path Mansoura Univ 3:307–318
11. Elliott NC, French BW, Reed DK, Burd JD, Kindler SDM (1994) Host species effects on parasitization by a Syrian population of *Diaeretiella rapae* (M'Intosh). (Hymenoptera: Aphididiidae). Can Entomol 126:1515–1517
12. Hafez M (1965) Characteristics of the open empty mummies of the cabbage aphid, *Brevicoryne brassicae* L. indicating the identify of emerged parasites. Agric Res Rev 43:85–88
13. Pike KS, Stary P, Miller T, Allison D, Graf G, Boydston L, Miller R, Gillespie R (1999) Host range and habitats of the aphid parasitoid *Diaeretiella rapae* (Hymenoptera: Aphididae) in Washington State. Environ Entomol 28:61–71
14. Saleh AAA (2000) Ecological and biological studies on certain aphid parasites at Mansoura district. MSc thesis, Faculty of Agriculture, Mansoura University, 85 pp
15. Saleh AAA (2004) Mass production and field application of some aphid natural enemies. PhD thesis, Faculty of Agriculture, Mansoura University, 161 pp
16. El-Heneidy A, Gonzalez D, Abdel-Awal M, Adly D (2006) Performance of certain exotic aphid parasitoid species towards cereal aphids under laboratory, field cage and open wheat field conditions in Egypt. Egypt J Biol Pest Control 16:67–72
17. Saleh AAA, Hashem MS, Abd-Elsamed AA (2006) *Aphidius colemani* Viereck and *Diaeretiella rapae* (M'Intosh) as parasitoids on the common reed aphid, Hyalopterus pruni (Geoffroy) in Egypt. Egypt J Biol Pest Control 16:93–97

18. Saleh AAA, Gatwary WGT (2007) Seasonal abundance of the oleander aphid *Aphis nerii* Boyer de Fonscolombe (Homoptera: Aphididae) in relation to the primary and hyper parasitoid on Dafla in Egypt. J Product Dev 12:709–730

19. Saleh AAA, Desuky WMH, Mohamed NE (2009) Studies on some parasitoids of the cowpea aphid *Aphis craccivora* Koch. (Homoptera: Aphididae) in Egypt. Egypt J Biol Pest Control 19:11–16

20. Vinson SB, Iwantsch BGF (1980) Host suitability for insect parasitoids. Ann Rev Ent 25:397–419

21. Elliott NC, Burd JD, Kindlers D, Lee JH (1995) The temperature effects on development of three cereal aphid parasitoids (Hymenoptera: Aphidiidae) Great Lakes. Entomologist 28:199–204

22. Hosseini GA, Fathipour Y, Talebi AA (2003) A comparison of stable population parameters of cabbage aphid *Brevicoryne brassicae* and its parasitoid *Diaeretiella rapae*. Iran J Agric Sci 34:4, Pe785–Pe791

23. Abdel-Samad SSM (2002) Bioagents for controlling aphids in wheat fields to minimize pesticide pollution. PhD thesis, Faculty of Agriculture, Ain Shams University, 129 pp

24. Saleh AAA, Gatwary WGT, Mohamed NE (2009) Effect of temperature, relative humidity and some biological aspects on performance parasitoids of the oleander aphid *Aphis nerii* Boyer De Fonscolombe. J Agric Res 87:983–998

25. Abou-Fakhr EM, Kawar NS (1998) Complex of endoparasitoids of aphids (Homoptera, Aphididae) on vegetables and other plants. Lebanon Entomol Obozrenie 77:753–763

26. Ragab ME, Abou El-Naga AM, Ghanim AA, Saleh AA (2002) Ecological studies on certain aphid parasitoids especially those of *Aphis craccivora* Koch. J Agric Sci Mansoura Univ 27:2619–2620

27. Saleh AAA (2008) Ecological and biological studies of *Diaeretiella rapae* (M'Intosh) (Hymenoptera: Aphidiidae) the parasitoid of some aphid species in Egypt. Egypt J Biol Pest Control 18:33–38

28. Zaki FN, El-Shaarawy MF, Farag NA (1999) Release of two predators and two parasitoids to control aphids and white flies. Anz Schadlingsk 72:19–20

29. Maghraby HMM (2012) Studies on the parasitoid *Diaeretiella rapae* on some aphid species in Sharkia governorate. MSc thesis, Faculty of Agriculture, Moshtohor, Benha University, 222 pp

30. Farrell JA, Stufkens MW (1990) The impact of Aphidius rophopalosiphi (Hymenoptera: Apidiidae) on population of the rose grain aphid (*Metopolophium dirhodum*) (Homoptera: Aphididae) on cereals in canKrbury. NZ Bull Entomol Res 80:377–383

31. Freuler J, Fischer S, Mittaz C, Terretaz C (2001) Role of banker plants to reinforce the action of *Diaeretiella rapae*, the main parasitoid of cabbage aphid. Revue Suisse Vitic Arboric Hortic 33:329–335

32. Ibrahim A, Fayad YH (1984) Rate of parasitism in certain species of aphids infesting some cultivated and uncultivated plants in Egypt. Ann Agric Sci Moshtohor 21:1079–1085

33. El-Maghraby MMA (1993) Seasonal abundance of the cruciferous aphid *Brevicoryne brassicae* L. (Homoptera: Aphididae) in relation to the primary and hyper parasitoids on cauliflower in Zagazig region Egypt. Zagazig J Agric Res 20:1627–1639

34. Okasha YAH (1998) Studies on certain parasitoids attacking the cabbage aphid. MSc thesis, Faculty of Agriculture Menoufiya University, 111 pp

35. Abdel-Megid JE (1999) The cabbage aphid, *Brevicoryne brassicae* (L.) (Homoptera: Aphididae) and it's associated parasitoids on cauliflower plantations at Zagazig district. J Agric Sci Mansoura Univ 24:7741–7752

36. Bahana J, Karuhize G (1986) The role of *Diaeretiella rapae* (M'Intosh) (Hymenoptera: Braconidae) in the population control of cabbage aphid, *Brevicoryne brassicae* (L.) (Hemiptera: Aphididae) in Kenya. Insect Sci Appl 7:605–609

37. Thakur JN, Rawat US, Pawar AD, Sidhu S (1989) Natural enemy complex of the cabbage aphid, *Brevicoryne brassicae* L. in Kull Valley, Himachal Pradesh. J Biol Control 3:69

38. Abdel-Samad SSM (1996) Studies on natural enemies of certain insects attacking leguminous crop. MSc thesis, Faculty of Agriculture, Ain Shams University, 94 pp
39. Stary P (1970) Biology of aphid parasites (Hymenoptera: Aphidiidae) with respect to integrated control. Entomologica 6, 643 pp
40. Vidal S (1997) Factors influencing the population dynamics of *Brevicoryne brassicae* in under sown Brussels sprouts. Biol Agric Hort 5:285–295
41. Ibrahim AMA, Afifi AI (1994) Aphidius colemani Viereck and *Aphidius picipes* (Nees) as a parasitoid on the mealy plum aphid, *Hyalopterus pruni* (Geoffroy) on peach in Egypt. Egypt J Biol Pest Control 1:45–56
42. Megahed HEA (2000) Studies on aphids. PhD thesis, Faculty of Agriculture, Zagazig University, 206 pp
43. Kavallieratos NG, Lykoressis SGP, Stathas SA, Athanassiou CG (2001) The aphidiinae (Hymenoptera: Icheumonoidea: Braconidae) of Greece. Phytoparasitica 29:306–340
44. Steenis MJV (1995) Evaluation of four aphidiine parasitoids for biological control of *A. gossypii*. Entomol Exp Appl 75:151–157
45. Albert R (1995) Biological control of the cotton aphid on cucumbers. Gartenbau Mag 4:32–34
46. Ohta I (2003) Parasitism of *Lysiphlebus japonicus* Ashmead on the cotton aphid, *Aphis gossypii* Glover. Proc Kansai Plant Prot Soc 45:33–35
47. Selim AA, El-Refai SA, El-Gantiry A (1987) Seasonal fluctuations in the population of *Aphis craccivora* Koch., *Myzus persicae* (Sulz.) *Aphis gossypii* (Glov.) and their parasites. Ann Agric Sci Ain Shams Univ 32:1837–1848
48. Ragab ME (1996) Biology and efficiency of *Trioxys angelicae* Hal. (Hymenoptera: Aphidiidae), a newly recorded parasitoid of *Aphis craccivora* Koch. (Homoptera: Aphidiidae). Egypt J Biol Control 6:7–11
49. Kant R, Minor MA, Trewick SA (2012) Reproductive strategies of *Diaeretiella rapae* (Hymenoptera: Aphidiinae) during fluctuating temperatures of spring season in New Zealand. Biocontrol Sci Technol 22:1–9
50. Nematollahi MR, Fathipour Y, Talebi AA, Karimzadeh J, Zalucki MP (2014) Parasitoid- and hyperparasitoid-mediated seasonal dynamics of the cabbage aphid (Hemiptera: Aphididae). Environ Entomol 43:1542–1551
51. Herakly FA, Abou El-Ezz A (1970) Seasonal abundance and natural enemies of the cabbage aphid, *Brevicoryne brassicae* L. Agric Res Rev 48:119–122
52. Horn DJ (1989) Secondary parasitism and population dynamics of aphid parasitoids (Hymenoptera: Aphidiidae). J Kansas Entomol Soc 62:203–210
53. Mackauer M, Volki W (1993) Regulation of aphid populations be aphidiid wasps: does parasitoid foraging behaviour or hyper parasitism limit impact? Oecologia 94:339–350
54. Vaz LAL, Tavares MT, Lomônaco C (2004) Diversity and size of parasitic Hymenoptera of *Brevicoryne brassicae* L. and *Aphis nerii* Boyer de Fonscolombe (Hemiptera: Aphididae). Neotrop Entomol 33:225–230
55. May RM (1973) Stability and complicity in model ecosystems. Princeton University Press, New Jersey, p 235
56. Saleh AAA, Khedr MMA (2014) Performance of the aphid parasitoid, *Diaeretiella rapae* (M'Intosh) towards certain aphid species in Egypt. J Entomol 11:127–141
57. Liu SS, Xu ZH, Li ZJ (1990) Preliminary investigation of the parasites of aphids on cruciferous vegetables in Hangzhou. Chin J Biol Control 6:5–8
58. Saleh AAA, El-Sharkaw HM, El-Santel FS, Abd-El-Salam RA (2017) Studies on some parasitoids of aphid *Aphis gossypii* Glover, (Homoptera: Aphididae) on cucumber plants in Egypt. Egypt Acad J Biol Sci 10:19–30
59. Carnevale AB, Bueno VHP, Sampaio MV (2003) Parasitism and development of *Lysiphlebus testaceipes* (Cresson) (Hymenoptera: Aphidiidae) on *Aphis gossypii* Glover and *Myzus persicae* (Sulzer) (Hermiptera: Aphididae). Neotrop Entomol 32:293–297
60. Saleh AAA, Desuky WMH, Hashem HHA, Gatwary WGT (2009) Evaluation the role of aphid parasitoid *Diaeretiella rapae* (M'Intosh) (Hymenoptera: Aphidiidae) on cabbage aphid *Brevicoryne brassicae* L. (Homoptera: Aphididae) in Sharkia district. Egypt J Biol Pest Control 19:151–155

61. Rodrigues SMM, Bueno VHP, De Bueno JS, Filho S (2001) Development and evaluation of an open rearing system for the control of *Aphis gossypii* Glover (Aphididae) by *Lysiphlebus testaceipes* in greenhouses. Neotrop Entomol 30:433–436

62. AI-Hag E, Al-Rokaibah AA, Zaitoon AA (1996) Natural enemies of cereal aphids in sprinkler—irrigated wheat in central Saudi Arabia. Bull Fac Agric Cairo Univ 47:649–663

63. Giustina W, Deriu P, Foessel P (1982) Role of specific natural enemies in the control of maize aphid populations in the Paris area preliminary results. Bull Srop 10:12–22

64. Bueno VHP, Souza MD (1992) Ethnology and life span of *Diaeretiella rapae* (M'Intosh). (Hymenoptera: Aphididiidae). Rev Agric (Piracicaba) 87:49–54

65. El-Batran LA, Awadallah SS, Fathy HM (1996) On some predators and parasitoids of the cabbage aphid *Brevicoryne brassicae* (L.) in Mansoura district. Egypt J Biol Pest Control 6:35–38

66. Heimpel GE, Jervis MA (2005) Does floral nectar improve biological control by parasitoids? In: Wäckers FL, van Rijn PCJ, Bruin J (eds) Plant-provided food for carnivorous insects: a protective mutualism and its applications. Cambridge University Press, Cambridge, pp 267–304

67. Lee JC, Heimpel GE (2002) Nectar availability and parasitoid sugar feeding. In: van Driesche RG (ed) 1st international symposium on biological control of arthropods. Forest Health Technology Enterprise Team, Morgantown, West Virginia, pp 220–225

68. Tena A, Pekas A, Wäckers FL, Urbaneja A (2013) Energy reserves of parasitoids depend on honeydew from non-hosts. Ecol Entomol 38:278–289

69. Yann-Davi V (2015) Floral resource subsidies for the enhancement of the biological control of aphids in oilseed rape crops. PhD thesis, Faculty of Agriculture, Lincoln University, 123 pp

70. Araj SE, Wratten S, Lister A, Buckley HL, Ghabeish I (2011) Searching behavior of an aphid parasitoid and its hyperparasitoid with and without floral nectar. Biol Control 57:79–84

71. Saleh AAA, Salem HEM, Gatwary WGT (2009) The role of primary parasitoids and hyperparasitoids associated with oleander aphid *Aphis nerii* Boyer de Fonscolombe (Homoptera: Aphididae). Bull Soc Entomol Egypt 86:115–129

72. Kean J, Wratten S, Tylianakis J, Barlow N (2003) The population consequences of natural enemy enhancement, and implications for conservation biological control. Ecol Lett 6:604–612

73. Jervis MA, Ellers J, Harvey JA (2008) Resource acquisition, allocation, and utilization in parasitoid reproductive strategies. Annu Rev Entomol 53:361–385

74. Jamont M, Crépellière S, Jaloux B (2013) Effect of extrafloral nectar provisioning on the performance of the adult parasitoid *Diaeretiella rapae*. Biol Control 65:271–277

75. Ayal Y (1987) The foraging strategy of *Diaeretiella rapae*. J Anim Ecol 56:1057–1068

76. Hopkinson JE, Zalucki MP, Murray DAH (2013) Honeydew as a source of nutrition for *Lysiphlebus testaceipes* (Cresson) (Hymenoptera: Braconidae): effect of adult diet on lifespan and egg load. Aust J Entomol 52:14–19

77. Volkl GP, Stechmann DH (1998) Parasitism of the black bean aphid (*Aphis fabae*) by *Lysiphlebus fabarum* (Hermiptera: Aphidiidae): the influence of host plant and habitat. J Appl Entomol 122:201–206

78. Murphy ST, Volkl W (1996) Population dynamics and foraging behaviour of *Diaeretus leucopterus* (Hymenoptera: Braconidae), and its potential for the biological control of pine damaging *Eulachnus* spp. (Aphididae). Bull Entomol Res 86:397–405

79. Kolaib MO (1991) Effect of temperature on *Pachyneuron aphidis* Bouche, a hyperparasitoid of *Diaeretiella rapae* (M'Intosh). Alex Sci Exch 12:30–42

80. Wilson GB, Lambdin PL (1987) Suitability of *Brevicoryne brassicae* L and *Myzus persicae* (Homoptera: Aphididae) as hosts of *Diaeretiella rapae*. Entomol News 98:140–146

81. Sandra EH, Scott JC, Mark DH (2004) Effects of variation among plant species on the interaction between a herbivore and its parasitoid. Ecol Entomol 29:44–51

82. Araj S, Wratten SD (2015) Comparing existing weeds and commonly used insectary plants as floral resources for a parasitoid. Biol Control 81:15–20

83. Tylianakis JM, Didham RK, Wratten SD (2004) Improved fitness of aphid parasitoids receiving resource subsidies. Ecology 85:658–666

84. Saleh AAA, Ali SAM, Abd-Elsamed AA, Elsayed AAA (2014) Development of the parasitoid *Diaeretiella rapae* (M'Intosh) reared on certain aphid species in relation to heat unit requirement. J Entomol 11:319–329

85. Kant R, Minor M, Sandanayaka M, Trewick S (2013) Effects of mating and oviposition delay on parasitism rate and sex allocation behaviour of *Diaeretiella rapae* (Hymenoptera: Aphidiidae). Biol Control 65:265–270

86. Srivastava M, Singh R (1988) Bionomics of *Trioxys indicus*, an aphidiid parasitoid of *Aphis craccivora*. Impact of host-extract on the oviposition response of the parasitoid. Biol Agric Hort 5:169–176

87. Debach P (1964) Biological control of insect pest and weeds, 844 pp

88. Archer TL, Murray CL, Eikenbary RD, Starks KJ, Morrison RD (1973) Cold storage of *Lysiphlebus testaceipes* mummies. Environ Entomol 2:1104–1108

89. Hofasvang T, Hagvar E (1977) Cold storage tolerance and super cooling points of mummies *Ephedrus cerasicola* Stary and *Aphidius colemani* Viereck (Hymenoptera: Aphidiidae). Norw J Entomol 24:1–6

90. Shalaby FF, Rabasse JM (1979) Effect of conservation of the aphid parasite *Aphidius matricariae* Hal. (Hymenoptera: Aphidiidae) on adult longevity, mortality and emergence. Ann Agric Sci Moshtohor 11:59–73

91. Scopes NEA, Biggerstaff SM, Goodall DE (1973) Cool storage of some parasites used for pest control in glasshouses. PL Path 22:189–193

92. Lyon JP (1968) Remarques preliminaries sur les possibilites d'utilisation pratique d'Hymenopteres parasites pour. Ann Epiphy 19:113–118

93. Gornall AG, Bardawilb CD, David MM (1949) Determination of serum protein by means of bruit reduction. J Biochem 177:751–766

94. Knight JA, Anderson S, Rawla JM (1972) Chemical basis of the sulfo-phospho-vanillin reaction for estimating total serum lipids. Clin Chem 18:199–202

95. Lee YP, Takahash T (1966) An improved colorimetric determination of amino acids with use of ninhydrin. Anal Biochem 14:71–77

96. DuBois M, Gilles KA, Hamilton JK, Rebers PA, Smith F (1956) Colorimetric method for determination of sugars and related substances. Anal Chem 28:350–356

97. Vanson SB, Iwantsch GF (1980) Host suitability for insect parasitoids. Annu Rev Entomol 25:397–419

98. Crompton M, Birt LM (1967) Changes in the amounts of carbohydrates, phosphagen and related compounds during the metamorphosis of the blowfly, *Lucilia cuprina*. J Insect Physiol 13:1575–1595

99. Sequeira R, Mackauer M (1993) The nutritional ecology of a parasitoid wasp *Ephedrus californicus* Baker (Hymenoptera: Aphidiidae). Can Entomol 125:423–430

100. Carver M, Sulivan DJ (1988) Encapsulative defense reactions of aphids (Hemiptera: Aphididae) to insect parasitoids and aphelinidae (minireview). In: Niemczyk E, Dixon AFG (eds) Ecology and effectiveness of aphidophaga: proceedings of an international symposium. SPB Academic Publishing, Hague, pp 299–303

101. Kant R, Sandanayaka WRM, He XZ, Wang Q (2008) Effect of the host age on searching and oviposition behaviour of *Diaeretiella rapae* (Hymenoptera: Aphidiidae). NZ Plant Prot 61:355–361

102. Downer RC (1978) Functional role of lipid in insects. In: Rockstein M (ed) Biochemistry of insects. Academic Press, London, pp 58–93

103. Van Alphen JJM, Bernstein C, Driessen G (2003) Information acquisition and time allocation in insect parasitoids. Trends Ecol Evol 18:81–87

104. Pierre JS (2011) Neuroeconomics in parasitoids: computing accurately with a minute brain. Oikos 120:77–83

105. Sinha TB, Singh R (1979) Studies on the bionomics of *Trioxys indicus* (Hymenoptera: Aphidiidae): effect of population densities on sex ratio. Entomophaga 24:289–294

106. Sinha TB, Singh R (1980) Studies on the bionomics of *Trioxys indicus* Subba Rao and Sharma (Hymenoptera: Aphidiidae): a parasitoid of *Aphis craccivora* Koch (Homoptera: Aphidiidae) the area of discovery of parasitoid. Zeit Angew Entomol 89:173–178

107. Acheampong S, Stark JD (2004) Can reduced rates of pymetrozine and natural enemies control the cabbage aphid *Brevicoryne brassicae* L. (Homoptera: Aphididae), on broccoli? Int J Pest Manag 50:275–279

108. Fathipour Y, Gharalari AH, Talebin AA (2004) Some behavioral characteristics of *Diaeretiella rapae* (Hym. Aphidiidae), parasitoid of *Brevicoryne brassicae* (Hom. Aphididae). Iran J Agric Sci 35:393–401

109. El-Naggar EM, Abou El-Naga AM, Ghanim AA, Saleh AAA (2008) Mass production and field application of some aphid natural enemies. Egypt J Agric Res 86:623–624

110. White TCR (2013) Experimental and observational evidence reveals that predators in natural environments do not regulate their prey: they are passengers, not drivers. Acta Oecol 53:73–87

111. Nematollahi MR, Fathipour Y, Talebi AA, Karimzadeh J, Zalucki MP (2014) Parasitoid and hyperparasitoid-mediated seasonal dynamics of the cabbage aphid (Hemiptera: Aphididae). Environ Entomol 43:1542–1551

112. Saleh AAA (2013) Efficacy of the aphid parasitoid *Diaeretiella rapae* (M'Intosh) to control *Brevicoryne brassicae* L., *Aphis craccivora* (Koch) and *Aphis nerii* Boyer at Sharkia governorate, Egypt. J Agric Res 92:21–31

113. Ralec A, Ribulé A, Barragan A, Outreman Y (2011) Host range limitation caused by incomplete host regulation in an Aphid parasitoid. J Insect Physiol 57:363–371

114. Abdul Rehman A, Powell W (2010) Host selection behavior of aphid parasitoids (Aphidiidae: Hymenoptera). J Plant Breed Crop Sci 2:299–311

115. Ragab ME, Ghanim AA (1997) Effect of different parasite/host ratio between *Trioxys angelicae* Hal, (Hymenoptera: Aphidiidae) and its host Aphis craccivora Koch. (Homoptera: Aphdidae) on the percentages of parasitism and population development. J Agric Sci Mansoura Univ 27:2619–2620

116. Saleh AAA (2012) Efficacy of the aphid parasitoid *Diaeretiella rapae* (M'Intosh) to control *Brevicoryne brassicae* L., *Aphis craccivora* (Koch) and *Aphis nerii* Boyer at Sharkia governorate, Egypt. J Agric Res 91:21–31

117. Zahra T, Talebi AA, Rakhshani E (2011) The foraging behavior *Diaeretiella rapae* (Hymenoptera: Braconidae) on *Diuraphis noxia* (Hymenoptera: Aphidiidae). Arch Biol Sci Belgrade 63:225–232

118. Chau A, Mackauer M (2001) Preference of the aphid parasitoid *Monoctonus paulensis* (Hymenoptera: Braconidae, Aphidiinae) for different aphid species: female choice and offspring survival. Biol Control 20:30–38

119. Gently GL, Barbosa P (2006) Effects of leaf epicuticular wax on the movement, foraging behavior, and attack efficacy of *Diaeretiella rapae*. Entomol Exp Appl 21:115–122

120. Brown WL, Eisner TE, Whittaker RH (1970) Allomones and kairomones: transspecific chemical messengers. Bioscience 20:21–22

121. Zhang WQ, Hassan SA (2003) Use of the parasitoid *Diaeretiella rapae* (McIntoch) to control the cabbage aphid *Brevicoryne brassicae*. J Appl Entomol 127:522–526

Predacious Insects and Their Efficiency in Suppressing Insect Pests

Nabil El-Wakeil and Nawal Gaafar

Abstract Using beneficial organisms, insect predators should help to decline the population density of pest organisms (insects or mites). Understanding the natural enemy capability to manage different insect pests may lead to sager insect control, comprising pesticide reduction and keeping the environment clean. The abundance of predators could help controlling and reducing the pest populations and thus preventing insect pest outbreaks. Interactions between multiple predator species should modify the strength of prey suppression. These interactions had been distinguished by different predation levels. There are various mechanisms may provide the greatest benefit for biocontrol agents. A suggested strategy for more efficient conservation biological control is containing collection natural enemies, preservation them and releasing the preserved biocontrol agents on target crops. Some of predators have a restricted tolerance to the prey feeding and abiotic factors; therefore, mass rearing and field application of insect predators are considered one of the main aspects in succeeding the biological control programs. In a case study, efficacy of insect predators against some aphid species in different crops was shown that insect predators played a significant role in controlling these aphid species. We tried to present the status and potential of insect predators throughout researching the abundance, the mass production and the field application of them.

Keywords Insect predators · Predacious activity · Biological control · Sustainable agriculture

1 Introduction

Crop production has to rise by 7% by 2050 to confront the human increase [1]. This production increase should be accomplished by conserving the environment and limits the use of insecticides and undesirable chemicals. Using selective pesticides at optimal concentrations and time will protect the population of insect predators [2]. Biological control is a natural practice that shows a vital role in pest suppressing

N. El-Wakeil (✉) · N. Gaafar
Pests and Plant Protection Department, National Research Centre, Dokki, Cairo, Egypt
e-mail: nabil.elwakeil@yahoo.com

© Springer Nature Switzerland AG 2020
N. El-Wakeil et al. (eds.), *Cottage Industry of Biocontrol Agents and Their Applications*, https://doi.org/10.1007/978-3-030-33161-0_4

133

pests [3–5]. Predation is a vital arranging force in natural populations (generally, the term "predator" is used equivalently to "biological control agent"). Mass application of predators could be effective in organic and conventional agriculture [6]. It is approved that integrated biological control is the preferred approach for realizing the supportable agriculture, especially in the sustainable agriculture programs [7].

Mass production and application of insect predator industry in a certain country be governed by some factors namely the qualified researchers or farmers, the infrastructure and the market. Abundance of the significant predators in the nature, possibility of mass producing these insect predators and field application for applying the biological control programs [8–11].

The history of predaceous animals generally may be considered from Egyptian records that cats were showed as useful in rat's management. With the development of stylish cultivation, the Chinese growers placed nests of predaceous ants in citrus trees where the ants fed on infested insects. Most famous predators work in all agricultural environments are naturally occurring ones, which offer a brilliant regulation of many insects with little or no support from humans [5, 12, 13, 14], this is due to eliminate or reduce of usage of pesticides [15] and promote the sustainable and organic agriculture to be more distributed [16].

There are many factors may be affected the insect predation; these factors could be divided into five key groups (a) prey density, (b) predator density, (c) ecological characteristics, (d) prey traits (defense performances), and (e) predator characteristics (attack practices) [17, 18]. Two of these factors, prey and predator density, are unavoidable features of every predator-prey interaction, so that the basic mechanisms of predation will arise from these universal parameters. However, the other factors are constant, so that the minor works will be shown by the effects of the lesser variables: characteristics of environment, prey and predator characteristics [19].

Mass rearing of natural enemies generally comprises producing of millions of insects (host and natural enemy) for controlling the target insect pests. Insect production that could be achieved at a suitable cost/benefit ratio to mass produce a relatively large supply of insects for field application may help to obtain an acceptable insect management. Normally, mass rearing is begonnen from research scale or intermediate-sized rearing, upon which basic research about the target (often an agricultural pest) and the predators are conducted. Therefore, to produce two predator species have to be reared, the pest (host/prey) and the natural enemy. The potential for rearing a large number of predaceous insects increased as artificial diets began to be developed since the 1960s [20, 21].

One of the vital goals for entomologists is developing methods which suppress insect species without harming the environment or other organisms. An ideal method would be to augment insect to control insects. Augmentative releases of lacewings are released into profitable crops with aphid infestations [22, 23]. Nevertheless, the results are obtained by a combination of what is released, and the higher level of naturally occurring predaceous insects in different fields. In Egypt, there is a combined biocontrol program of *Helicoverpa armigera* by using parasitoids and predators [24]. We recommended such control type which deserves more attention as an achievable approach in IPMs.

There are many predaceous insects in nature, which help to avoid some insect pests from population outbreaks. These predators certainly eat/prey their preys. We aimed in this chapter to discuss potential of using insect predators to sustain their population through mass production and the field application of these natural enemies for managing different insect pests in the numerous target crops, vegetables and fruit orchards. In the end of this chapter, there is one of our case studies, which aimed to study efficacy of insect predator species against many aphid species in different crops.

2 Multiple Predator Species Interaction

Every host/prey species regularly faces various predator species, the influences of each may be non-additive [25–27]. Along with being interesting in their own right, these predator effects have received heightened interest with respect to the different biological control program for some of insect pests [28, 29]. Multiple predator species may interact with others and thus either support or decline their potency and generally prey parameters. Effects of multiple predators is depending on combining ways of these predators at different relative densities as mentioned by Griffin et al. [30], who suggested that these effects of multiple predators are not always sensitive to species relative abundance, but given that changes in predator comparative abundance are recurrently detected in nature.

Predator traits play a vital role for suppressing or reducing prey population [26]. According to Soluk [31] a major deficiency of studies which concentrate on the combined predator effects is that they are implemented at single prey densities. Prey accessibility could determine the outcome of interactions, for example when gratified predators become less aggressive, so reducing predation risk [32]. Therefore, a wide range of prey densities are required to explore the nature of interactions. Furthermore, such experimental data may contribute to disentanglement difficulties in hypothesis by other studies and eventually encourage understanding the effect of multiple predators on prey population [33, 34].

Interactions between multiple predator species should modify the strength of prey suppression. These interactions had been distinguished by different predation levels of such predators [25, 35, 36]. Positive effects are assumed to result from mechanisms such as interspecific resource-use complementarity [37], and conflicted escape behavior of prey [38]. Negative effects are attributed to intraguild predation (IGP) [39, 40] or interspecific interference competition [41].

For explaining the nature of effect of multiple predators, it is significant to determine the ecosystem models and community parameters [25, 42], and effects of predator species population could be predicted on reductions and losses on ecosystem functioning [43]. Many of researcher's work has inspected the multiple effects with respect to interactions between predator pairs [25], and gradually, mixtures of three or more predator species [30, 44]. For understanding this mechanism, multiple effects differ obviously with the relative richness, which lead to stronger effects, where populations of predator species are most even, making a positive relationship in a two-predator system comparing to the relative abundance of one species. This interpretation may result in a speedily saturating relationship between relative abundance [26, 45].

The qualified abundances of predator species fluctuate constantly in place and time in response to local ecological conditions and location relative to the center of species' respective biogeographic ranges. Researchers are exploring the effects of predator species' relative abundances on the functioning of predator guilds [36, 46], who mentioned that they affect prey suppression in crop fields.

From a predator's nutritional ways and the expected relationships among prey species, a general closed-form multi-species functional response for describing predators changing between multiple prey species [33]. Who mentioned that functional response could be divided into five main types, thus providing a natural explanation for the known complications of investigating prey switching in the fields?

Interactions among predators behavior on the same prey have been understood to play a vital role in determining their population structure. These studies with multiple predators have revealed that the effectiveness of such interactions may not be expected and is dependent mostly on individual active-behavioral characteristics and density [34]. Although consumption of predators and their forage are recorded, there has been practically research on effects of multiple predators; as well as their importance in environmental systems and population dynamics, expected more attention with respect to biocontrol of insects [47]. Principally, conservation of biocontrol agents aims to increase the efficiency of diverse natural enemies [48–50]. Approachability of prey could regulate the outcome of interactions among predators and their prey [32].

3 Does Intraguild Predation Interrupt the Biocontrol Agents Activities?

There are two categories of biocontrol agents: generalists and specialist. Specialists attack or infect one or as a maximum few hosts or prey. Generalist agents are likely to feed on various host species, as well as they could be reproduced on alternative food sources, hosts or prey. Conversely, these generalist predators may attack another biological control agent, when they feed on the same insect hosts (Intraguild Predation) [51, 52].

Intraguild predation (IGP) is known as a predator is preying on another natural enemy either predator or parasitoid when both biocontrol agents share the same host species. This means that intraguild predation can decline the resource competition and individual fitness of the competitor might be directly improved for similar resources [53, 54]. Generally, a lot of predators are generalists that mean they prey a wide range of t host types, which could rise IGP chance. There are many IGP examples associated with biocontrol agents either used in insectary, greenhouse or field. For example, adults of *Orius majusculus* feed on *Bemisia tabaci* nymphs parasitized by *Encarsia formosa*, while pupae were preyed upon less than the nymphs [55]. It has been advised that parasitized whiteflies have a tendency to tremendous and become muddy, which makes the parasitized whitefly nymphs more noticeable to the searching predators, then it has been directly preyed.

3.1 Types of Intraguild Predation

Intraguild predation has two types; unidirectional or bidirectional. Unidirectional IGP happens in case of one of the interacting biocontrol agents is considered the intraguild predator and another natural enemy is the intraguild prey. For example, *Orius laevigatus* has been presented to feed on both *Neoseiulus cucumeris* and *Iphiseius degenerans*, but neither of the predatory mites feeds on *O. laevigatus*. While another type known as a bidirectional IGP occurred when both of predators are capable of feeding on each other. For example, IGP is a common existence when the predatory mites, *Neoseiulus cucumeris* and *Amblyseius swirskii* both inhabit similar zones in the crop-plant parts [56]. This kind of IGPs is of interest to biocontrol applicators because use of both predator species must be evaded, if IGP decreases the insect control levels due to mutual exclusion of predators comparing to releasing one predator only, even if each predator share to pest management [40].

3.2 Some Examples

Rosenheim et al. [51] studied lacewing survivorship in the existence of another predator; their results showed that IGP was a main source of predator mortality. In spite of the fact that the trophic web was too complicated to delineate distinct trophic levels within the predatory arthropod community. Perdikis et al. [57] mentioned that two predators were found on different places on the plants, with *Nesidiocoris tenuis* manipulating mostly the upper part, whereas *Macrolophus pygmaeus* were habitually saw on the 5th to 7th leaf. A high proportion of the dead nymphs found with their body liquids totally sucked indicating that they had been preyed by *N. tenuis*. However, large 4th instar *M. pygmaeus* nymphs were much less susceptible to *N. tenuis* than younger. The behavior of *N. tenuis* was influenced by *M. pygmaeus* incidence, but at a rate similar to that when 2 individuals of *N. tenuis* were together surrounded.

3.3 Intraguild Impact on Biological Control

It has been suggested that IGP can negatively influence or interrupt biocontrol programs and could be an effective factor in determining the abundance and distribution of biocontrol agents [51, 58, 59]. Though, there is a discussion on whether IGP between natural enemies affect a positive or negative or no effect that could be accompanying with the natural enemies types involved. Lucas and Rosenheim [60] reported that one of the key aspects affecting the extent of IGP is the density of extraguild prey. Most laboratory and field mesocosm studies have reported a decrease in IGP intensity as extraguild prey density increases. Nevertheless, a rising of extraguild prey population leads to a growth in the intraguild predator density [61].

4 Climate Change Effects on Predator–Prey Interactions

In Egypt, there are different weather conditions according to the four seasons. These weather conditions play a significant role on insect pest population as well as the related activities of associated natural enemies [62]. The effects of climate change on environmental systems have more attention which affect the species composition and functions of different predator populations [63–66]. The climate change affected the food webs to become rewired or entirely disassemble because of the disability of heat sensitive species to bear or familiarize for climate change [67].

Schmitz and Barton [68] mentioned that predacious activity of some insect predators is either become more active or become disrupted according to changing climate. They also discusses climate change effects on predator–prey interactions and reported that these effects depended on the of predator and prey tolerances to changing abiotic conditions [63].

The global climate change played a prospective role to interrupt the prevailing programs of conservation biological control [68]. Climate change could modify abiotic conditions such as temperature, rainfall, humidity and wind that in sequence may alter the life-cycles of some insect predatory and their prey species which affect on the behavior and power of interactions [65, 66].

5 Pillars of Biological Control Industry

There are main pillars of biocontrol industry; abundance, mass production and finally field application. These three main objects will be presented as the followings.

5.1 Abundance of Natural Enemies

Insect predators belong to several orders (Heteroptera, Neuroptera, Diptera and Coleoptera) and found in most of the ecological fields over the world. Larvae and/or adults are the insatiable predators of exposed eggs, small larvae of beetle and lepidopterous pests. It also preys on slow-moving, soft-bodied arthropods such as thrips, whitefly, scales, aphids, jassids, mealy bugs, and mites.

Orius albidipennis, O. laevigatus, Chrysoperla carnea, Coccinella undcimpunctata, Mantis religiosa, Labidura riparia, Scymnus interruptus, Stethorus punctillum and Paederus alferii associating with corn borers S. cretica, C. agamemnon and O. nubilalis in maize fields were recorded in Egypt [69–71]. EL-Heneidy et al. [72] surveyed the predator population frequently in clover then in cotton fields, they mentioned that the predators began to increase in the fields of clover during March–May until be got their peaks of abundance during late April and May. In June, the predators wander to the cotton fields and achieve the maximum abundance in the latter fields during June and July. In cotton fields, El-Heneidy et al. [73]; Abou-Elhagag [74] recorded these predators in Egypt C. undcimpunctata, Scymnus sp. Paederus alferii, Orius spp., Chrysopa carnea and true spider. The dignified predators reached their peaks during July and Early of August. Orius species was the main predator followed by spider. El-Heneidy et al. [75] calculated directly the predatory species accompanying cotton pests in Uganda. They recorded immature and adults stages of several predators; syrphidae, coccinellids, anthocoris, staphylinidae, ants, labidurida, and true spiders. El-Heneidy et al. [76] found that predators associated cotton insect pests were more dominant in early cultivation than those in the late one, as well as in the mid of season than the end of the season.

The famous aphidophagous predatory species such as C. undecempunctata, Scymnus interruptus, Paederus alferii, Orius spp., Chrysopa carnea and true spider were documented preying Rhopalosiphum padi, R. maidis, Schizaphis graminum in wheat fields [77]. Nesidiocoris tenuis is considered a wide distributed predator in the Mediterranean region, was recorded for the first time in Egypt associated with T. absoluta in aubergine and tomato cultivations in Giza, Qaluobia, and Fayoum Governorates [78]. Sayed [79] recorded C. septempunctata, Andrallus spinidens, Rhynocoris fuscipes, Componotus sp., and Mantis religiosa on Pigeonpea plants and stated that these active predators played a pivotal role in reducing the population of insect pests. An assortment of survey attempts of some predaceous insects was shown in Table 1.

5.2 Mass Production of Some Insect Predators

The mass production of biocontrol agents depends on the capability of insectaries to produce and commercially market a highly reliable and objectively convenient supply of natural enemies. Accomplishing our purposes needs effective, identical

Table 1 Abundance of predaceous insects in Egypt (cited from Saleh et al. [7])

Crop	Natural enemy	Pest	Reference
Maize	*Mantis religiosa, Labidura riparia, Orius albidipennis, O. laevigatus, C. carnea, C. undcimpunctata, Scymnus interruptus, Paederus alferii*	*Sesamia cretica, Chilo Agamemnon, Ostrinia nubilalis*	Fayad et al. [69], Ragab et al. [69, 70]
Wheat	*C. undcimpunctata, Scymnus sp. Paederus alferii, Orius spp., C. carnea*	*Rhopalosiphum padi, R. maidis, Schizaphis graminum, Sitobion avenae*	EL-Heneidy and Abdel-Samad [77]
Cotton fields	*C. undcimpunctata, Scymnus sp., Paederus alferii, Orius albidipennis, O. laevigatus, C. carnea, Stethorus punctillum*	*Pectinophora gossypiella*	El-Heneidy et al. [73], Abou-Elhagag [74]
Mulberry trees	*Orius* spp., *Coranus* sp., *C. undecimpunctata, Cydonia* sp., *Mantis religiosa*	*Brevipalpus* sp., *Panonychus ulmi, Icerya aegyptiaca, I. purchase, Ceroplastes rusci, Saissetia oleae*	Hendawy et al. [80]
Pigeonpea (*Cajanus cajan*)	*C. septempunctata, Andrallus spinidens, Rhynocoris fuscipes, Componotus* sp., *Mantis religiosa*	*Aphis fabae, Oxyrachis tarandus, Melanoplus bivittatus, S. indica*	Sayed [79]
Tomatoes	*Nesidiocoris tenuis*	*T. absoluta*	El-Arnaouty and Kortam [78]

mass-rearing processes: (1) using of promising, nutritive media, (2) automatic and space effectual raring programs, (3) consistent storage techniques, and (4) assessment of quality of biocontrol agents [81]. In every step of these steps, generally it depends on the environmental, physiological and behavioral aspects. In Chrysoperla production which has contributed to the reasonable and applied improvements in mass production. Though, the active market of biocontrol agents and the right way for applying is being serious problems in need of more considerations [82].

Mass rearing of natural enemies especially the insect predator plays a significant role in rising the greenhouse industry over the world as well as in achieving a high quality agricultural production [83]. Sattar and Abro [84] mass reared *Chrysoperla carnea* adults for Integrated Pest Management Programs of some target insect pests. Rearing procedures for *C. carnea* have been developed to have its mass production in insectary for mass rearing in order to use in the fields. In this technique, for avoiding the cannibalistic predator, the rearing larvae were kept separately in hard gelatin capsules (500 mg) instead of using the anesthesia or vacuum sucker. After 10 days of development the larvae, then being collecting the molded pupae, which placing these pupae into cages for adult emergence and egg laying. The collected eggs will be used for field application [20, 21, 85].

Rearing of larvae classifies the most economically phase in mass rearing of *Chrysopesrla* principally because all three larval instars are predaceous. Many of mass rearing companies have produced lacewing and other insect predators using factitious hosts i.e. lepidopteran eggs: *Sitotroga* or *Corcyra*, which is comparatively propitious versus artificial diets [86, 87] (Fig. 1a, b). The recent work has intensive on rising the quantity and quality of mass produced predators by establishing abundance of prey needed and augmenting prey with artificial diet [88] in case of needed.

Nutritional necessities of adult predators face many practical problems for mass-production and selling of these predators. *Chrysoperla* adults need native hosts to sustain their egg production; these characteristics obscure mass production rearing procedures. Conversely, adult of *Chrysoperla* has an advantage comparing to other predators because they feed on honeydew and nectar [89]. From early experiment results, it could be concluded that nutrition of lacewings yielded quietly low-priced and effective artificial media that maintain the greatest rates of oviposition [88, 90, 91]. This effective diet affords a satisfactory illustration for mass rearing of some insect predators. Lacewing was mass reared as shown in Table 2.

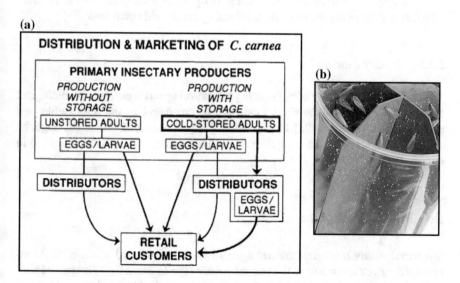

Fig. 1 **a** Strategy for distributing mass-produced *Chrysoperla carnea* (cited from Tauber et al. [87]). **b** Mass production of *Chrysoperla carnea* in cylinder

Table 2 Mass production of lacewings

Natural enemy	Host	Crop	Strategy and technique	References
Chrysoperla carnea	Aphids, new larvae, artificial diets	Wheat, cotton, maize	The cannibalistic larvae are kept separately in gelatin capsules to avoid using anesthesia or vacuum sucker	Ashfaq et al. [20, 21], Sattar and Abro [84]

5.2.1 Long-Term Storage

Storing of different insect predator species for a long time is a vital factor in the cost-effective mass production and marketing of biocontrol agents [92]. Storage abilities provide insectaries the opportunity to supply of more biological control agents for usage during high season periods. Furthermore, an effective preservation or storage strategies may provide alternative procedures for supply the natural enemies for a long-term with convenient, price [86, 87, 88, 93, 94].

Significant new developments in mass production of insect predators, mechanized production systems, long-term storage, and high quality agents could decrease the cost and easy for field application. Additionally, these procedures should be more effective and motiving for more mass releasing under field conditions [95].

5.2.2 Quality Control

The identical mass rearing of high quality biocontrol agents is essential for biological control application and farmers' awareness of biocontrol as the responsible pest management approaches [96]. Quality control of produced natural enemies might be different commercially, because there is no fixed quality control tactics applied in our regions [88, 97].

5.3 Field Application

In general, many biological control agents have been released in different developmental stages according to the natural enemy type in some target crops and fruit orchards; for example the lacewing *Chrysoperla carnae* had been released either egg or larval stages. While as the coccinellids is released as larvae or adults; however the nymph or adult stages of the anthocoris predatory bug are released in order to manage different insect eggs or new hatching larvae. Different insect predators were mass released as shown in Table 3.

Table 3 Examples of field application of insect predators in different crops

Crop	Natural enemy	Pest	Findings	Reference
Cotton	*C. undcimpunctata, Scymnus* sp. *Paederus alferii, Orius* sp., *C. carnea, S. punctillum*	*H.armigera, E. insulana, P. gossypiella*	Significant effects	El-Heneidy et al. [73, 75, 76]
Faba bean	*C. carnea, C. undecimpunc-tata*	*Aphis gossypii, Aphis fabae*	100% reduction in *Aphis gossypii*	Zaki et al. [98]
	Harmonia axyridis	*A. craccivora*	Significant reduction	El-Arnaouty et al. [99]
Green pepper	*Chrysoperla carnea*	*Myzus persicae*	Successful control of aphids	El-Arnaouty et al. [100]
Ornamental plant	*Cryptolaemus montrouzieri*	*Ferrisia virgata*	Reduction % reached to 99.99	Attia and El-Arnaouty [101]
Ornamental shrubs	*C. montrouzieri*	*Planococcus citri*	Reduction rates reached to 100%	Afifi et al. [102]

In Egypt, there are many researchers have inspected several species of insect predators in cotton fields such as *Scymnus interruptus, C. undcimpunctata, Paederus alferii, Orius albidipennis, O. laevigatus, C. carnea* [73, 75, 76]. Those predators have played a significant role in reducing population of many cotton insect pests in different cotton fields in northern and southern Egypt. El-Arnaouty and Sewify [103] controlling some aphid species such as *Aphids gossypii* by releasing the eggs and larvae of lacewings *C. carnae* in cotton fields. They mentioned that *C. carnae* succeeded to achieve a major declining of *A. gossypii* population.

Zaki et al. [98] carried out several experiments by releasing ladybeetle *Coccinella undecimpunctata* and lacewings *Chrysoperla carnea* against different aphid species such as *Aphis gossypii* and *A. fabae* under greenhouse conditions. They stated that 100% reduction in population of *A. gossypii* were achieved by twice releases of *C. carnea* post twelve days; while as an individual release of *C. undecimpunctata* caused 99.97% decreasing in aphid population.

Applying *Harmonia axyridis* for managing *Aphis craccivora* population was released on faba bean plants in faba fields. Results presented that predator like *H. axyridis* is able to manage aphids and provided a pivotal reduction in aphid population [99]. Releasing eggs or the 2nd larval instar of *Chrysoperla carnea* on pepper plants caused reducing *Myzus persicae* population [100]; who mentioned that a positive control of aphids could be achieved in the greenhouses where applied the predator in order to reduce the aphid infestation.

For controlling the striped mealybug *Ferrisia virgata* on the ornamental plant *Acalypha macrophylla*, the coccinellid predator *Cryptolaemus montrouzieri* was released by Attia and El-Arnaouty [101] in Egypt. This predator were distributed once in Fall October and the researchers have recorded that the reduction percent of *F. virgata* reached to 90% for, 75% and 68% for crawlers, nymphs and adults, respectively; while as these percents increased in the 11th week post release to reach 100%, 89% and 95% for the same developmental stages. Afifi et al. [102] applied the *C. montrouzieri* for managing the citrus mealybug, *Planococcus citri* on some ornamental shrubs such as *Codiaeum variegatum*. After one month post field application, numbers of egg masses, nymphs and adults of *P. citri* reduced to reach 41.5, 42.3 and 57.5%, respectively. While after 2 months, the corresponding rates were 80.6, 86.5 and 91.5%. The reduction rates reached to 100% for all stages of this pest after 3 months of releasing the predator, https://www.ncbi.nlm.nih.gov/pubmed/20464943.

6 Insect Predator Preservation

Crop fields (plant structure) differ greatly in insect pest populations during the different year seasons compared with fruit orchards (permanent crops). Paradoxically, the orchards have naturally the higher natural enemy diversity [50, 104, 105], and preserving high natural enemy population in the field crops might be very difficult. Therefore, there are various mechanisms may provide the greatest benefit for biological control agents. In many natural environments, separating the biodiversity effects from the ecological influences that produce it may be a key task to suppress the important insect pests for ecosystem services [106]. A larger sympathetic of the interactions among diversity components as well as effects of mechanisms fundamental biodiversity may improve efforts to support biological control in different agroecosystems [107]. Introducing predators for classical biological control program, in case of low diversity of natural enemies may provide solutions to an applied problem [108, 109]. A suggested strategy for more efficient conservation biological control is containing collection natural enemies, preservation them and releasing the preserved natural enemies on target crops [50].

Conservation strategies should be prolonged over bigger zones to be effective playing an essential role for suppressing the insect pests [45, 110]. Fiedler et al. [111] stated that habitat management isn't considered as a stand-alone practice and should be viewed holistically for the several ecosystem facilities.

6.1　Nectar Resources

Most predators necessitate sugar sources. Nectar nurturing could increase predator survival to 20-fold [112], enhance fecundity, increasing general reproductive fitness by providing also floral supplements for natural enemies [113, 114].

6.2　Pollen Sources

Vandekerkhove and De Clercq [115] mentioned that nymphs of mirid predator *Macrolophus pygmaeus* which fed on plan pollen survived to 80%; this method is considered a preventative release strategy similar to the one suggested by De Cocuzza et al. [116] for *Orius* species.

6.3　Shelter

For protecting the predators in abnormal conditions, shelter provides a suitable condition for overwintering and aestivation or any disturbances in fields or orchards [117].

6.4　Weed as Shelter and Food

Generally, in the interval time between the agricultural seasons (main crops), the biocontrol agents escape to the field borders (where are weeds) searching for food and safe place. Researchers acknowledge role of weed introduction and their effects on communities of biological control agents [118–120]. Effects of different weed groups on biological control agents in many agricultural systems were investigated [121]. In organic farming, it generally harbors 3 times more abundant seed densities and 1.5 times more weed species compared with conventional farms [122]. Herbicides used in intensive agroecosystems leads to reduce the weed abundance, which it is considered the microhabitat of insects as well as the associated natural enemies, especially in case of absent the main crops [120].

7　Case Study: Aphids and Their Associated Predators

There are many aphid species infest various crops, vegetables and fruit orchards over the world. For example cotton aphid, *Aphis gossypii*, is one of the principal insect pests of commercially in cotton fields especially in the summer months (June–October) [123]. Cereal aphids, *Rhopalosiphum padi*, *R. maidis* and *Metopolophium dirhodum* infest wheat plants during the winter and maize and sorghum during the summer season [70, 71, 124, 125, 126]. Faba aphids *Aphis craccivora* and *A. fabae* infest faba beans [99, 100, 127], potato/artichoke aphid *Myzus persicae* infests artichoke as confirmed by El-Wakeil and Saleh [128] which caused a serious damage for these crops [129].

The case study demonstrated that the aphid species and their associated predator species were highly varied among the most important crops. Efficacy of insect predator species against many aphid species in different crops i.e., wheat, faba bean, corn and potato/artichoke was shown that insect predators played a significant role in controlling many aphid species. Among the observed predators were coccinellid, syrphid, anthorid and chrysopid species. *Coccinella septempunctata*, *C. undecimpunctata*, *Harmonia axyridis*, *Episyrphus balteatus*, *Orius* spp. and *Chyrsoperla carnea*. The density of the most abundant species (thirteen species) varied across the four crops. *C. septempunctata* was the most abundant species in faba bean and potato/artichoke crops. In contrast, three species dominated the predator guild in wheat crops; namely, *E. balteatus*, *C. septempunctata*, and *C. carnea*. *H. axyridis* had higher densities in bean and in potato/artichoke than the other crops. *Coccinella septempunctata* was the most abundant predator in many crops.

This population variation might be due to the inherent traits of these crops including leaf and steam constructions, volatile components and sap conformation [130]. Aphid abundance was regulated by several aspects such as abiotic ecological conditions (micro-climate), plant volatiles, plant structure, and multiplicity of host plants [131, 132]. Results obtained during this study show that *C. septempunctata* and *H. axyridis* was more abundant in potato/artichoke and in faba bean than in wheat and corn, these results are similar which obtained by Vandereycken et al. [133]. The corn plants being higher than the other cultures the water is dropped between maize plants affected negatively the predator communities; such result was confirmed by Hodek et al. [134].

Aphid populations might suffer periods of rapid growth regardless of the development of generalist predator populations such as lacewings and lady beetle. The most important predators are lacewing and lady beetle species, which played a significant role with other predator species in reducing the aphid densities. Either larvae or/and adults are known to be the effective predators of aphids; there are many of natural enemies which affected by habitat features at different landscape scales [135]. Coccinellids and Lacewing are reared by insectaries commercially and then released widely in different biological control programs against many of aphid species (Table 4). There are other predators which play a very important role in managing numerous aphid species; predatory true bugs (Hemiptera) *Orius* spp. as well as *Lygus* bugs [45, 136]. Finally, the lady beetle species are considered the efficient predators for controlling aphids on a range of field crops (Fig. 2a, b, c), vegetables and fruit trees [99, 125, 133, 134, 137, 138].

Table 4 Aphid species and associated predator species obtained in wheat, maize, faba bean, and potato/artichoke crops

Crop	Aphid species	Predator	Family
Wheat	*Rhopalosiphum padi* *Sitobion avenae* *Metopolophium dirhodum*	*Coccinella septempunctata* *Coccinella undecimpunctata* *Harmonia axyridis* *Hippodamia undecimnotata* *Propylea quatuordecimpunctata* *Scymnus interruptus*	Coccinellids
Maize	*Rhopalosiphum maidis*	*Chrysoperla carnea*	Chrysopids
Faba bean	*Aphis fabae* *Aphis craccivora*	*Syrphus ribesii* *Episyrphus balteatus* *Metasyrphus corollae*	Syrphids
Potato/artichoke	*Myzus persicae* *Aphis nasturtii*	*Orius albidipennis* *Orius insidiosus* *Orius laevigatus*	Anthorids

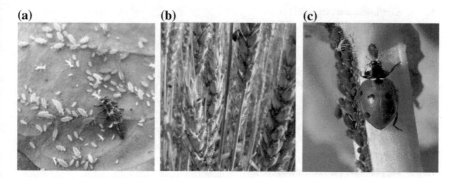

Fig. 2 Coccinellid larvae prey potato aphids (**a**); coccinellid adults prey wheat aphids (**b**); bean aphids (**c**)

Understanding the natural enemy capability to manage different insect pests may lead to sager insect control, comprising pesticide reduction and keeping the environment clean. The abundance of generalist predators could help controlling and reducing the pest populations and thus preventing insect pest outbreaks [109, 126, 132, 139]. Some of predators have a restricted tolerance to the prey feeding and abiotic factors; lacewing larval food consumption affected the fecundity of Chrysopid adults [140–142]. Therefore, mass rearing and field application of insect predators are considered one of the main aspects in succeeding the biological control programs.

8 Conclusions and Recommendation

Agriculture provides of 13% of the Egyptian gross domestic product (GDP) and more than 30% of occupation chances. To face the population increase, agricultural production should be increased by 7%; which should be accomplished throughout the sustainable agriculture system which conserves the ecological alternatives and confines the use of insecticides in the different programs for insect pest management. Consequently, using insect predators in programs of biological control might be a foremost part of sustainable agriculture. About 20 species of insect predators listed in this book chapter have been considered by Egyptian biocontrol researchers (see Table 5). These predators prey most all the economic agricultural insect pests of

Table 5 Alphabetical list of insect predators listed in this chapter

1.	*Andrallus spinidens*
2.	*Chrysoperla carnea*
3.	*Coccinella septempunctata*
4.	*Coccinella undcimpunctata*
5.	*Componotus montrouzieri*
6.	*Componotus sp.*
7.	*Coranus sp.*
8.	*Cryptolaemus montrouzieri*
9.	*Cryptolaemus sp.*
10.	*Cydonia sp.*
11.	*Episyrphus balteatus*
12.	*Harmonia axyridis*
13.	*Hippodamia undecimnotata*
14.	*Labidura riparia*
15.	*Macrolophus pygmaeus*
16.	*Mantis religiosa*
17.	*Metasyrphus corollae*
18.	*Nesidiocoris tenuis*
19.	*Orius albidipennis*
20.	*Orius insidiosus*
21.	*Orius laevigatus*
22.	*Paederus alferii*
23.	*Propylea quatuordecimpunctata*
24.	*Rhynocoris fuscipes*
25.	*Scymnus interruptus*
26.	*Syrphus ribesii*
27.	*Stethorus punctillum*

field crops, vegetables, fruit orchards, ornamentals and greenhouse crops. It has been focused in this chapter on richness of predators in our environment and the challenges of mass rearing and field application of the most significant predators.

As declared in this chapter the biocontrol production is influenced by many factors, such as infrastructure, personnel, tools, transportation and market, beside presence of the mentioned predators in the local environment and tries of mass rearing and releasing. In Egypt, almost of these elements are found, what is still need is the support of decision makers in private and public sectors for having the infrastructure as well as some equipment. There is an increasing consciousness to the risks of agricultural chemicals and the need to the biocontrol for preserving the environment in Egypt. Egypt has a great market of biological products represented mostly by the agricultural private sector which distribute these bio- or organic products of vegetables and fruit to Arab and European countries. The following is an alphabetical list of the existed insect predators and recorded throughout this chapter (Table 5).

References

1. Aune JB (2012) Conventional, organic and conservation agriculture: production and environmental impact. In: Lichtfouse E (ed) Agroecology and strategies for climate change, vol 8, Sustainable agriculture reviews. Springer Science + Business Media B.V, Dordrecht/New York
2. Anon (2014) Recommendations adopted to combat agricultural pests, 296 pp. Agricultural Pesticides Committee, Egyptian Ministry of Agriculture and Land reclamation, Media Support Centre Press, Dekerness, Dakahila, Egypt (in arabic)
3. DeBach P (1964) Biological control of insect pests and weeds. Chapman and Hall, London
4. DeBach P (1974) Preface. In: DeBach P, Rosen D (eds) Biological control by natural enemies. Cambridge University Press, Cambridge
5. DeBach P, Rosen D (1991) Biological control by natural enemies, 2nd edn, 440 pp. Cambridge University Press, Cambridge. ISBN 0-521-39191-1
6. Mahr S (2017) The role of biological control in sustainable agriculture. University of Wisconsin–Madison. http://www.entomology.wisc.edu/mbcn/fea405.html. 28 Sept 2017
7. Saleh MME, El-Wakeil NE, Elbehery H, Gaafar N, Fahim S (2019) Biological pest control for sustainable agriculture in Egypt. In: Negm AM, Abu-Hashim M (eds) Sustainability of agricultural environment in Egypt: part II. Springer Publisher. ISBN 978-3-319-95356-4, © Springer Nature Switzerland AG 2019-Soil-Water-Plant Nexus, Hdb Env Chem 77:145–188. https://doi.org/10.1007/978-3-319-95357-1
8. Gurr G, Wratten S (2000) Biological control: measures of success, 429 pp. Springer Science + Business Media, Dordrecht
9. van Lenteren JC (2000) Success in biological control of arthropods by augmentation of natural enemies. In: Wratten S, Gurr G (eds) Biological control: measures of success. Springer Science + Business Media, Dordrecht, pp 77–103
10. van Lenteren JC (2003) Commercial availability of biological control agents. In: van Lenteren JC (ed) Quality control and production of biological control agents: theory and testing procedures. CABI, Oxon, UK, pp 167–179
11. van Lenteren JC (2012) The state of commercial augmentative biological control: plenty of natural enemies, but a frustrating lack of uptake. Biocontrol 57:1–20
12. Tawfik MFS (1997) Biological control for insect pests (in Arabic), 2nd edn., 757 pp. Academic Bookshop, Cairo

13. Doutt RL (1964) The historical development of biological control. In: DeBach P (ed) Biological control of insect pest and weeds. Chapman and Hall, London
14. El-Wakeil NE, Abd-Alla AM, El Sebai TN, Gaafar NM (2015) Effect of organic sources of insect pest management strategies and nutrients on cotton insect pests (Chap 2). In: Gorawala P, Mandhatri S (eds) Frame of book Agricultural research updates, vol 10, pp 49–81. ISBN: 978-1-63482-745-4
15. El-Wakeil N, Gaafar N, Sallam A, Volkmar C (2013) Side effects of insecticides on natural enemies and possibility of their integration in plant protection strategies. In: Trdan S (ed) Agricultural and biological sciences "insecticides—development of safer and more effective technologies". Intech, Rijeka, Croatia, pp 1–54
16. Bianchi FJ, Booij CJ, Tscharntke T (2006) Sustainable pest regulation in agricultural landscapes: a review on landscape composition, biodiversity and natural pest control. Proc Biol Sci 273:1715–1727
17. Leopold A (1933) Game management. Charles Scribner's Sons, New York, NY
18. Holling CS (1961) Principles of insect predation. Annu Rev Entomol 6:163–182
19. Holling CS (1959) Some characteristics of simple types of predation and parasitism. Can Entomologist 91:385–398
20. Ashfaq M, Abida N, Gulam MC (2002) A new technique for mass rearing of green lacewing on commercial scale. Pak J Appl Sci 2:925–926
21. Ashfaq M, Nasreen A, Cheema GM (2004) Advances in mass rearing of *Chrysoperla carnea* (Stephen) (Neuroptera: Chrysopidae). South Pac Stud 24:47–53
22. van Lenteren JC, Manzaroli G (1999) Evaluation and use of predators and parasitoids for biological control of pests in greenhouses. In: Albajes R, Gullino ML, van Lenteren JC, Elad Y (eds) Integrated pest and disease management in greenhouse crops. Kluwer, Dordrecht, The Netherlands, pp 183–201
23. van Lenteren JC, Bueno VHBP (2003) Augmentative biological control of arthropods in Latin America. Biocontrol 48:123–139
24. El-Wakeil NE, Vidal S (2005) Using of *Chrysoperla carnea* in combination with *Trichogramma* species for controlling *Helicoverpa armigera*. Egypt J Agric Res 83:891–905
25. Sih A, Englund G, Wooster D (1998) Emergent impacts of multiple predators on prey. Trends Ecol Evol 13:350–355
26. Casula P, Wilby A, Thomas MB (2006) Understanding biodiversity effects on prey in multi-enemy systems. Ecol Lett 9:995–1004
27. Letourneau D, Jedlicka J, Bothwell S, Moreno C (2009) Effects of natural enemy biodiversity on the suppression of arthropod herbivores in terrestrial ecosystems. Ann Rev Ecol Evol Syst 40:573–592
28. Hochberg M (1996) Consequences for host population levels of increasing natural enemy species richness in classical biological control. Am Nat 147:307–318
29. Stiling P, Cornelissen T (2005) What makes a successful biocontrol agent? A meta-analysis of biological control agent performance. Biol Control 34:236–246
30. Griffin JN, Toscano BJ, Griffen BD, Silliman BR (2015) Does relative abundance modify multiple predator effects? Basic Appl Ecol 16:641–651
31. Soluk DA (1993) Multiple prey effects: predicting combined functional response of stream fish and invertebrate predators. Ecology 74:219–225
32. Vance-Chalcraft HD, Soluk DA (2005) Estimating the prevalence and strength of non-independent predator effects. Oecologia 146:452–460
33. van Leeuwen E, Brännström A, Jansen VAA, Dieckmann U, Rossberg AG (2013) A generalized functional response for predators that switch between multiple prey species. J Theor Biol 328:89–98
34. Lampropoulos PD, Ch Perdikis D, Fantinou AA (2013) Are multiple predator effects directed by prey availability? Basic Appl Ecol 14:605–613
35. Schmitz OJ (2007) Predator diversity and trophic interactions. Ecology 88:2415–2426
36. Schmitz OJ (2009) Effects of predator functional diversity on grassland ecosystem function. Ecology 90:2339–2345

37. Griffin JN, De la Haye KL, Hawkins SJ, Thompson RC, Jenkins SR (2008) Predator diversity and ecosystem functioning: density modifies the effect of resource partitioning. Ecology 89:298–305

38. Losey JE, Denno RF (1998) Positive predator–predator interactions: enhanced predation rates and synergistic suppression of aphid populations. Ecology 79:2143–2152

39. Finke DL, Denno RF (2005) Predator diversity and the functioning of ecosystems: the role of intraguild predation in dampening trophic cascades. Ecol Lett 8:1299–1306

40. Gagnon A, Heimpel G, Brodeur J (2011) The ubiquity of intraguild predation among predatory arthropods. PLoS ONE 6(11):e28061. https://doi.org/10.1371/journal.pone.0028061

41. Griffen BD, Byers JE (2006) Partitioning mechanisms of predator interference in different habitats. Oecologia 146:608–614

42. McCoy MW, Stier AC, Osenberg CW (2012) Emergent effects of multiple predators on prey survival: the importance of depletion and the functional response. Ecol Lett 15:1449–1456

43. Duffy JE, Cardinale BJ, France KE, McIntyre PB, Thebault E, Loreau M (2007) The functional role of biodiversity in ecosystems: incorporating trophic complexity. Ecol Lett 10:522–538

44. Griffin JN, Byrnes JE, Cardinale BJ (2013) Effects of predator richness on prey suppression: a meta-analysis. Ecology 94:2180–2187

45. Finke DL, Snyder WE (2008) Niche partitioning increases resource exploitation by diverse communities. Science 321:1488–1490

46. Crowder DW, Northfield TD, Strand MR, Snyder WE (2010) Organic agriculture promotes evenness and natural pest control. Nature 466:109–112

47. Werling BP, Lowenstein DM, Straub CS, Gratton C (2012) Multi-predator effects produced by functionally distinct species vary with prey density. J Insect Sci 12:1–7

48. Straub CS, Finke DL, Snyder WE (2008) Are the conservation of natural enemy biodiversity and biological control compatible goals? Biol Control 45:225–237

49. Tylianakis JM, Romo CM (2010) Natural enemy diversity and biological control: making sense of the context-dependency. Basic Appl Ecol 11:657–668

50. El-Wakeil NE, Saleh MME, Gaafar N, Elbehery H (2017) Conservation biological control practices (Chap 3) in frame of book biological control of pest and vector insects. Intech Open Access. ISBN 978-953-51-5041-1

51. Rosenheim JA, Wilhoit LR, Armer CA (1993) Influence of intraguild predation among generalist insect predators on the suppression of an herbivore population. Oecologia 96:439–449

52. Polis GA, Myers CA, Holt RD (1989) The ecology and evolution of intraguild predation: potential competitors that eat each other. Ann Rev Ecol Syst 20:297–330

53. Lucas E (2005) Intraguild predation among aphidophagous predators. Eur J Entomol 102:351–364

54. Lucas E (2012) Intraguild interactions. In: Hodek I, van Emden HF, Honek A (eds) Ecology and behaviour of the ladybird beetles (Coccinellidae), pp 43–373. Wiley Blackwell

55. Jakobsen L, Enkegaard A, Brodsgaard HF (2004) Interaction between two polyphagous predators, *Orius majusculus* and *Macrolophus caliginosus*. Biocontrol Sci Technol 14:17–24

56. Janssen A, Sabelis MW, Magalhães S, Montserrat M, Van Der Hammen T (2007) Habitat structure affects intraguild predation. Ecology 88:2713–2719

57. Perdikis D, Lucas E, Garantonakis N, Giatropoulos A, Kitsis P, Maselou D, Panagakis S, Lampropoulos P, Paraskevopoulos A, Lykouressis D, Fantinou A (2014) Intraguild predation and sublethal interactions between two zoophytophagous mirids, *Macrolophus pygmaeus* and *Nesidiocoris tenuis*. Biol Control 70:35–41

58. Rosenheim JA, Limburg DD, Colfer RG (1999) Impact of generalist predators on a biological control agent, *Chrysoperla carnea*: direct observations. Ecol Appl 9:409–417

59. Martinou AF, Raymond B, Milonas PG, Wright DJ (2010) Impact of intraguild predation on parasitoid foraging behaviour. Ecol Entomol 35:183–189

60. Lucas E, Rosenheim JA (2011) Influence of extraguild prey density on intraguild predation by heteropteran predators: a review of the evidence and a case study. Biol Control 59:61–67

61. Rosenheim JA (2007) Intraguild predation: new theoretical and empirical perspectives. Ecology 88:2679–2680

62. El-Wakeil NE, Volkmar C (2011) Effect of weather conditions on frit fly (*Oscinella frit*, Diptera: Chloropidae) activity and infestation levels in spring wheat in central Germany. Gesunde Pflanzen 63:159–165

63. Schmitz OJ, Post E, Burns CE, Johnston KM (2003) Ecosystem responses to global climate change: moving beyond color-mapping. Bioscience 53:1199–1205

64. Tylianakis JM, Didham RK, Bascompte J, Wardle DA (2008) Global change and species interactions in terrestrial ecosystems. Ecol Lett 11:1351–1363

65. Traill LW, Lim MLM, Sodhi NS, Bradshaw C (2010) Mechanisms driving change: altered species interactions and ecosystem function through global warming. J Anim Ecol 79:937–947

66. Tylianakis JM, Binzer A (2014) Effects of global environmental changes on parasitoid–host food webs and biological control. Biol Control 75:77–86

67. Sheldon KS, Yang S, Tewksbury JJ (2011) Climate change and community disassembly: impacts of warming on tropical and temperate montane community structure. Ecol Lett 14:1191–1200

68. Schmitz OJ, Barton BT (2014) Climate change effects on behavioral and physiological ecology of predator–prey interactions: implications for conservation biological control. Biol Control 75:87–96

69. Fayad YH, Hafez M, El-Kifl AH (1979) Survey of the natural enemies of the three corn borers *Sesamia cretica*, *Chilo agamemnon* and *Ostrinia nubilalis* in Egypt. Agric Res Rev 57:29–33

70. Ragab ZA, Awadallah KT, Farghaly HT, Ibrahim AM, El-Wakeil NE (2001) Population dynamics of corn pests and their associated predators in sorghum varieties grown in El-Giza Governorate in Egypt. Egypt J Appl Sci 16:652–666

71. Ragab ZA, Awadallah KT, Farghaly HT, Ibrahim AM, El-Wakeil NE (2001) Seasonal abundance of certain corn pests and their associated predators in maize varieties grown in E-Beheira Governorate in Egypt. Egypt J Appl Sci 16:298–312

72. EL-Heneidy AH, Abbas MST, El-Dakruory MSI (1978–1979) Seasonal abundance of certain predators in untreated Egyptian clover and cotton fields in Fayoum Governorate, Egypt. Bull Soc Entomol Egypt 62:89–95

73. El-Heneidy AH, Abbas MS, Khidr AA (1987) Comparative population densities of certain predators in cotton fields treated with sex pheromones and insecticides in Menoufia Governorate, Egypt. Bull Soc Entomol Egypt, Econ Ser 16:181–190

74. Abou-Elhagag GH (1998) Seasonal abundance of certain cotton pest and their associated natural enemies in Southern Egypt. Assiut J Agric Sci 29:253–267

75. EL-Heneidy AH, Sekamatte B, Mwambu N, Nyamutale C, Soroti PO (1996) Integrated pest management approach in cotton agro-ecosystem in Uganda. 1. Basic field data. Afr Crop Sci J 4:1–13

76. EL-Heneidy AH, Ebrahim AA, Gonzalez D, Abdel-Salam NM, Ellington J, Moawad GM (1997) Pest-predator-interactions in untreated cotton fields at three plant growth stages. 2. Planting date impact. Egypt J Agric Res 75:137–155

77. EL-Heneidy AH, Abdel-Samad SS (2001) Tritrophic interactions among Egyptian wheat plant, cereal aphids and natural enemies. Egypt J Pest Control 11:119–125

78. El-Arnaouty SA, Kortam MN (2012) First record of the mired predatory species, *Nesidiocoris tenuis* Reuter (Heteroptera: Miridae) on the tomato leafminer, *Tuta absoluta* (Meyrick) (Lepidoptera: Gelechiidae) in Egypt. Egypt J Biol Pest Control 22:223–224

79. Sayed HE (2016) Ecological and biological studies on some destructive and beneficial insects on tomato plants in Egypt, 372 pp. PhD thesis, Faculty of Science, Al-Azhar University, Egypt

80. Hendawy AS, Saad IAI, Taha RH (2013) Survey of scale insects, mealy bugs and associated natural enemies on mulberry trees. Egypt J Agric Res 91:1447–1458

81. Ruberson JR, Nechols JR, Tauber MJ (1999) Biological control of arthropod pests, PI'. 417–448. In: Ruberson JR (ed) Handbook of pest management. Marcel Dekker, New York

82. Slininger PJ, Behle RW, Jackson MA, Schisler DA (2003) Discovery and development of biological agents to control crop pests. Neotrop Entomol 32:183–195

83. van Lenteren JC, Roskam MM, Timmer R (1997) Commercial mass production and pricing of organisms for biological control of pests in Europe. Biol Control 10:143–149

84. Sattar M, Abro GH (2011) Mass rearing of *Chrysoperla carnea* (Chrysopidae) adults for integrated pest management programs. Pak J Zool 43:483–487

85. El-Arnaouty SA, Abdel-Khalek S, Hassan H, Shahata M, Game M, Mahmoud N (1998) Mass rearing of aphidophagous predators, *Chrysoperla carnea* and *Harmonia axyridis* on *Ephestia kuehniella* eggs. In: Regional symposium for applied biological control in Mediterranean countries, Cairo, Egypt, 25–29 Oct 1998

86. Wang R, Nordlund DA (1994) Use of *Chrysoperla* spp. (Neuroptera: Chrysopidae) in augmentative release programs for control of arthropod pests. Biocontrol News Inf 15:51N–57N

87. Tauber MJ, Tauber CA, Daane KM, Hagen KS (2000) Commercialization of predators: recent lessons from green lacewings. Am Entomol 46:26–38

88. Chang YF, Tauber MJ, Tauber CA (1996) Reproduction and quality of F1 offspring in *Chrysoperacamea*: differential influence of quiescence, artificially-induced diapause, and natural diapause. J Insect Physiol 42:521–528

89. Hagen KS (1987) Nutritional ecology of terrestrial insect predators. In: Slansky Jr. F, Rodriguez JG (eds) Nutritional ecology of insects, mites, and spiders, pp 517–533. Wiley, New York

90. Albuquerque GS, Tauber CA, Tauber NJ (1994) *Chrysoperla extenza* (Neuroptera: Chrysopidae): life history and potential for biological control in Central and South America. Biol Control 4:8–13

91. Anonymous (1997) Mass-reared insects get fast-food. USDA Agric Res 5–7

92. Osman MZ, Selman BJ (1993) Storage of *Chrysoperla carnea* Steph. (Neuroptera, Chrysopidae) eggs and pupae. J Appl Entomol 115:420–424

93. Tauber MJ, Tauber CA, Gardescu S (1993) Prolonged storage of *Chrysoperla carnea* (Neuroptera: Chrysopidae). Environ Entomol 22:843–848

94. Tauber MJ, Albuquerque GS, Tauber CA (1997) Storage of nondiapausing Chrysoperla externa adults: influence on survival and reproduction. Biol Control 10:69–72

95. Jones SL, Ridgway RL (1976) Development of methods for field distribution of eggs of the insect predator *Chrysopa carnea* Stephens, 5 pp. USDA, ARS–S–124

96. Bigler F (ed) (1992) Quality control of mass reared arthropods. In: Proceedings, 5th workshop of the IOBC global working group, Wageningen, The Netherlands, 25–28 Mar 1991. Swiss Federal Research Station for Agronomy, Zurich

97. O'Neil RJ, Giles KL, Obrycki JJ, Mahr DL, Legaspi JC, Katovich K (1998) Evaluation of the quality of four commercially available natural enemies. Biol Control 11:1–8

98. Zaki FN, El-Shaarawy MF, Farag NA (1999) Release of two predators and two parasitoidds to control aphids and whiteflies. J Pest Sci 72:19–20

99. El-Arnaouty SA, Beyssat-Arnaouty V, Ferran A, Galal H (2000) Introduction and release of the coccinellid *Harmonia axyridis* for controlling *Aphis craccivora* on faba beans in Egypt. Egypt J Biol Pest Control 10:129–136

100. El-Arnaouty SA, Gaber N, Tawfik MFS (2000) Biological control of the green peach aphid *Myzus persicae* by *Chrysoperla carnea* (stephens) *sensu lato* (Chrysopidae) on green pepper in greenhouses in Egypt. Egypt J Biol Pest Control 10:109–116

101. Attia AR, El-Arnaouty SA (2007) Use of the coccinellid predator *Cryptolaemus montrouzieri* against the striped mealybug, *Ferrisia virgata* on the ornamental plant, *Agalypha macrophylla* in Egypt. Egypt J Biol Pest Control 17:71–76

102. Afifi AI, El Arnaouty SA, Attia AR, Abd Alla AEL-M (2010) Biological control of citrus mealybug, *Planococcus citri* using coccinellid predator, *Cryptolaemus montrouzieri*. Pak J Biol Sci 13:216–222

103. El-Arnaouty SA, Sewify GH (1998) Apilot experiment of using eggs and larvae of *Chrsoperla carnae* against *Aphids gossypii* on cotton in Egypt. Acts Zoll Fenica 209:103–106

104. Tscharntke T, Bommarco R, Clough Y, Crist TO, Kleijn D, Rand TA et al (2007) Conservation biological control and enemy diversity on a landscape scale. Biol Control 43:294–309

105. Tscharntke T, Sekercioglu CH, Dietsch TV, Sodhi NS, Hoehn P, Tylianakis JM (2008) Landscape constraints on functional diversity of birds and insects in tropical agroecosystems. Ecology 89:944–951

106. Shanker C, Katti G, Padmakumar AP Padmavathi C, Sampathkumar M (2012) Biological control, functional biodiversity and ecosystem services in insect pest Management. In: Venkateswarlu et al (eds) Crop stress and its management: perspectives and strategies. https://doi.org/10.1007/978-94-007-2220-0_14, © Springer Science + Business Media B.V

107. Crowder DW, Jabbour R (2014) Relationships between biodiversity and biological control in agroecosystems: current status and future challenges. Biol Control 75:8–17

108. Prasad R, Snyder WE (2006) Polyphagy complicates conservation biological control that targets generalist predators. J Appl Ecol 43:343–352

109. Perdikis D, Fantinou A, Lykouressis D (2011) Enhancing pest control in annual crops by conservation of predatory Heteroptera. Biol Control 59:13–21

110. Barbosa P (ed) (1998) Conservation biological control. Academic, New York

111. Fiedler AK, Landis DA, Wratten SD (2008) Maximizing ecosystem services from conservation biological control: the role of habitat management. Biol Control 45:254–271

112. Bianchi FJJA, Wackers FL (2008) Effects of flower attractiveness and nectar availability in field margins on biological control by parasitoids. Biol Control 46:400–408

113. Winkler K, Wackers F, Bukovinszkine-Kiss G, van Lenteren J (2006) Sugar resources are vital for Diadegma semiclausum fecundity under field conditions. Basic Appl Ecol 7:133–140

114. Winkler K, Wackers FL, Termorshuizen AJ, van Lenteren JC (2010) Assessing risks and benefits of floral supplements in conservation biological control. Biocontrol 55:719–727

115. Vandekerkhove B, De Clercq P (2010) Pollen as an alternative or supplementary food for the mirid predator *Macrolophus pygmaeus*. Biol Control 53:238–242

116. De Cocuzza GE, Clercq P, Lizzio S, Van de Veire M, Degheele D, Tirry L, Vacante V (1997) Life tables and predation activity of *Orius laevigatus* and *O. albidipennis* at three constant temperatures. Entomol Exp Appl 85:189–198

117. Griffiths GJK, Holland JM, Bailey A, Thomas MB (2008) Efficacy and economics of shelter habitats for conservation biological control. Biol Control 45:200–209

118. Jabbour R, Crowder DW, Aultman EA, Snyder WE (2011) Entomopathogen biodiversity increases host mortality. Biol Control 59:277–283

119. Jabbour R, Zwickle S, Gallandt ER, McPhee KE, Wilson RS, Doohan D (2013) Mental models of organic weed management: comparison of New England US farmer and expert models. Renew Agric Food Syst. http://dx.doi.org/10.1017/S1742170513000185

120. Chisholm P, Gardiner M, Moon E, Crowder DW et al (2014) Exploring the toolbox for investigating impacts of habitat complexity on biological control. Biol Control 75:48–57

121. Roschewitz I, Gabriel D, Tscharntke T, Thies C (2005) The effects of landscape complexity on arable weed species diversity in organic and conventional farming. J Appl Ecol 42:873–882

122. José-María L, Sans FX (2011) Weed seedbanks in arable fields: effects of management practices and surrounding landscape. Weed Res 51:631–640

123. Solangi GS, Mahar GM, Oad FC (2008) Presence and abundance of different insect predators against sucking insect pest of cotton. J Entomol 5:31–37

124. Lang A (2003) Intraguild interference and biocontrol effects of generalist predators in a winter wheat field. Oecologia 134:144–153

125. El-Wakeil NE, Volkmar C, Sallam AA (2010) Jasmonic acid induces resistance to economically important insect pests in winter wheat. Pest Manage Sci 66:549–554

126. El-Wakeil NE, Volkmar C (2012) Effect of jasmonic acid application on economically insect pests and yield in spring wheat. Gesunde Pflanzen 64:107–116

127. El-Wakeil NE, El-Sebai TN (2009) Role of biofertilizer on faba bean growth, yield, and its effect on bean aphid and the associated predators. Arch Phytopathol Plant Prot 42:1144–1153

128. El-Wakeil NE, Saleh SA (2009) Effects of neem and diatomaceous earth against *Myzus persicae* and associated predators in addition to indirect effects on artichoke growth and yield parameters. Arch Phytopathol Plant Prot 42:1132–1143

129. Werling BP, Gratton C (2010) Local and broadscale landscape structure differentially impact predation of two potato pests. J Appl Ecol 20:1114–1125

130. Webster B, Bruce T, Dufour S, Birkemeyer C, Birkett M, Hardie J, Pickett J (2008) Identification of volatile compounds used in host location by the black bean aphid, *Aphis fabae*. J Chem Ecol 34:1153–1161

131. Goffreda JC, Mutschler MA, Tingey WM (1988) Feeding behavior of potato aphid affected by glandular trichomes of wild tomato. Entomol Exp Appl 48:101–107

132. Alhmedi A, Haubruge E, Bodson B, Francis F (2007) Aphidophagous guilds on nettle (*Urtica dioica*) strips close to fields of green pea, rape and wheat. Insect Sci 14:419–424

133. Vandereycken A, Brostaux Y, Joie E, Haubruge E, Verheggen FJ (2013) Occurrence of *Harmonia axyridis* (Coccinellidae) in field crops. Eur J Entomol 110:285–292

134. Hodek I, Van Emden HF, Honěk A (2012) Ecology and behaviour of the ladybird beetles (Coccinellidae), 561 pp. Wiley-Blackwell

135. Caballero-López B, Bommarco R, Blanco-Moreno JM, Sans FX, Pujade-Villar J, Rundlöf M, Smith HG (2012) Aphids and their natural enemies are differently affected by habitat features at local and landscape scales. Biol Control 63:222–229

136. Snyder WE, Snyder GB, Finke DL, Straub CS (2006) Predator biodiversity strengthens herbivore suppression. Ecol Lett 9:789–796

137. Vandereycken A, Durieux D, Joie E, Francis F, Haubruge E, Verheggen FJ (2015) Aphid species and associated natural enemies in field crops: what about the invasive ladybird *Harmonia axyridis*? Entomol Faunistique 68:3–15

138. Chakraborty A, Kumar K, Rajadurai G (2014) Biodiversity of insect fauna in okra (*Abelmoschus esculentus*) ecosystem. Trends Biosci 7:2206–2211

139. Alhmedi A, Haubruge E, D'Hoedt S, Francis F (2011) Quantitative food webs of herbivore and related beneficial community in non-crop and crop habitats. Biol Control 58:103–112

140. Zheng Y, Km Daane, Hagen KS, Mittler TE (1993) Influence of larval food consumption on the fecundity of the lacewing *Chrysoperla camea*. Ent Exp Appl 67:9–14

141. Lopez-Arroyo JI, Tauber CA, Tauber MJ (2000) Storage of lacewing eggs: post-storage hatching and quality of subsequent larvae and adults. Biol Control 18:165–171

142. Gaafar N (2002) Effects of some neem products on *Helicoverpa armigera* and their natural enemies *Trichogramma* spp. and *Chrysoperla carnea*. MSc, Faculty of Agricultural Science, at the Georg-August University Goettingen, Germany

Mass Production of Predatory Mites and Their Efficacy for Controlling Pests

Faten Momen, Shimaa Fahim and Marwa Barghout

Abstract Biological control has long been recognized as an effective and an environmentally safe pest management method. Several species of Mesostigmata and Prostigmata predatory mites are effective biological control agents against several phytophagous mites and insects also nematodes. Based on predator feeding habits, biological characteristics of it can be affected by nutritional value of the prey. Acari predators are mass reared on a wide scale of natural food sources like mites belonging to families Eriophyidae, Tetranychidae, Tarsonemidae, Tenuipalpidae and Tydeidae, also insects like whiteflies, mealybugs and thrips, or on factitious food like storage mites and insect eggs. Hence, a cost for predator's mass production is a fundamental precondition; this chapter discussed the rearing method on various diets referring to the advantages, the cost and benefit as well as the abundance of these predators and the possibility of their application and success, especially in Egypt.

Keywords Predacious mites · Phytophagous mites · Eriophyidae · Tetranychidae · Tarsonemidae · Tenuipalpidae and Tydeidae · Life table parameters

1 Introduction

Mites are common in abundant places and found in different habitats as in soil, stored product, ornamentals and field crops. Some of them are phytophagous give rise to direct injury to plant by feeding or indirectly by transmitting disease agents. Members of most species are highly host specific, and many form extremely dense populations causing characteristic injury both locally and generally on host through their feeding activities.

Mites of the family Tetranychidae are strictly phytophagous and are represented in all region of Egypt. Spider mites can undoubtedly cause severe crop loss [1, 2]. Mites of the family Eriophyidae including the rust, gall and bud mites constitute a large and strictly phytophagous group represented throughout Egypt on a wide

F. Momen (✉) · S. Fahim · M. Barghout
Pests and Plant Protection Department, National Research Centre, Dokki, Cairo, Egypt
e-mail: fatmomen@yahoo.com

© Springer Nature Switzerland AG 2020
N. El-Wakeil et al. (eds.), *Cottage Industry of Biocontrol Agents and Their Applications*, https://doi.org/10.1007/978-3-030-33161-0_5

variety of hosts [3, 4]. Also some eriophyids are vectors of severely deleterious plant viruses [5].

Acaricides have been most common method of mites control in Egypt, leading to many problems such as environmental pollution, development of resistant mite strains, rising costs and disturbing the natural enemies [6, 7]. In order to save the high costs of importing Acaricides and to replace those rendered unable by mite resistance, the attention has been focused to use biological control. According to [8], Acari are the second biggest group (after Hymenoptera) of natural enemies used commercially in the biological control of the arthropod pests. The previous author also listed the predatory mites (30 species) that represented about 13.1% of all arthropod natural enemies used in pest control worldwide. However, Acari predators is characterized by some characteristics such as the easily mass rearing of them, they may capable of control various pest species, the ability of their releasing by a mechanical ways; they are comparatively small in size and do not spread over a wide area. However, the successful biological control of spider, tarsonemid and eriophyid mites implies the evaluation of mass-reared predator effectiveness which depends not only on geographical origin, but also on food experienced in the mass rearing [9] and some other phenomena associated with predators. Predatory mites of the Mesostigmata and Prostigmata groups are playing an important role in biological control.

The Mesostigmata predators are encountered in soil and litter, on aerial parts of plants, nests or galleries of insects, mammals, and birds, where they feed on small insects, nematodes, collembola or on phytophagous and mycophagous mites [10–12]. However, many are also able to feed on fungi, while others have evolved to feed on pollen and nectar [13]. Prostigmata predators are also found in diverse habitats, including soils and overlaying litter layers as well as aerial parts of plants. Predators of the genus *Agistemus* were play an important role in biological control of economic pests [14, 15].

The objective of this chapter is to give a highlight about the most dominant predatory mites associated with vegetable crops and fruit trees recorded in Egypt. A highlight is concentrated also on selected predatory species from various groups of mites for augmentation release for rapid biological control. The mass production of various predatory mites and their potential of biological control in Egypt are also included. Studied mass production includes (a) Predatory mites of some mesostigmatic group, (b) Predatory mites of some prostigmatic group, and (c) Role of these mites in attacking various pests by natural and factitious prey.

2   Abundance of Predatory Mites

There are many predatory mites in the environment, and they can reduce some insect/mite pests from population outbreaks. Predatory mites inhabiting soil or vegetation certainly consume their prey. Many of the predatory Mesostigmata have been studied because of their potential to be used as biological control agents of agricultural pests or of parasites. Few studies have been conducted to validate the potential of laelapid species (Family Laelapidae) in field experiments. Most of these studies refer to the use of *Gaeolaelaps aculeifer*. They are widespread in soil, litter, manure, decaying woods and galleries of bark beetles [16, 17].

Predatory mites of the family Phytoseiidae are successfully used as bio-control agents in controlling of phytophagous mites, thrips and whitefly [18]. Several researches had been studied the biology of these predators and concluded that there were many food types of the phytoseiids including small arthropods, nematodes, fungi and pollen [13]. Another family, namely Ascidae has many predators like: *Gamasellodes adrianae*, *Gamasellodes bicolor*, and *Gamasellodes claudiae* were confirmed experimentally under laboratory conditions to predate on nematodes [19, 20]. Species of the genus *Proctolaelaps* (Family Melicharidae) seem to be quite variable in relation to feeding habits. Five species of that genus were recorded from Egypt. The dominant *Proctolaelaps* species have been shown to develop and reproduce on nematodes. *Agistemus exsertus* is one of the most important species of the family Stigmaeidae (Acari: Prostigmata) in Egypt, since it was associated with many economic pests and able to attack tetranychoid and eriophyoid mites as well as scale insects and white fly in field crops and orchards.

3 Mass Production of Predatory Mites

Successful biological control mostly implies production of billions of predatory mites and evaluation for their efficiency [21], taking into consideration a cost-effective method. For Example traditional phytoseiid mites rearing systems are tritrophic (plant materials, natural prey and the predator) that has several disadvantages, laborious and expensive [22]. So, turning mass rearing system viable by strategy requires not only finding an effective and abundant species, but also finding economic method to produce them with inexpensive cost and this strategy basically depending on understanding feeding habits of the predatory mites in general and its behavior. [23]. Based on feeding habits, phytoseiid mites were classified as specialized predators on *Tetranychus*/eriophyid/tydeid species, selective predators of tetranychid mites, generalist predators and generalist but pollen feeders [13].

Generalist predators can be mass cultured on different prey species like phytophagous mites, various insect species including mealybugs, whiteflies and scale crawlers and also can feed on pollen. Some of these predators have been commercially produced around the world and used as biological control agents in greenhouses like, *Amblyseius swirskii*, *A. andersoni*, *Neoseiulus barkeri*, *N. cucumeris* and *N. californicus* [24, 25]. Extensive attention goes to *Neoseiulus californicus*; although it was having a feeding habit as selective predators of tetranychid mites; however it can also consume other mite species, small insects or pollen when the primary prey is unavailable. In an effort to allow the production of predators at low cost, factitious food items have been evaluated. Ex: alternative food sources that they would not normally found in their natural habitat but can develop and reproduce. In addition,

such different items of food can be useful as food supplements to support predator populations after release in the crop [26]. Predatory mites inhabiting soil like
Rhodacaridae, Laelapidae, Ascidae and Melicharidae were mass cultured on some
natural and factitious foods such as the onion thrips *Thrips tabaci* and insect eggs
[27].

4 Field Application of Predatory Mites

The predatory phytoseiid mites are success as biological control agents for various mite pests. Additionally, the success of some phytoseiids, such as *Phytoseiulus
persimilis* against spider mites, is due to the great population growth rate of *P. persimilis* comparative to the spider mites [28]. However, there were other predators
that have a parallel or somewhat lower population growth rate compared to their
prey, an example of that is *N. cucumeris* that used in controlling of thrips [29]. In
this case, *N. cucumeris* persevere in field is a result of its feeding on another food
such as pollen [30]. Moreover that the reproduction of some phytoseiid mites can
be enhanced when consumed the pollen [31]. In such cases, the pollen consumption
supports the persistence of predatory immature that has difficulties with large prey
stages [32].

4.1 Examples Successful Usage of Phytoseiids as Bio-Control Agents Against Pests

There were several cases of the efficacious usage of the predatory phytoseiids as
bio-control agents of several destructive pests. Two examples were summarized as
followed:

4.1.1 *Tetranychus urticae*

The two-spotted spider mite, *T. urticae*, is extremely polyphagous mite pest and
cosmopolitan tetranychid species. This mite pest is one of the greatest severe pest
species infested many fruit tree, cotton, vegetables and a variety of greenhouse crops.
The feeding of *T. urticae* cause a small spots on leaves as a results of chlorophyll
reduction, webbing, dry leaf-fall, up to necrosis in young leaves and stems, or even
the plant death in a severe mite-infestation [33]. Additionally, the mite feeding causes
a reduction of photosynthesis, carbon dioxide absorption and transpiration as a result
of the leaf tissue injure. Such effects were observed in many plants such as strawberry,
apple, peach, tomato and cotton [34–36] lead to reduction in the crop yields such as
that detected in cotton [37], soybean [38, 39] and others. In other cases, *T. urticae*

can feed directly on tomato fruit causing discoloration of the tomato, which may have a bad effect on the marketability of the tomato [40].

Mainly at high temperatures, *T. urticae* has rapid development. At 30–32 °C, the total life cycle only takes 8–12 days. The female can oviposition about 90–110 eggs during a lifetime of about 30 days; therefore the mite populations can be raised quickly during the summer, or under greenhouse conditions [18, 41]. However, the females can deposit unfertilized eggs that become males [42].

Biological control: One of the most successful biological control examples is the use of *P. persimilis* in controlling of *T. urticae*. This predator is used in the management of *T. urticae* on various crops, since 1960s [43], till now.

Neoseiulus californicus has used effectively in the management of *T. urticae* [44, 45]. Moreover, *N. californicus* was providing efficacious control of other tetranychid mites such as *Panonychus ulmi* and *Oligonychus perseae* in several countries [i.e. 46–48]. In the USA, [49] stated that there were other phytoseiid predators such as *Neoseiulus fallacies* which proved to be an effective bio-control agent against *T. urticae* on strawberries and on apple in Oregon [50]. Also, *Euseius victoriensis* [51] and *Neoseiulus womersleyi* in Australia [52] showed efficiency in the control of *T. urticae*.

4.1.2 Polyphagotarsonemus latus

The broad mite, *P. latus*, is a serious tarsonemid mite pest in tropical and subtropical regions and has been collected from about 60 different plant families [53]. It has a large host range including cotton, potato, tomato, eggplant, peppers, apple, pear, cantaloupe, mango, guava, avocado, citrus, coffee, tea and others [54]. In addition, broad mites infest many ornamentals such as African violet, lantana, begonia, marigold, azalea, dahlia, zinnia, impatiens, jasmine, verbena, gerbera and chrysanthemums [55]. This destructive mite pest attacks newest leaves and plant buds. It's feeding making the terminal leaves and flower buds malformed. The toxic saliva of the mite leads to hardened and twisted of the leaf edges [55].

The mite infestation cause flowers deformation, fruits blistering and finally yield reduction. In a young pepper plant, 10 mites are able to cause a considerable reduction in the plant height and leaf surface [56]. On fruit trees, the shaded parts of the fruits are generally damaged as a result of mite feeding and become discolored. In heavy infestations, premature fruit may fall. Additionally, the damaged fruit is not acceptable in the markets [54]. Adult females deposit several eggs ranged from 30 to 76 eggs in a period from 8 to 13 days. The unmated females laid eggs that become males. Generally, the mated females lay 4 female eggs for every male egg [54].

Biological control: A number of predatory mites have been reported to be effective in the controlling of *P. latus*. *Neoseiulus californicus* successfully control *P. latus* on pepper [25, 57] in addition to the tarsonemid mite pest *Phytonemus pallidus* on strawberry [24]. *Neoseiulus barkeri* has been found to be an important biological control agent for the management of *P. latus* in protected crops [58]. Also, *N. cucumeris* [59] and *E. victoriensis* [51] have been mentioned as noticeable predators effectively used in the biological control of *P. latus*. *Amblyseius swirskii* [60, 61] were reported to control the broad mite successfully in greenhouses. Recently, [62] indicated that *A. largoensis* is an effective control agent for the broad mite.

5 Pillars of Biological Control Industry

5.1 Abundance of Predatory Aerial Mites

5.1.1 Family Phytoseiidae

Phytoseiid species have been recorded from all continents except Antarctica. Phytoseiids are primarily plant-inhabiting mites, with more than 2077 species (representing about 85% of species) described from diverse plant species. The phytoseiids have a typical Gamasina life history, usually with five life stages: egg, larva, protonymph, deutonymph and adult. Females must mate in order to produce progeny. Multiple-mated females produce more eggs [63–65].

Some phenomena associated with phytoseiid species provide additional food to the predators when the preferred prey are low like cannibalism and intraguild predation (predator–predator dynamics) and it is important to be understood to increase their overall fitness [66, 67]. These studies have centered mainly on the ability and frequency of predation of immatures by adults of its own (cannibalism) or of different (intraguild predation) species [68].

Collection of some predatory mites, including family Phytoseiidae were surveyed in Egypt and presented in Table 1. Twenty five phytoseiid species were recorded during a survey conducted in some governorates in Egypt; this survey included some crops, fruit trees and weeds.

According to [13] predators belonging to the genus *Phytoseiulus* are specialist predators on *Tetraychus* sp. (ex: *P. persimilis–P. macropilis–P. longipis*) (specialist predators Type 1a). Also other phytoseiid mites were recorded to be specific predator on tydeid mites (*Paraseiulus talbi–Proprioseiopsis cabonus*).

Life table parameters of the specialist phytoseiid predator *Phytoseiulus macropilis* cultured on various stages of *Tetranychus urticae* in Egypt had been shown in Table 2.

Life table parameters of the specialist phytoseiid predators *Paraseiulus talbi* and *Proprioseiopsis cabonus* cultured on various tydeid mites in Egypt and Italy were presented in Table 3.

Yet other group of specialist phytoseiid predators on eriophyoid mites was recorded from Egypt. This group included *Proprioseiopsis lindquisti*, *P. badryi*, *Amblyseiella denmarki* and *Typhlodromus transvaalansis*. All species were recorded to feed on various Eriophyoid mites. These pests included: *Aceria olive*, *Aculops*

Table 1 Incidence of some phytoseiid species in Egypt

Species	Crop/weed	Locality
Neoseiulus barkeri	Solanym lycopersicum–Solanym melongena	Fayoum–Tamai–Tobhar
Proprioseiopsis kadii	Mangifera indica–P. persica	Ismailia–Al-Tal Al-Kber
Neoseiulus californicus	Cucumis sativus	Fayoum–Sennours
Cydnoseius negevi	S. lycopersicum–S. melongena	Ismailia–Abou–Souer–Fayed
Proprioseiopsis badryi	S. lycopersicum–M. indica	Fayoum–Tobhar–Tamai
Typhlodromips swirskii	S. lycopersicum	Fayoum–Sennours
Amblyseius largoensis	S. lycopersicum	Ismailia–Abou-Souer
Proprioseiopsis lindquisti	Musca paradisiacal	Fayoum–Atsa
Typhlodromus transvaalens	S. lycopersicum	Ismailia–Fayed
Proprioseiopsis cabonus	S. melongena–M. indica	Fayoum–Atsa
Typhlodromus balanites	Pluchea dioscoridis	Fayoum–Sennours
Typhlodromus rhenanus	S. lycopersicum–S. melongena	Fayoum–Atsa
Neoseiulus arundonaxi	Arondo donax	Fayoum
Typhlodromus mangifera	M. indica	Giza
Amblyseius denmarki	M. paradisiacal	Fayoum–Sennours
Euseius finlandicus	Piper nigrum	Fayoum–Tobhar
Euseius scutalis	P. nigrum	Fayoum–Tobhar
Phytoseiulus persimilis	L. sculentum	Fayoum–Demo
Phytoseius finitimus	Eichhornia crassipes	Kewesna
Euseius yousefi	A. donax	Tokh
Amblyseius zaheri	E. crassipes	Manfalout
Paraseiulus talbi	M. indica	Manfalout
Typhlodromus athiasae	Ceratophyllum demersum	Mostorud
Amblyseius deleoni	Conyze dioscoridis	Farscor
Amblyseius zaheri	E. crassipes	Kewesna

lycopersici, Aculus fockeui and *Cisaberoptus kenyae*. The main hosts of these erio-phyoid mites are mango, tomato, pear and olive leaves.

Life table parameters of specialist phytoseiid predators on eriophyoid mites cultured on various these pests in Egypt have been shown in Table 4.

In their classification [13] divided family Phytoseiidae to 4 groups, the 3rd group is the generalist predatory mites which can able to feed and sustain oviposition on various kind of foods included the factitious prey. Many phytoseiids in this group was reared and mass cultured on natural, alternative and factitious prey. Natural prey included *T. urticae, B. tabaci, Thrips tabaci,* eriophyid mites and pollen grains, while factitious prey included various insect eggs and acarid mites. Type IV of generalist phytoseiid mites included also predators preferred pollen grains to other foods, but sometimes eriophyid mites could be the best and this group included predatory mites

Table 2 Life table parameters of *Phytoseiulus macropilis* a specific predatory mite on *Tetranychus urticae* used in biological control and mass cultured on various stages of the pest in Egypt

Predatory mite species	Prey	Egg to adult (days)	♀Longevity (days)	Exp. conditions	Fecundity		R_o	r_m	T	λ	Authority
					Total eggs/♀	Daily eggs/♀ /day					
Phytoseiulus macropilis (Banks)	*T. urticae* (Eggs)	5.7	57.2	20 °C	35.5	0.8	13.00	0.13	20.00	1.10	Ali [69]
			42.9	25 °C	68.3	1.9	86.28	0.36	2.10	1.40	
			29.2	28 °C	78.8	3.3	88.90	0.47	8.80	1.60	
			36.7	30 °C	46.2	1.5	54.70	0.43	9.20	1.50	
			28.0	32 °C	24.7	1.1	24.00	0.29	10.80	1.30	
	T. urticae (Nymphs)	5.40	29.25	30 °C, 70% R.H.	41.50	1.71	28.24	0.31	10.73	1.36	Abdel-Khalek and Momen [70]
	T. urticae (Eggs) 0–24 h	6.08	28.92		66.80	2.71	52.13	0.32	12.00	1.38	
	T. urticae (Eggs) 72–96 h	5.50	33.54		82.27	2.83	58.41	0.36	11.06	1.44	

Table 3 Life table parameters of two specific predatory phytoseiid mites on tydeid mites cultured on various tydeid species in Egypt and Italy

Phytoseiid species	Prey/pollen	Egg to adult (days)	♀ Longevity (days)	Exp. conditions	Fecundity		R_o	r_m	T	λ	Authority
					Total eggs/♀	Daily eggs/♀/day					
Paraseiulus talbii (Athias-Henriot)	Tydeus caudatus	12.15	35.21	20 °C	35.21	–	17.45	0.089	31.81	1.09	Camporese and Duso [71]
	Colomerus vitis	10.16	43.20	27 °C	5.14	–	2.39	0.03	29.40	1.03	
	T. caudatus	6.66	50.55	–	32.50	–	17.50	0.165	17.33	1.179	
	P. ulmi	10.63	3.66	–	–	–	–	–	–	–	
	Eotetranychus carpini	8.25	7.25	–	0.24	–	–	–	–	–	
	Tydeus californicus	8.80	49.80	25 °C, 70 ± 5% R.H. 16:8 h. (L:D)	33.4	0.79	15.72	0.15	17.79	1.17	Zaher et al. [72]
	Coccus acuminatum	10.00	53.20		24.8	0.68	13.58	0.12	21.44	1.13	

(continued)

Table 3 (continued)

Phytoseiid species	Prey/pollen	Egg to adult (days)	♀ Longevity (days)	Exp. conditions	Fecundity Total eggs/♀	Daily eggs/♀/day	R_o	r_m	T	λ	Authority
	C. kenyae	–	–		–	–	–	–	–	–	
	O. mangiferus	–	–		–	–	–	–	–	–	
Proprioseiopsis cabonus (Schicha and Elshafie)	Neoapolorryia aegyptiaca El Bagoury and Momen	9.93	34.87	30 ± 1 °C, 70 ± 5% R.H.	47.62	2.03	35.95	0.212	16.84	1.23	Momen [73]
	Lorryia aegyptiaca	11.07	29.87		40.80	1.84	24.64	0.174	18.38	1.19	
	A. olive	6.78	28.22		30.39	1.25	17.32	0.19	14.46	1.21	
	A. lycopersici	8.05	21.80		16.95	1.17	9.83	0.168	13.53	1.18	
	C. kenyae	7.27	24.32		23.32	1.28	12.87	0.18	13.81	0.21	

Table 4 Life table parameters of some specific predatory phytoseiid mites mass cultured on various eriophyoid mites and used in biological control in Egypt

Phytoseiid species	Prey/pollen	Egg to adult (days)	♀Longevity (days)	Exp. conditions	Fecundity Total eggs/♀	Daily eggs/♀/day	R_o	r_m	T	λ	Authority
Proprioseiopsis lindquisti (Schuster and Pritchard)	Aceria olivi	6.38	32.00	28 ± 2 °C, 70–75% R.H.	35.69	–	–	–	–	–	Momen [74]
	P. dactylifera (Pollen)	6.61	26.23		1.85	–	–	–	–	–	
	A. dioscoridis	–	–		34.08	–	–	–	–	–	
	C. kenyae	–	–		19.08	–	–	–	–	–	
Proprioseiopsis badryi (Yousef and El-Borolossy)	A. dioscoridis	7.41	25.38	27 °C	30.75	–	18.51	0.170	17.17	1.19	Abou-Awad et al. [75]
	T. urticae	8.00	22.80		–	–	–	–	–	–	
	E. orientalis	8.30	31.70		–	–	–	–	–	–	
	A. dioscoridis	7.5	26.05	30 ± 1 °C, 75 ± 5% R.H.	30.44	1.37	18.87	0.20	14.39	1.22	Momen et al. [76]
	A. olivi	6.78	28.22		30.39	1.25	17.32	0.19	14.46	1.21	
	A. lycopersici	8.05	21.80		16.95	1.17	9.83	0.168	13.53	1.18	
	C. kenyae	7.27	24.32		23.32	1.28	12.87	0.18	13.81	0.21	
Amblyseiella denmarki (Zaher and El-Borolossy)	A. dioscoridis	4.00	23.64	27 ± 1 °C, 70 ± 5% R.H.	41.23	2.72	25.33	0.378	8.54	1.45	Momen et al. [77]
	A. olivi	4.00	23.14		45.57	2.91	35.74	0.419	8.52	1.52	
	C. kenyae	4.00	19.54		14.76	1.08	9.40	0.255	8.78	1.29	
	A. fockeui	8	39.3	28 ± 2 °C	49.40	1.56	35.56	0.216	16.52	1.24	Momen [78]

of the genus *Euseius*. Two species were recorded from Egypt: *E. scutalis* and *E. yousefi*.

Life table parameters of some generalist phytoseiid predators cultured on natural prey in Egypt were shown in Table 5. These species included: *Neoseiulus barkeri, Cydnoseius negevi, Typhlodromus mangifera, Amblyseius zaheri, Amblyseius deleoni*. Life table parameters of some generalist phytoseiid predators cultured on factitious foods had been presented in Table 6. Factitious foods proved to be an excellent alternative prey for certain generalist phytoseiid mites. Factitious food used in this text included: *Aleuroglyphus ovatus, Carpoglyphus lactis, Suidasia medanensis, Tyrophagous putrescentiae, E. kueniella, Artemia franciscana* and pollen grains.

5.1.2   Family Stigmaeidae

Most of Stigmaeidae members are free-living predators on plant leaves and branches and are belonging to two genera, *Agistemus* and *Zetzellia*. Stigmaeid mites have five life stages as the members of the family Phytoseiidae. There are also three quiescent periods after larval, protonymphal and deutonymphal stages. Unmated females produce only male egg (arrhenotokous) [100].

Collection of some areal predatory mites including family Stigmaeidae was surveyed in Egypt and presented in Table 7. Three stigmaeid species were recorded during a survey conducted at some governorate in Egypt. Survey included some crops, fruit trees and weeds.

Life table parameters of some generalist stigmaeid predators which have been cultured on natural and factitious foods shown in Table 8. Predatory mites of the family Stigmaeidae are classified to 2 categories, specialist and generalist. A specialist stigmaeid mite like *Agistemus olive* can be mass cultured on various eriophoid mites' infested olive, mango and pear leaves. Generalist stigmaeid mites were mass cultured on natural and factitious food. Natural foods included tetranychid, eriophyid, scale insects and small arthropods, while factitious food included *E. kuehniella, Galleria mellonella*.

It is well known that *Agistemus exsertus* is an egg predator since the species prefer to feed on egg stage to active stages.

5.2   Abundance of Predatory Soil Mites

5.2.1   Family Laelapidae

For the control of pests that spend part of their life in the soil, four predatory mites of the family Laelapidae have also been used, namely *Androlaelaps casalis, Gaeolaelaps aculeifer, Stratiolaelaps miles* and *Stratiolaelaps scimitus*.

Laelapidae have five developmental stages: egg, larva, protonymph, deutonymph and adults. The reproduction in this family can be by haplodiploidy, arrhenotoky

Table 5 Life table parameters of generalist predatory phytoseiid mites cultured on natural foods and used in biological control in Egypt

Phytoseiid species	Prey/pollen	Egg to adult (days)	♀ Longevity (days)	Exp. conditions	Fecundity Total eggs/♀	Daily eggs/♀/day	R_o	r_m	T	λ	Authority
Neoseiulus californicus	*T. urticae* (Eggs)	9.76	31.6	28 ± 2 °C, 70 ± 5% R.H.	44.00	1.73	32.10	0.17	19.35	1.19	El-Laithy and El-Sawi [79]
	T. urticae (Nymphs)	8.05	35.8		64.70	2.20	47.57	0.23	16.57	1.26	
	Aceria dioscoridis	7.35	39.2		32.95	1.21	25.04	0.21	15.22	1.23	
	Phoenix dactylifera L. Date palm (Pollen)	9.63	17.5		–	–	–	–	–	–	
Typhlodromus mangifera	*Aceria mangifera*	7.32	22.27	30 °C, 55% R.H.	37.0	2.30	23.72	0.176	17.96	1.19	Abou-Awad et al. [80]
	Metaculus mangiferae	7.62	20.61		25.70	1.74	16.21	0.156	17.83	1.17	
	Cisaberoptus kenyae	7.44	20.18		28.8	2.03	18.59	0.166	17.6	1.18	
	Oligonychus mangiferae	7.60	23.18		23.9	1.40	16.41	0.153	18.27	1.16	
Typhlodromips swirskii	*Aceria ficus* (Cotte)	8.0	21.2	29 °C	28.2	–	21.25	0.155	19.74	1.16	Abou-Awad et al. [3]
	Rhyncaphytoptus ficifoliae Keifer	9.3	24.5		20.4	–	15.62	0.122	22.37	1.13	

(continued)

Table 5 (continued)

Phytoseiid species	Prey/pollen	Egg to adult (days)	♀ Longevity (days)	Exp. conditions	Fecundity Total eggs/♀	Daily eggs/♀/day	R_o	r_m	T	λ	Authority
Typhlodromips swirskii	*Thrips tabaci*	6.35	39.05	–	40.18	1.65	27.11	0.211	15.64	1.234	Abou-Elella [81]
	A. lycopersici	7.00	24.60	–	35.4	1.7	26.78	0.23	13.96	1.26	Momen and Abdel-Khalek [4]
	Aculus fockeui	6.57	28.70	28 °C, 70% R.H.	43.00	1.91	28.90	0.24	13.78	1.27	Momen [78]
	A. mangiferae	9.87	23.32	25 °C, 60% R.H.	34.08	1.92	23.82	0.157	20.10	1.17	Abou-Awad et al. [80]
	M. mangiferae	9.62	23.90		29.72	1.72	19.58	0.144	20.52	1.155	
	Cisaberoptus kenyae	8.79	26.10		30.30	1.62	22.26	0.146	21.20	1.157	
	O. mangiferus	9.39	24.50		20.90	1.2	14.62	0.133	20.14	1.142	
Amblyseius zaheri Yousef and El-Borolossy	*T. urticae* (Nymphs)	5.07	17.14	27 ± 1 °C, 70 ± 5% R.H.	27.57	2.07	18.26	0.300	9.67	1.35	Rasmy et al. [82]
	E. orientalis (Nymphs)	6.38	17.13		15.77	1.42	8.79	0.200	10.83	1.22	
	A. dioscoridis	5.00	30.21		39.79	2.16	28.67	0.316	10.58	1.37	

(continued)

Table 5 (continued)

Phytoseiid species	Prey/pollen	Egg to adult (days)	♀ Longevity (days)	Exp. conditions	Fecundity Total eggs/♀	Fecundity Daily eggs/♀/day	R_o	r_m	T	λ	Authority
	Phoenix dactylifera (Pollen)	5.38	21.38		45.38	2.91	33.80	0.338	10.40	1.40	
	Parlatoria ziziphus	5.46	23.46		18.23	1.14	11.74	0.218	11.28	1.24	
	Bemisia tabaci	7.69	20.15		15.92	1.15	9.60	0.167	13.46	1.18	
Amblyseius deleoni (Muma and Denmark)	T. urticae (Eggs)	–	–	–	–	–	–	–	–	–	Rasmy et al. [83]
	T. urticae (Nymphs)	5.92	32.31		45.31	1.87	25.46	0.255	12.65	1.29	
	E. orientalis	6.22	28.23		40.54	1.85	24.07	0.241	13.16	1.27	
	A. dioscoridis	5.39	28.08		41.15	2.10	24.53	0.280	11.40	1.32	
	P. ziziphus	6.07	23.38		8.23	0.91	3.53	0.119	10.56	1.12	
	B. tabaci	7.15	13.31		9.23	1.47	4.69	0.153	10.09	1.16	
	P. dactylifera (Pollen)	6.23	29.23		41.00	1.87	20.91	0.225	13.49	1.25	
	R. communis (Pollen)	7.46	34.54		12.00	0.50	5.33	0.105	15.94	1.11	
Neoseiulus barkeri (Hughes)	Tegolophus hassani	12.31	31.60	$15 \pm 1\,°C$, 50% R.H.	27.70	1.15	19.04	0.111	–	1.11	Metwally et al. [84]

(continued)

Table 5 (continued)

Phytoseiid species	Prey/pollen	Egg to adult (days)	♀ Longevity (days)	Exp. conditions	Fecundity Total eggs/♀	Daily eggs/♀/day	R_o	r_m	T	λ	Authority
		8.84	28.00	25 ± 1 °C, 70% R.H.	45.50	2.07	32.49	0.167	20.90	1.17	
		5.61	25.00	31 ± 1 °C, 80% R.H.	54.20	2.71	36.70	0.201	17.90	1.23	
	T. urticae (Nymphs)	13.27	32.20	15 ± 1 °C, 70% R.H.	18.40	0.79	15.23	0.098	27.80	1.090	
		8.69	27.80	25 ± 1 °C, 70% R.H.	40.57	1.87	24.87	0.142	22.70	1.10	
		6.38	25.20	31 ± 1 °C, 80% R.H.	49.20	2.39	33.00	0.158	18.60	1.20	
Neoseiulus cucumeris (Oudemans)	T. urticae	4.00	31.56	27 ± 2 °C, 70 ± 5% R.H.	48.69	2.37	35.06	0.378	9.41	1.46	Abou El-Elela and Abou-Elella [85]
	A. dioscoridis	4.38	30.35		42.81	2.17	29.27	0.332	10.16	1.39	
	T. tabaci	5.94	32.56		50.56	2.04	38.35	0.280	13.04	1.32	
Cydnoseius negevi (Swirski and Amitai)	A. ficus	8.90	36.23	–	35.77	–	30.87	0.23	15.04	1.26	Abou-Awad et al. [75]
	T. urticae (Eggs)	9.00	29.37		25.64	–	23.07	0.16	19.16	1.18	
	B. tabaci	6.07	32.20	28 ± 1 °C, 75 ± 5% R.H. 28 ± 1 °C, 75 ± 5% R.H.	21.2	–	18.02	0.23	12.25	1.26	Momen et al. [86]
	Insulaspis pallidula	7.70	16.82		10.20	–	8.14	0.14	14.84	1.15	

Table 5 (continued)

Phytoseiid species	Prey/pollen	Egg to adult (days)	♀ Longevity (days)	Exp. conditions	Fecundity		R_o	r_m	T	λ	Authority
					Total eggs/♀	Daily eggs/♀/day					
	Phoenicoccus marlatti	7.80	14.37		6.40	–	3.61	0.049	13.59	1.09	
	R. communis	6.33	37.00		32.90	–	29.04	0.271	12.38	1.312	
	B. tabaci	6.07	32.20		21.2	–	18.02	0.23	12.25	1.26	
Euseius scutalis (Athias-Henriot)	R. communis (Pollen)	6.25	21.37	28 ± 1 °C, 75 ± 5% R.H.	37.06	2.03	24.91	0.338	9.52	–	Momen and Abdel-Khalek [87]
	E. orientalis	6.44	23.12		21.81	1.07	14.65	0.25	10.44	1.29	
	A. ficus	6.38	24.23		26.08	1.20	12.51	0.218	11.59	1.243	
	Rhyncaphytoptus ficifolia	7.00	22.57		24.14	1.20	12.02	0.215	11.52	1.240	
	Icerya aegyptiaca	6.31	22.62		20.50	1.05	13.34	0.25	10.31	1.28	
Euseius yousefi (Zaher and El-Borolossy)	A. dioscoridis	7.30	44.50	27 °C, 70 ± 5% R.H.	63.80	–	47.46	0.23	16.66	1.26	Momen [88]
	T. urticae	6.20	25.18		50.50	–	38.82	0.29	12.40	1.34	
	P. dactylifera (Pollen)	7.00	43.30		57.60	–	37.00	0.21	16.85	1.23	
	Zea mays L. (Pollen)	13.20	3.20		–	–	–	–	–	–	
	R. communis (Pollen)	7.90	29.3	30 ± 1 °C	57.13	–	43.28	0.258	14.05	1.29	Momen [89]
	Helianthus annuus (Pollen)	7.20	30.10		28.06	–	11.79	0.158	15.57	1.17	

Table 6 Life table parameters of generalist predatory phytoseiid mites cultured on factitious foods and used in biological control

Phytoseiid species	Prey/pollen	Egg to adult (days)	♀ Longevity (days)	Exp. conditions	Fecundity Total eggs/♀	Daily eggs/♀/day	R_o	r_m	T	λ	Authority
Typhlodromips swirskii (Athias-Henriot)	Suidasia medanensis	5.01	–	–	–	1.71	–	0.222	–	1.249	Midthassel et al. [90]
	T. latifolia	7.44	34.45	23 °C	29.00	–	19.71	0.158	19.00	–	Nguyen et al. [91]
	Carpoglyphus lactis	7.00	35.90		29.03	–	20.58	0.175	17.33	–	
	The basic artificial diet (AD1)	7.57	22.56		9.94	–	6.46	0.104	22.08	–	
	The basic artificial diet (AD2)	7.26	42.07		38.26	–	27.21	0.181	18.30	–	
	Ephestia kuehniella	8.21	44.12	23 ± 1 °C, 65 ± 5% R.H.	31.64	–	21.78	0.169	18.22	–	Nguyen et al. [92]
	Artemia franciscana	7.67	37.79		37.71	–	28.33	0.176	19.03	–	
	AD1-G1	7.65	40.32		34.50	–	24.71	0.174	18.49	–	
	AD2-G2	7.36	49.14		34.10	–	23.23	0.160	19.65	–	
Amblyseius zaheri Yousef and El-Borolossy	E. kuehniella (Eggs)	12.90	38.30	27 ± 2 °C, 70-75% R.H.	41.10	1.50	23.50	0.128	24.60	1.136	Momen and El-Laithy [93]
Neoseiulus cucumeris	Apple (Pollen)	7.17	93.9	25 °C, 65: 70% R.H.	76.72	1.53	38.79	0.148	24.64	1.16	Nar et al. [94]

(continued)

Table 6 (continued)

Phytoseiid species	Prey/pollen	Egg to adult (days)	♀ Longevity (days)	Exp. conditions	Fecundity Total eggs/♀	Daily eggs/♀/day	R_o	r_m	T	λ	Authority
	Birch (Pollen)	7.7	98.9		67.33	1.31	26.88	0.126	26.09	1.13	
	Christmas cactus (Pollen)	7.06	73.3		80.15	1.92	28.87	0.555	21.63	1.16	
	Horse-chestnut (Pollen)	6.3	81.1		89.23	2.15	49.83	0.180	21.69	1.19	
	Maize (Pollen)	8.21	97.8		69.59	1.19	20.96	0.101	29.57	1.10	
	Tulip (Pollen)	5.93	98.4		89.48	2.06	38.99	0.167	21.92	1.18	
	Carpoglyphus lactis	5.1	67.3	25 ± 1 °C, 90 ± 5% R.H.	53.3	–	31.05	0.212	16.20	–	Ji et al. [95]
	T. latifolia (Pollen)	6.67	52.64	25 °C, 70 ± 5% R.H.	45.45	1.51	32.97	0.185	18.95	–	Nguyen et al. [96]
	Artificial diet	9.00	71.56		25.78	0.52	17.16	0.090	31.74	–	
Iphiseiodes zuluagai	T. putrescentiae	3.90	25.40	25 °C, 88 ± 7% R.H. 12:12 h	13.20/10 days	1.3/10 days	7.1	0.11	18.6	1.11	De Albuquerque and de Moraes [97]
Neoseiulus baraki Athias-Henriot	T. putrescentiae	5.8	26.1		39.4	–	18.67	0.18	15.8	1.20	Domingos et al. [98]

(continued)

Table 6 (continued)

Phytoseiid species	Prey/pollen	Egg to adult (days)	♀ Longevity (days)	Exp. conditions	Fecundity		R_o	r_m	T	λ	Authority
					Total eggs/♀	Daily eggs/♀/day					
	E. kuehniella	11.80	43.00	27 °C, 75% R.H.	50.40	1.70	32.88	0.139	25.03	1.14	Momen and El-Laithy [93]
	Aleuroglyphus ovatus	11.66	45.47	20 °C	10.67	0.65	6.81	0.050	23.69	1.05	Xia et al. [99]
		7.86	34.57	24 °C	30.64	1.53	20.14	0.136	22.07	1.14	
		4.98	23.72	32 °C	20.52	1.47	11.44	0.165	14.75	1.80	

Table 7 Incidence of stigmaeid species in Egypt

Species	Crop/weed	Locality
Agistemus exsertus	C. dioscoridis–M. indica	Fayoum–Tobhar
Agistemus olivi	C. dioscoridis–S. lycopersicum	Ismailia–Fayed
Agistemus vulgaris	C. sativus–S. lycopersicum	Fayoum–Tobhar–Tamai

or thelytoky [110]. Experimental releases of *G. aculeifer* and *S. miles* have shown reduction of respectively 80.5 and 61% of the *Frankliniella occidentalis* populations in bean crops [111]; who reported to provide better thrips control combining with *N. cucumeris* than when released in isolation [112].

Collection of some predatory mites in Egypt, including family Laelapidae, six laelapid species was recorded during a survey in some governorates. Survey included crops and weeds (Table 9).

In Egypt, many genera of the family Laelapidae were recorded from different habitat, includes *Hypoaspis, Laelaspis, Cosmolaelaps, Ololaelaps, Androlaelaps* and *Geolaelaps*.

The life history and life table parameters of some predatory laelapid mites cultured on natural and factitious foods in Egypt were presented in Table 10. Members of the family Laelapidae can play an important role in biological control. *Thrips tabaci* and nematodes are the major pests and can be consumed by various laelapid species. Natural foods includes free living nematodes like *Meloidogyne javanica, Tylenchulus semipenetrans* and *Rhabditella muscicola* as well as *T. tabaci* and *Phthorimaea operculella* while factitious food were included *Entomobrya musatica*, acarid mites ex: *Caloglyphus Rodriguez, Rhizoglyphus robini* and *T. putrescentiae, Drosophila melanogaster* and *Musca domestica*.

5.2.2 Family Ascidae

Species of the ascid mites have been described from a wide range of habitats. Most of the species were described from soil, grasses, mosses, or dead organic matter on soil surface. The most important genera are included *Antennoseius, Arctoseius, Asca, Gamasellodes, Protogamasellus*.

Arctoseius cetratus has reported to prey on eggs and first instar larvae of *Drosophila melanogaster* (Drosophilidae) and of *Lycoriella auripila* (Sciaridae) [117].

Gamasellodes adriannae was reported as a voracious predator of nematodes [118]. Walter and Lindquist [119] indicated that *Protogamasellus mica* was able to develop to adulthood on the nematode *Acrobeloides* sp., while *Protogamasellus minutus* and *Protogamasellus primitivus similis* were reported to develop and reproduce on the nematodes *Acrobeloides* sp. and *Rhabditis* sp. [120]. *Protogamasellus minutus* was also reported by [121] to develop and reproduce on the collembolan *Lepidocyrtinus incertus* as well as on the *Rhizoglyphus robini* (Acaridae) and *T. putrescentiae*.

Table 8 Life table parameter of predatory stigmaeid mites (Acari: Stigmaeidae) cultured on some natural and factitious prey

Stigmaeid species	Prey/pollen	Plant	Egg to adult (days)	♀Longevity (days)	Exp. conditions	Fecundity Total eggs/♀	Daily eggs/♀/day	R_o	r_m	T	λ	Authority
Agistemus exsertus	T. urticae	–	10.80	–	27–29 °C, 70–80% R.H.	90.50	–	57.91	0.15	26.21	1.16	Abou-Awad and El-Sawi [101]
	E. orientalis		11.70			61.95		35.32	0.14	24.90	1.15	
	Brevipalpus pmcher [C. and F.]		10.80			67.3		31.75	0.12	28.30	1.13	
	B. tabaci		10.40			60.12		33.23	0.13	27.69	1.14	
	E. kuehniella		13.00	31.2	27 °C, 70 ± 5% R.H.	97.78	–	61.25	0.196	20.99	1.22	Momen [102]
	P. ziziphus		15.00	23.6		75.27	–	43.71	0.174	21.70	1.19	
	S. littoralis (Eggs)	–	10.6	21.2	27 °C	68.9	–	44.07	0.21	17.89	1.23	Momen and El-Sawi [103]
	Agrotis ipsilon (Eggs)		10.0	23.8		49.9		30.378	0.18	18.91	1.20	
	P. operculella (Eggs)		13.4	11.1		4.9		2.29	0.03	20.78	1.04	
	Chrysomphalus ficus (Eggs)	–	10.8	32.4	27 °C, 70–75% R.H.	70.9	2.5	45.5	0.17	22.64	1.18	El-Sawi and Momen [14]
	C. ficus (Crawlers)		10.5	21.0		15.3	1.1	7.7	0.098	20.58	1.11	

(continued)

Table 8 (continued)

Stigmaeid species	Prey/pollen	Plant	Egg to adult (days) ♀	♂	♀Longevity (days)	♂	Exp. conditions	Total eggs/♀	Daily eggs/♀/day	R_o	r_m	T	λ	Authority
	Parlatoria blanchardi (Eggs)		15.5		4.0			–	–	–	–	–	–	
	T. urticae	Eggplant leaves	25.5		40.2		20 °C	26.72	1.02	20.94	0.08	36.66	1.08	Al-Shammery [104]
			19.2		32.4		25 °C	60.00	2.42	29.46	0.11	28.46	1.12	
			16.3		28.6		30 °C	47.7	2.84	36.97	0.13	27.23	1.14	
			15.0		28.2		35 °C	31.6	2.40	15.37	0.10	26.05	1.11	
A. exsertus	*Galleria mellonella* (Eggs)		9.97		40.00		30 °C, 70 ± 5% R.H.	131.47	3.65	92.30	0.2384	18.98	1.2692	Momen [105]
	T. urticae (Eggs)		11.30		36.53			97.93	3.10	57.291	0.2084	19.41	1.2318	
	Pulvinaria psidii		20.1	15.4	23.4	12.0	26 °C	25.67	–	–	–	–	–	Salwa [106]
			11.31	7.54	10.03	5.6	30 °C	8.65		–	–	–	–	
Zetzellia mali	*T. urticae* (Eggs)		12.18		10.00		28 ± 1 °C, 65 ± 5% R.H.	1	1.4	7.25	0.0146	12.88	1.15	Khodayari et al. [107]
Agistemus olivi	*A. oleae*		23.81		37.11		20 °C	23.75	1.5	–	–	–	–	Abou-Awad et al. [108]

(continued)

Table 8 (continued)

Stigmaeid species	Prey/pollen	Plant	Egg to adult (days)	♀Longevity (days)	Exp. conditions	Fecundity Total eggs/♀	Daily eggs/♀ /day	R_o	r_m	T	λ	Authority
			12.90	27.96	25 °C	59.25	3	–	–	–	–	
			8.90	22.48	30 °C	119.37	6.92	–	–	–	–	
	T. hassani		25.33	40.25	20 °C	37.05	1.30	–	–	–	–	
			13.31	29.91	25 °C	58.19	2.75	–	–	–	–	
			8.68	25.12	30 °C	116.67	6.08	–	–	–	–	
	A. mangiferae		10.77	25.35	30 ± 1 °C, 75 ± 5% R.H.	123.70	6.12	92.779	0.263	17.188	1.301	Momen [15]
	A. fockeui		10.40	28.53		116.07	4.85	83.376	0.249	17.720	1.283	
	A. lycopersici		13.07	25.53		73.67	3.69	35.36	0.178	20.001	1.195	
Agistemus vulgaris	T. urticae		12.56	18.84	28 ± 2 °C, 70 ± 5% R.H.	28.43	1.43	12.825	0.1363	18.633	1.141	Mohamed [109]
	O. sayedi		12.09	23.06		53.75	2.84	24.384	0.1666	19.18	1.181	
	A. lycopersici		10.75	13.15		21.93	2.03	9.816	0.1489	15.341	1.161	

Table 9 Incidence of soil predatory laelapid mites in Egypt

Species	Host/habitat	Locality
Hypoaspis reticulatus	*E. crassipes* (roots)–*C. dioscoridis* (roots)–*A. donax* (roots)–*S. marianum* (roots)	Tokh Kewesna El-Bagoor
Ololaelaps chanti	*S. marianum* (roots) *C. demersum* (roots)	Bosh–El-Wasta El-Badary
Cosmolaelaps keni	*L. sculentum* (soil and roots)	Fayoum–Tamai
Geolaelaps aculeifer	*L. sculentum* (soil and roots)	Fayoum–Abshawi
Androlaelaps aegypticus	*S. lycopersicum* (roots)	Ismailia–Fayed
Laelaspis astronomicus	*S. marianum* (roots)	Bosh–El-Wasta

Collection of survey some predatory mites, including family Ascidae, Ten Ascid species were recorded during a survey conducted in some governorates in Egypt (Table 11). Survey included some crops, orchards and weeds.

In Egypt, Predatory ascid mites were able to feed and sustain oviposition on various kinds of foods. Foods included fungi ex: *Fusarium solani*, *Aspergillus niger*, *A. flavus* and *F. oxysporum* acarid mites, Collembola and free-living nematode were used to culture various ascid species. The life history of some predatory ascid mites cultured on natural and factitious foods in Egypt (Table 12).

5.2.3 Family Rhodacarida

Three genera of the family Rhodacaridae were reported by various authors, included *Rhodacarellus*, *Multidentorhodacarus*, *Protogamasellopsis* sp. Nematodes have been reported many times to be preyed by rhodacarids [129, 130]. Thelytoky seems common in rhodacarids [20, 131]. Only one species was recorded from Egypt namely *Protogamasellopsis denticus*. This species was able to feed on insect and mite foods. Insects prey included: *Bactrocera zonata*, *Phothorimaea operculella*, *Agrotis ipsilon* and *Agrotis ipsilon* while mites prey included: *Rhizoglyphus robini* and *T. putrescentiae*.

The life history and life table parameters of *Protogamasellopsis denticus* cultured on natural and factitious foods in Egypt (Table 13).

5.2.4 Family Melicharidae

Proctolaelaps is the most important genus in family Melicharidae, and can play an important role in biological control. *Proctolaelaps bickleyi*, *P. bickleyi*, *P. deleoni* were developed and reproduced on various kinds of food such as Nematodes, mites, pollen and fungi ex: *Botrytis fabae*, *Rhizoctonia solani*, *Aceria guerreronis*, *Tyrophagus putrescentiae*, *Rhizopus* aff. *stolonifer*, *Steneotarsonemus furcatus*, *T. urticae*

Table 10 Selected species of predatory mites of the family Laelapidae used in biological control and mass cultured on natural and factitious foods in Egypt

Laelapid species	Prey	Egg to adult (days)	♀Longevity (days)	Exp. conditions	Fecundity Total eggs/♀ 18 °C	Daily eggs/♀/day 25 °C	Daily eggs/♀/day 30 °C	Authority
Hypoaspis zachvkinae	*Musca domestica* (Eggs)	8.0	22.7	18 °C 25 °C 30 °C	7.3	45.8	45.8	Ahmed [113]
	Rhizoglyphus robini	10.0	18.3		9.0	32.4	63.2	
	Free-living nematodes	8.2	17.7		5.2	26.7	45.0	
	Drosophila melanogaster	9.6	20.3		3.3	24.1	45.4	
Laelaspis imitatus	*T. putrescentiae*	6.92	21.55	25 °C 70% R.H.	17.33	0.80		Ezz El-Dein [114]
		6.07	20.65	35 °C	18.27	0.89		
	(Collembola) *Entomobrya musatica*	10.21	23.66	25 °C	4.63	0.19		
		–	–	35 °C	–	–		
	Rhabditella muscicola	6.26	20.9	25 °C	23.87	1.18		
		4.87	17.8	35 °C	17.53	1.02		

Predatory mite species	Prey	Egg to adult (days)	♀Longevity (days)	Exp. conditions	Fecundity Total eggs/♀	Daily eggs/♀	R_o	r_m	T	λ	Authority
Cosmolaelaps simplex	*Caloglyphus Rodriguez*	10.3	–	26 °C, 65% R.H.	–	–	20.46	0.135	22.24	–	Al-Rehiayani and Fouly [115]

(continued)

Table 10 (continued)

Predatory mite species	Prey	Egg to adult (days)	♀Longevity (days)	Exp. conditions	Fecundity Total eggs/♀	Daily eggs/♀	R_o	r_m	T	λ	Authority
	Meloidogyne javanica	12.6	–		–	–	13.62	0.123	21.07	–	
	Tylenchulus semipenetrans	13.9	–		–	–	13.10	0.127	20.22	–	
Cosmolaelaps qassimensis	T. putrescentiae	11.94	20.89	26 ± 1 °C, 70 ± 5% R.H.	–	–	30.99	0.119	28.69	–	Fouly and Abdel-Baky [116]
	C. rodriguez	12.85	20.55		–	–	27.12	0.105	31.17	–	
	M. incognita	14.95	20.72		–	–	17.82	0.087	32.84	–	
Cosmolaelaps keni	Thrips tabaci (nymph)	13.33		28 ± 2 °C, 70 ± 10% R.H	94.11	1.85	70.301	0.169	25.198	1.183	Lamlom [27]
	Agrotis ipsilon (eggs)	8	35.67		107.11	4.47	82.583	0.163	27.144	1.176	
	Tuta absoluta (eggs)	9.26	59.42		94.22	3.10	69.348	0.247	17.194	1.279	
	Phthorimaea operculella (eggs)	8.17	42.66		97.44	4.05	75.812	0.289	14.962	1.335	

Table 11 Incidence of soil predatory Ascid mites in Egypt

Species	Host/habitat	Locality
Lasioseius africanus	S. lycopersicum (Debris and soil)	Fayoum–Abshway
Lasioseius aegypticus	S. melongena (Debris and soil)	Ismailia–Al-Kassasen
Lasioseius zahri	Prunus persica (Debris)	Ismailia–Abou-Souer
Lasioseius athiasae	M. indica (Debris)	Ismailia–Al-Tal Al-Kber
Blattisocius tarsalis	M. indica (Debris and soil)	Fayoum–Atsa
Cheiroseius egypticus	Prunus persica (Debris)	Ismailia–Al-Tal Al-Kber
Lasioseius lindquisti	A. donax (leaves)	Tokh
Cheiroseius egypticus	Eichhornia crassipes (roots)	El-Badary–Bosh
Cheiroseius nepalensis	C. dioscoridis (leaves)	Sheben El-Kom
Cheiroseiulus crassipes	M. indica (Debris and soil)	Mostorud

and coconut pollen [133, 134]. *Proctolaelaps cyinchuanensis* was observed to feed on *Suidasia pontifi* [135]. Most *Proctolaelaps* species studied are able to develop and reproduce on several food items, indicating the polyphagous habit of members of this genus.

Collection of some predatory mites, including family Melicharidae in Egypt was surveyed. Five species were recorded during a survey at some governorate in Egypt. Survey included some crops, orchards and weeds (Table 14).

The life history and fecundity of various species of the genus *Proctolaelaps* cultured on natural and factitious foods in Egypt have been presented in Table 15.

5.3 Mass Production Predatory Mites

5.3.1 Mass Production of Aerial Predatory Mites—Family Phytoseiidae

Since its introduction approximately 40 years ago, phytoseiids have gained recognition for their importance as natural enemies of thrips, whiteflies and spider mites [140]. Zhang [141] recorded that phytoseiid mites (as a minimum 20 species) were obtainable at a commercial scale and have been mainly used on greenhouse crops.

Worldwide, the farmers use *P. persimilis* in the management of *T. urticae* and other tetranychid mite pests both in greenhouses and on field crops [8]. However,

Table 12 Life history of some predatory mites of the family Ascidae cultured on natural and factitious foods in Egypt

Ascid species	Diets	Female				Temp. °C	Source
		Life cycle (days)	Oviposition period (days)	Longevity (days)	Fecundity (eggs/female)		
Lasioseius parberlesei	*T. urticae* eggs	7.5	17.2	23.4	12	25	Afifi [120]
Cheroseius nepalensis	*Thaminidium elegans*	5	–	17.6	–	25	El-Bishlawy, Shahira [122]
Lasioseius aegypticus	*R. robini*	13.2	14.1	31.5	38.6	25	Ahmed [123]
	T. putrescentiae	14.2	13.4	29	24.8	25	
Protogamasellus minutus	*R. robini*	12.4	59	70.4	56.7	25	Afifi et al. [121]
	T. putrescentiae	13.8	52.5	62.3	50.3	25	
	Rhabditis sp.	15.3	17.5	31.3	10.4	25	
	L. incertus	19.3	24.2	42.8	19.1	25	
Protogamasellus primitives similis	*Fusarium solani*	14.9	36.4	47.3	13.9	25	Nawar and Nasr [124]
	A. flavus	16.8	34.2	46.1	12.1	25	
Lasioseius athiasae	*R. robini*	8.2	35.4	43.4	44.6	25	Nasr et al. [125]
	T. putrescentiae	9.1	27.9	37.7	32.5	25	
	Free-living nematodes	9.5	25.7	35.8	28.7	25	
	Aspergillus niger	10.7	10.8	11.8	11.5	25	
	F. oxysporum	10.5	14.7	26.4	16.1	25	

(continued)

Table 12 (continued)

Ascid species	Diets	Female Life cycle (days)	Oviposition period (days)	Longevity (days)	Fecundity (eggs/female)	Temp. °C	Source
Lasioseius bispinosus	*R. robini* immatures	33.2	85.6	144.6	38.61	20	Nawar et al. [126]
		15.7	51.4	90.4	71.45	25	
		6.8	20.4	38	45.67	30	
	T. putrescentiae immatures	25.8	78.6	119.8	33.4	20	
		12.9	50.8	73.8	69	25	
		6.6	19.2	29	40	30	
Lasioseius zaheri	*T. putrescentiae*	8.3	33.4	41.5	48.6	25	Nawar and El-sherif [127]
	F. oxysporum	10.4	14	25.7	10.8	25	
	A. flavus	10.6	11.2	22.9	9	25	
	Meloidogyne incognita	17.9	37.3	54.1	25		Mowafi [128]
		11.1	46	57.4	34.7		
		10.3	6.6	11.9	8.2		
	R. scanica	15.1	27	35.3	20.4		
		8	47.5	51.8	38.5		
		7.1	30.8	41	38.5		
Lasioseius aegypticus	*T. putrescentiae*	7.25	11.6	16.63	16.1	25	Ezz El-Din [114]
		20.68	11.24	17.23	13.9	15	
	Collembola	12.8	10.17	15.95	8.7	25	
		21.0	11.53	17.79	6.43	15	
	Free-living nematode	5.88	8.74	13.07	22.6	25	
		19.4	10.7	16.3	16.9	15	

Table 13 *Protogamasellopsis denticus* a predatory mite of the family Rhodacaridae used in biological control and mass cultured on natural and factitious foods in Egypt

Diets	Female			Fecundity (eggs/female)	Temp. °C	Source
	Life cycle (days)	Oviposition period (days)	Longevity (days)			
Bactrocera zonata	10.22	39.61	49.44	93.27	30	Mahmoud [132]
Phothorimaea operculella	17.94	61.27	72.05	84.24	30	
Agrotis ipsilon	13.05	103.83	114.11	68.76	30	
Sitotroga cerealella	11.83	34.05	39.94	41.44	30	
Rhizoglyphus robini	7.00	36.38	43.61	113.68	30	
Tyrophagus putrescentiae	9.00	27.61	31.64	71.69	30	

Table 14 Incidence of soil predatory Melicharid mites in Egypt

Species	Host/habitat	Locality
Proctolaelaps bickleyi	*M. indica* (Debris)	Banha–Tokh
Proctolaelaps pygmaeus	*S. lycopersicum* (Debris and soil)	Kaluob–Mostorud
Proctolaelaps orientalis	*Prunus persica* (Debris)	El-Badary
Proctolaelaps naggarii	*Pluchea dioscoridis* (Leaves)	Sheben El-Kom
Proctolaelaps aegyptiaca	*P. persica* (Debris)	Sheben El-Kom

procedures of the mass production of phytoseiids such as *P. persimilis/N. californicus* based on bean plants growing in greenhouse for the production of spider mite while, the predatory mite was introduces latterly. A clean spider mite culture, without any predators, is required for phytoseiids mass rearing. Infested bean leaves from the spider mite culture are used to infest bean plants growing in greenhouse to offer continuous source of spider mite as prey. Latterly, predators are introduced to bean plants that are heavily infested with the prey. The plants are collected when it reached to the highest predator density [142]. The introduction of predatory mite (*P. persimilis*) into the infested bean leaves needed ideal timing to obtain the highest production of spider mite without causing plant damage as a result of mite infestation [143]. Predator harvesting usually exposes the predators to food shortage stress and a lot of them are lost due to unsuitable collection techniques.

Table 15 Life history of some species of the family Melicharidae used in biological control and mass cultured on natural and factitious foods in Egypt

Melicharid species	Diets	Female				Fecundity (eggs/female)	Temp. °C	Source
		Life cycle (days)	Oviposition period (days)	Longevity (days)				
Proctolaelaps orientalis	*M. domestica* larvae	6.7	12.3	20.5		30.4	25	Gomaa [136]
	Free-living nematodes	6.8	14.3	19.3		27.7	25	
	R. robini eggs	9.3	7.7	15.3		11	25	
	M. domestica larvae	9.5	12.1	17.6		25.4	18	
	Free-living nematodes	9.2	14	20.3		26.6	18	
	R. robini eggs	10.1	12.1	17.9		19.6	18	
	F. oxysporum	6	13.3	17.5		31.1	18	
	A. niger	5.8	13	17.5		30.3	18	
	P. notatum	8	7.7	11.4		11.9	18	
Proctolaelaps aegyptiaca	*Aceria ficus*	7.27	24.09	36.18		55.82	27	Abou-Elela [137]
	T. urticae nymphs	10.30	24.08	37.40		41.50	27	
	E. orientalis nymphs	10.0	30.20	38.50		30.50	27	
Proctolaelaps bickleyi	*A. ficus*	4.50	8.69	10.73		57.92	27	
Proctolaelaps naggarii	Free-living nematodes	9.53	16.11	21.60		45.60	25	Mahmoud [138]
		9.15	13.12	18.66		54.18	30	
Proctolaelaps bickleyi	*R. scanica*	6.6	12.1	22.3		13.1	30	Nasr et al. [139]
	R. solani	11.1	5.9	12		6.1	30	
	A. niger	16.8	3.6	8.3		4.5	30	
	Botrytis fabae	16.4	4.5	10.8		4.8	30	

Any rearing systems should offered more environmental conditions regulation and better controls that keep predatory mites save from the severe losses. Numerous enclosed systems have been suggested to rear the predatory mite (*P. persimilis*) on its prey (*T. urticae*) using different types of enclosure [144, 145]. A rearing system consists of a plastic foam piece or sponge placed in the center of a tray full of water was designated by [144] for the mass production of a number of species of predatory mite for example, *P. persimilis, P. macropilis, Typhlodromus pyri, N. fallacis* and *N. cucumeris*. A fence is made by surrounding the foam piece with wet tissue paper. In order to keep a continuous saturation, one side of the wet tissue paper needs to touch the water.

However, modifications that required developing those rearing systems with the aim of get high predator production have proven difficult. Ramakers and Van Lieburg [146] described an enclosed system used for the production of predatory mites and consist of a cube rearing cage occupying approximately ½ m². These rearing cages have two doors; upper door on the upper part of the cage and another lower door on the cage bottom. In addition, collection cups were placed at the cage top for collecting the predatory mites. In this method bean plants were grown in trays and infested with *T. urticae*. Lately, the bean plants are cut from the soil base and introduced into the rearing cages where the predacious mites are introduced latterly. Fresh infested bean plants are introduced through the upper door. After being exposed to the predacious mites for a week, the plants are taken away through the lower door.

The gravid females of the predatory mites tend to stay close to spider mites, while young adult predators have a tendency to move to the upper parts of the rearing cage. Therefore, the predatory mites accumulated in the collection cup are daily collected. This small rearing cage was able to a continuous production of more than14.000 predatory mites each week.

Providing easily produced frozen, live, lyophilized insects and mites as primary food sources (factitious food) instead of natural prey that support the development and reproduction of predators may make the Cost-effective rearing system economical. For example astigmatid mites and eggs of many lepidopterans have been considered useful for mass rearing predatory mites. Some mites of Astigmatina have been considered suitable factitious food for several Mesostigmata species for many years, and it can be economically produced in large numbers in relatively small containers on flour, bran or similar substrates [147]. This usually renders the rearing process less expensive than those using phytophagous mites/thrips/nematoda as natural foods, due to reduced requirements for space, labor and maintenance costs [100].

It was found that *N. cucumeris* and *N. barkeri* had been successfully developed and reproduced on *Tyrophagus casei* (Acari: Acaridae) as factitious food by [147]. Also [148] verified that the storage mites, *Carpoglyphus lactis* (Acari: Carpoglyphidae) was an excellent factitious food for *A. swirskii* and *N. cucumeris* for development and reproduction. Xia et al. [99] proved that *N. barkeri* developed and reproduced well on *A. ovatus* especially at high temperature. Midthassel et al. [90] verified that *A. swirskii* consumed and strongly attack eggs and active stages of astigmatid mite *Suidasia medanensis* (Acari: Suidasiidae). Barbosa and de Moraes [149] found that *N. barkeri* was mass reared successfully on *Thyreophagus* sp. and it was the most suitable astigmatid mite for development and reproduction of the predator. Massaro et al. [150] noted that *Amblyseius tamatavensis*, which distributed in Brazil and used in controlling the whitefly, *Bemisia tabaci*, a serious pest of various crops in many countries and was successfully mass-reared on some astigmatine species.

5.3.2 Mass Production of Soil Predatory Mites

Predatory soil mites can be easily reared on insect eggs or acarid mites. Stock colonies of astigmatids and predators were always reared in plastic containers similarly to those described beyond [98–151], consisting of plastic pots (12 cm high, 7.5 cm diameter), each containing holes for ventilation (2 cm diameter) closed with a polyester screen of 0.2 mm mesh. The diet of astigmatid mite was consisted of a mixture of 30–50% of brewer's yeast and 70% of wheat bran. The diet was refreshed every week. This method doesn't require a host plant, which causes more time and space restraints.

Several studies have shown that eggs of the Mediterranean flour moth *Ephestia kuehniella* Zeller constitute a nutritionally superior food for various insect predators [152], predatory phytoseiid and laelapid mites [91, 92, 153]. Conversely, the continuous use of *E. kuehniella* eggs as a factitious prey in mass rearing system has disadvantage, is their high cost (EUR 500 for 1 kg) in the Egyptian market. This has begun in a search for cheaper alternative prey, like another insect eggs or astigmatid mites. Eggs of the peach fruit fly *Bactrocera zonata* (Saunders) (Diptera: Tephritidae) have been proposed as a potential source of factitious prey for *N. barkeri*, *Amblyseius largoensis* (Muma) and *Proprioseiopsis kadii* (El-Halawany and Abdel-Samad), whereas experiments were tested only for the 1st generation [154]. Eggs of *B. zonata* are produced by the billion in mass-rearing facilities for rearing predatory insect purposes; market prices of their eggs are competitive with those of *E. kuehniella* since the cost needed for producing 285.000 eggs is approximately 0.25 USD [154].

6   Using Predatory Mites Against Different Agricultural Pests

Mites of the family Phytoseiidae are predators generally related to the phytophagous mite and/or insect pests in the field. Therefore, field applications were conducted to evaluate the efficacy of these predators as successful biological control agents against several agricultural pests infested many crops growing in the greenhouses/field. *Phytoseiulus persimilis*, *A. swirskii*, *N. californicus*, *P. macropilis*, *N. cucumeris* and *Neoseiulus zaheri* are the most phytoseiid species used in releasing program to suppress *T. urticae*/some insect pests.

El-Saiedy [155] *N. californicus* and *P. persimilis* were used successfully for controlling *T. urticae* on strawberry, where their release resulting in reduction percentages of *T. urticae* infestation (71.78 and 97.20%). *Neoseiulus californicus* and *P. macropilis* were released against *T. urticae* infesting 2 cultivars of cucumber that grown in plastic houses. Both predators showed the maximum percentages of reduction of *T. urticae* (97.36 and 97.51%) and (87.14 and 92.50%) in both cultivars [156]. *Neoseiulus cucumeris* gave a higher reduction in infestation of *Aphis gossypii* and *T. stabaci* than *N. californicus* when both released on 2 cultivars of eggplant in open field [157].

Elmoghazy et al. [44], a high performance was given by *N. californicus* and *A. swirskii*, to control *T. urticae* on 2 cultivars of *Vicia faba* in open field at Beheira Governorate. *Neoseiulus californicus* clearly decrease the population density of *T. urticae* (87.22 and 74.22% redaction of mite/leaf) comparing to *T. swirskii* (57.49 and 41.5% redaction of mite/leaf) on both cultivars, respectively. Releasing of *N. californicus* in addition to *A. swirskii* can give good results in controlling *T. urticae*. Furthermore, the release of *P. persimilis* and *N. californicus* resulted in reduction of *T. urticae* population on sweet pepper growing in greenhouses during 2 successive seasons [158]. Under both open field and plastic low tunnels conditions, releasing *P. persimilis* and *A. swirskii* to control *T. urticae* on two watermelon and muskmelon cultivars were gave an excellent results. The highly percentage of reduction was attained by *P. persimilis*, followed by *T. swirskii* on all tested plants [159]. Kame et al. [160] studied 2 sweet pea cultivars and under high tunnel of plastic-net condition, releasing *P. persimilis* and *N. californicus* gave the best results in reducing the population of *T. urticae* as compared to *Typhlodromus negevi* and *E. scutalis*, probably because both species are pollen feeder.

7 Performances of Predatory Mites in Controlling Pests After Long-Term Feeding on Factitious Foods

It is important for a predator keeps its potential after long-term rearing on alternative food to control natural prey, so the efficacy of factitious food for predatory mites deserved to be investigated by authors. The effective factitious food supply sufficient nutritional requirements to the predators and lead to continuous production of high quality progeny with no reverse effect on the predator performance [161]. *Neoseiulus barkeri* is predatory mite which distributed worldwide and has been used to manage pests in plastic houses since the method of its mass-rearing were developed.

In Denmark, [162] reared *N. barkeri* on storage mites (*Acarus* spp.) in large numbers at a relatively small cost and was utilized to successfully control *T. tabaci*, on cucumber plants in glasshouses. Ling et al. [163] found that the storage mite, *A. ovatus* suitable substitute food for *N. barkeri* and used for mass production and it was also exhibited a strong predatory ability on the spider mite *Panonychus citri* in Ganzhou as well as [164] reported that *N. barkeri* fed on *A. ovatus* could efficiently manage *P. citri*. *Amblyseius swirskii* has been proved to be an effective biological control agent against whiteflies, thrips, broad mites in several plastic houses crops [61, 165, 166].

A study by [92] explored the potential of *E. kuehniella* eggs on development and reproduction of different life stages of *A. swirskii* for five successive generations. Moreover, the study reported that *A. swirskii* did not lose their power to control its natural prey first instars of *F. occidentalis* after six tested generations. Only the predation rates of the predators of generation six were slightly lower than the predation rates of the predators of generation one. Likewise, [167] verified that rearing *N. californicus* on almond pollen for multiple generations (up to 20) had an influence on development and positively affected fecundity and accordingly higher potential to control *T. urticae*.

8 Conclusion

This chapter summarized the role of some predatory mites inhabiting soil or vegetation as biological control agents and it has been focused on the abundance of these predacious mites in Egypt as well as their mass production and the most important species used in controlling pests in field. It was mentioned that twenty five phytoseiid, three stigmaeid, six laelapid, ten ascid, five melicharid, and only one Rhodacarid species namely *Protogamasellopsis denticus* were recorded in Egypt. Also, Predatory phytoseiid mites are successfully used as bio-control agents in controlling of phytophagous mites, thrips and whitefly. *Agistemus exsertus* is one of the most important species of the family Stigmaeidae and able to attack pests of the families Tetranychidae and Eriophyidae as well as scale insects and whitefly in field crops and orchards.

The chapter concentrated on the method of mass production of aerial and soil predacious mites that can be reared traditionally by introducing natural or by providing factitious foods. It was explained in detail the rearing system taking into consideration the cost, advantage and disadvantage.

References

1. van der Geest LPS (1985) Pathogens of spider mites. In: Helle W, Sabelis MW (eds) Spider mites. Their biology, natural enemies & control. World Crop Pests 1B:247–258
2. Gerson U (2008) The Tenuipalpidae: an under-explored family of plant feeding mites. Syst Appl Acarol 2:83–101
3. Abou-Awad BA, El-Sawaf BM, Abdel-Khalek AA (2000) Impact of two eriophyoid fig mites, *Aceria ficus* and *Rhyncaphytoptus ficifoliae*, as prey on postembryonic development and oviposition rate of the predacious mite *Amblyseius swirskii*. Acarologia 4:367–371

4. Momen FM, Abdel-Khalek A (2008) Effect of the tomato rust mite *Aculops lycopersici* (Acari: Eriophyidae) on the development and reproduction of three predatory phytoseiid mites. Int J Trop Insect Sci 1:53–57
5. Oldfield GN, Proeseler G (1996) Eriophyoid mites as vectors of plant pathogens. In: Eriophyoid mites their biology, natural enemies and control. World Crop Pests 6:259–275
6. van Leeuwen T, Vontas J, Tsagkarakou A, Dermauw W, Tirry L (2010) Acaricides resistance mechanisms in the two-spotted spider mite *Tetranychus urticae* and other important Acari: a review. Insect Biochem Mol Biol 40:563–572
7. Uddin N, Alam Z, Miah UR, Mian HI, Mustarin EK (2015) Toxicity of pesticides of *Tetranychus urticae* Koch (Acari: Tetranychidae) and their side effects on *Neoseiulus californicus* (Acari: Phytoseiidae). Int J Acarol 41:688–693
8. van Lenteren JC (2012) The state of commercial augmentative biological control: plenty of natural enemies, but a frustrating lack of uptake. Biol Control 57:1–20
9. Bigler F (1989) Quality assessment and control in entomophagous insects used for biological control. J Appl Entomol 108:390–400
10. Krantz GW (1978) A manual of acarology, 2nd edn. Oregon State University Bookstores, Corvallis, p 509
11. Halliday RB (2006) New taxa of mites associated with Australian termites (Acari: Mesostigmata). Int J Acarol 32:27–38
12. Moreira GF, de Morais MR, Busoli AC, de Mraes GJ (2015) Life cycle of *Cosmolaelaps jaboticabalensis* on *Frankliniella occidentalis* and two factitious food sources. Exp Appl Acarol 65:219–226
13. McMurtry JA, de moraes JG, Sourassou FN (2013) Revision of the lifestyles of phytosiid mites and implication for biological control strategies. Syst Appl Acarol 4:297–320
14. El-Sawi SA, Momen FM (2006) *Agistemus exsertus* Gonzalez (Acari: Stigmaeidae) as a predator of two scale insects of the family Diaspididae (Homoptera: Diaspididae). Arch Phytopathol Plant Prot 39:421–427
15. Momen FM (2012) Influence of life diet on the biology and demographic parameters of *Agistemus olive* Romeih, a specific predator of eriophyid mites (Acari: Stigmaeidae and Eriophyidae). Trop Life Sci Res 23:25–34
16. Lindquist EE, Kantz GW, Walter DE (2009) Order Mesostigmata. In: Krantz GW, Walter DE (eds) A manual of acarology, 3rd edn. Texas Tech University Press, Lubbock, pp 124–232
17. Castilho RC, de Moraes GJ, Halliday B (2012) Catalogue of the mite family Rhodacaridae Oudemans, with notes on the classification of the Rhodacaroidea. Zootaxa 3471:1–69
18. Helle W, Sabelis MW (1985) Spider mites: their biology, natural enemies and control, vol 1B. Elsevier, Amsterdam, p 458
19. Walter DE (2003) The genus *Gamasellodes* (Acari: Mesostigmata: Ascidae) New Australian and North American species. Syst Appl Acarol (Spec Publ Lond) 15:1–10
20. Walter DE, Ikonen EK (1989) Species, guilds, and functional groups: taxonomy and behavior in nematophagous arthropods. J Nematol 21:315–327
21. Collier T, Van Steenwyk R (2004) A critical evaluation of augmentative biological control. Biol Control 31:245–256
22. Hoy MA (2009) Agricultural acarology: introduction to integrate mite management. CRC Press, Boca Raton
23. van Lenteren JC (2003) Quality control and production of biological control agents: theory and testing procedures. CABI Publishing, Wallingford, UK
24. Easterbrook MA, Fitzgerald JD, Solomon MG (2001) Biological control of strawberry tarsonemid mite *Phytonemus pallidus* and *Tetranychus urticae* on strawberry in the UK using species of *Neoseiulus* (*Amblyseius*). Exp Appl Acarol 25:25–36
25. Jovicich E, Cantliffe DJ, Osborne LS, Stoffella PJ, Simonne EH (2008) Release of *Neoseiulus californicus* on pepper transplants to protect greenhouse-grown crops from *Polyphagotarsonemus latus* infestations. In: Mason PG, Gillespie DR, Vincent C (eds) Proceedings of 3rd international symposium on biological control of arthropods, Christchurch, pp 347–353

26. Wade MR, Zalucki MP, Wratten SD, Robinson KA (2008) Conservation biological control of arthropods using artificial food sprays: current status and future challenges. Biol Control 45:185–199

27. Lamlom M (2017) Mites associated with some plants of the family solanaceae at Fayoum and Beheira governorates. M.Sc., Faculty of Agriculture, Cairo University, 176 pp

28. Sabelis MW, van der Meer J (1986) Local dynamics of the interaction between predatory mites and two-spotted spider mites. In: Metz JAJ, Diekmann O (eds) Dynamics of physiological structured populations. Lecture notes in biomathematics. Springer-Verlag, Berlin, pp 1–24

29. van Rijn PCJ, Mollema C, Steenhuis-Broers GM (1995) Comparative life-history studies of *Frankliniella occidentalis* and *Thrips tabaci* (Thysanoptera, Thripidae) on cucumber. Bull Entomol Res 85:285–297

30. van Rijn PCJ, Sabelis MW (1990) Pollen availability and its effect on the maintenance of populations of *Amblyseius cucumeris*, a predator of thrips. Int Symp Crop Prot Meded Fac Landbouwwet Rijksuniv Gent 55:335–342

31. Sabelis MW, van Rijn PCJ (1997) Predation by insects and mites. In: Lewis T (ed) Thrips as crop pests. CAB-International, Wallingford, UK, pp 259–354

32. van Rijn PCJ, van Houten YM (1991) Life history of *Amblyseius cucumeris* and *Amblyseius barkeri* (Acarina: Phytoseiidae) on a diet of pollen. In: Dusbabek F, Bukva V (eds) Modern acarology, vol 2. Academia, Prague and SPB Academic Publishing BV, The Hague, pp 647–654

33. van der Geest LPS (1985) Aspects of physiology. World crop pests: In: Helle W, Sabelis MW (eds) Spider mites. Their biology, natural enemies and control, vol 1A. Elsevier, Amsterdam, The Netherlands, pp 171–182

34. Mobley KN, Marini RP (1990) Gas exchange characteristics of apple and peach leaves infested by European red mite and two spotted spider mite. J Am Soc Hort Sci 115:757–761

35. Nihoul P, Hance T, Impe GV, Marecha B (1992) Physiological aspects of damage caused by spider mites on tomato leaflets. J Appl Entomol 113:487–492

36. Bondada BR, Oosterhuis DM, Tugwell NP, Kim KS (1995) Physiological and cytological studies of two spotted spider mite, *Tetranychus urticae* Koch, injury in cotton. Southwest Entomol 20:171–180

37. Wilson LJ (1993) Spider mites (Acari: Tetranychidae) affect yield and fiber quality of cotton. J Econ Entomol 86:566–585

38. Singh OP (1988) Assessment of losses to soybean by red spider mite in Madhya Pradesh. Agric Sci Digest (Karnal) 8:129–130

39. Suekane R, Degrande PE, de Melo EP, Bertoncello TF, Sde LJI, Kodama C (2012) Damage level of the two-spotted spider mite *Tetranychus urticae* Koch (Acari: Tetranychidae) in soybeans. Rev Ceres 59:77–81

40. Meck ED, Walgenbach JF, Kennedy GG (2012) Association of *Tetranychus urticae* (Acari: Tetranychidae) feeding and gold fleck damage on tomato fruit. Crop Prot 42:24–29

41. Helle W, Sabelis MW (1985) Spider mites: their biology, natural enemies and control, vol 1A. Elsevier, Amsterdam, p 405

42. Feiertag-Koppen CCM (1976) Cytological studies of the two-spotted spider mite *Tetranychus urticae* Koch (Tetranychidae, trombidiformes). I: meiosis in eggs. Genetica 46:445–456

43. Hussey NW, Parr WJ, Gould HJ (1965) Observations on the control of *Tetranychus urticae* Koch on cucumbers by the predatory mite *Phytoseiulus riegeli* Dosse. Entomol Exp et Appl 8:271–281

44. Elmoghazy MME, El-Saiedy EMA, Romeih AHM (2011) Integrated control of the two spotted spider mite *Tetranychu surticae* Koch (Tetranychidae) on faba bean *Vicia faba* (L.) in an open field at Behaira Governorate, Egypt. Int J Environ Sci Eng 2:93–100

45. Greco NM, Liljesthröm GG, Ottaviano MFG, Cluigt N, Cingolani MF, Zembo JC, Sánchez NE (2011) Pest management plan for the two-spotted spider mite, *Tetranychus urticae*, based on the natural occurrence of the predatory mite *Neoseiulus californicus* in strawberries. Int J Pest Manag 57:299–308

46. Hoddle MS, Aponte O, Kerguelen V, Heraty J (1999) Biological control of *Oligonychus perseae* (Acari: Tetranychidae) on avocado. I. Evaluating release timings, recovery and efficacy of six commercially available phytoseiids. Int J Acarol 25:211–219

47. Hoddle MS, Robinson L, Virzi J (2000) Biological control of *Oligonychus perseae* (Acari: Tetranychidae) on avocado: III. Evaluating the efficacy of varying release rates and release frequency of *Neoseiulus californicus* (Acari: Phytoseiidae). Int J Acarol 26:203–214

48. Jolly RL (2000) The predatory mite *Neoseiulus californicus*: its potential as biological control agent for the fruit tree red spider mite, *Panonychus ulmi*. BPC Conf Brighton Pest Control 1:487–490

49. Croft BA, Coop LB (1998) Heat units, release rate, prey density and plant age effects on dispersal by *Neoseiulus fallacis* (Acari: Phytoseiidae) after inoculation into strawberry. J Econ Entomol 91:94–100

50. Croft BA, Pratt DA, Luh HK (2004) Low-density release of *Neoseiulus fallacis* provide for rapid dispersal and control of *Tetranychus urticae* (Acari: Phytoseiidae, Tetranychidae) on apple seedlings. Exp Appl Acarol 33:327–339

51. James DG (2001) History and perspectives of biological mite control in Australian horticulture using exotic and native phytoseiids. In: Halliday RB, Walter DE, Proctor HC (eds) Acarology: proceedings of the 10th international congress. CSIRO Publishing, Melbourne, pp 436–443

52. Waite GK (1988) Integrated control of *Tetranychus urticae* in strawberries in south-east Queensland. Exp Appl Acarol 5:23–32

53. Gerson U (1992) Biology and control of the broad mite, *Polyphagotarsonemus latus* (Banks) (Acari: Tarsonemidae). Exp Appl Acarol 13:163–178

54. Pena JE, Campbell CW (2005) Broad mite. EDIS. http://edis.ifas.ufl.edu/CH020

55. Baker JR (1997) Cyclamen mite and broad mite. Ornamental and turf insect information notes (2 May 2016)

56. Rodriguez H, Montoya A, Miranda I, Rodriguez Y, Ramos M (2011) Influence of the phenological phase of two pepper cultivars on the behaviour of *Polyphagotarsonemus latus* (Banks). Rev Prot Veg 26:73–79

57. Pena JE, Osborne L (1996) Biological control of *Polyphagotarsonemus latus* (Banks) (Acarina: Tarsonemidae) in greenhouses and field trials using introduction of predacious mites (Acarina. Phytoseiidae). Entomophaga 41:279–285

58. Fan YQ, Petitt FL (1994) Biological control of broad mite, *Polyphagotarsonemus latus* (Banks), by *Neoseiulus barkeri* Hughes on pepper. Biol Control 4:390–395

59. Weintraub PG, Kleitman S, Mori R, Shapira N, Palevsky E (2003) Control of broad mites (*Polyphagotarsonemus latus* (Banks)) on organic greenhouse sweet peppers (*Capsicum annuum* L.) with the predatory mite, *Neoseiulus cucumeris*. Biol Control 26:300–309

60. Tal C, Coll M, Weintraub PG (2007) Biological control of *Polyphagotarsonemus latus* (Acari: Tarsonemidae) by the predaceous mite *Amblyseius swirskii*. IOBC/WPRS Bull 30:25–36

61. van Maanen R, Vila E, Janssen A (2010) Biological control of broad mite (*Polyphagotarsonemus latus*) with the generalist predator *Amblyseius swirskii*. Exp Appl Acarol 52:29–34

62. Rodriguez H, Montoya A, Miranda I, Rodriguez Y, Depestre TL, Ramos M, Badii-Zabeh, MH (2015) Biological control of *Polyphagotarsonemus latus* (Banks) by the predatory mite *Amblyseius largoensis* (Muma) on sheltered pepper production in Cuba. Rev Prot Veg 30:70–76

63. Momen FM (1993) Effect of single and multiple copulation on fecundity, and sex ratio of the predacious mite, *Amblyseius barkeri*. Anz Schädlingsk Pflanzen Umweltschutz 66:148–150

64. Momen FM (1997) Copulation, egg production and sex ratio in *Cydnodromella negevi* & *Typhlodromus athiasae* (Acari: Phytoseiidae). Anz Schädlingsk Pflanzen Umweltschutz 70:34–36

65. Abdel-Khalek A, Fahim SF (2018) Influence of multiple mating and food deprivation on reproduction, longevity and sex ratio of *Amblyseius largoensis*. Biosci Res 15:437–442

66. Schausberger P (2004) Ontogenetic isolation favours sibling cannibalism in mites. Anim Behav 67:1031–1035

67. Zannou ID, Hanna R, de Moraes GJ, Kreiter S (2005) Cannibalism and interspecific predation in a phytoseiid predator guild from Cassava fields in Africa: evidence from the laboratory. Exp Appl Acarol 37:27–42

68. Momen FM (2010) Intra- and interspecific predation by *Neoseiulus barkeri* and *Typhlodromus negevi* (Acari: Phytoseiidae) on different life stages: predation rates and effects on reproduction and juvenile development. Acarina 18:81–88

69. Ali FS (1998) Life tables of *Phytoseiulus macropilis* (Banks) (Gamasida: Phytoseiidae) at different temperatures. Exp Appl Acarol 22:335–342

70. Abdel-Khalek A, Momen F (2009) Mating and prey stage affecting life history, reproduction and life table of the predacious mite *Phytoseiulus macropilis* (Banks) (Acari: Phytoseiidae). Arch Phytopathol Plant Prot 42:751–765

71. Camporese P, Duso C (1995) Life history and life table parameters of the predatory mite *Typhlodromus talbii*. Entomol Exp Appl 77:149–157

72. Zaher MA, El-Borolossy MA, Ali FS (2001) Morphological and biological studies on *Typhlodromus talbii* Athias-Henriot (Gamasida: Phytoseiidae). Insect Sci Appl 21:43–54

73. Momen FM (2011) Life tables and feeding habits of *Proprioseiopsis cabonus*, a specific predator of tydeid mites (Acari: Phytoseiidae and Tydeidae). Acarina 19:103–109

74. Momen FM (1999) Biological studies of *Amblyseius lindquisti* a specific predator of eriophyid mites (Acari). Acta Phytopathol Entomol Hung 34:245–251

75. Abou-Awad BA, El-Sherif AA, Hassan MF, Abou-Elela MM (1998) Laboratory studies on development, longevity, fecundity and predation of *Cydnoseius negevi* (Swirski and Amitai) (Phytoseiidae) with two mite species as prey. J Plant Dis Prot 105:429–433

76. Momen FM, Metwally AM, Nasr AK, Abdallah AA, Saleh KM (2014) Life history of *Proprioseiopsis badri* feeding on four eriophyid mite species. Phytoparasitica 42:23–30

77. Momen FM, Rasmy AH, Zaher MA, Nawar MS, Abou-Elella GM (2004) Dietary effect on the development, reproduction and sex-ratio of the predatory mite *Amblyseius denmarkeri* Zaher & El-Borolossy (Acari: Phytoseiidae). Int J Trop Insect Sci 24:192–195

78. Momen FM (2009) Potential of three species of predatory phytoseiid mites as biological control agents of the peach silver mite, *Aculus fockeui* (Acari: Phytoseiidae and Eriophyidae). Acta Phytopathol Entomol Hung 44:151–158

79. El-Laithy AYM, El-Sawi SA (1998) Biology and life table parameters of the predatory mite *Neoseiulus californicus* fed on different diet. J Plant Dis Prot 105:532–537

80. Abou-Awad BA, Metwally AM, Al-Azzazy MM (2010) *Typhlodromips swirskii* (Acari: Phytoseiidae) a predator of eriophyid and tetranychid mango mites in Egypt. Acta Phytopathol Entomol Hung 45:135–148

81. Abou-Elella GM (2003) *Thrips tabaci* (Lind.) as suitable prey for three predacious mites of the family Phytoseiidae (Acari: Phytoseiida). J Agric Sci Mans Univ 28:6933–6939

82. Rasmy AH, Momen FM, Zaher MA, Abou-Elella GM (2003) Influence of diet on life history and predatory capacity of *Amblyseius zaheri* Yousef & El-Borolossy (Acari: Phytoseiidae). Insect Sci Appl 23:31–34

83. Rasmy AH, Momen FM, Zaher MA, Nawar MS, Abou-Elella GM (2002) Dietary influence on life history and predation of the phytoseiid mite, *Amblyseius deleoni* (Acari: Phytoseiidae). In: Acarid phylogeny and evolution: adaptation in mites & ticks, pp 319–323

84. Metwally AM, Abou-Awad BA, Al-Azzazy MMA (2005) Life table and prey consumption of the predatory mite *Neoseiulus cydnodactylon* Shehata and Zaher (Acari: Phytoseiidae) with three mite species as prey. J Plant Dis Prot 112:276–286

85. Abou El-Elela MM, Abou-Elella GM (2001) Laboratory studies on development and oviposition of *Neoseiulus cucumeris* (Oudemans) (Acari: Phytoseiidae) fed on various preys. Egypt J Biol Pest Control 11:115–118

86. Momen FM, Abdel-Khalek A, El-Sawi S (2009) Life tables of the predatory mite *Typhlodromus negevi* feeding on prey insect species and pollen diet (Acari: Phytoseiidae). Acta Phytopathol Entomol Hung 44:353–361

87. Momen FM, Abdel-Khalek A (2008) Influence of diet on biology and life-table parameters of the predacious mite *Euseius scutalis* (A.H.) (Acari: Phytoseiidae). Arch Phytopathol Plant Prot 41:418–430

88. Momen FM (2001) Biology of *Euseius yousefi* (Acari: Phytoseiidae) life tables and feeding behaviour on different diets. Acta Phytopathol Entomol Hung 36:411–417

89. Momen FM (2004) Suitability of the pollen grains, *Ricinus communis* and *Helianthus annuus* as food for six species of phytoseiid mites (Acari: Phytoseiidae). Acta Phytopathol Entomol Hung 39:415–422

90. Midthassel A, Leather SR, Baxter IH (2013) Life table parameters and capture success ratio studies of *Typhlodromips swirskii* (Acari: Phytoseiidae) to the factitious prey *Suidasia medanensis* (Acari: Suidasidae). Exp Appl Acarol 11:69–78

91. Nguyen DT, Vangansbeke D, Lü X, De Clercq P (2013) Development and reproduction of the predatory mite *Amblyseius swirskii* on artificial diets. Biol Control 58:369–377

92. Nguyen DT, Vangansbeke D, De Clercq P (2014) Artificial and factitious foods support the development and reproduction of the predatory mite *Amblyseius swirskii*. Exp Appl Acarol 2:181–194

93. Momen FM, El-Laithy AY (2007) Suitability of the flour moth *Ephestia kuehniella* (Lepidoptera: Pyralidae) for three predatory phytoseiid mites (Acari: Phytoseiidae) in Egypt. Int J Trop Insect Sci 27:102–107

94. Nar BR, Goleva I, Zebitz CPW (2014) Life tables of *Neoseiulus cucumeris* exclusively fed with seven different pollens. BioControl 59:195–203

95. Ji J, Zhang Y, Wang J, Lin J, Sun L, Chen X, Ito K, Saito Y (2015) Can the predatory mites *Amblyseius swirskii* and *Amblyseius eharai* reproduce by feeding solely upon conspecific or heterospecific eggs (Acari: Phytoseiidae)? Appl Entomol Zool 50:149–154

96. Nguyen DT, Vangansbeke D, De Clercq P (2015) Performance of four species phytoseiid mites on artificial and natural diets. Biol Control 80:56–62

97. De Albuquerque FA, de Moraes GJ (2008) Perspectives for mass rearing of *Iphiseiodes zuluagai* Denmark & Muma (Acari: Phytoseiidae). Neotrop Entomol 3:328–333

98. Domingos CA, Melo JWS, Gondim MGC, de Moraes GJ, Rachid HLM, Peter S (2010) Diet-dependent life history, feeding preference and thermal requirements of the predatory mite *Neoseiulus baraki* (Acari: Phytoseiidae). Exp Appl Acarol 50:201–215

99. Xia B, Zou Z, Li P, Lin P (2012) Effect of temperature on development and reproduction of *Neoseiulus barkeri* fed on *Aleuroglyphus ovatus*. Exp Appl Acarol 56:33–41

100. Gerson U, Smiley RL, Ochoa R (2003) Mites (Acari) for pest control. Blackwell Science, Oxford, p 539

101. Abou-Awad BA, EL Sawi SA (1993) Biology and life table of the predacious mite, *Agistemus exsertus*. Anz Schädlingsk Pflanzen Umweltschutz 66:101–103

102. Momen FM (2001) Effects of diet on the biology and life tables of the predacious mite *Agistemus exsertus* (Acari: Stigmaeidae). Acta Phytopathol Entomol Hung 36:173–178

103. Momen FM, El- Sawi SA (2006) *Agistemus exsertus* (Phytoseiidae) predation on insects: life history and feeding habits on three different insect eggs. Acarologia 47:211–217

104. Al-Shammery KA (2011) Plant pollen as an alternative food source for rearing *Euseius scutalis* (Acari: Phytoseiidae) in Hail, Saudi Arabia. J Entomol 8:365–374

105. Momen FM (2011) Natural and factitious prey for rearing the predacious mite *Agistemus exsertus* Gonzales (Acari: Stigmaeidae). Acta Phytopathol Entomol Hung 46:267–275

106. Salwa MES (2012) A new diet for reproduction of the predacous mite *Agistemus exsertus* Gonzalez (Acari: Stigmaeidae). J Appl Sci Res 8:2321–2324

107. Khodayari S, Kamali K, Fathipour Y (2008) Biology, life table and predation of *Zetzellia mali* (Stigmaeidae) on *Tetranychus urticae* (Acari: Tetranychidae). Acarina 16:191–196

108. Abou-Awad BA, Hassan MF, Romeih AH (2010) Biology of *Agistemus olivi*, a new predator of eriophid mites infesting olive trees in Egypt. Arch Phytopathol Plant Prot 43:1–8

109. Mohamed OMO (2014) Biological aspects of the predaceous mite, *Agistemus vulgaris* Soliman and Gomaa and life table parameters on three host phytophagous mite species, (Acari: stigmaeidae). Egypt Acad J Biol Sci 7:165–171

110. Norton RA, Kethley JB, Johnston DE, O'Connor BM (1993) Phylogenetic perspectives on genetic systems and reproductive modes of mites. In: Wrensch DL, Ebbert MA (eds) Evolution and diversity of sex ratio of insects and mites. Chapman & Hall Publications, New York, pp 8–99

111. Berndt O, Meyhöfer R, Poehling HM (2004) The edaphic phase in the ontogenesis of *Franklin-iella occidentalis* and comparison of *Hypoaspis miles* and *Hypoaspis aculeifer* as predators of soil dwelling thrips stages. Biol Control 30:17–24

112. Wiethoff W, Poehling H, Meyhöfer R (2004) Combining plant- and soil-dwelling predatory mites to optimize biological control of thrips. Exp Appl Acarol 34:239–261

113. Ahmed WGO (1992) Studies on certain predaceous mites species in Sharkia and Giza Governorates. M.Sc. thesis, Faculty of Agriculture, Zagazig University, 157 pp

114. Ezz El-Dein SA (2003) Studies on some soil predacious mite associated with some field crop. M.Sc. thesis, Faculty of Science, Al-Azhar University (Girls), 148 pp

115. Al-Rehiayani SM, Fouly AH (2005) *Cosmolaelaps simplex* (Berlese), a polyphagous predatory mite feeding on root-knot nematode *Meloidogyne javanica* and citrus nematode *Tylenchulus semipenetrans*. Pak J Biol Sci 1:168–174

116. Fouly AH, Abdel-Baky NF (2015) Influence of prey types on the biological characteristics of *Cosmolaelaps qassimensis* (Acari: Laelapidae). J Entomol 1:21–29

117. Binns ES (1972) *Arctoseius cetratus* (Sellnick) (Acarina: Ascidae) phoretic on mushroom sciarid flies. Acarologia 14:350–356

118. Beaulieu F, Weeks AR (2007) Free-living mesostigmatic mites in Australia: their roles in biological control and bioindication. Aust J Entomol Agric 47:460–478

119. Walter DE, Lindquist EE (1989) Life history and behavior of mites in the genus *Lasioseius* (Acari: Mesostigmata: Ascidae) from grassland soils in Colorado, with taxonomic notes and description of a new species. Can J Zool 67:2797–2813

120. Afifi AM, Hassan MF, Nawar MS (1986) Notes on the biology feeding habits of *Protogamasellus minutus* Hafez, El-Badry & Nasr. Bull Soc Entomol Egypt 66:251–259

121. Afifi AM (1977) Studies on some soil predacious mites. M.Sc. thesis, Faculty of Agriculture, Cairo University, 104 pp

122. El-Bishlawy SMO (1978) Ecological and biological studies on mites associated with weeds, with special reference to lawn grasses. Ph.D. dissertation, Faculty of Agriculture, Cairo University, 148 pp

123. Ahmed MA (1984) Biology of some soil fauna, feeding on soil microorganism. M.Sc. thesis, Faculty of Agriculture, Zagazig University, 109 pp

124. Nawar MS, Nasr AK (1988) Biology of the ascid mite *Protogamasellus primitivus similis* Genis, Loots & Ryke with description of immature stages (Ascidae). Bull Soc Entomol Egypt 68:85–94

125. Nasr AK, Nawar MS, Mowafi MH (1990) Biological studies and feeding habits of *Lasioseius athiasae* Nawar & Nasr (Mesostigmata: Ascidae) in Egypt. Bull Soc Entomol Egypt 39:75–88

126. Nawar MS, Rakha MA, Ali FS (1990) Laboratory studies on the predaceous mite, *Lasioseius bispinosus* Evans (Acari: Mesostigmata: Ascidae) on various kinds of food substances. Bull Soc Entomol Egypt 69:247–255

127. Nawar MS, El-Sherif AA (1992) Biological studies and description of developmental stages of *Lasioseius zaheri* Nasr (Acari: Ascidae). Ann Agric Sci Moshtohor 30:581–589

128. Mowafi MA (1993) Studies on some important economic predacious mites in Egypt. Ph.D. dissertation, Faculty of Agriculture, Al-Azhar University, 130 pp

129. Walter DE, Hunt HW, Elliott ET (1988) Guilds or functional groups? An analysis of predatory arthropods from a shortgrass steppe soil. Pedobiologia 31:247–260

130. Castilho RC, de Moraes GJ, Silva ES, Silva LO (2009) Predation potential and biology of *Protogamasellopsis posnaniensis* Wiśniewski and Hirschmann. Biol Control 48:164–167

131. Walter DE, Oliver JH (1989) *Geolaelaps oreithyiae*, n. sp. (Acari: Laelapidae), a thelytokous predator of arthropods and nematodes, and a discussion of clonal reproduction in the Mesostigmata. Acarologia 30:291–303

132. Mahmoud R (2019) Taxonomy and behavior of some predacious gamasid mites used in biological control. Ph.D. dissertation, Cairo University, Faculty of Agriculture

133. Lawson-Balagbo LM, Gondim MGC Jr, de Moraes GJ, Hanna R, Schausberger P (2007) Life history of the predatory mites *Neoseiulus paspalivorus* and *Proctolaelaps bickleyi*; candidates for biological control of *Aceria guerreronis*. Exp Appl Acarol 43:49–61

134. Galvão AS, Gondim MGC Jr, de Moraes GJ (2011) Life history of *Proctolaelaps bulbosus* feeding on the coconut mite *Aceria guerreronis* and other possible food types occurring on coconut fruits. Exp Appl Acarol 53:245–252

135. Navasero MM, Hirao GA, Santiago DR, Navasero MV, Raros LC (2004) Laboratory rearing technique for the predatory *Proctolaelaps yinchuanensis* Xue, Sui & Yi (Ascidae: Gamasida: Acari) using *Suidasia pontifica* Oudemans (Suidasiidae: Acaridae: Acari). Philip Entomol 18:180

136. Gomaa WO (1998) Biological studies on some species of mesostigmatic mites with special reference to their chemical analysis together with preys. Ph.D. dissertation, Faculty of Agriculture, Cairo University, 299 pp

137. Abou-Elela MM (1999) Biological studies on some predacious mites associated with fruit trees and its debris. Ph.D. dissertation, Faculty of Agriculture, Cairo University, 185 pp

138. Mahmoud AM (1999) Ecological studies on certain soil fauna in Dakahlia governorate. M.Sc. thesis, Faculty of Agriculture, Al-Azhar University, 86 pp

139. Nasr AK, Nawar MS, Mowafi MH (1990) Biological studies on *Proctolaelaps bickleyi* Bram (Acari: Gamasida: Ascidae). Bull Soc Entomol Egypt 39:89–100

140. Bjornson S (2008) Natural enemies of mass-reared predatory mites (Family: Phytoseiidae) used for biological pest control. J Exp Appl Acarol 46:299–306

141. Zhang QZ (2003) Mites of greenhouses identification, biology and control. CABI, Oxon, UK

142. Gilkeson LA (1992) Mass rearing of phytoseiid mites for testing and commercial application. In: Andersonand TE, Leppla NC (eds) Advances in insect rearing for research and pest management. West View Press, Boulder, CO, USA, pp 489–506

143. Overmeer WPJ (1985) Rearing and handling. In: Helle W, Sabelis MW (eds) Spider mites: their biology, natural enemies, and control. Elsevier, Amsterdam, The Netherlands, pp 161–170

144. Morales-Ramos JA, Rojas MG, Cahn D (2012) System and methods for production of predatory mites. Patent No. US 8,327,797 B1

145. Morales-Ramos JA, Rojas MG (2014) A modular cage system design for continuous medium to large scale in vivo rearing of predatory mites (Acari: Phytoseiidae). Psyche 2014:1–8

146. Ramakers PMJ, Van Lieburg MJ (1982) Start of commercial production and introduction of *Amblyseius mekenziei* Sch. & Pr. (Acarina: Phytoseiidae) for the control of *Thrips tabaci* in glasshouses. Med Fac Landbow Rijksuniv Gent 47:541–545

147. Schliesske J (1981) On the technique for mass rearing of predatory mites under controlled conditions. Med Fac Landbow Rijksuniv Gent 46:511–517

148. Bolckmans KJF, van Houten YM (2006) Mite composition, use thereof, method for rearing the phytoseiid predatory mite *Amblyseius swirskii*, rearing system for rearing said phytoseiid mite and methods for biological pest control on a crop. WO patent WO/2006/057552

149. Barbosa MFC, de Moraes GJ (2015) Evaluation of astigmatid mites as factitious food for rearing four predaceous phytoseiid mites (Phytoseiidae). Biol Control 91:22–26

150. Massaro M, Martin JP, de Moraes GJ (2016) Factitious food for mass production of predaceous phytoseiid mites commonly found in Brazil. Exp Appl Acarol 70:411–420

151. Cavalcante ACC, Santos VLV, Rossi LC, de Moraes GJ (2015) Potential of five Brazilian populations of Phytoseiidae (Acari) for the biological control of *Bemisia tabaci* (Insecta: Hemiptera). J Econ Entomol 108:29–33

152. De Clercq P, Bonte M, Van Speybroeck K, Bolckmans K, Deforce K (2005) Development and reproduction of *Adalia bipunctata* (Coleoptera: Coccinellidae) on eggs of *Ephestia kuehniella* (Lepidoptera: Phycitidae) and pollen. Pest Manag Sci 61:1129–1132

153. Navarro-Campos C, Wackers FL, Pekas A (2016) Impact of factitious foods and prey on the oviposition of the predatory mites *Gaeolaelaps aculeifer* and *Stratiolaelaps scimitus* (Acari: Laelapidae). Exp Appl Acarol 70:69–78

154. Momen FM, Nasr AK, Metwally AM, Mahmoud YA, Saleh KM (2016) Performance of five species of phytoseiid mites (Acari: Phytoseiidae) on *Bactrocera zonata* eggs (Tephritidae) as a factitious food. Acta Phytopathol Entomol Hung 51:123–132

155. EI-Saiedy EMA (2003) Integrated control of red spider mite *Tetranychus urticae* Koch on strawberry plants. Ph.D. dissertation, Faculty of Agriculture, Cairo University, Egypt, 170 pp

156. Hassan MF, Ali FS, Hussein AM, Mahgob MH (2007) Control measures of *Tetranychs urticae* Koch on two cucumber cultivars in plastic houses. Acarines J Egypt Soc Acarol 1:11–15

157. El-Kholy MY, El-Saiedy EMAK (2009) Biological control of *Thrips tabaci* (lind.) and *Aphis gossypii* (Glover) using different predatory Phytoseiid mites and the biocide vertimec on eggplant at Behaira governorate. Egypt Acad J Biol Sci 2:13–22

158. Hassan AS (2013) Biological and chemical control of two spotted spider mite and important insects infesting sweet pepper in green houses in Egypt. Ph.D. dissertation, Faculty of Agriculture, Cairo University, Egypt

159. Abou-Awad BA, Afia SI, El-Saiedy E (2017) Population dynamics of Tetranychid mite and its predator on watermelon and muskmelon and effect of mite feeding on the phytochemical components of the host plants. Biosci Res 14:879–886

160. Kame MS, Afia SI, El Saiedy E (2018) Biological control of *Tetranychus urticae* (Acari: Tetranychidae) using four predatory mites (Acari: Phytoseiidae) on two sweet pea cultivars. Biosci Res 15:185–191

161. Cohen AC (2004) Insect diets: science and technology. CRC Press, Boca Raton

162. Hansen LS (1988) Control of *Thrips tabaci* (Thysanoptera: Thripidae) on glasshouse cucumber using large introductions of predatory mites *Amblyseius barkeri* (Acarina: Phytoseiidae). Entomophaga 33:33–42

163. Ling P, Xia B, Li PX (2008) Functional response of *Amblyseius barkeri* (Acarina: Phytoseiidae) on *Panonychus citri* (Acari: Tetranychidae). Acta Arachnol Sin 17:29–34

164. Shu C, Zhong L, Li AH (2007) A preliminary report on the effect of control *Panonychus citri* by releasing *Amblyseius barkeri*. China Plant Prot 27:23–24

165. Nomikou M, Janssen A, Schraag R, Sabelis MW (2002) Phytoseiid predators suppress populations of *Bemisia tabaci* on cucumber plants with alternative food. Exp Appl Acarol 27:57–68

166. Messelink GJ, van Steenpaal SE, Ramakers PM (2006) Evaluation of phytoseiid predators for control of western flower thrips on greenhouse cucumber. BioControl 51:753–768

167. Khanamani M, Fathipour Y, Talebi AA, Mehrabadi M (2017) Quantitative analysis of long-term mass rearing of *Neoseiulus californicus* (Acari: Phytoseiidae) on almond pollen. J Econ Entomol 110:1442–1450

Microorganisms for Controlling Insect Pests

Production and Application of *Bacillus thuringiensis* for Pest Control in Egypt

Hussein S. Salama

Abstract The leading role in *Bacillus thuringiensis B.t.* research in Egypt has been taken by the National Research Centre (NRC). The present book chapter is an account of studies that have been carried out on this pathogen in Egypt and its possible role as a biological control agent. These studies included bioassay of various cultures of *B.t.* to detect the potent strains versus different insect species. The introduction of agro-industrial byproducts in the fermentation media have been explored for economic commercial production of the pathogen. Attempts were made to develop feeding stimulants and bait formulations aiming to overcome short environmental persistence. Novel approaches including the use of chemical additives with *B.t.* formulations were adopted to enhance potency against insects through biochemical reactions that occurred in the insect midgut. Reference was given to the mode of action of *B.t.* and its distribution. The joint action of *B.t.* varieties and its integration with other microbial and chemical control agents was highlighted. Investigations on the pathogen effect on various development stages of Lepidoptera are given. Pilot scale production of *B.t.* has been described. Studies dealing with the large scale field application of *B.t.* are given. Reference has been given to the current status of genetically modified technology (GM) in Egypt.

Keywords Fermentation media · Agro industrial byproducts · Adjuvants · Chemical additives · Feeding stimulants · Pilot production · GM technology

1 Historical Overview

At present *Bacillus thuringiensis* (*B.t.*) accounts for perhaps 80–90% of the total world microbial insecticide market. *B.t.* appeared to be a simple, spore forming infectious bacterium, easy to grow on laboratory media. The discovery of *B.t.* as reported by Frankenhuyzen [1] dated back to 1901 and then, it was isolated by Berliner [2, 3] from diseased larvae of the flour moth *Anagasta kuehniella* in Germany and he named it *B.t.* During the next two decades field testing of *B.t.* continued

H. S. Salama (✉)
Pests and Plant Protection Department, National Research Centre, Dokki, Cairo, Egypt
e-mail: hsarsalama@hotmail.com

© Springer Nature Switzerland AG 2020
N. El-Wakeil et al. (eds.), *Cottage Industry of Biocontrol Agents and Their Applications*, https://doi.org/10.1007/978-3-030-33161-0_6

against Lepidopterous insects in Europe and U.S.A. Weiser [4] reported Sporeine as the first product used in 1938 in France and its production continued during 1950s in USSR, Chechoslovakia, France and Germany. Steinhaus [5] encouraged the commercialization of B.t. leading to the production of Thuricide in 1957.

The discovery of the *kurstaki* isolate HD-1 by Dulmage [6] showed its high activity against some agricultural pests more than other isolates and therefore it was adopted for commercial production. During this time, an international system for standardizing the potency of commercial products based on biological units was established to substitute the reliance on spore counts which did not relate to the total insecticidal activity [7–10]. Becker and Margalit [11] referred to the discovery of B.t. subspecies *israelensis*; and similar results were also recorded by Goldberg and Margalit [12]. Margalit and Dean [13] opened a new chapter in the biological control of mosquito and blackfly larvae and it was characterized as serotype H-14. Following that, a hundred local strains of this serotype have been isolated in many countries, but no isolate has yet been found that has better mosquitocidal effect than the original B.t. *israelensis* isolate. However, mention should be made of B.t. subspecies *morrisoni* (serotype H 8a:8b) that was isolated in Philippines by Padua et al. [14]. Although it belongs to a different subspecies and serotype, but it proved to be as effective as B.t. *israelensis*.

Again, Keller and Langenbruch [15] reported that a novel B.t. strain (BI 256-82) was isolated and identified by Huger and Krieg [16] from a dead pupa of the yellow mealworm *Tenebrio molitor* which showed activity against Coleoptera. It was described as B.t. subspecies *tenebrionis* and represented a new B.t. pathotype [17]. In the following years, further strains with coleopteran activity appeared, but still *tenebrionis* is the most important commercial subspecies used in the control of coleopteran pests and four companies have placed its products on the market (M-one, Trident, Ditera and Novodor).

In Egypt, the leading role in research and development of B.t. has been taken by the National Research Centre (NRC). The efforts of the NRC team led to the establishment of scientific cooperation with the USDA (1977–1985) and Agriculture Canada (1985 onwards) to develop research in this area and to combat the key lepidopteran pests of field, oilseed and vegetable crops. The present chapter covers the expanding knowledge of the production, mode of action and application of this pathogen in Egypt. Emphasis has been given to Lepidoptera, the most destructive group of plant pests.

2 Mode of Action Research

When the insect ingests B.t. crystal containing either class Cry I or Cry II proteins, many steps lead to the death of the insect larvae. Firstly, the crystal dissolves in the alkaline midgut of the insect generating 130–140 kDa Cry I proteins or 71 kDa Cry II proteins.

Recently Höfte and Whitely [18] referred to a system of classification of the δ-endotoxins in relation to their insecticidal properties and molecular relationships. They described four major classes of δ-endotoxins (Cry I, II, III, IV) and a cytolysin (Cyt) found in the crystals of the mosquito active strains. The different δ-endotoxins belonging to each of the four Cry classes contain a single family of proteins [19, 20]. These proteins are active against Lepidoptera (Cry I), Lepidoptera and Diptera (Cry II), Coleoptera (Cry III) and Diptera (Cry IV).

In addition to this large variety of crystal proteins, some strains synthesize heat stable toxins designated B-exotoxins [21]. These exotoxins presumably contribute to the overall toxicity of a *B.t.* strain. The mode of action of *B.t.* involves biochemical changes in the haemolymph of the treated insect. Salama et al. [22] and Salama [23] reported that the level of proteins in *S. littoralis* markedly decreased during larval treatment with *B.t.* Other quantitative changes in the amino acids were detected. It is known that amino acids play an important role in the osmo-regulation of insects and abnormal changes in their concentration may lead to fatal consequences. In addition, a disruption in the regulation of potassium, sodium, magnesium and zinc cues occurred leading to general paralysis.

Also, Salama [23] referred to investigation of Boctor and Salama [24] who "determined the differences between lipids in larvae of *S. littorals* as compared to those treated with *B.t.*" Salama and Sharaby [25] investigated the histopathological changes caused by *B.t.* var. *entomocidus* HD-635 in the cotton bollworm, *Heliothis armigera*, using electron microscopy. Most of the changes that occurred on the fourth day after treatment with *B.t.* were mainly localized in the midgut, where the epithelium was greatly affected losing its integrity; the peritrophic membrane and microvilli were degenerated; and the musculosa was also affected. This coincides with the observations of Sutter and Raun [26] and Reese et al. [27] with other insect species. Other associated effects were observed in the integument, nerve ganglion, fat body cells, tracheoles and Malpighian tubules. In the integument, the exo- and endocuticles were clumped with an obvious separation from each other.

An obvious degeneration of the nerve cells surrounding the second abdominal nerve ganglion as well as the neurilemma of the nerve fibres occurred. Vacuolization of the fat body cells, degeneration of their nuclei and destruction of the membranous sheath surrounding these cells occurred. Tracheoles showed excessive cellular hypertrophy with disintegration of its mitochondria. The Malpighian tubules showed a reduction in their lumen, with nuclei degeneration and nuclear chromatin clumping. Uric acid crystals were released in the lumen of the tubules and a rupture was observed in some parts of the microvilli. A rapid phagocytosis occurred in the haemocytes and this agree with Cheung et al. [28] who reported that the plasmatocytes and granular haemocytes phagocytosed the bacteria. The counts of haemocytes were at a minimum and bacterial numbers at a maximum on the fourth day after feeding the larvae on *B.t.* contaminated diet".

3 Fermentation Media and Production of *B.t.* in Egypt

3.1 Fermentation Technologies

For the purpose of producing *B.t.* in Egypt, studies have been designed and continued for many years to find low priced fermentation ingredients universally available and of high potential for δ-endotoxin production. Achievements by Salama et al. [29–31] and Salama and Morris [32] open up the possibility of utilization of cheap agro-industrial byproducts for production of endotoxin from *B.t.* [23].

So, "novel approaches were pursued in the development of practical media for δ-endotoxin production of *B.t.* subspp. *kurstaki* and *entomocidus*. Several agro-industrial byproducts, including cottonseed meal, fish meal, beef blood, slaughterhouse residues, corn steep liquor, and sorter liquor, were investigated for their abilities to support toxin production by these varieties [29]". "In addition, fodder yeast and a variety of low-priced plant proteins available in Egypt, exemplified by such leguminous seeds as horse beans, kidney beans, lima beans, soybeans, chick peas, lentils, and peanuts were incorporated in fermentation media as sole sources of proteins for biosynthesis of the endotoxins. The cotton pests, *S. littoralis*, *S. exigua*, and *H armigera*, were used as test insects for biological assays of spore-δ-endotoxin formulations derived from the novel fermentation media. Fodder yeast, beef blood, and slaughterhouse residues were among the byproducts yielding good sporulation titers and potent spore-δ-endotoxin preparations. Formulations of subsp. *kurstaki* produced from media containing these nutrients killed 80–100% of larvae of *H. armigera* when tested at 500 μg/mL diet. Most of the formulations derived from fermentations using leguminous seeds as sole sources of protein also contained high levels of spores and endotoxin. For example, LC50 values determined against *S. littoralis* of spore-endotoxin preparations of subsp. *entomocidus* grown in media containing kidney beans, chick peas, or peanuts were 93.4, 93.4, and 110.0 μg/mL diet, respectively".

"Accordingly these findings open up the way for utilization of a variety of agro-industrial byproducts for the production of endotoxins from *B.t.* Furthermore, the introduction of leguminous seeds as the major protein source in the fermentation media have given promising results with potential application in commercial media [29].

In a parallel line, "Salama et al. [30] devised a simplified approach to recycle sweet whey in production of spore-δ-endotoxin complex from certain entomopathogenic varieties of *B.t.* The process suggested aimed at the protection of the environment through dual channels namely biological oxygen demand (BOD) reduction of the byproduct under investigation and its incorporation in a microbial fermentation for production of pollution-free biological insecticides. The sweet whey could be used successfully for endotoxin production as complete fermentation media both as such and with simple treatments. In Egypt more than 500,000 ton of whey are produced annually as a byproduct of cheese industry from buffalo milk. This includes sweet whey and a larger proportion of salted whey that contains more than 10% sodium chloride resulting from manufacturing types of cheese. Up till now all these amounts of the liquor virtually are run to waste. Successful utilization of whey in production of δ-endotoxins from *B.t.* highly active against some major cotton pests, namely *S. littoralis, S. exigua* and *H. armigera* was achieved. Such approaches aimed at increasing the availability and reducing production costs of biological means of pest control, therefore, contributing to protection of environment against pollution with chemical insecticides".

Furthermore, Salama et al. [31] "evaluated a number of newly-devised fermentation media with respect to their ability to support sporulation and biosynthesis of endotoxins by strains of *B.t.* that are biologically active against *S. littoralis, H. armigera*, and *S. exigua*. Fodder yeast from dried cells of *Saccharomyces cerevisiae* could be used as a complete mono-component medium for production of highly active spore-δ-endotoxin complexes from *B.t.*, vars. *entomocidus, kurstaki* and *galleriae*. Highest sporulation titers were obtained at 2% fodder yeast concentration with endotoxin yields ranging between 7 and 9 g/L of medium. Ground horse beans and kidney bean seeds could also be used successfully as complete media for sporulation and endotoxin production. Extracts of potato tubers and sweet potato roots were efficient media for active endotoxin production from *B.t.* var. *kurstaki*, although they obtained yields more much lower than those produced in fodder yeast media. The utilization of fish meal, cotton seed meal, and residues of chicken from the slaughter-house residues as media for the production of endotoxins active against *S. littoralis*, was not successful. On the other hand, minced citrus peels, ground seeds of dates, and wheat bran could be successfully used in combination with fodder yeast as media for production of endotoxins, active against *H. armigera* and *S. exigua*. Re-utilization of culture supernatants in a second fermentation cycle after supplementation with some nutrients gave promising results with some of the strains tested".

"This direction emphasises on developing new fermentation media, basing on cheap ingredients that could substitute expensive components, thus contributing to reducing the production costs of *B.t.* In this concern Dulmage [6] reported the successful utilization of partially defatted cotton seed flour in production of spore-endotoxin preparation from twelve varieties of *B.t.* Later the same author developed a novel medium for endotoxin production, based on the use of defatted soybean meal as the sole source of protein [33]. Nagamma et al. [34] reported a new solid medium for *B.t.* fermentation, containing ground nut cake, tamarind kernel powder, and agar. Their medium supported the fermentation of high sporulation yields that was pathogenic to larvae of *Cadra cautella*. On the other hand, unsuccessful attempts to use coconut water, a byproduct resulting from coconut dehydration factories, were reported by Fernandez et al. [35]. The results obtained in our work could successfully introduce the use of low-price fodder yeast as a mono-component medium for production of highly active spore-δ-endotoxin complexes against a number of major cotton insects. Furthermore, in combination with a variety of agricultural byproducts, the fodder yeast medium could support the biosynthesis of large yields of such active endotoxins from various strains of *B.t.*"

4 Pilot Scale Production of *B.t.*

In Egypt, where applied *B.t.* research has been going for many years and where several insect pests of field crops, oil seed crops and vegetables have become endemic, it was necessary to take the decision of trying to establish a pilot production line of *B.t.* formulations based upon the locally available media ingredients and with the assistance of a local industrial company. So, the pilot scale production of the promising potent *B.t.* strains was carried out in a mobile fermentation unit located in the Sugar and Distillery Company in Hawamdia, Giza [36]. The unit structure is shown in (Fig. 1).

The Sugar and Distillery Company is known to produce fodder yeast and Molasses that could be used as cheap local byproducts for making of the fermentation media. In the course of *B.t.* fermentation, various choices were evaluated with respect to the components of the fermentation medium. The potency of the endotoxin preparations obtained using different choices was evaluated. The results indicate that the potency of the local *B.t.* product is comparable to that of Dipel 2× particularly on using modified-fodder yeast molasses medium as were used by Foda et al. [36] as shown in Table 1.

This was repeatedly employed and gave consistently good results.

4.1 *Sequence of Fermentation Steps*

The fermentation steps began with preparation of the inoculum. Sterile tubes containing *B.t.* culture dried and mounted on sterile filter paper were used as inoculums source. They were used to inoculate 50 mL of the inoculation medium placed in 500 mL conical flasks which were then incubated on a rotary shaker 150 rpm at a 28 °C for two days. The contents of the flasks were used as seed cultures for 5 L flasks, each containing one liter of the same medium. The flasks were incubated on the shaker for 48 h.

4.2 *Seeding of the Paddle Wheel Fermenter of the Mobile Units and the Fermentation Conditions*

The sterile fermentation medium (1000 L) placed in the paddle wheel fermenter were inoculated with the actively growing culture of the inoculum at an inoculum size ranging between 0.5 and 1.0% (v/v).

Strong aeration at a rate of about 60 m³/h was supplied to the fermenter with continuous agitation. The fermentation process was allowed to proceed at an adjusted temperature 30 °C for three days on average basis. In some instances, the fermentation period was extended for extra 24 h. Samples were withdrawn periodically from

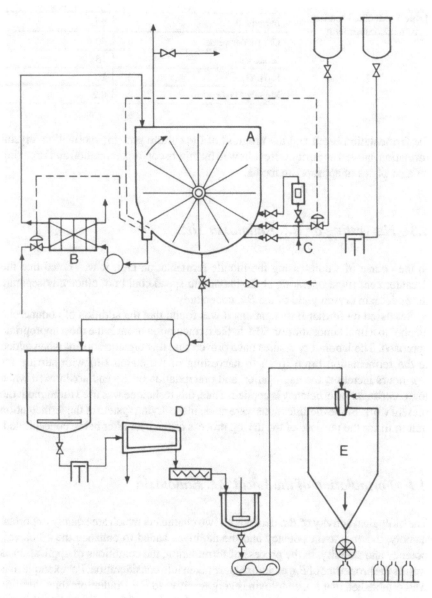

Fig. 1 Diagrammatic sketch of the fermentation unit for *B.t.* production in the Sugar and Distillery Company, Hawamdia, Egypt. A = paddle-wheel fermenter; B = fedder water chiller; C = compressed air set; D = decanter centrifuge; E = spray dryer

Table 1 Modified-fodder
yeast molasses medium

Ingredient	g/L
Dry fodder yeast	40
Molasses	15
K_2HPO_4	1
$MgSO_4 \, 7H_2O$	0.2

the fermentation broth and the progress of the culture growth; sporulation, crystal formation, as well as purity were followed by microscopic examination and streaking on agar plates of appropriate media.

4.3 Harvesting of B.t. Endotoxin Yield

In the course of studies using the mobile fermentation unit, it was noted that the decanter centrifuge operating at the maximum speed could not efficiently separate the endotoxin crystal yield of the *B.t.* understudy.

So, based on further investigations it was found that the addition of commercial alcohol to a final concentration 50% in the fermentation broth is the most appropriate approach. The laboratory studies have proven that this organic solvent when added to the fermentation broth (prior to harvesting of the endotoxin) with stirring for few hours increases the aggregation and precipitation of the endotoxin even when low centrifugation efficiency is applied. Thus, this technique was used throughout the present work. Some modifications were made in the drying system of the fermentation unit to fit for the purpose of the drying process of the beer after being concentrated.

4.4 Formulation of the Local B.t. Endotoxin

The biological activity of *B.t.* depends on two characters which are equally important namely, the endotoxin potency and the additives added to enhance the efficiency, potency and stability. In the process of formulation, the conditions of application as well as the properties of *B.t.* endotoxins are taken into consideration. For example, it is well established that *B.t.* endotoxin is very sensitive to UV-irradiation; thus, addition of UV-protectant is a significant component of the formulation. Regarding diluents and carriers, local mixture of minerals of small particle size may be selected. In some formulations, molasses in the wettable powders are included. However, this adjuvant is recommended to be tank-mix. So, the *B.t.* can be prepared as water-dispersible powder taking into consideration the following factors:

1. Flowability, wettability, dispersibility and suspensibility to promote stable suspensions with a consequent over plant distribution.

2. Addition of UV-protectant.
3. Carriers and adjuvants free of moisture.
4. Deactivation of the acidic active sites on the carrier to a pka higher than 3.3, to ensure shelf-life stability.
5. Since foaming is undesirable, adjuvants with antifoaming activity were included.

4.5 Agerin (B.t. Product) Production from Genetic Engineering Research Institute (AGERI)

Few years later, AGERI produced the biopesticide Agerin. Research efforts and collaboration of scientists from AGERI and the University of Wyoming led to the development of a biological pesticide derived from active strain of *B.t.* from Egypt. This strain showed high activity against various insects (Lepidoptera, Coleoptera and Diptera). AGERI collaborated with a private sector to establish "BIOAGRO" International. This company was responsible for commercialization of research results in AGERI [37].

Agerin is a *B.t.* subspecies *aegyptia*. It is a wettable powder 6% containing 32×10^6 IU/mg of *B.t. aegyptia*. It was introduced by Bioagro International—Egypt. It was used in Egypt by permission from AGERI-Agricultural Research Centre, Ministry of Agriculture. The recommended rate of application is 500 g/feddan. At present, Agerin is widely used all over the country to control lepidopteran insects on cotton and other crops.

5 Persistence and Field Stability of *B.t.*

The low field persistence of *B.t.* is a major problem regulating its effective use for pest control. Salama et al. [38] studied the persistence of different formulations of *B.t.* spores after spray application in cotton cultivations in Egypt. They found an obvious reduction after one day of weathering. The decay in the spore's viability is progressively correlated with the time of exposure in the field. The spores half-life of the tested *B.t.* formulations ranged between 61 and 256 h and cannot be correlated with the temperatures attained on the surface of sunny cotton leaves. Ultraviolet radiation seemed to be the dominant factor affecting the viability of spores.

A measure of the viable spores of *B.t.* preparations at various intervals after spray application in cotton cultivations was carried out during two successive seasons. Salama and Zaki [39] determined the mortality of neonate larvae of *S. littoralis* in correlation with the field persistence and subsequent decay of spores of *B.t.* preparations on the cotton plant leaves. Various preparations of *B.t.* differ in their degree of protection from sunlight and their efficacy against *S. littoralis*. The decay in the

spore viability showed to be proportional to larval mortality, but this does not necessarily mean that insect intoxication is only correlated to the spore's viability. The
same authors reported that sunlight seems to be the main factor affecting the spore's
viability. They also mentioned that the use of adjuvants or phagostimulants such as
Coax, extracts of cotton leaves or Jews mallow increased the activity of *B.t.* versus
S. littoralis and they can compensate for subsequent spores decay through increased
ingestion but they did not act as protectants from sunlight inactivation.

In this concern the half-life of *B.t. entomocidus* was 109 hours (h) in May and
this decreased to 89 h in June, when the number of hours of sunlight reached a
maximum. In the last week of July, the spore's half-life decreased further to 61 h
at temperatures ranging between 33 and 45 °C on exposed cotton leaves. Relative
humidity fluctuation showed no obvious correlation with the spore viability.

The effect of PH and temperature on the activity of *B.t. kurstaki* was assessed
against *H. armigera* [40]. He found that storing *B.t.* formulation for 2–4 days at PH
values 3–13 caused a decrease in the potency against this species. The decrease in the
potency of *B.t.* exposed to 55° was more obvious at PH 11. Morris and Moore [41]
found that one day of direct sunlight in May can inactivate over 90% of Dipel spores
on potted white spruce and the trees themselves in the dark can inactivate 78% of the
spores in 14 days. So, UV radiation seems to be the main factor affecting the spore
viability. Hamed and Hassanein [42] also reported that after 6 days of exposure to
sunlight, *B.t.* treated cotton foliage killed 7% of *S. littoralis* compared with 27% on
treated cotton foliage held in the shade.

The short persistence of *B.t.* as caused by solar U.V. can be minimized by adding
substances with high degree of U.V. absorbance [43]. Evaluation of some protective
materials revealed that lingnosulphonate (orzan) was strongly protective particularly
when combined with molasses. Ragaei [40] reported that congo red, starch, encapsulated *B.t.*, charcoal, chitin, oxybenzone, peptonized milk 5%, Brewer's yeast, egg
albumen singly or combined with Brewer's yeast led to high protection against U.V.

5.1   Isolation of Mutants Resistant to Physical and Chemical Factors

In another approach and in order to prolong *B.t.* persistence in the field, attempts
were made by Salama et al. [44, 45] to isolate mutants that can be resistant to
physical factors, namely U.V. and high temperature resistant mutants responsible
for endotoxin decay. Preliminary mutagenesis studies were made using *B.t.* vars.
kurstaki HD-251 and *entomocidus* HD-635 to obtain autotrophic mutants as a means
for elucidation of possible relation between the endotoxin potencies versus some
lepidopterous insects and certain nutritional markers [44, 45].

Thus, wild type cultures of var. *entomocidus* HD-635 and var. *kurstaki* HD-251
were mutagenized with *N*-methyl-*N*-nitrosoguanidine and a number of auxotrophic
mutants were isolated. The auxotrophic mutants were compared with their wild type

strains with respect to their growth, sporulation titers and δ-endotoxin potencies in standard media. Although the auxotrophic mutants were partially characterized, and irrespective to their different genetic lesions. No specific correlations could be, yet, detected between mutation toward auxotrophicity in general and endotoxin production. In another series of mutational experiments, emphasis was placed on the isolation of mutants that are resistant to physical factors, namely UV-resistant and high temperature resistant mutants, that are prevalent in the field and which are possibly responsible for endotoxin decay. The ultimate goal of the obtainment of such types of mutants is to find out whether it is possible to extend the endotoxin longevity in the field through selection and application of such types of mutants resistant to physical factors. In addition, resistance to common antibiotics, e.g., penicillin G, streptomycin and chloramphenicol were selected to study the possible interrelationships between endotoxin potency, longevity and resistance markers. Several mutants were obtained in this respect and they were compared to their wild type strains.

6 Environmental Safety

Utilization of *B.t.* is safe as reported by Matter [46] and it has no in vivo or in vitro toxicity, histopathological changes or haemolytic activity on mammalian cells. Risks of cytogenetic damage to human by exotoxin in amounts normally used to control pests were considered negligible.

6.1 Effect of **B.t.** on Insect Parasitoids and Predators

The application of *B.t.* against insect pests in the field necessitated to explore its possible effect on the parasites and predators of these pests. It is expected that the combination of *B.t.* with parasites or predators is one of the most effective strategies in integrated pest management programs. So, investigations have been made by Salama et al. [47] to explore the relation between the host insect *S. littoralis*, the pathogen *B.t.* and parasites or predators. "The parasite *Microplitis demolitor* showed to be affected from certain aspects when its host larvae were fed on a diet containing *B.t.*, these aspects include reduction in percentage of emergence and reproductive potential. Also, the predator *Chrysopa carnea* was affected in terms of larval duration and rate of food consumption, when the larvae were fed on a treated host, such as *S. littoralis* or *Aphis durantae*. Similar effects were obtained with *Coccinella undecimpunctata*. These results indicate that *B.t.* may cause few biological effects to the parasites and predators exposed to it". Further studies by "Salama et al. [47] indicated that the adult predators *Paederus alferii* were not affected unlike the predator larvae of other species such as *C. carnea* and *Coccinella undecimpunctata*. Exposure of the predator adults to treated host larvae may give little chance for them to be affected by the changes that occurred in the host larvae".

Also, it appears that *B.t.* (Dipel) had no detrimental effect on the predator complex of *S. littoralis* under field conditions. In the case of *Orius* and *C. undecimpunctata* larvae and which feed on egg masses or aphids, no obvious effect was observed. The effects observed in the following counts however may be correlated to developmental retardation or low rate of prey consumption when sprayed with *B.t.* and as previously reported by Salama et al. [47] and Salama and Zaki [48, 49]. With the other predators *Coccinella* adults and *C. carnea* which feed mainly on eggs and larvae of *S. littoralis* and *Aphis gossypii*, the population was reduced immediately after spraying with *B.t.* This is related to the drop in the larval population of the main host prey (*S. littoralis* larvae) caused by *B.t.* and possibly due to low rate of consumption and retardation of development. Since the persistence of *B.t.* is short under the field conditions, a rebuild in the population of the predators that were indirectly affected may occur within a short period.

Salama and Zaki [50] found that feeding of the adult parasite of *Trichogramma evanescens* on a honey solution containing 500 mg of *B.t.* var. *galleriae* HD-129 for 4–5 days had no effect on their longevity, productivity or their capability to parasitize the host eggs of *S. littoralis* or *A. kuehniella*. After spraying the freshly laid host eggs with *B.t.* at the same concentration, a significant decrease was observed in the percentage of parasitism. No deleterious effects were observed in the development of the immature stages of the parasite and the percentage of its emergence, as affected by *B.t.*, when applied to the host eggs before or after parasitism.

7   Strain Survey and Potency Bioassay

Salama [23] reported differences in the susceptibility of different insect species to various strains of *B.t.* Bioassay of some *B.t.* preparatious versus *A. kuehniella*, *P. interpunctella* and *Achroia grisella* were made by Afify [51–53], Afify and Merdan [54], Afify et al. [55] to evaluate the response of *A. ypsilon*, *Laphygma exigua* and *Prodenia litura* to some *B.t.* preparations. Further screening tests were attempted by Afify et al. [56] and Soliman et al. [57] against some lepidopterans species.

Salama et al. [58] screened 29 cultures of *B.t.* belonging to 14 serotypes with respect to their activities against *S. littoralis*, *S. exigua*, *H. armigera*, *Pectinophera gossypiella*, *Earias insulana* and *A. ypsilon*. "In a further attempt, 17 varieties of *B.t.* obtained from Dulmage [33]; who were screened for their potency against *S. littoralis* and most of the endotoxin preparatious showed a low activity. Only *B.t.* var. *entomocidus* showed a high potential activity versus *S. littoralis* at a concentration of 500 µg/mL of the diet and the potency was 65.520 IU/mg for the first instars. The ability of this strain to produce potential to δ-endotoxin against *S. littoralis* was repeatedly shown on different media with different composition [59]. On the other hand, cultures belonging to varieties *kurstaki* and *aizawai* were highly effective against *H. armigera*. A high activity was reported for the vars. *finitimus*, *dendrolimus*, *subtoxicus*, *entomocidus* and *aizawai against P. gossypiella*. The larvae of *E. insulana* were susceptible to vars. *entomocidus*, *thompsoni*, *sotto*, *galleriae* and *aizawai*.

The larvae of *A. ypsilon* showed susceptibility to endotoxins of vars. *thuringiensis, tolworthi, finitimus, sotto* and *dendrolimus*. The LC$_{50}$ values were determined for different strains versus different insect species".

"In addition, the spore-δ-endotoxin preparations of other 25 cultures of *B.t.* belonging to 16 varieties and 13 serotypes were bioassayed with respect to their activities against the boll-worms *P. gossypiella, E. insulana* and *A. ypsilon* [60]. The endotoxin preparations were obtained from cultures aerobically grown in fodder yeast media using lactose-complexing acetone precipitation procedure. A high activity against *P. gossypiella* was recorded for some cultures belonging to vars. *finitimus, dendrolimus, subtoxicus, entomocidus and aizawai*. The larvae of *E. insulana* were highly susceptible to the vars. *entomocidus, thompsoni, sotto, galleriae* and *aizawai*. On the other hand, the larvae of *A. ypsilon* showed susceptibility to the endotoxins of some cultures of vars. *thuringiensis, tolworthi, finitimus, sotto* and *dendrolimus*. Data obtained on the potency of the active *B.t.* varieties indicate that the most potent varieties were *finitimus, entomocidus* and *tolworthi* for *P. gossypilla, E. insulana and A. ypsilon*, respectively".

7.1   Effect of Exposure to Sublethal Concentrations of B.t.

The larvae of insects are exposed in the field to sublethal concentrations of the endotoxin after application and this exposure may be continuos or intermittent. This is natural since the toxin application may not cover the whole plant.

The necessity to elucidate the possible interactions occurring between the endotoxin and the target insect under these circumstances received the attention of some workers in Egypt. So, Abul-Nasr and Abdallah [61], Abdallah and Abul-Nasr [62, 63] studied the effect of sublethal doses of some commercial preparations of *B.t.* on the biology and reproduction of *S. littoralis* and they recorded reduction in egg production and shortening in moth longevity with irregular oviposition. "The amount of food consumed by the sublethally infected larvae with *B.t.* was less than the untreated ones. Afify and Matter [64], Afify et al. [65] were working with *A. kuehniella* and Soliman et al. [57] working with *Pieris rapae*, reported similar effects of sublethal doses of *B.t.* Matter and Zohdy [66] also reported retardation in the larvae of *H. armigera* fed on a diet containing different concentrations of Bactospeine". "Salama et al. [58], Salama and Sharaby [67] investigated the effect of exposure time of *S. littoralis, S. exigua* and *H. armigera* and *A. ypsilon* to low endotoxin concentrations of *B.t.* They found retardation in larval development, reduction in egg production of the moths and fertility of their eggs; together with a significant reduction in pupal weight and appearance of deformities in both pupae and moth population. The percentage of larvae that survived and succeeded to pupate increased with the decrease in the toxin concentrations and with the decrease in exposure' time. The reduction of the pupal weight significantly, increased with the increase in either toxin concentration or the duration of exposure. The longevity of the moths was not affected by larval treatment".

8   Distribution of *B.t.* as a Soil Microorganism

"The distribution of *B.t.* and *Bacillus cereus* in the soil of Egyptian governorates was surveyed [32, 68]. High *bacillus* counts were detected in most fertile soils while sandy soils had low counts. Some isolates were found to be active against *S. exigua* and *H. armigera*, but none of the isolates exhibited significant activity against *S. littoralis*". *B.t.* strains were also isolated from insects such as *P. interpunctella* and *P. gossypiella* [69, 70].

Studies by Abdel-Ghany [71] showed no correlation between the occurrence of *B.t.* in the soil, its type, PH value, salinity, organic matter content, total nitrogen, organic carbon and available copper concentration. In this concern, DeLucca et al. [72] reported that the distribution of *B.t.* was not affected by soil PH. Also, Anwar et al. [73] found *B.t.* in soils with pH values of 5.3–7.95; they also found that copper level in the soil had significant association with *B.t.* index. Morris et al. [74] reported that *B.t.* was found most frequently in organic-rich soil samples from six different types of Canadian soil.

9   Effect of *B.t.* on Eggs, Prepupae, Pupae and Adult of Lepidoptera

B.t. is normally used against the larvae of lepidopterous species. Investigations were made by Ali and Watson [75] and Potter et al. [76] to evaluate the effect of *B.t.* on the adult and egg stages of *H. virescens*. Similarly "Salama [77] evaluated the effect of *B.t. galleriae* on the moth and eggs of *S. littoralis*, aiming to get some results that could be of value in suppressing the moth population of this pest. The longevity and egg production were significantly affected when the moths were fed on a sucrose diet containing 17.2×10^3 IU/mg of the tested formulation or higher rates. At lower concentrations ($4.4–8.7 \times 10^3$ IU/mg), the females gave normal egg production but the longevity was adversely affected. Investigations show that a formulation with spores alone affects the egg production and hatching while a formulation with active crystals alone had no effect on egg production. Sprays of *B.t.* combined with sucrose on flowering cotton plants seem to affect the biology of the moths released on it. Egg masses sprayed with *B.t.* hatched normally, but the survival of the hatched larvae was reduced when treatment was made shortly before egg hatching. Based on the results obtained, *B.t.* showed a promising effect in the control of the moth *S. littoralis* as well as an ovicide-larvicide agent".

"The prepupal stage of *Spodoptera littoralis* was also affected by Dipel as mentioned by Salama and Zaki [78] at high concentration (5%), when sprayed or kept in soil treated with the pathogen. The emerged moths showed a short longevity associated with low egg production and low fertility. Increase in malformed individuals was observed. Treatment of the pupae of *S. littoralis* with *B.t.* during different ages showed an increased effect when treatment was applied, immediately after pupation".

So, it is recommended that Dipel may be provided with irrigation water or sprayed on the soil when the population of these immature stages reach the maximum. Under these conditions, repeated application may be required and so the stages which escape the effect of spraying will be further affected. "It is worth mentioning that the drift of *B.t.* that fall on the soil during spray application of cotton cultivation may also affect the prepupal, pupal stages of *S. littoralis* in the soil". Abdallah et al. [79] found that larvae of *S. littoralis* living in the soil treated with *B.t.*, pupated and gave adults which deposited eggs with low hatching rate.

10 Joint Action of *B.t.* Varieties and Integration with Microbial and Chemical Control Agents

10.1 Activity of Combination of **B.t.** Varieties

Salama et al. [38] explored the activity of different combinations of *B.t.* varieties versus *S. littoralis*, *S. exigua* and *H. armigera* in search for the possibilities of enhancing their effectiveness. They reported that some varieties of *B.t.* were found to synergies the action of others versus the tested species, while others did not and with no apparent correlation with the serotype or potential activity of the single variety. These results though preliminary, but they showed a kind of interactions or competition that may occur between the different varieties combined together at a post-harvest stage. In the growing stage, however, it has been demonstrated that the activity of some varieties when grown together decreased as with the vars. *entomocidus* and *kurstaki* HD-1, and the activity of the mixture was drastically reduced compared to the potential activity caused by both of them if used singly.

10.2 Activity of Interaction of **B.t.** and Chemical Insecticides

Salama et al. [80] stated that the potential activity after combination of *B.t.* with chemical insecticides has to be determined as a base for any pest management system. There is some contradiction in the literature in this respect. Abdallah [81], Altahtawy and Abaless [82, 83] carried out investigations in this concern.

"Salama et al. [80] conducted some studies on the effects of chemical insecticides of different chemical groups on sporulation yields of *B.t.* var. *entomocidus*. Among the carbamates, tested, carbaryl exhibited a more deleterious effect on the sporulation process of *B.t.* than Methomyl. Within the organophosphorus group phoxim inhibited sporulation yields less than profenofos. The pyrethroid group, represented by fenvalerate, cypermethrin and permethrin, generally had less deleterious effects on sporulation yield of *B.t.* than did the carbamate and organophosphorus compounds.

The pyrethroids and most organophosphorus compounds tested potentiated the activity of *B.t.* applied against *S. littoralis*. The carbamates, diflubenzuron, and a combination of methomyl and diflubenzuron (disa) showed an additive effect when jointly applied with *B.t.* varieties. The mild effect of pyrethroids on sporulation processes of *B.t.*, compared to effects of other classes of chemical insecticides, suggested little or no interference with the ecology and perpetuation of this useful bacterium at the site of application. Synergistic interactions suggest that application of pyrethroids with *B.t.* may be a safe and effective means for controlling *S. littoralis*".

10.3   Activity of Combination of **B.t.** and Viral Diseases

Salama et al. [84] found that combinations of *B.t.* HD-129 and the nuclear polyhedrsis virus of *S. littoralis* (SLNPV) showed antagonistic effect while combinations of *B.t.* HD-635 and SLNPV showed an additive effect. The median lethal dose of *B.t.* varieties decreased with the increase of the viral dose in the mixture.

11   Enhancement of *B.t.* Potency for Field Application

The dosages of *B.t.* needed for effective insect control in the field required repeated application due to its short persistence. So, the incorporation of adjuvant with this pathogen was practiced to achieve high efficacy to overcome the short persistence or to extend the spectrum of activity of various strains. In this concern, various approaches were adopted.

11.1   Feeding Stimulants

Studies by Salama et al. [85, 86] were made to find feeding eliciting materials of some cotton pests in Egypt and to determine the feasibility of using these materials in a bait formulation to increase the effect of *B.t.* Laboratory experiments revealed that the petroleum ether extract of some host plants such as cotton, Jews mallow, castor oil, clover and sweet potato plants are active feeding stimulants to either the cotton leafworm *S. littoralis* or *exigua*. "The LC50 of the tested *B.t.* formulations significantly decreased when the extracts of the forementioned plants were supplemented with the pathogen incorporated into the diet at 0.1% concentration. The extracts of either cotton leaves or jews mallow contain a volatile and non volatile fractions which also proved to be active feeding stimulants leading to the increase in the effectiveness of *B.t.* The components of the volatile fraction of cotton leaves as identified by thin layer chromatography, are ∞-pinene, ∞-terpineol, citronellol, humulene, (−) nerol, ∞-bisabolol, trans-B-ocimene, camphene, linalool, L-caryophyllene, caryophyllene oxide and phenyl ethyl alcohol. Some of these components had a high effectiveness in increasing the potency of *B.t. entomocidus* versus *S. littoralis*, while others had no effect. Cotton seed flour served also as a feeding stimulant when combined with sucrose. Sucrose and maltose had also a similar effect".

On the base of these findings, 14 field experiments were carried out in 2, successive years 1980 and 1981 by Salama et al. [86] to determine the role of some feeding stimulants, increasing the effect of *B.t.* formulations versus *S. littoralis* and *S. exigua* under the local environmental conditions. "The bacterial preparations used were *B.t. entomocidus* HD-635, *B.t. kurstaki* HO-251 and *B.t. galleriae* HD-129. The results obtained generally showed that the potency of *B.t.* preparations was increased when they were applied in combination with some feeding adjuvants such as extracts of cotton leaves, jews mallow and cotton seed flour combined with sucrose".

El-Nockrashy et al. [87] attempted to develop some preparations to be combined with *B.t.* to make it more acceptable and be to ingested by the target insect. Theses preparations were obtained from cottonseed or soybean. So, those containing cotton-seed kernels extracted with 70% ethanol and then acetone–hexane water led to more acceptability of the meal as compared to the commercial adjuvant "Coax". When gossypol content increased, there was a decrease in the efficiency of the formulations. The addition of soybean flour was also highly efficient.

11.2 Biochemical Approaches

Charles and Wallis [88], Smirnoff [89], Burges [90] reported that the efficacy of *B.t.* increased against some insect species after the addition of boric acid, chitinase or P-amino-salysilic acid to the pathogen fed to the insect, with no explanation of the mode of action of these additives. These findings are mostly caused due to the attack of subunits of the crystal to the midgut lining and thus causing paralysis via leakage of the gut contents into the haemocoel [91].

Accordingly, new approaches have been followed by Salama et al. [45, 68] aiming to enhance the endotoxin effect through optimizing conditions inside the insect gut required for release of intoxicating fragments. This procedure was based on the incorporation of some selected essentially safe nontoxic compounds with the endotoxin resulting in its potentiation. Among the chemicals selected were inorganic salts, nitrogenous compounds including amino acids and some salts, aromatic compounds, protein solubilizaing agents and lipid emulsifying agents.

11.2.1 Effect of Inorganic Salts

Screening reveals that potassium carbonate, potassium bicarbonate and potassium dibasic phosphate increased the activity of *B.t. kurstaki* HD-1 against *S. littoralis* 2.5–4.5 times. Since potassium carbonate is naturally present in the midgut, it is expected that potassium ions leaking from the damaged epithelium lining of the gut into the haemolymph as caused by endotoxin and so enhancing insect paralysis and death.

With calcium salts, the data show the drastic effect of calcium carbonate (0.25%) and calcium oxide (0.1%) in enhancing the potency of tested endotoxin versus the target insect, with 4.3 and 14.7 fold increases, respectively. It is worthy to note that these particular compounds are of alkaline nature. Since, it has been reported that the low PH of the gut juice of *S. litura* is a main factor contributing to the weak susceptibility of this species to many *B.t.* preparations [92], it is expected that the addition of such alkaline compounds will change the PH of the gut, being more alkaline and thus enhancing the endotoxin breakdown and release of toxic fragments [93].

The results of Salama et al. [45, 68] show the remarkable effect of zinc sulphate (0.1%) in enhancing the potency of the endotoxin of *B.t.* var. *kurstaki* HD-1 with 16 fold increase. "The mode of action of this salt may be correlated to its effect on the proteolytic enzymes present in the insect midgut. In this view, Dixon and Webb [94] reported that the divalent cations are generally known either as activators or co-factors for many proteolytic enzymes. From the data, it appears also that most of the ammonium salts tested was ineffective in potentiating the activity of the tested endotoxin. Ammonium dibasic phosphate (1%) however was markedly effective since its incorporation in the diet increased the insecticidal activity of *B.t.* var. *kurstaki* HD-1."

"A drastic increase in the potency of *B.t.* (19 and fivefold), Salama et al. [45, 68] occurred when calcium oxide (0.05%) and calcium hydroxide (0.01%) were used. Also calcium carbonate (0.25%), calcium sulphate (1%) and calcium acetate (1%) potentiated the effectiveness of *B.t.* 13, 6 and 6 times, respectively. Zinc sulphate (0.05%) showed a remarkable potentiution of *B.t.* 24 fold increase". Endo and Nishitsuji-Uwo [95] found that injection of the midgut juice and some salt solutions as well as buffers and alkali into the haemocoel of *B. mori*. caused paralysis symptoms in the normal larvae in absence of *B.t.*

11.2.2 Effect of Nitrogenous Compounds

"The data obtained by Salama et al. [45, 68] reveal that the incorporation of acetamide (1%) or L-tryptophane (0.5%) into the larval diet containing *B.t.* caused 21.7 and 13 fold increase in its potency versus the target insect. The LC50's were drastically reduced, being 15 and 28 compared with 360 µg in the control. L-arginine (0.1%) and sodium nitrate (1%) also showed to be potential in increasing the effectiveness of *B.t.* and the factors of increase were 3.6 and 3.9, respectively". Feeding of the larvae on a diet combined with amino acids, may lead to the leakage of amino acids to the haemocoel in an unorganized and abrupt manner. It is expected that the regulatory mechanisms of the insect would be challenged to reset the natural composition of the haemolymph possibly through specific reactions, e.g., deamination, decarboxylation and catabolic activities, in addition to the enhancement of excretion mechanisms for such undesired nitrogenous compounds. The destructive effect caused by the endo-toxin to the midgut may minimize the rate of excretion mechanism which results in reserving the amino acids in the circulating blood and thus increased intoxica-tion. The alteration of the haemolymph composition may interfere with the normal physiological processes. According to Wigglesworth [96], the insect haemolymph contains amino acids with its nitrogen and its interference may cause increase in the insect susceptibility to the toxin action.

11.2.3 Effect of Aromatic Compounds

Increase in the activity of *B.t.* endotoxin fed to the larvae of *S. littoralis* occurred after incorporation of 0.5% picric acid. Salama et al. [45, 68] Also, sodium benzoate (0.5%) potentiated the effectiveness of *B.t.* and the factor of increase was 2.5 fold.

11.2.4 Effect of Protein Solubilizing Agents

"The incorporation of disodium glycerophosphate (0.05%) and dipotassium hydro-gen phosphate (1%) into the larval diet containing *B.t.* Salama et al. [45, 68] led to 5.9 and 5.6 fold increase in its potency versus the target insect, With sodium thio-glycollate (1%) and urea (0.5%), the potency of *B.t.* was elevated with factors of 16.4 and 2.7, respectively". Nickerson [93] assumed that the disulphide bonds of protein molecule may prevent dissolution of endotoxin and thus leading to reduction of these bonds to sulfhydryl group and increasing endotoxin solubility in the insect gut causing mortality.

11.2.5 Effect of Lipid Emulsifying Agents

The incorporation of Tween at 0.5% into the larval diet containing *B.t.* caused an elevation in its potency against the target insect with factors of 3, 9 and 5.2 for Tweens 40, 60 and 80, respectively [45, 68]. This potential effect was associated with a decrease in the LC50 values. Tweens are complexes of nonionic surface acting agents, the basic structure being hexahydric alcohols, alkene oxides and fatty acids. Due to its safety, it has been used as an adjuvant in the field spraying of *B.t.* [97]. It is assumed that these compounds, may affect the permeability barrier of damaged gut epithelium and so leading to loss of permeability of the membranes followed by infiltration of endotoxin into the haemocoel, paralysis and then death.

The significance of the foregoing findings with respect to the toxicity enhancing effects of some of the tested compounds lies in the fact that all of them except amino acids are low priced, all are nontoxic to humans or animals which add to their feasibilities in application. The results were based on experiments on natural host plants. These results would undoubtedly be of interest to both research work and industry. The expected results should contribute to the significant reduction of cost production of endotoxin preparations and to fill the gaps now present with respect to spectra of activity of various strains of *B.t.* and consequently would render the commercial production of bacterial insecticides most economic and valuable through the incorporation of these additives at a postharvest stage.

12 Case Studies: Field Application of *B.t.*

A series of large scale field experiments was carried out to evaluate the effect of locally produced strains of *B.t.*, compared to the available commercial products Dipel 2X, Dipel ES, as well as the recommended chemical insecticides against soybeans and cotton lepidopterous pests. For this purpose, an area of 120 feddans cultivated with soybeans in Menoufia Governorate was selected. Also, an area of 70 feddans cultivated with cotton plants (*Gossypium barbadense*) was included in the experiments with the participation of representatives from Egyptian Ministry of Agriculture. The target insects were *A. ypsilon, S. littoralis, P. gossypiella and E. insulana*. Baits based on *B.t.* were used against the greasy cutworm *A. ypsilon*, while sprays of *B.t.* (locally produced and available commercial products) were applied against other insects at intervals of 2–3 weeks or whenever spraying is required. The chemical insecticides applied were Hostathione, Denet, Lannate and Malathion. Adjuvants such as molasses and some inorganic salts were combined with *B.t.* treatments. By the end of the season, the yield of either soybeans or cotton was evaluated in different treatments for comparison. The results obtained indicate the efficacy of *B.t.* products in controlling the target insects, and the effect is almost equal to that of chemical insecticides. The yield of cotton after spray applications with *B.t.* was almost equal to that obtained with conventional chemical insecticides (860 compared with 892 kg/feddan) but this is compared to 261 kg/feddan for the control. The yield of soybeans was 1.5 ton/feddan in areas treated with *B.t.* compared with 1.44 ton/feddan in those areas treated with chemical insecticides. In the control, the average yield was 0.83 ton/feddan.

Since the greasy cutworm *A. ypsilon* infests other crops mainly vegetables during the winter season, some field experiments were also carried out to control this insect throughout the agricultural cycle. At early summer this insect pest migrates to some important crops as cotton and corn causing destruction to the seedlings and young plants. The damage caused by larvae living few centimeters below or just above the soil, is almost inflicting upon the plant seedlings between the root and the stem causing them to wither. This has emphasized the use of baits in controlling the larvae of this pest. Hostathion 40% EC was the chemical insecticide recommended in baits application against this pest (Pest Control Program, Ministry of Agriculture—Egypt). The role of bait formulations of *B.t.* in suppressing *A. ypsilon* larval population was explored.

In the light of the results obtained it appears that:

- The different tested materials caused appreciable percentages of reduction in the larval population of *A. ypsilon* and a higher yield was obtained as compared with the check in the different vegetable crops.
- The effective threshold rate of application for *B.t.* products in baits was 250 g/feddan.
- *B.t.* applied in baits led to a better control than if it was applied as spray (potato experiment). The site of baits application was the ground below plant foliage, ensure shady conditions that protect rapid deterioration of *B.t.* toxins [98, 99].
- The incorporation of some inorganic salts (adjuvants) as calcium oxide and calcium sulphate has significantly accelerated the biotic efficacy of the pathogen (as judged from eggplant and chilies experiments). Salama et al. [86] found that the addition of calcium oxide or calcium sulphate to *B.t.* treated diet increased the potency of the pathogenic bacterium against *S. littoralis* by 19 or 6 fold, respectively. Salama et al. [98] working on *A. ypsilon* attacking horse bean cultivations, found that baits with Dipel 2X alone may be less effective than those with Hostathion unless a chemical adjuvant is involved.
- Regarding the biological baits treatments (one application) on different crops, a significant efficacy was obtained in potato cultivations (92.3% reduction in larval density, and about 2.3 fold increase in the yield). They were almost as effective as Hostathion bait treatment in potato, lentils and peas cultivations. In eggplant, chilies and okra, Hostathion-bait treatment surpassed biological treatment when applied once and was almost equitoxic when all treatments were applied twice [99, 100]. These fruitful prospects should encourage the utilization of baits based on microbial pathogens in *A. ypsilon* control program.

13 Transgenic Plants

To cope and catch up with the recent advances in *B.t.* research, GM technology has been also developed in Egypt but very cautiously. This is to overcome the problem of poverty and to improve quality of life of small scale farmers who cannot adopt control measures of insect pests since the costs are high and to protect them from exposure to hazards of insecticides [101]. Assem [102] reported the approval for commercialization of genetically modified (GM) *B.t.* corn hybrid (Ajeeb YG) was given by the Egyptian Ministry of Agriculture in 2008 and as decided by the National Biosafety Committee (NBC) and the Seed Registration committee. This hybrid produces a protein toxic to certain lepidopterous insects. This protein is identified as δ-endotoxin or Cry 1 Ab protein. In this concern, South Africa issued the first conditional release permits for GM crops in 1997. This was followed by Egypt in 2008 and then Burkina–Faso approved the release of *B.t.* cotton. In 2008, Egypt planted 700 ha in 2010/2011. According to EL-Banna [103] the benefits from planting 2000 ha in 2010 was of the order of US$550.000. However on 8th March 2012, the Ministry approval was suspended temporarily as a result of some information regarding health concern [102].

In this concern, Saker et al. [104] reported that the transgenic tomato (CV. Money maker) over expressing *B.t.* (Cry 2 Ab) gene was produced using *Agrobactetium* mediated transformation method. Obvious effects were judged by the mortality of *H. armigera* and the potato tuber, *Phthorimaea operculella* when fed on *B.t.* tomato.

14 Conclusion

It appears that *B.t.* is a desirable agent for pest control and ideal for use in Egypt and other developing countries because its possible low production cost and lack of toxicity. The possibility of increasing regional production using inexpensive material and agro-industrial byproducts is a particularly attractive option. Novel approaches can be adopted to enhance the potency of *B.t.* preparations by using feeding stimulants and safe and cheap chemical additives leading to biochemical reactions in the midgut of the treated insects [105–108]. These additives were able to increase the potency and to extend spectrum level of activity of *B.t.* preparations by many folds. It was found that *B.t.* varieties can be successfully integrated with pyrethroids, most organophosphorus compounds and some biological control agents within an integrated system of control. No deleterious effects were reported on parasites and predators of the target lepidopteran insects. It was found that *B.t.* can affect the various developmental stages of lepidopteran insects other than the larvae.

Field application of the local *B.t.* product against lepidopteran insects in cotton, soybean, vegetables and some other field crops indicate that the yield was almost equal to that obtained with conventional *B.t.* products and chemical insecticides and economically profitable. Research for development of GM crops is on-going in a number of research institutions and universities in Egypt, but the developed plants did not reach the stage of commercial release due to lack of national legislation on Biotech crops. However the approval for commercialization of genetically modified (GM) *B.t.* corn hybrid was made for a short period but this approval was suspended temporarily on March 2012 as a result some information regarding health concern.

Acknowledgements The author is greatly indebted to Dr. I. Shehata and Miss Dalal Ali, in NRC for their sincere efforts and help during the preparation of this book chapter.

References

1. Frankenhuyzen KV (1993) The challenge of *Bacillus thuringiensis*. In: Entwistle P, Cory J, Bailey M, Higgs S (eds) *Bacillus thuringiensis*, an environmental biopesticide: theory and practice, pp 1–35
2. Berliner E (1911) Uber die Schlaffsucht der Mehlmottenraupe. Z Gesamte Getreidewesen Berlin 3:63–70

3. Berliner E (1915) Uber die Schlaffsucht der Mehlmottenraupe. Z ang Ent 2:29–56
4. Weiser J (1986) Impact of *Bacillus thuringiensis* on applied entomology in Eastern Europe and in the Soviet Union. In: Krieg A, Huger AM (eds) Mitteilungen aus der Biologischen Bundesanstalt für Land – und Forstwirtschaft Berlin-Dahlem, vol 233. Paul Parey, Berlin, pp 37–50
5. Steinhaus EA (1951) Possible use of *Bacillus thuringiensis* as an aid in the biological control of the alfalfa caterpillar. Hilgardia 20:359–381
6. Dulmage HT (1970) Production of the spore-δ-endotoxin complex by variants of *Bacillus thuringiensis* in two fermentation media. J Invertebr Pathol 16:385–389
7. Menn JJ (1960) Bioassay of a microbial insecticide containing spores of *Bacillus thuringiensis*. J Insect Pathol 2:134–138
8. Krieg A (1965) Über die vivo-titration von insektenpathogenen, speziell von *Bacillus thuringiensis*. Entomophaga 10:3–20
9. Burges HD (1967) The standardization of products based on *Bacillus thuringiensis*. In: van der Laan (ed) Insect pathology and microbial control. North Holland Publication Co, Amsterdam, pp 306–314
10. Dulmage HT, Rhodes RA (1971) Production of pathogens in artificial media. In: Burges HD, Hussey NW (eds) Microbial control of insects and mites. Acad Press, London, pp 507–540
11. Becker N, Margalit J (1993) Use of *Bacillus thuringiensis israelensis* against mosquitoes and blackflies. In: Entwistle P, Cory J, Bailey M, Higgs S (eds) *Bacillus thuringiensis*. An environmental biopesticide: theory and practice. Wiley, pp 147–170
12. Goldberg LH, Margalit J (1977) A bacterial spore demonstrating rapid larvicidal activity against *Anopheles sergenti, Uranotaenia unguiculata, Culex univattatus, Aedes aegyptii* and *Culex pipiens*. Mosq News 37:355–358
13. Margalit J, Dean D (1985) The story of *Bacillus thuringiensis israelensis*. J Am Mosq Control Assoc 1:1–7
14. Padua LE, Ohba M, Aizawa K (1984) Isolation of *Bacillus thuringiensis* strain (serotype 8a: 8b) highly and selectively toxic against mosquito larvae. J Invertebr Pathol 44:12–17
15. Keller B, Langenbruch G (1993) Control of coleopteran pests by *Bacillus thuringiensis*. In: Entwhistle P, Cory J, Bailey M, Higgs S (eds) *Bacillus thuringiensis*. An environmental biopesticide theory and practice. Wiley, pp 171–191
16. Huger AM, Krieg A (1989) Über zwei typen parasporaler kristalle beim käferwirksamen stamm BI. 256-82 von *Bacillus thuringiensis* subsp. *tenebrionis*. J Appl Entomol 108:490–497
17. Krieg A, Huger AM, Langenbruch GA, Schnetter W (1983) *Bacillus thuringiensis* var. *tenebrionis*: ein neuer gegenüber larven von Coleopteran wirksamer Pathotyp. Z ang Ent 96:500–508
18. Höfte H, Whitely H (1989) Insecticidal crystal proteins of *Bacillus thuringiensis*. Microbiol Rev 53:242–255
19. Agaisse H, Lereclus D (1995) How does *Bacillus thuringiensis* produce so much insecticidal crystal protein? J Bacteriol 177:6027–6032
20. Lereclus D, Agaisse H, Gominet M, Chaufaux J (1995) Overproduction of encapsulated insecticidal crystal proteins in a *Bacillus thuringiensis* spo0A mutant. Biotechnology (N Y) 13:67–71
21. Lecadet M, De Barjac H (1981) *Bacillus thuringiensis* beta-exotoxin. In: Davidson E (ed) Pathogenesis of invertebrate microbial diseases. Allanheld Osmun and Co., New Jersey, pp 293–316
22. Salama HS, Sharaby A, Ragaei M (1983) Chemical changes in the haemolymph of *Spodoptera littoralis* as affected by *Bacillus thuringiensis*. Entomophaga 28:331–337
23. Salama HS (1984) *Bacillus thuringiensis* Berliner and its role as a biological control agent in Egypt. Z ang Ent 98:206–220
24. Boctor I, Salama H (1983) Effect of *Bacillus thuringiensis* on the lipid content and composition of *Spodoptera littoralis* larvae. J Invertebr Pathol 41:381–384
25. Salama HS, Sharaby A (1985) Histopathological changes in *Heliothis armigera* infected with *Bacillus thuringiensis* as detected by electron microscopy. Insect Sci Appl 4:503–511

26. Sutter GR, Raun ES (1967) Histopathology of European corn borer larvae, treated with *Bacillus thuringiensis*. J Invertebr Pathol 9:90–103
27. Reese J, Yonke T, Fairchild M (1972) Fine structure of the midgut epithelium in larvae of *Agrotis ypsilon*. J Kansas Entomol Soc 45:242–251
28. Cheung P, Grula EA, Burton RL (1978) Haemolymph responses in *Heliothis zea* to inoculation with *Bacillus thuringiensis* or *Micrococcus lysodeikticus*. J Invertebr Pathol 3:148–156
29. Salama HS, Foda MS, Dulmage HT, El-Sharaby A (1983) Novel fermentation media for production of δ-endotoxins from *Bacillus thuringiensis*. J Invertebr Pathol 41:8–19
30. Salama HS, Foda MS, El-Sharaby A, Selim M (1983) A novel approach for whey recycling in production of bacterial insecticides. Entomophaga 28:151–160
31. Salama HS, Foda MS, Selim MH, El-Sharaby A (1983) Utilization of fodder yeast and agro-industrial byproducts in production of spores and biologically active endotoxins from *Bacillus thuringiensis*. Zentralblatt Mikrobiol 138:553–563
32. Salama HS, Morris O (1993) The use of *Bacillus thuringiensis* in developing countries. In: Entwistle P, Cory J, Bailey M, Higgs S (eds) *Bacillus thuringiensis*. An environmental biopesticide: theory and practice. Wiley, pp 237–253
33. Dulmage HT (1971) Production of δ-endotoxin by eighteen isolates of *Bacillus thuringiensis* serotype 3, in 3 fermentation media. J Invertebr Pathol 18:353–358
34. Nagamma MV, Ragnathan AN, Majumder SK (1972) A new medium for *Bacillus thuringiensis* Berliner. J Appl Bact 35:367–370
35. Fernandez WL, Ocampo TA, Perez DC (1974) Coconut water in three media reduces cell yield of *Bacillus thuringiensis* var. *thuringiensis*. Philipp Agric 28:273–279
36. Foda MS, Salama HS, Fadel M (1993) Local production of *Bacillus thuringiensis* in Egypt. Advantages and constraints. In: Salama H, Morris O, Rached E (eds) The biopesticide *Bacillus thuringiensis* and its application in developing countries, pp 149–165
37. Madkour MA (2000) Egypt: biotechnology from laboratory to the marketplace: challenges and opportunities. In: Persley HJ, Lantin MM (eds) Agriculture biotechnology and the poor: proceeding of an international conference, Washington DC, 21–29 Oct
38. Salama HS, Foda S, Zaki F, Khalafallah A (1983) Persistance of *Bacillus thuringiensis* Berliner spores in cotton cultivations. Z ang Ent 95:321–326
39. Salama HS, Zaki F (1985) Application of *Bacillus thuringiensis* Berliner and its potency for control of *Spodoptera littoralis* (Boisd.). Z ang Ent 99:425–431
40. Ragaei M (1985) Studies on the effect of some environmental and chemical factors on the potency of *Bacillus thuringiensis* against some cotton pests. M.Sc. thesis, Cairo University, Egypt
41. Morris O, Moore A (1975) Studies on the protection of insect pathogens from sunlight inactivation II. Preliminary field trials. Report Cc-x-113, Chemical Control Research Institute, 34 pp (8)
42. Hamed A, Hassanein F (1985) Persistance and virulence of *Bacillus thuringiensis* under sunny and shady conditions. Bull Entomol Soc Egypt Econ Ser 14:73–77
43. Morris O (1983) Protection of *Bacillus thuringiensis* from inactivation by sunlight. Can Entomol 115:1215–1227
44. Salama HS, Foda S, Selim M (1984) Isolation of *Bacillus thuringiensis* mutants resistant to physical and chemical factors. Z ang Ent 97:139–145
45. Salama HS, Foda S, Selim M (1984) Mutation in relation to sporulation and potency of *Bacillus thuringiensis* vs. cotton pests. Z ang Ent 97:29–36
46. Matter M (1993) *Bacillus thuringiensis* and environmental safety. In: Salama HS, Morris O, Rached E (eds) The biopesticide *Bacillus thuringiensis* and its applications in developing countries, pp 257–265
47. Salama HS, Zaki FN, Sharaby AF (1982) Effect of *Bacillus thuringiensis* Berl. on parasites and predators of the cotton leafworm *Spodoptera littoralis* (Boisd.). Z ang Ent 94:498–504
48. Salama HS, Zaki FN (1983) Interaction between *Bacillus thuringiensis* Berliner and the parasites and predators of *Spodoptera littoralis* in Egypt. Z ang Ent 95:425–429

49. Salama HS, Zaki FN (1984) Impact of *Bacillus thuringiensis* Berl. on the predator complex of *Spodoptera littoralis* (Boisd.) in cotton fields. Z ang Ent 97:485–490
50. Salama HS, Zaki FN (1985) Biological effects of *Bacillus thuringiensis* on the egg parasitoid *Trichogramma evanescens*. Insect Sci Appl 6:145–148
51. Afify AM (1964) Bioassay of three bacterial insecticides on the base of *Bacillus thuringiensis* Berliner, using its original host *Anagasta kuehniella* as a test insect. Bull Soc Entomol Egypt 48:103–109
52. Afify AM (1965) Studies on the susceptibility of certain stored products insects to bacterial insecticides: 1-tests with "Bakthane L-69". Bull Soc Entomol Egypt 49:59–64
53. Afify AM (1968) Bioassay of "Biospore 2902" using two species of lepidopterous larvae of different susceptibility levels. J Invertebr Pathol 10:283–286
54. Afify AM, Merdan AI (1969) On tracing the response of some Egyptian cotton worms in different larval ages to *Bacillus thuringiensis* Berliner. Z ang Ent 63:263–267
55. Afify A, Hafez M, Merdan A (1969) Preliminary investigations on the virulence of 13 *Bacillus* preparations against 3 Egyptian noctuids. Anz Schadlingsk Pflanz Unwelt 42:54–57
56. Afify AM, El-Sawaf S, Habib E, Hammad SM (1970) Pathogenicity tests of Biotrol BTB process 183, on *Anagasta kuehniella* Zeller. Z ang Ent 65:29–37
57. Soliman A, Afify A, Abdel-Rahman H, Atwa W (1970) Effectiveness of different components of *Bacillus thuringiensis* against three larval stages of *Pieris rapae*. Anz Schadlingsk Pflanz Umwelt 43:161–165
58. Salama HS, Foda MS, El-Sharaby A (1981) Potency of spore endotoxin complexes of *Bacillus thuringiensis* against some cotton pests. Z ang Ent 91:388–398
59. Salama HS, Foda MS (1982) A strain of *Bacillus thuringiensis* var. *entomocidus* with high potential activity on *Spodoptera littoralis*. J Invertebr Pathol 39:110–111
60. Salama HS, Foda MS (1984) Studies on the susceptibility of some cotton pests to various strains of *Bacillus thuringiensis*. J Plant Dis Prot 91:65–70
61. Abul-Nasr S, Abdallah MD (1969) Lethal and sublethal action of *Bacillus thuringiensis* Berliner on the cotton leafworm *Spodoptera littoralis* (Biosd.). Bull Entomol Soc Egypt Econ Ser 4:151–160
62. Abdallah MD, Abul-Nasr S (1970) Feeding behavior of the cotton leafworm *Spodoptera littoralis* (Boisd.) sublethally infected with *Bacillus thuringiensis* Berliner. Bull Entomol Soc Egypt Econ Ser 4:161–170
63. Abdallah MD, Abul-Nasr S (1970) Effect of *Bacillus thuringiensis* Berliner on reproductive potential of the cotton leafworm (Lep., Noctuidae). Bull Entomol Soc Egypt Econ Ser 4:171–176
64. Afify AM, Matter MM (1969) Retarded effect of *Bacillus thuringiensis* Zell. Entomophaga 14:447–456
65. Afify A, Hafez M, Matter M (1971) The retarding effect of *Bacillus thuringiensis* on larval development of flour moth *Anagasta kuehniella* with a new method of determining the duration of instars. Acta Entomol Bohemoslov 68:6–14
66. Matter M, Zohdy N (1981) Biotic efficiency of *Bacillus thuringiensis* and a nuclear polyhedrosis virus on larvae of the American bollworm *Heliothis armigera*. Z ang Ent 92:336–343
67. Salama HS, Sharaby A (1988) Effects of exposure to sublethal levels of *Bacillus thuringiensis* on the development of the greasy cutworm *Agrotis ypsilon* (Hbn). Z ang Ent 106:396–401
68. Salama HS, Foda S, Zaki F, Ragaei M (1986) On the distribution of *Bacillus thuringiensis* and closely related *Bacillus cereus* in Egyptian soils and their activity against Lepidopterous cotton pests. Z ang Zool 73:257–265
69. Abdel-Rahman H (1966) Study of the pathogenicity of crystalline inclusion of *Bacillus thuringiensis*. Ain Shams Sci Bull 10:89–95
70. Abul-Nasr S, Ammar E, Merdan A, Farrag S (1979) Infectivity tests on *Bacillus thuringiensis* and *B. cereus* isolated from resting larvae of *Pectinophora gossypiella*. Z ang Ent 88:60–69
71. Abdel Ghany N (2006) Studies on the potential activity and molecular characterization of *Bacillus thuringiensis* isolated from Lepidopterous cotton insects and soil in Egypt. M.Sc. thesis, Ain Shams University, Egypt

72. Delucca A, Simonsen JG, Larson A (1981) *Bacillus thuringiensis* distribution in soils of the United States. Can J Microbiol 27:865–870

73. Anwar M, Sohel A, Sirajul H (1997) Abundance and distribution of *Bacillus thuringiensis* in the agricultural soil of Bangladesh. J Invertebr Pathol 70:221–225

74. Morris O, Converse V, Kanagaratnam P, Cote J (1998) Isolation, characterization and culture of *Bacillus thuringiensis* from soil and dust from grain storage bins and their toxicity for *Mamestra configurata*. Can Entomol 130:515–537

75. Ali A, Watson T (1982) Effect of *Bacillus thuringiensis* var. *kurstaki* on tobacco budworm adult and egg stages. J Econ Ent 75:596–598

76. Potter M, Jensen M, Watson T (1983) Influence of sweet bait *Bacillus thuringiensis* var. *kurstaki* combinations on adult tobacco budworm. J Econ Entomol 75:1157–1160

77. Salama HS (1985) Control of *Spodoptera littoralis* through moth and eggs treatment with *Bacillus thuringiensis*. Insect Sci Appl 6:49–53

78. Salama HS, Zaki F (1986) Effects of *Bacillus thuringiensis* Berliner on prepupal and pupal stages of *Spodoptera littoralis*. Insect Sci Appl 7:747–749

79. Abdallah M, Zaazou H, El-Tantawi M (1974) Wirkung eines *Bacillus thuringiensis* preparats und eines Juvenile hormone – Analogous über dem erdboden auf *Spodoptera littoralis*. Anz Schadlingsk Pflanz Umwelt 47:170–172

80. Salama HS, Foda MS, Zaki FN, Moawad S (1984) Potency of combinations of *Bacillus thuringiensis* and chemical insecticides on *Spodoptera littoralis* (Lepidoptera: Noctuidae). J Econ Entomol 77:885–890

81. Abdallah M (1969) The joint action of microbial and chemical insecticides on the cotton leafworm *Spodoptera littoralis*. Bull Entomol Soc Egypt Econ Ser 3:201–217

82. Altahtawy M, Abaless I (1972) Thuricide 90 Ts flowable, a recent approach to the biological control of *Spodoptera littoralis*. Z ang Ent 72:299–308

83. Altahtawy M, Abaless I (1973) An integrated control trial of *Spodoptera littoralis* (Boisd.) using *Bacillus thuringiensis* associated with insecticides. Z ang Ent 74:255–263

84. Salama HS, Moawed S, Zaki F (1987) Effects of nuclear polyhedrosis virus—*Bacillus thuringiensis* combinations on *Spodoptera littoralis* (Roisd.). Z ang Ent 104:22–27

85. Salama HS, Foda S, El-Sharaby A (1983) Biological activity of mixtures of *Bacillus thuringiensis* against some cotton pests. Z ang Ent 95:69–74

86. Salama HS, Foda S, Sharaby A (1985) Role of feeding stimulants in increasing the potency of *Bacillus thuringiensis* vs. *Spodoptera littoralis*. Entomol Gener 10:111–119

87. El-Nockrashy S, Salama HS, Taha F (1984) Developemt of bait formulations for control of *Spodoptera littoralis*. Z ang Ent 103:313–319

88. Charles C, Wallis R (1964) Enhancement of the action of *Bacillus thuringiensis* var. *thuringiensis* on *Porthetria dispar* (Linn.) in laboratory tests. J Insect Pathol 6:423–429

89. Smirnoff WA (1974) The symptoms of infection by *Bacillus thuringiensis* and chitinase formulation in larvae of *Choristoneura fumiferana*. J Invertebr Pathol 23:397–399

90. Burges HD (1977) Control of the waxmoth *Galleria mellonella* on beecomb by H-serotype v *Bacillus thuringiensis* and the effect of chemical additives. Apidologie 8:155–168

91. Couch TL, Ross D (1980) Production and utilization of *Bacillus thuringiensis*. Biotechnol Bioeng 22:1297–1304

92. Narayanan K, Govindarajan R, Jayaraj S (1976) Role of alkali components and gut microflorae of *Papilio demoleus* L. and *Spodoptera litura* F. in the mode of action of *Bacillus thuringiensis* Berliner. Madras Agric J 64:344–346

93. Nickerson KW (1980) Structure and function of *Bacillus thuringiensis* protein crystal. Biotechnol Bioeng 22:1305–1333

94. Dixon M, Webb EC (1964) Enzymes. Academic Press Inc., New York, pp 67–70

95. Endo Y, Nishitsuji-Uwo (1980) Mode of action of *Bacillus thuringiensis* δ-endotoxins: histopathological changes in the silkworm midgut. J Invertebr Pathol 36:90–103

96. Wigglesworth VB (1972) The principles of insect physiology. English Language Book Society, Chapman Hall, England

97. Patti H, Carver GR (1974) *Bacillus thuringiensis* investigation for control of *Heliothus* spp. on cotton. J Econ Entomol 67:415–418

98. Salama HS, Foda S, El-Sharaby A, Matter M, Khalafallah M (1981) Development of some lepidopterous cotton pests as affected by exposure to sublethal levels of endotoxins of *Bacillus thuringiensis* for different periods. J Invertebr Pathol 38:220–229

99. Salama HS, Salem S, Zaki F, Matter M (1990) Control of *Agrotis ypsilon* on some vegetable crops in Egypt using the microbial agent *Bacillus thuringiensis*. Anz Schadlingsk Pflanz Umwelt 63:147–151

100. Salama HS, Salem S, Matter M (1991) Field evaluation of the potency of *Bacillus thuringiensis* on lepidopterous insects infesting some field crops in Egypt. Anz Schadlingskd Pflanz Umwelt 64:150–154

101. Pilcher CD, Rice M, Obrycki J, Lewis L (1997) Field and laboratory evaluation of transgenic *Bacillus thuringiensis* corn on secondary Lepidopteran pests (Lepidoptera: Noctuidae). J Econ Entomol 90:669–678

102. Assem SK (2014) Opportunities and challenges of commercializing biotech products in Egypt: Bt. maize, a case study. In: Wambugu F, Kammanga D (eds) Biotechnology in Africa: emergence, initiatives and future, pp 37–47

103. El-Banna H (2011) Terza Giornata Mondiale del Mais, Bioenergy, Italy. Maize production in Egypt. Al-Ahram Agriculture, Cremona, 18–20 Mar 2011

104. Saker M, Salama HS, Salama M, El-Banna A, Abdel-Ghany N (2011) Production of transgenic tomato plants expressing Cry 2 Ab gene for the control of some lepidopterous insects endemic in Egypt. J Genetic Eng Biotechnol 9:149–155

105. Salama HS, Foda S, Sharaby A (1984) Novel biochemical avenues for enhancing *Bacillus thuringiensis* endotoxin potency against *Spodoptera litoralis*. Entomophaga 29:171–178

106. Salama HS, Foda S, Sharaby A (1986) Possible extension of the activity spectrum of *Bacillus thuringiensis* strains through chemical additives. J Appl Ent 101:304–313

107. Salama HS (1993) Enhancement of *Bacillus thuringiensis* for field application. In: Salama HS, Morris O, Rached E (eds) The biopesticide *Bacillus thuringiensis* and its application in developing countries. Al-Ahram Press, Cairo, pp 105–116

108. Osman GEH, Already R, Assaeedi ASA, Organji SR El-Ghareeb D, Abulreesh HH, Althubiani AS (2015) Bioinsecticide *Bacillus thuringiensis* a comprehensive review. Egypt J Biol Pest Control 25:271–288

Isolation, Mass Production and Application of Entomopathogenic Fungi for Insect Pests Control

Mohamed Abdel-Raheem

Abstract **Background** Entomopathogenic fungi have played a uniquely important role in the history of microbial control of insects. *Beauveria bassiana*, commonly known as white muscardine fungus attacks a wide range of immature and adult insects. *Metarhizium anisopliae* a green muscardine fungus is reported to infect 200 species of insects and arthropods. **Isolation** Both of these entomopathogenic fungi are soil borne and widely distributed. These fungi have been documented to occur naturally in over 750 species of host insects. Nevertheless there are studies on isolation of these fungi from insect cadavers. The insect cadavers were put on a wetted filter paper in a Petri-dish and incubated at 24 ± 1 °C for 7 days. The new fungal generation was isolated from the surface of insect cadaver and cultured on PDA medium in Petri-dishes. Fungal cultures were purified weekly until pure cultures were obtained. **Mass Production** Fungi were grown and maintained on: Peptone medium, Potato dextrose agar and Rice Grains, *B. bassiana* and *M. anisopliae* isolates were propagated on wetted rice. Two kilograms wetted rice were washed in boiled water for 10 min and put in thermal bags. These bags were autoclaved at 120 °C for 20 min, then infected by isolates and incubated at 25 ± 2 °C for 15 days. The Conidia were harvested by distilled water and filtered through cheese cloth to reduce mycelium clumps and Tween 80% was added. Field **Application** Using of *B. bassiana* and *M. anisopliae* were the most effect to decrease the total number of *C. vittata*. The fungus *B. bassiana* was the most effect to controlling the tortoise beetle, *C. vittata* than the Fungus *M. anisopliae*. Also, Using of *B. brongniartii* and *N. rileyi*, against the potato tuber moth *Phthorimaea operculella* under semi-field conditions, the corresponding LC_{50} recorded 1.20×10^6 and 9.7×10^5 spores/mL, respectively. *Sitophilus granaries* treated with the last concentrations of *B. carterii* the larval mortality were significant decrease to 29.61, 15.81 and 0.613% as compared to zero in the control. Cabbage worm, *Pieris rapae* was treated with three isolates of *B. bassiana*, *M. anisopliae* and *V. lecanii* under laboratory and field conditions. *B. bassiana* was the best fungus against *P. rapae* followed by *M. anisopliae* and *V. lecanii*. Also, Sugar beet fly, *Pegomyia mixta* treated with *B. bassiana*, and *M. anisopliae*.

M. Abdel-Raheem (✉)
Pests & Plant Protection Department, National Research Centre, Dokki, Giza, Egypt
e-mail: abdelraheem_nrc@yahoo.com

© Springer Nature Switzerland AG 2020
N. El-Wakeil et al. (eds.), *Cottage Industry of Biocontrol Agents and Their Applications*, https://doi.org/10.1007/978-3-030-33161-0_7

Keywords Isolation · Mass production · Application · Entomopathogenic fungi · Insect

1   Introduction

Entomopathogenic fungi have played a uniquely important role in the history of microbial control of insects. Entomopathogenic fungi were the first to be recognized as microbial diseases in insects [1]. *Beauveria bassiana*, commonly known as white muscardine fungus attacks a wide range of immature and adult insects. *Metarhizium anisopliae* a green muscardine fungus is reported to infect 200 species of insects and arthropods. Both of these entomopathogenic fungi are soil borne and widely distributed. These fungi have been documented to occur naturally in over 750 species of host insects [2–4]. Soil is considered to be the natural environment of these fungi because they deposit their infectious spores there and remain in the soil for some duration of their life cycle. Therefore, it was determined that soil is the most appropriate place to determine their occurrence [5]. The occurrence and distribution of insect pathogenic fungi in agricultural field soils have been extensively investigated in previous studies [6–11]. Nevertheless there are studies on isolation of these fungi from insect cadavers as well [12–17]. Wild isolates are still of great importance due to their potential unique characteristics in biological control of insect pests. The presence of certain entomopathogenic fungal species can be considered as an indicator of their ability to survive in that environment. This information is useful for the selection of biocontrol agents because the indigenous dominant species are generally the most suitable candidates [18]. Factors such as the geographical location, climate, habitat, altitude, and pH of the soil or organic matter impact the presence of fungal species, and the response of each species to these conditions varies [18].

Use of *Galleria*-bait technique is a common method for isolation of entomopathogenic fungi from soil [19–22]. Although identification of the common entomopathogenic fungi like (*Beauveria bassiana*, *Metarhizium anisopliae*, and *Verticillium lecanii*) looks easy even to non-mycologists because they have basic diagnostic characters making them easily identifiable. It must be remembered that these species have other complex molecular, morphological and pathobiological traits [23, 24]. We have aimed to discuss the following objectives:

1. Isolation and abundance the entomopathogenic fungi
2. Mass Production of Entomopathogenic Fungi
3. Field applications of entomopathogenic fungi against some insect pests.

2   Mode of Action of Entomopathogenic Fungi

With most fungi disease development can be divided into nine steps as given by Roberts [25]:

1. Attachment of the infective unit (e.g. conidium or zoospore) to the insect epicuticle.
2. Germination of the infective unit on the cuticle.
3. Penetration of the cuticle, either directly by germ tubes or by infection pegs from appressoria.
4. Multiplication in the yeast phase (hyphal bodies) in the haemocoel.
5. Production of toxic metabolites.
6. Death of the host.
7. Growth in the mycelial phase with invasion of virtually all host organs.
8. Penetration of hyphae from the interior through the cuticle to the exterior of the insect.
9. Production of new infective units (conidia).

Many authors use entomopathogenic fungi as a biological control agent against Insects pests like [12, 26–41].

3 Isolation of Entomopathogenic Fungi

3.1 Isolation from Soil

The entomopathogenic fungi were isolated from soil samples using the *Galleria*-bait technique based on Zimmermann [22]. Soil samples were collected from the soil at 0–20 cm-depth in a circle of 1 m-diameter around tree trunks. A weight of approximately 1.5 kg of soil was collected in plastic bags and transferred in a cooled box to the laboratory. Larvae of the greater wax moth, *G. mellonella* were distributed in the cups at a rate of three larvae/cup and wetted to approximately 15% water content.

After seven days of incubation at 25 °C, infected larvae were transferred to 5 cm-diameter Petri dishes furnished with wetted filter paper and left for development and sporulation. Numbers of positive samples (samples containing entomopathogenic fungi) and numbers of infected *G. mellonella* larvae in each sample were recorded [42].

3.2 Identification and Maintenance of Obtained Fungi

Extracted entomopathogenic fungi were identified morphologically based on the morphological characteristics of reproductive structures according to Humber [24] and Salem et al. [43]. The white muscardine fungus *B. bassiana* was the highest abundant forming 25.03% of total collected samples followed by the green muscardine fungus, *M. anisopliae* (17.76%) then the metallic pink fungus *V. lecanii* (14.49) [42]. Percentages of infected *G. mellonella* larvae in collected samples ranged from

1.11 to 17.78%. In a survey study conducted by Asensio et al. [44], positive soil samples containing *B. bassiana* or *M. anisopliae* comprised less than 2% of total samples collected from El Behaira, Kafr Elshaikh or Aswan.

Beauveria bassiana was the most frequent entomopathogenic fungus in the soils followed by *M. anisopliae* in Spain [45]. In China, *B. bassiana* was more abundant than *M. anisopliae* in soils [46]. *M. anisopliae* had two periods of occurrence; the first was from January to March while the second was from mid April to May with a peak of 50% positive samples and a minimum of 10%. Percentages of *G. mellonella* larvae infected with *M. anisopliae* in collected soil samples were between 1.1 and 6.67%. *V. lecanii* occurred from mid January to March recording 10–40% positive samples and 1–12.2% infected *G. mellonella* larvae [42].

3.3 Relationship Between Fungus and Plant Cover

Beauveria bassiana was found in soil under all kinds of fruit trees recording 10–40% of collected samples. This fungus was found mostly under Mango trees (50% of samples) then under Pomegranates (40%). Its occurrence was between 10 and 30% for the rest of fruit kinds. *M. anisopliae* was found under seven fruit kinds. Its distribution looked almost even among the fruit kinds. Its range of occurrence was between 10 and 30% of collected samples. *V. lecanii* was found under 6 kinds of fruits with no clear relationship between the fungus type and the fruit kind [42]. No clear relationship between the distribution of entomopathogenic fungi and the kind of fruit tree could be determined as mentioned by Saleh et al. [39]. Charnley [46] mentioned the organic content and temperature among the factors affecting fungal abundance and activity in the soil.

3.4 Isolation from Insects Cadavers

Collected cadavers from the field were kept in Petri-dishes (9 cm diameter × 1.5 cm) at 24 °C. Incubated dishes were inspected daily to observe the fungal growth that was purified and used to confirm the disease cycle (Koh's postulates), then stored on slant of PDA artificial media at 4 °C until used in subsequent experiments [12].

3.5 Isolation of New Fungal Generation from the Infected Insects

The infected insects covered with fungal mycelium were collected from the field and carried to the laboratory. Different stages from *Cassida vittata* and *Scrobipalpa*

Table 1 The entomopathogenic fungi isolated from insect cadavers from different regions in Kafr El-Sheikh [12]

Fungi	Original host	Host plant	Location	Date
B. bassiana	S. ocellatella (4th larvae)	Sugar-beet	Abu Ghalab	May 2001
	C. vittata (larvae-adults)	Sugar-beet	Abu Ghalab	April 2001
M. anisopliae	C. vittata (larvae-adults)	Sugar-beet	Beiala	April 2003

ocellatella were found dead and totally covered with fungal mycelium [12]. The insect cadavers were put on a wetted filter paper in a Petri-dish and incubated at 24 ± 1 °C for 7 days. The new fungal generation was isolated from the surface of insect cadaver and cultured on PDA medium in Petri-dishes. Fungal cultures were purified weekly until pure cultures were obtained [12]. Two isolates of *B. bassiana* and *M. anisopliae* were isolated from infected cadavers of *C. vittata* and *S. ocellatella* from two regions (Beiala and Abu Ghalab) in Kafr El-Sheikh Governorate, Egypt as given in Table 1 [12].

4 Mass Production of Entomopathogenic Fungi

Fungi were grown and maintained on the following media:

4.1 Peptone Medium

This medium consists of: 10 g peptone, 40 g dextrose, 2 g yeast extract, 15 g agar and 500 mL chloramphenicol and completed to one liter by distilled water. The medium was autoclaved at 120 °C for 20 min, and poured in Petri-dishes (9 cm diameter × 1.5 cm) then inoculated with the entomopathogenic fungi and kept at 25 ± 2 °C and 85 ± 5 R.H. The fungal isolates were re-cultured every 15–30 days and kept at 4 °C.

4.2 Potato Dextrose Agar

The medium consisting 250 g potatoes, 20 g agar and 1000 mL distilled water was autoclaved at 120 °C for 20 min, and poured in Petri-dishes (9 cm diameter × 1.5 cm). The fungal isolates were re-cultured every 14–30 days and isolates were kept at 4 °C.

To restore the virulence of the isolates they were passed through their natural host or through the wax moth larvae *G. mellonella*.

4.3 Rice Grains

Beauveria bassiana and *M. anisopliae* isolates were propagated on wetted rice. Two kilograms wetted rice were washed in boiled water for 10 min and put in thermal bags. These bags were autoclaved at 120 °C for 20 min, then infected by isolates and incubated at 25 ± 2 °C for 15 days. The Conidia were harvested by distilled water and filtered through cheese cloth to reduce mycelium clumps and Tween 80% was added [24].

5 Application of Entomopathogenic Fungi for Insect Pests Control

5.1 Bioassay of Isolated Fungi Against Insects

5.1.1 Larvae of *G. mellonella*

G. mellonella is usually used as a test insect for assessment of virulence of ento-mopathogenic microorganisms including fungi [47]. Fungal concentrations around 1×10^7 spores/mL are frequently used in these bioassayes. Target insect may be directly used if available. Gupta [47] compared the virulence of five strains of *B. bassiana* against the larvae of *G. mellonella*. Percentages of mortality among *G. mellonella* larvae exposed for 10 days to serial concentrations were determined. At the lowest concentration, the mortalities ranged from 14.4 to 99.2 for *G. mellonella*, 19.4–90.4% for *M. anisopliae* and 10.4–84% for *V. lecanii*. At the highest concentration, the tested fungi reached their maximum effect after application being 100, 98.4 and 97.6% [39].

5.1.2 *Scrobipalpa ocellatella* and *Cassida vittata*

Using of *B. bassiana* and *M. anisopliae* were the most effect to decrease the total number of *C. vittata* (Vill.) The fungus *B. bassiana* was the most effect to controlling the tortoise beetle, *C. vittata* than the Fungus *M. anisopliae* (Figs. 1, 2 and 3; Tables 2, 3, 4 and 5).

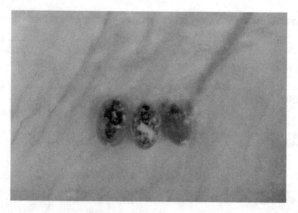

Fig. 1 Adult of *C. vittata* infected with *B. bassiana* [12]

Fig. 2 Adult of *S. ocellatella* infected with *M. anisopliae* [12]

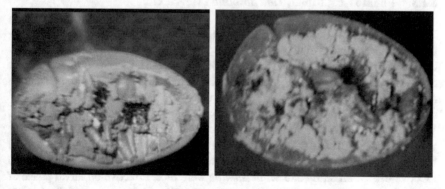

Fig. 3 Adult of *C. vittata* infected with *M. anisopliae* [12]

Table 2 Susceptibility of different stages of tortoise beetle *C. vittata* to entomopathogenic fungus *B. bassiana* at different concentrations [12]

Treated stages	%Mortality at indicated concentrations					LC_{50}	Fiducial limits 95%	Slope ± SE
	10^6	5×10^6	10^7	5×10^7	10^8			
1st instar larvae	50	75	85	95	100	1.08×10^6	6.35×10^5 to 1.6×10^6	1.09 ± 0.121
2nd instar larvae	40	55	80	90	100	2.24×10^6	2.92×10^5 to 5.42×10^6	1.1 ± 0.2
3rd instar larvae	40	50	65	90	100	1.49×10^6		0.79 ± 0.31
4th instar larvae	80	80	85	80	100	1.03×10^5	0.2×10^5 to 6.02×10^5	2.33 ± 0.73
5th instar larvae	50	60	70	70	85	1.04×10^6	2.16×10^5 to 2.38×10^6	0.43 ± 0.08
Pupa	50	65	65	75	95	1.32×10^6	0.013×10^6 to 6.04×10^6	0.59 ± 0.17
Adult	80	85	85	95	100	0.4×10^5	0.005×10^5 to 3.13×10^5	2.55 ± 0.76

5.1.3   Tomato Leaf Miner, *Tuta absoluta* (Meyrick)

Efficacy of 3 concentrations of *B. bassiana*, *M. anisopliae* and *V. lecanii* were pre-pared at concentrations of (1×10^5; 1×10^6; 1×10^7) and were tested on *T. absoluta* larvae (Neonate "newly hatched", 2nd and 3rd instar) [48]. Eggs of *T. absoluta* were exposed to *B. bassiana*, *M. anisopliae* and *V. lecanii* to evaluate their effect on hatch-ability. The estimated LC_{50} of *V. lecanii* were 3.25×10^5 spores/mL, 5.47×10^5 and 3.28×10^5 for neonate, 2nd instar and 3rd instar *T. absoluta* larvae, respectively. While the LC_{50} values of *B. bassiana* and *M. anisopliae* were (0.28×10^5 and 0.11×10^5), (0.45×10^5 and 0.46×10^5) and (0.32×10^5 and 0.27×10^5 conidia/mL) for neonate, 2nd instar and 3rd instar *T. absoluta* larvae, respectively. According to LC_{50} values, *B. bassiana* and *M. anisopliae* were most effective on larval phase of *T. absoluta* than *V. lecanii* [48]. Daily mortality (%) of larval phase of *T. absoluta* fed in the newly hatched 2nd instar and 3rd instar larva with leaves treated with *V. lecanii*.

Table 3 Susceptibility of different stages of tortoise beetle *C. vittata* to entomopathogenic fungi *M. anisopliae* at different concentrations [12]

Treated stages	%Mortality at indicated concentration					LC_{50}	Fiducial limits 95%	Slope ± SE
	10^6	5×10^6	10^7	5×10^7	10^8			
1st instar larva	60	70	70	95	100	3.68×10^6		1.4 ± 0.5
2nd instar larva	60	55	55	60	75	2.45×10^5	6.86×10^5 to 1.2×10^6	0.16 ± 0.08
3rd instar larva	60	65	70	85	100	8.49×10^5		0.69 ± 0.22
4th instar larva	60	60	60	70	100	9.8×10^5		0.52 ± 0.49
5th instar larva	55	55	65	75	90	1.22×10^6	0.03×10^6 to 5.79×10^6	0.5 ± 0.14
Pupa	45	50	55	60	70	3.65×10^6	6.32×10^5 to 8.72×10^6	0.3 ± 0.08
Adult	60	75	70	95	100	2.27×10^5	0.9×10^5 to 1.12×10^6	2.15 ± 0.76

Table 4 Susceptibility of different stages of sugar beet mining moth *S. ocellatella* to entomopathogenic fungus *B. bassiana* at different concentrations [12]

Treated stages	%Mortality at indicated concentrations					LC_{50}	Fiducial limits 95%	Slope ± SE
	10^6	5×10^6	10^7	5×10^7	10^8			
4th instar larva	60	75	60	60	70	0.4×10^5		0.04 ± 0.14
Pupa	75	90	95	100	100	2.99×10^5	0.9×10^5 to 5.63×10^5	1.2 ± 0.19

The results revealed that when neonate larvae fed on *V. lecanii* the pathogen effect was evident by the third day of evaluation after exposure in the three concentrations (10^5; 10^6; 10^7 spores/mL) with recorded mortality (13.3, 13.3, 33.3%) respectively [48]. Thereafter, the values of the corrected mortalities of neonate larvae increased gradually from the 4th day after exposure until the last day (10th) with mortality 88.8% for the first concentration 10^5 spores/mL. For the second concentration (10^6 spores/mL) the mortality values were increased and reached its maximum in

Table 5 Susceptibility of different stages of sugar beet mining moth *S. ocellatella* to entomopathogenic fungi *M. anisopliae* at different concentrations [12]

Treated stages	%Mortality at indicated concentrations					LC_{50}	Fiducial limits 95%	Slope ± SE
	10^6	5×10^6	10^7	5×10^7	10^8			
4th instar larva	70	70	75	90	95	2.27×10^5	6.14×10^5 to 6.53×10^5	0.52 ± 0.09
Pupa	20	20	35	70	75	1.92×10^7	6.14×10^6 to 9.6 $\times 10^7$	0.88 ± 0.18

the 8th day of exposure (100% reduction). The mortality values reached its maximum in the 7th day and in the 5th day for the second and third concentration (10^6 and 10^7 spores/mL) and recorded 100% reduction for the two concentrations [48]. The higher effective concentration of *V. lecanii* on neonate larvae of *T. absoluta* was 10^7 spores/mL followed by 10^6 spores/mL while the other concentrations (10^5 spores/mL) showed moderate effect [48].

For the second instar larva of *T. absoluta* the pathogen effect was evident by the fifth day for (10^5 spores/mL), by the 4th day for (10^6 spores/mL) and by the 3rd day for (10^7 spores/mL) of evaluation after exposure with corresponding mortalities (53.3, 40.0, and 6.6%) for the three concentrations, respectively [48]. The values of the corrected mortalities of 2nd instar larvae increased gradually until the 9th day after exposure to record 88.8% mortality for the first concentration (10^5 spores/mL) [48]. The mortality values reached its maximum in the 9th day, 8th day and in the 7th day for the three concentrations and recorded 100% reduction. Daily mortality (%) of larval phase of *T. absoluta* fed in the newly hatched "neonate", 2nd instar and 3rd instar larva with leaves treated with *B. bassiana*. The higher effective concentration of *B. bassiana* on the 2nd instar larvae of *T. absoluta* resulted from the two concentrations 10^7 and 10^6 conidia/mL evenly, followed by 10^7 and 10^6 conidia/mL, respectively [48].

When 3rd instar larvae fed on *B. bassiana* the pathogen effect was evident by the 3rd day of evaluation after exposure in the first concentration (10^5 conidia/mL) with recorded mortality (20.0%). While the pathogen effect was evident by the 2nd day of evaluation after exposure in the 1st, 2rd and 3th concentrations (10^5, 10^6 and 10^7) with recorded mortalities (6.7%) for the three concentrations [48].

5.1.4 The Potato Tuber Moth *Phthorimaea operculella* (ZELLER)

The LC_{50} of the target pest 1×10^6 and 8.7×10^5 spores/mL after *P. operculella* treated with *B. brongniartii* and *N. rileyi*, respectively. Under semi-field conditions, the corresponding LC_{50} recorded 1.20×10^6 and 9.7×10^5 spores/mL respectively

[49]. The effect of the entomopathogenic fungi against *P. operculella*, the number of eggs laid/female recorded significant reduction to 42 ± 1.7 and 33 ± 8.9, after being treated with *B. brongniartii* and *N. rileyi* as compared to the control, respectively. The percentage of emerged adults was significantly decreased by 4% as compared with the control [49].

5.1.5 Cabbage Worm, *Pieris rapae* L.

Three isolates of *B. bassiana*, *M. anisopliae* and *V. lecanii* were evaluated against cabbage worm, *P. rapae* L. under laboratory and field conditions. *B. bassiana*, *M. anisopliae* and *V. lecanii* have not effect on cabbage worm, *Pieris rapae* L. after two days from treatment [50]. Mortalities are occurred in the 3rd day. The percent of mortalities are increased gradually and reached to the maximum in the 10th day from treatment. The percent of mortalities ranged between 70–80%, 60–75.7% and 40.0–55.2% with *B. bassiana*, *M. anisopliae* and *V. lecanii*, respectively, in the 10th day after treatment. This means that *B. bassiana* isolation is more effective than *M. anisopliae* and *V. lecanii* [50]. The % mortalities with all concentrations of *B. bassiana* isolation were 70, 72.4 and 80%, respectively.

Also, mortalities are occurred in the 3rd day. The percent of mortalities are increased gradually and reached to the maximum in the 10th day from treatment. Data also showed a positive correlation between concentrations of fungi and the percentage of *P. rapae* L. Pupae mortality. The percent of mortalities ranged between 65–76%, 58–70.1% and 42.0–60% with *B. bassiana*, *M. anisopliae* and *V. lecanii*, respectively, in the 10th day after treatment. This means that *B. bassiana* isolation is more effective than *M. anisopliae* and *V. lecanii*. The percent of mortalities with all concentrations of *B. bassiana* isolation were 65, 70 and 76%, respectively [50].

5.1.6 Whitefly, *Bemisia tabaci* (Genn.)

There are no effect for *V. lecanii*, *M. anisopliae* and *B. bassiana* to *B. tabaci* after 3 days from treatment. The percent of mortalities were increased gradually and reached to the maximum in the 7th day from treatment. With the all concentrations, the percent of mortalities were increased with increase of concentrations [51].

The % mortalities ranged between 68.3–100%, 56.2–90.5% and 62.7–97.3% with *V. lecanii*, *M. anisopliae* and *B. bassiana*, respectively, in the 7th day after treatment. This mean that *V. lecanii* isolation is more effective than *M. anisopliae* and *B. bassiana*. The percent of mortalities with all concentrations of *V. lecanii* isolation were 68.3, 90.4 and 100%, respectively [51]. The number of *B. tabaci* per leave was decreased compared with control after the 2nd application. That the percent of reduction by *V. lecanii M. anisopliae* and *B. bassiana* after the third application were 57.3, 90.4 & 100, 50.2, 70.0 & 75.5 and 55.5, 79.5 & 90% with C_1, C_2 and C_3, respectively [51]. There were no significant differences between the concentrations and controls after the first application all plots. After the 3rd application there were highly significant differences among all concentrations. The L.S.D was two, three after the third application. The statistical analysis confirmed that the third concentration (1×10^9) was the highly toxic compared the first and the second concentrations in all treatment [51].

The percent of reduction by *V. lecanii M. anisopliae* and *B. bassiana* after the third application were 55.2, 85.2 & 100, 50.0, 71.0 & 76.2 and 56.2, 78.3 & 92% with C_1, C_2 and C_3, respectively [51]. The % of reduction by *V. lecanii*, *M. anisopliae* and *B. bassiana* after the third application were 55.2, 85.2 & 100, 50.0, 71.0 & 76.2 and 56.2, 78.3 & 92% with C_1, C_2 and C_3, respectively [51].

5.1.7 *Rhyzopertha dominica*, *Sitophilus oryzae* and *Oryzaephilus surinamensis*

R. dominica LC_{50} and LC_{90} values confirmed that *R. dominica* was more susceptible to *M. anisopliae* than *B. bassiana*, where LC_{50} and LC_{90} were 1.2×10^5, 2.3×10^5 conidia/g and 2.7×10^5, 0.01×10^6 conidia/g, respectively. The LT_{50}s were calculated as 4.6 and 6.6 days for the two respective fungi, indicating the superiority of *B. bassiana* over *M. anisopliae* [52].

5.1.8 Sitophilus oryzae

The % cumulative mortality of *R. dominica* maximum mortality percentages adult were recorded 11 days after treatment with *B. bassiana* and *M. anisopliae*, where the mortality percentages were 25.2, 38.1, 45.2 and 50.3% for *B. bassiana* at the tested concentrations of 0.12×10^6 conidia/g, 0.22×10^6 conidia/g, 0.32×10^6 conidia/g, and 0.42×10^6 conidia/g, respectively, and 63.0, 72.3, 75.3, and 79.3% for *M. anisopliae* at the tested concentrations of 0.12×10^6 conidia/g, 0.22×10^6 conidia/g, 0.32×10^6 conidia/g, and 0.42×10^6 conidia/g, respectively.

The cumulative mortality percentages started with low levels on the 3rd day after treatment. Then the Percent mortality of *S. oryzae* was increased gradually, till the maximum was recorded on the 11th day of observation, where the mortality percentages were recorded 23, 32, 40 and 54% for *B. bassiana* respectively. The *S. oryzae* was more susceptible to infection with *B. bassiana* than *M. anisopliae*. LC_{50} and LC_{90} values proved that *S. oryzae* was more susceptible to *M. anisopliae* than *B. bassiana*, where the LC_{50} and LC_{90} of, *B. bassiana* and *M. anisopliae* were 4.3×10^5 and 0.08×10^6 conidia/g, and 1.6×10^5 and 7.1×10^5 conidia/g, respectively. LT_{50}s were calculated as 8.8 and 6.3 days for the two respective fungi, indicating that the conidial spores of *B. bassiana* were less efficient on *S. oryzae* than those of *M. anisopliae* [51].

5.1.9 *Oryzaephilus surinamensis*

The same effective was found when the entomopathogenic fungi were used against *O. surinamensis*. The concentrations 0.42×10^6 conidia/g of fungi, induced increase the mortalities rates. The highest cumulative mortalities were recorded after 11 days of treatment when mortality percentages were recorded 25, 35, 43 and 51% for *B. bassiana* at the tested concentrations of 0.12×10^6 conidia/g, 0.22×10^6 conidia/g, 0.32×10^6 conidia/g, and 0.42×10^6 conidia/g, respectively. The mortality percentage of *O. surinamensis* treated with *B. bassiana* and *M. anisopliae* differed significantly among different concentrations. Also mortality percentage differed significantly between *O. surinamensis* treated with *B. bassiana* and with *M. anisopliae* [51].

6 Application in the Field of Isolated Fungi Against Many Insects

6.1 Scrobipalpa ocellatella *and* C. vittata

The application of *B. bassiana* and *M. anisopliae* blast spores in sugar-beet fields was applied in Kafr El-Sheikh (Abu Kalab), Egypt. Sugar-beet plants were sprayed with the fungal suspensions to control *C. vittata*. Blastspores were grown on agitated bio malt liquid medium. Blastspores of *B. bassiana* and *M. anisopliae* were applied to sugar-beet plants by using 10 L. Knapsack sprayer with a concentration of 10^8 spores/mL. Another untreated plot was used as control. In the treated and control plots, eggs, larvae, pupae and adults were counted on 25 plants. The spraying was repeated for three times in week, three weeks and five weeks [12].

A. Using *M. anisopliae* for controlling *C. vittata* in the field

Using of *M. anisopliae* in the field application against the insect of *C. vittata*. Number of eggs, larvae, pupa and adult was counted and recorded in treatment and control before the beginning of the treatment and every two weeks after the application. Before spraying the average number of eggs in treatment was 0.6/plant and 0.26/plant in control, the average number of larvae in treatment was 1.12/plant and 1.2/plant in control, the average number of pupa in treatment was 0/plant and 0/plant in control, the average number of adult in treatment was 0.6/plant and 0.6/plant. After two weeks from the application the average number of eggs in treatment were 2.64/plant and in control was 5.76 egg/plant, the average number of larvae in treatment was 3.32/plant and in control was 5.24/plant, average number of pupae in treatment was 2.0/plant and in control was 2.08/plant, the average number of adult in treatment was 3.28/plant and in control was 4.28/plant.

After 4 weeks after treatment the average number of eggs in treatment was 1.84/plant and in control was 3.44/plant, the average number of larvae in treatment was 3.96/plant and in control was 7.6/plant, the average number of pupae in treatment was 1.12/plant and in control was 1.64/plant, the average number of adult in treatment was 2.36/plant and in control was 4.84/plant. After six weeks after treatment the average number of eggs was 4.84/plant in treatment and 4.92/plant in control, the average number of larvae was 5.52/plant in treatment and 10.4/plant in control, the average number of pupae was 0.92/plant in treatment and 1.5/plant in control, the average number of adult was 1.68/plant in treatment and 2.92/plant in control [12].

B. Using *B. bassiana* for controlling *C. vittata* in the field

Using of *B. bassiana* in the field application against the insect of *C. vittata*. Number of eggs, larvae, pupa and adult was counted and recorded in treatment and control before the beginning of the treatment and every two weeks after the application. Before spraying the average number of eggs in treatment was 6.0/plant and 5.8/plant in control, the average number of larvae in treatment was 4.8/plant and 5.2/plant in control, the average number of pupa in treatment was 1.9/plant and 2.0/plant in control, the average number of adult in treatment was 4.6/plant and 4.2/plant in control [12]. After two weeks from the application the average number of eggs in treatment were 1.88/plant and in control was 3.44 egg/plant, the average number of larvae in treatment was 3.2/plant and in control was 7.6/plant, average number of pupae in treatment was 0.88/plant and in control was 1.6/plant, the average number of adult in treatment was 2.6/plant and in control was 4.8/plant [12].

After four weeks after treatment the average number of eggs in treatment was 3.24/plant and in control was 4.92/plant, the average number of larvae in treatment was 4.2/plant and in control was 10.4/plant, the average number of pupae in treatment was 1.04/plant and in control was 1.5/plant, the average number of adult in treatment was 1.6/plant and in control was 2.9/plant [12]. After 6 weeks after treatment the average number of eggs was 1.8/plant in treatment and 3.64/plant in control, the average number of larvae was 2.84/plant in treatment and 7.7/plant in control, the average number of pupae was 3.2/plant in treatment and 7.2/plant in control, the average number of adult was 2.7/plant in treatment and 7.9/plant in control [12].

6.2   The Potato Tuber Moth, Phthorimaea operculella (ZELLER)

Application of the fungi occurred was accomplished at the rate of 1×10^8 spores/mL. Applications were made by a sprayed at the sunset. Four applications were made at 4-week intervals during crop growing season. Control plots were left without any treatments. Examinations of 40 plants/plot/treatment were carried out just before the first application and seven days after last application to calculate the average reduction percentages in the target insect infestation percentages which were calculated in each treatment. The agricultural practices followed the recommendations of the Ministry of Agricultural. Twenty tubers were taken from the first 5 rows in each treatment and in the control [49].

Under field conditions the yields weight of potatoes were significantly increased to 25.45 ± 55.66 and 29.67 ± 61.11 ton/feddan in plots treated with *B. brongniartii* and *N. rileyi* respectively as compared to 17.88 ± 55.43 ton/feddan in the control. The weight of potatoes were significantly increased to 25.97 ± 67.91 and 29.94 ± 54.98 ton/feddan as compared to 12.27 ± 45.09 in the control [49].

6.3 Sitophilus granarius

The effect of the grain weevil *S. granarius* were significantly decreased after oil extract treatments, the percentage of larval mortality, 88.14, 39.13 and 2.21 after *N. Sativa* treatments at the concentrations of 5, 0.5 and 0.05%. The corresponding concentrations of *T. distichum* gave the larval mortality 57.18, 20.26 and 4.41% of the *S. granarius*, respectively. When *B. carterii* treated with the last concentrations the larval mortality were significant decrease to 29.61, 15.81 and 0.613% as compared to zero in the control [52].

The oviposition deterrent effect in the means number of eggs were significantly decreased to 13.0 ± 7.8, 37 ± 7.9 and 45 ± 3.8 eggs/female as when treated with 3% of the oil *N. sativa*, *T. distichum and B. carterii*, respectively compared to 198 ± 4.9 in the control [52].

The half-life period of the target insect pest after oil extract treatments which show that the LC_{50} obtained 2.1×10^5, 3.2×10^5, and 3.7×10^5 as compared to 8.8×10^5 sores/mL. [52]. Accumulative mortality of *S. granarius* during the first week of rice seeds exposed to treated foam with *N. sativa*, *T. distichum*, and *B. carterii* on *S. granarius* oils. After seven days from treated rice seeds the accumulative mortality was 78.15, 50.17, and 30.1, respectively compared with 2.1 in untreated [52].

6.4 Pieris rapae

There are significant difference between 1st (C_1) and 2nd (C_2) spores concentrations and control after the first application in all parts, the differences appear gradually after the second and third application. On the other hand the third concentration (C_3) in *B. bassiana* was the best concentration against *P. rapae* L. followed by the third concentration in *M. anisopliae* and the third concentration in *V. lecanii*. This means that the third concentration in *B. bassiana* was the best concentrations against *P. rapae* [50].

There are significant difference also between 1st (C_1) and 3rd (C_2) spores concentrations and control after the first application in all parts, the differences appear gradually after the second and third application. On the other hand the third concentration (C_3) in *B. bassiana* was the best concentration against *P. rapae* followed by the third concentration in *M. anisopliae* and the third concentration in *V. lecanii*.

The percent of reduction in all treatment was ranged between 48.5 and 75.8% in all concentrations [50].

B. bassiana, M. anisopliae and V. lecanii isolates are promising agents for Cabbage worm control in the field [50].

6.5   Sugar Beet Fly, Pegomyia mixta

P. mixta started to appear during the second week of January. In mid-February, the first application was carried out using two entomopathogenic fungi B. bassiana, and M. anisopliae [51]. The number of infested/10 plants was 70% before the first spray by B. bassiana and reached to 10% after seven days from treatment and reached to zero% infested after 14 days from first treatment. Also, the number of infested leaves 11 before the first spray by B. bassiana and reached to 4 after seven days from treatment and reached to zero infested after 14 days from first treatment [53]. The number of infested egg patches/plant 13 before the first spray by B. bassiana and reached to 5 after seven days from treatment and reached to zero infested after 14 days from first treatment.

The number of infested larvae/plant 22 before the first sprays by B. bassiana and reached to 4 after seven days from treatment and reached to zero infested after 14 days from first treatment. The number of infested/10 plants was 80% before the first spray by M. anisopliae and reached to 20% after seven days from treatment and reached to 10% infested after 14 days from first treatment. Also, the number of infested leaves 12 before the first spray by B. bassiana and reached to 6 after seven days from treatment and reached to 2% infested after 14 days from first treatment.

The number of infested egg patch/plant 12 before the first spray by B. bassiana and reached to 6 after seven days from treatment and reached to 3% infested after 14 days from first treatment. The number of infested larvae/plant 21 before the first spray by B. bassiana and reached to 15 after seven days from treatment and reached to zero infested after 14 days from first treatment [53].

The number of infested/10 plants 60% before the first spray by B. bassiana and reached to 20% after seven days from treatment and reached to zero% infested after 14 days from first treatment. Also, the number of infested leaves 13 before the first spray by B. bassiana and reached to 3 after seven days from treatment and reached to zero infested after 14 days from first treatment.

The number of infested egg patches/plant 12 before the first spray by B. bassiana and reached to 4 after seven days from treatment and reached to zero infested after 14 days from first treatment [53]. The number of infested larvae/plant 25 before the first spray by B. bassiana and reached to 12 after seven days from treatment and reached to zero infested after 14 days from first treatment.

The number of infested/10 plants 70% before the first spray by Metarhizium anisopliae and reached to 30% after seven days from treatment and reached to Zero% infested after 14 days from first treatment. Also, the number of infested leaves 13 before the first spray by Metarhizium anisopliae and reached to 6 after seven days from treatment and reached to Zero% infested after 14 days from first treatment. The number of infested egg patches/plant 12 before the first spray by *M. anisopliae* and reached to 6 after seven days from treatment and reached to Zero% infested after 14 days from first treatment.

The number of infested larvae/plant 23 before the first sprays by *M. anisopliae* and reached to 14 after seven days from treatment and reached to zero infested after 14 days from first treatment. The entomopathogenic fungi *B. bassiana* and *M. anisopliae* can be used as a promising agent in pest control and integrated pest management programs instead of conventional pesticides to reduce the environmental pollution especially when the pests were under the economic threshold [53].

6.6 The Oliver Black Scale Insect, Saissetia oleae *(Oliver)*

Percent of sampled substrate infested with black scale Insect, *S. oleae* (Oliver) before and after treatment with *B. bassiana* reduction from 92.0% in the first week from treatment to 9.2% after seven weeks from treatment [54]. Also when we treated the trees with *M. anisopliae* reduction in population of the black scale Insect, *S. oleae* was 2.1% from 93.0% after seven weeks from treatment [54]. Also when we treated the trees with *V. lecanii* reduction in population of the black scale Insect, *S. oleae* was 9.5% from 90.0% after seven weeks from treatment. *M. anisopliae* caused the highest reduction in black scale Insect, *S. oleae* than *B. bassiana* and *V. lecanii* [54].

Percent of sampled substrate infested with black scale Insect, *S. oleae* (Oliver) before and after treatment with *B. bassiana* reduction from 84.0% in the first week from treatment to 8.3% after seven weeks from treatment [54].

Also when we treated the trees with *M. anisopliae* reduction in population of the black scale Insect, *S. oleae* was 1.0% from 86.0% after seven weeks from treatment.

Also when we treated the trees with *V. lecanii* reduction in population of the black scale Insect, *S. oleae* was 8.5% from 85.0% after seven weeks from treatment. *M. anisopliae* caused the highest reduction in black scale Insect, *Saissetia oleae* than *B. bassiana* and *V. lecanii* [54].

6.7 Anti-soft Scale Insect Pulvinaria tenuivalvata

The mean numbers of adults found on leaves before treatments, ranged from 58.6 to 68.2 scales/50 leaves, indicating a relatively uniform distribution of insect infestation. Three days after spraying, the treatments suppressed the levels of infestation to different degrees as compared to untreated control. *Beauveria bassiana* (10^8 spores/mL), Petroleum ether fraction of *Cressa cretica* with a rate of (50 mL/L) and significantly lowered the percentage of infestation to 39.9% and 44.6%, respectively, although they didn't reach KZ oil (15 mL/L) activity (54.0%). Two weeks after treatments, *B. bassiana* (10^8 spores/mL), Petroleum ether fraction (50 mL/L) and KZ oil (15 mL/L) had almost similar activity, 65.8, 69.1 and 79.4% reduction in the infestation, respectively. Statistical analysis showed significant differences between tested compounds and control on the soft scale insect population.

The low susceptibility of the adult stage to the tested compounds may be attributed to the presence of protective scales which prevent the penetration of sprayings [55]. Mortality percentages increased when *Beauveria bassiana*, *Cressa cretica* with Petroleum ether fraction 5% and KZ oil 95% were used. The average calculated yield in the control was 21,710 kg/feddan, as compared to 26,750, 27,180 and 27,670 kg/feddan after using the *Beauveria bassiana*, *Cressa cretica* Petroleum ether fraction 5% and KZ oil 95%, respectively [55].

7 Conclusion and Future Prospects

Entomopathogenic Fungi are one of the most promising agents for the biological control of insects, where it permits the cost of production and preservation of public health. As the fungus *Beauveria bassiana* successfully used against different insects as well as the fungus *Metarhizium anisopliae* and especially the rank of sheaths wings also *Verticillium lecanii* against insects sucking mouth parts.

The production of Entomopathogenic Fungi in various forms of liquid and powder as well as the body of nanoparticles through laboratories and factories. The dissemination of this means of promising in Integrated Pest Management (IPM) programs. Spreading awareness among the farmers about the importance of using this method as one of the most important means of biological control of pests.

8 Summary

- *Beauveria bassiana*, commonly known as white muscardine fungus attacks a wide range of immature and adult insects. *Metarhizium anisopliae* a green muscardine fungus is reported to infect 200 species of insects and arthropods.
- Both of these entomopathogenic fungi are soil borne and widely distributed. These fungi have been documented to occur naturally in over 750 species of host insects.
- The entomopathogenic fungi were isolated from soil samples using the *Galleria*-bait technique based on Zimmermann [22].
- The entomopathogenic fungi were identified morphologically based on the morphological characteristics of reproductive structures according to Salem [43].
- The infected insects which covered with fungal mycelium were collected from the field and carried to the laboratory.
- Fungi were grown and maintained on peptone medium, Potato dextrose agar and Rice Grains.
- *B. bassiana* the fastest effect against larvae of *G. mellonella*, followed by *M. anisopliae* then *V. lacanii*.

- The fungus *B. bassiana* was the most effect to controlling the tortoise beetle, *C. vittata* than the Fungus *M. anisopliae*.
- The higher effective concentration of *B. bassiana* on the 2nd instar larvae of *T. absoluta* resulted from the two concentrations 10^7 and 10^6 conidia/mL evenly, followed by 10^7 and 10^6 conidia/mL, respectively [49].
- The effect of the entomopathogenic fungi against *P. operculella*, the number of eggs laid/female recorded significant reduction.
- *B. bassiana* isolation is more effective than *M. anisopliae* and *V. lecanii* against cabbage worm, *Pieris rapae* L.
- The percent of mortalities of Whitefly, *Bemisia tabaci* (Genn.) were increased gradually and reached to the maximum in the 7th day from treatment.
- *R. dominica* was more susceptible to *M. anisopliae* than *B. bassiana*.
- *B. bassiana* were less efficient on *S. oryzae* than those of *M. anisopliae*.
- The mortality percentage of *O. surinamensis* treated with *B. bassiana* and *M. anisopliae* differed significantly among different concentrations.

References

1. Ainsworth (1956) Agostino Bassi, 1773–1856. Nature 177:255–257
2. Hajek, Leger (1994) Interactions between fungal pathogens and insect hosts. Annu Rev Entomol 39:293–322
3. Inglis et al (2001) Use of hyphomycetous fungi for managing insect pests. In: Butt TM, Jackson C, Magan N (eds) Fungi as biocontrol agents: progress, problems and potential. CABI Publishing, Wallingford, pp 23–69
4. Shah, Pell (2003) Entomopathogenic fungi as biological control agents. Appl Microbiol Biotech 61:413–423
5. Medo, Cagan (2011) Factors affecting the occurrence of entomopathogenic fungi in soils of Slovakia as revealed using two methods. Biol Control 59:200–208
6. Chandler et al (1997) Sampling and occurrence of entomopathogenic fungi and nematodes in UK soils. Appl Soil Ecol 5:133–141
7. Bidochka et al (1998) Occurrence of the entomopathogenic fungi *Metarhizium anisopliae* and *Beauveria bassiana* in soils from temperate and near-northern habitats. Can J Bot 76:1198–1204
8. Ali-Shtayeh et al (2002) Distribution, occurrence and characterization of entomopathogenic fungi in agricultural soil in the Palestinian area. Mycopathologia 156:235–244
9. Klingen et al (2002) Effects of farming system, field margins and bait insect on the occurrence of insect pathogenic fungi in soils. Agric Ecosyst Environ 91:191–198
10. Keller et al (2003) Distribution of insect pathogenic soil fungi in Switzerland with special reference to *Beauveria brongniartii* and *Metarhizium anisopliae*. Biocontrol 48:307–319
11. Meyling, Eilenberg (2006) Isolation and characterization of *Beauveria bassiana* isolates from phylloplanes of hedgerow vegetation. Mycol Res 110:188–195
12. Abdel-Raheem (2005) Possibility of using the entomopathogenic fungi *Beauveria bassiana* and *Metarhizium anisopliae* for controlling the sugar-beet insects *Cassida vittata* Vill. and *Scrobipalpa ocellatella* Boh. in Egypt. Ph.D. Faculty of Agriculture, Cairo University, Cairo, 86 pp
13. Abdo et al (2008) Isolation of *Beauveria* species from Lebanon and evaluation of its efficacy against the cedar web-spinning sawfly, *Cephalcia tannourinensis*. Biocontrol 53:341–352

14. Brownbridge et al (2010) Association of entomopathogenic fungi with exotic bark beetles in New Zealand pine plantations. Mycopathologia 169(1):75–80

15. Glare et al (2008) *Beauveria caledonica* is a naturally occurring pathogen of forest beetles. Mycol Res 112(3):352–360

16. Santoro et al (2008) Selection of *Beauveria bassiana* isolates to control *Alphitobius diaperinus*. J Invertebr Pathol 97(2):83–90

17. Vu et al (2007) Selection of entomopathogenic fungi for aphid control. J Biosci Bioeng 104(6):498–505

18. Meyling, Eilenberg (2007) Ecology of the entomopathogenic fungi *Beauveria bassiana* and *Metarhizium anisopliae* in temperate agroecosystems: potential for conservation biological control. Biol Control 43:145–155

19. Veen, Ferron (1966) A selective medium for isolation of *Beauveria tenella* and of *Metarhizium anisopliae*. J Invertebr Pathol 8:268–269

20. Shimazu, Sato (1996) Media for selective isolation of an entomogenous fungus, *Beauveria bassiana* (Deuteromycotina: Hyphomycetes). Appl Entomol Zool 31:291–298

21. Zimmermann (1998) Suggestion for a standardized method for reisolation of entomopathogenic fungi from soil using the bait method. "Insect pathogens and insect parasitic nematodes". IOBC Bull 21(4):289

22. Zimmermann (1986) The *Galleria* bait method for detection of entomopathogenic fungi in soil. J Appl Entomol 102(2):213–215

23. Soper et al (1988) Isolation and characterization of *Entomophaga maimaiga* sp. nov., a fungal pathogen of gypsy moth *Lymantria dispar*, from Japan. J Invertebr Pathol 51:229–241

24. Humber (1996) Fungal pathogen of the Chrysomelidae and prospects for their use in biological control. In: Jolivet PH, Cox ML, Hsiao TH (eds) Biology of Chrysomilidae, vol IV. SPB Academic Publishing, The Hague

25. Roberts (1980) Fungi toxins. In: Bureges HD (ed) Microbial control of insects, mites, and plant diseases, vol 2. Academic Press, New York, pp 223–248

26. Abdel-Raheem (2013) Susceptibility of the red palm weevil, *Rhynchophorus ferrugineus* Oliver to some entomopathogenic fungi. Bull NRC 38(1):69–82

27. Abdel-Raheem, Lamya (2017) Virulence of three entomopathogenic fungi against white fly, *Bemisia tabaci* (Genn.) (Hemiptera: Aleyrodidae) in tomato crop. J Entomol 14(4):155–159

28. Zaki, Abdel-Raheem (2010) Using of entomopathogenic fungi and insecticide against some insect pests attacking peanuts and sugar beet. Arch Phytopathol Plant Prot 43(18):1819–1828

29. Ismail, Abdel-Raheem (2010) Evaluation of certain entomopathogenic fungi for microbial control of *Myzus persicae* (Zulzer) at different fertilization rates in potatoes. Bull NRC 35(1):33–44

30. Abdel-Raheem MA, Ragab ZA (2010) Using entomopathogenic fungi to control the cotton aphid, *Aphis gossypii* glover on sugar beet. Bull NRC 35(1):57–64

31. Abdel-Raheem (2011) Impact of entomopathogenic fungi on cabbage aphids, *Brevicoryne brassicae* in Egypt. Bull NRC 36(1):53–62

32. Abdel-Raheem et al (2011) Effect of entomopathogenic fungi on the green stink bug, *Nezara viridula* L. in sugar beet. Bull NRC 36(2):145–152

33. Sabry et al (2011) Efficacy of the entomopathogenic fungi, *Beauveria bassiana* and *Metarhizium anisopliae* on some pests under laboratory conditions. J Biol Pest Control 21(1):33–38

34. Abdel-Raheem et al (2011) Virulence of some entomopathogenic fungi on *Schistocerca gregaria* (Forskal). Bull NRC 36(4):337–346

35. Sabbour et al (2012) Pathogenicity of entomopathogenic fungi against olive insect pests under laboratory and field conditions in Egypt. J Appl Sci Res 8(7):3448–3452

36. Sabbour, Abdel-Raheem (2013) Repellent effects of *Jatropha curcas*, Canola and Jojoba seed oil, against *Callosobruchus maculatus* (F.) and *Callosobruchus chinensis* (L.). J Appl Sci Res 9(8):4678–4682

37. Sabbour, Abdel-Raheem (2014) Evaluations of *Isaria fumosorosea* isolates against the red palm weevil *Rhynchophorus ferrugineus* under laboratory and field conditions. Curr Sci Int 3(3):179–185

38. Abdel-Raheem M (2015) Insect control by entomopathogenic fungi & chemical compounds. Book, Lambert Academic Publishing, 76 pp. ISBN: 978-3-659-81638-3
39. Saleh et al (2016) Natural abundance of entomopathogenic fungi in fruit orchards and their virulence against *Galleria mellonella* larvae. Egypt J Biol Pest Control 26(2):203–207
40. Abdel-Raheem M et al (2016) Entomopathogenic fungi, 112 pp. ISBN: 978-3-659-91451-5
41. Abdel-Raheem et al (2016) Evaluation of some isolates of entomopathogenic fungi on some insect pests infesting potato crop in Egypt. Int J Chem Tech Res 9(8):479–485
42. Lacey (1997) Manual of techniques in insect pathology. Academic Press (Elsevier), NY, p 409
43. Salem et al (2015) Lab-field evaluation of some Egyptian isolates of entomopathogenic fungi *Metarhizium anisopliae* and *Beauveria bassiana* against sugar beet beetle *Cassida vittata* Vill. Swift J Agric Res 1(2):9–14
44. Asensio et al (2003) Entomopathogenic fungi in soils from Alicante province, Spanish. J Agric Res 1(3):37–45
45. Bing-Da, Xing-Zhong (2008) Occurrence and diversity of insect-associated fungi in natural soils in China. Appl Soil Ecol 100–108
46. Charnley (1997) Entomopathogenic fungi and their role in pest control. In: Wicklow DT, Soderstrom B (eds) The mycota IV. Environmental and microbial relationships. Springer, Berlin, Heidelberg, Germany, pp 185–201
47. Gupta (1994) Relationships among enzyme activities and virulence parameters in *Beauveria bassiana* infections of *Galleria mellonella* and *Trichoplusia ni*. J Invertebr Pathol 64:13–17
48. Abdel-Raheem et al (2015) Efficacy of three entomopathogenic fungi on tomato leaf miner, *Tuta absoluta* in tomato crop in Egypt. Swift J Agric Res 1(2):15–20
49. Sabbour M, Abdel-Raheem M (2015) Efficacy of *Beauveria brongniartii* and *Nomuraea rileyi* against the potato tuber moth, *Phthorimaea operculella* (zeller). Am J Innov Res Appl Sci 1(6):197–202
50. Abdel-Raheem et al (2016) Nano entomopathogenic fungi as biological control agents on cabbage worm, *Pieris rapae* L. (Lepidoptera: Pieridae). Der Pharma Chemica 8(16):93–97
51. Abdel-Raheem MA, Ismail IA, Abdel-Rahman RS, Farag NA, Abdel-Rhman IE (2015) Entomopathogenic fungi, *Beauveria bassiana* (Bals.) and *Metarhizium anisopliae* (Metsch.) as biological control agents on some stored product insects. J Entomol Zool Stud 3(6):316–320
52. Sabbour M, Abdel-Raheem M (2015) Toxicity of the fungus *Beauveria bassiana* and three oils extract against *Sitophilus granaries* under laboratory and store condition. Am J Innov Res Appl Sci 1(7):251–256
53. Abdel-Raheem et al (2016) Isolates, virulence of two entomopathogenic fungi as biological control agent on sugar beet fly, *Pegomyia mixta* in Egypt. Der Pharma Chemica 8:132–138
54. Abdel-Raheem et al (2017) Virulence of some entomopathogenic fungi against the oliver black scale insect, *Saissetia oleae* (Oliver) on olive trees in Egypt. J Pharm Chem Biol Sci 5(2):97–102
55. Abdel-Rahman et al (2017) The strategy of anti-soft scale insect *Pulvinaria tenuivalvata* (Newstead) infesting sugarcane. J Pharm Chem Biol Sci 5:125–132
56. Hussein et al (2012) Pathogenicity of *Beauveria bassiana* and *Metarhizium anisopliae* against *Galleria mellonella*. Phytoparasitica 40:117–126

Commercialization of Biopesticides Based on Entomopathogenic Nematodes

Mahmoud M. E. Saleh, Hala M. S. Metwally and Mokhtar Abonaem

Abstract Entomopathogenic nematodes (EPN) are microscopic organisms existing in the soil and kill insects with the aid of their symbiotic bacteria. EPNs are considered safe to mammals, environment, and non-target organisms. The importance in the commercial developments of EPNs is due to its ease of mass production and exemption from registration. The importance in the commercial developments of EPNs is due to its ease of mass production and exemption from registration. These nematodes are mass produced worldwide using in vivo or in vitro techniques. In vivo culture (culture in live insect host) is low technology, has low establish costs and the quality of these nematodes is high. In vitro solid production, i.e. growing the nematode and bacteria on two-dimensional arenas containing different media or on three-dimensional rearing system (crumbled polyurethane foam). This technique offers an intermediate level of technology and costs. In vitro liquid technique requires the largest establish funds and the nematode quality is decreased. Efficiency in EPN applications can be supported through improved formulations. Recently, extensive progress has been made in developing EPN formulations, particularly for foliar applications. Efficacy of nematodes can also be increased through discovery of new strains and species, strain improvement and developed application equipment or approaches.

Keywords Entomopathogenic nematodes · In vivo production · In vitro production · Formulation · Foliar application

1 Introduction

Entomopathogenic nematodes (EPNs) are non-segmented round worms that carry bacteria inside their bodies and inhabit the soil. EPN species classified under two genera *Steinernema* and *Heterorhabditis* and are symbiotically associated with certain enterobacteria [1, 2]. Both steinernematids and heterorhabditids pass through four

M. M. E. Saleh · H. M. S. Metwally (✉) · M. Abonaem
Pests and Plant Protection Department, National Research Centre (NRC), Cairo, Egypt
e-mail: halasayed@ymail.com

© Springer Nature Switzerland AG 2020   253
N. El-Wakeil et al. (eds.), *Cottage Industry of Biocontrol Agents and Their Applications*, https://doi.org/10.1007/978-3-030-33161-0_8

juvenile instars before the maturing. Only the third-instar infective juvenile can survive outside their host and move searching for a host insect. Infective juveniles (IJs) carry symbiotic bacteria (*Xenorhabdus* spp. for steinernematids and *Photorhabdus* spp. for heterorhabditids) in their intestines and use it to kill their hosts by releasing them in the insect blood [3]. Nematodes locate and invade suitable insect hosts through the natural openings (spiracles, mouth, or anus) or in case of heterorhabditids through the cuticle of certain insects [4].

After the nematode entry to the host's hemocoel, it releases the bacteria in the insect blood. Bacteria develop and cause a septicemia that kill the insect host within 48 h. After the bacteria break down the haemolymph, the nematodes start feeding and develop and complete their life cycle. At 18–28 °C, the life cycle lasted 6–18 days depending on the insect host and the nematode species. Invading IJs belonging to *Steinernema* develop into females or males and *Heterorhabditis* IJs develop into hermaphrodites and this only the first generation. Up to three progeny generations develop inside one host and the nematode reproduction continues until host nutrients are depleted. Nematodes become third-stage IJs that emerge from the cadaver, survive in the environment and search for new insect hosts. The IJs are eligible to tolerate stresses fatal to other developmental stages. Therefore, they can be formulated and maintained for several months.

The attributes that EPNs have made them excellent insect biocontrol agents as they can kill their insect host within 48 h; have a wide host range; have the ability to move searching for hosts; they can be in vivo or in vitro mass produced; and present no hazard vertebrates and most non-target invertebrates [5–7]. The mutualistic relationship between the nematodes and the bacteria make them act together as a potent insecticidal complex against a wide range of insect species [8]. Therefore, EPNs are used as biological control agents in many agroecosystems in many countries [9, 10]. From a commercial standpoint, the production of a durable IJ stage and the symbiotic association with lethal bacteria are the most attractive features of steinernematid and heterorhabditid nematodes. The nematodes mass production plays a key role in the commercially development of insect pests management. The recent development in the nematodes mass production and formulation has increased the interest in these biocontrol agents. Therefore, there are many nematode products commercially available and used successfully in many crop systems against soil-borne insect pests. Nowadays, foliar application against insect pests on leaf foliage is possible in suitable formulations and has been elaborated in a number of cases.

The present chapter focuses on isolation, mass production, formulation, as well as field application of EPNs for the management of insect pests in different crops, vegetables and fruit orchards.

2   Entomopathogenic Nematodes Isolation

Nematode isolation is the first step to start establishing EPNs culture. Nematodes can be isolated from the soil where they naturally inhabit. The nematodes occur naturally in most of the soil types. There are some factors increases the nematodes availability like the moisture and the presence of insects which are the elements that accompany the existence of plants. Here is the simplest and the most popular technique for collecting EPNs from soil. This technique based on two main steps, firstly collecting soil samples, and secondly isolating the nematodes from the soil samples.

2.1 Soil Samples Collection

Soil samples collected from under cultivated plants should be at the fine roots area and at a depth of 15–30 cm (Fig. 1a). If the cultivated plants are trees, the proper place to collect the samples is at last point of the tree shade at noon and from two sides of the tree. The recommended sample size is 750–1000 g soil. The soil samples could be collected in paper or plastic bags and should be marked with location details, date, and the associated vegetation (Fig. 1b). The collected samples should be kept at temperature between 10 and 20 °C during the transporting to the laboratory.

Fig. 1 Procedure of soil samples collection and isolation EPNs from the collected samples. **a** The fine root area where the soil sample should be collected and at depth of 15–30 cm. **b** Soil sample in a plastic bag and a hand shovel used for collecting the sample. **c** Increasing the moisture in the soil sample by adding water using a hand-sprayer. **d** Infected *Galleria mellonella* larva after 72 h in soil sample using insect-baiting technique

2.2 The Nematode Isolation from the Collected Soil Samples

This step based on the insect-baiting technique described by Bedding and Akhurst [11]. The collected soil should be moistened with water using a Hand-sprayer and be mixed gently (Fig. 1c). After that, the soil could be divided and added to plastic cups containing 7–10 *Galleria mellonella* last instar larvae. Then the cups should be covered by their lids after making small ventilation holes. The cups could be stored at room temperature or at 22–26 °C. The cups could be checked after 3–5 days to collect the dead larvae (Fig. 1d). The cadavers should be rinsed in water and placed individually in White-traps to collect the nematode progeny. Emerged nematodes (infective juveniles IJs) would be collected after 8–10 days. The infective juveniles (IJs) could be reproduced in vivo in *G. mellonella* larvae (see procedure below in this chapter).

3 Mass Production Approaches

Production process is vital for the success of EPNs in biological control. Production techniques comprise in vivo, and in vitro techniques (solid or liquid culture). In vivo production of EPNs is the suitable method for small scale field experiments. In vivo production is also suitable for niche markets and small farmers where a lack of capital and scientific expertise. In vitro approach is used when large scale production is required at reasonable quality and cost.

3.1 In Vivo Production of EPNs

Most EPNs intended for commercial application are produced in artificial media via solid or liquid fermentation. However, for laboratory research and small greenhouse or field trials, in vivo production of EPNS is the common method of propagation. This technique has managed to maintain itself as a cottage industry. In vivo production is probable to continue as small business projects for niche markets or in developing countries where labour is reasonably priced. Advances in this mechanization and host production have led to improvements in efficiency. In vivo culture is a two dimensional system that relies on production in trays and shelves [12, 13].

3.1.1 The Host Insect

The most common insect host used for EPNs in vivo production is the greater wax moth *Galleria mellonella* last instar larvae. Also, the yellow mealworm, *Tenebrio molitor*, has been used for in vivo nematode production [14]. Here we focus on mass rearing of *G. mellonella*.

The greater wax moth *Galleria mellonella* L. is a serious pest attacking bee hives and stored bee wax. However, it is widely used as a model to study the interactions between pathogens and their hosts [15]. *G. mellonella* last instar larvae are the most used as insect hosts for producing EPNs in vivo. They have some advantages make them widely mass reared to use them for the previous purposes. The main advantages of this insect include their fast life cycle, the ease of rearing on artificial diets, the high susceptibility EPNs. There are many authors studied different proposed diets for rearing *G. mellonella* in laboratory conditions [16–19]. They studied the proposed diets effects on larval weight, longevity and/or fecundity of *G. mellonella*. Their studies results showed that no significant difference between the bee wax and artificial diets of main constituents: wheat bran, corn flour, wheat flour, rice bran, yeast, milk powder, honey, glycerol, malt, pollen, bee wax, and sucrose. The weight of full grown larva ranged between 0.2 and 0.25 g. Eischen and Dietz [20] found that adding 5% pollen; bee wax or honey significantly increased the longevity of *G. mellonella* moths.

For improving native EPNs in vivo production in *G. mellonella* larvae, Metwally et al. [21] studied the differences among the natural food (bee-wax) and four low cost different diets for mass rearing *G. mellonella*. The evaluation among these diets was based on the diet cost, the food consumption, the larvae number and weight, the larval lipid content and the nematode productivity of produced larvae. They found two diets that were costly lower than the bee wax and had no adverse effects on the studied parameters. The first suggested diet consists of wheat flour (350 g), corn bran (200 g), milk powder (130 g), yeast powder (70 g), honey (100 mL) and glycerol (150 mL). The second diet was the same previous diet, but with adding sorbitol (150 mL) instead of glycerol.

Diet preparing

In our laboratory, *G. mellonella* is permanently mass-reared using an artificial diet consists of Wheat flour 30%, Wheatgerm 30%, Corn grits 10%, Brewer's yeast 5%, Milk powder 5%, Honey 5%, and Glycerol 15%. The diet preparing could be fulfilled by firstly, mixing the dry ingredients together in one container and thereafter adding the liquid components to the mixture and manually mixing. The resulting mixture (diet) could be stored in a refrigerator till use for feeding *G. mellonella* larvae in the rearing containers.

Rearing protocol

G. mellonella culture could be started by collecting contaminated bee wax from any hives. After that, the contaminated wax could be placed in glass jars or plastic boxes as rearing containers. The rearing containers could be closed by their lids which have a handmade metal screen window for allowing gaseous exchange (Fig. 2a). As oviposition sites, Filter paper or tissue strips could be placed under the lid on the container edges (Fig. 2b). The containers should be incubated in the dark at 28–30 °C.

Fig. 2 Rearing protocol of host insect *G. mellonella*. **a** *G. mellonella* larvae inside a rearing box. **b** *G. mellonella* female moth lay eggs on paper tissue as an oviposition site. **c** *G. mellonella* deposited eggs on paper strip, then places on a fresh diet. **d** *G. mellonella* last instar larvae feed on artificial diet

The larvae feed and develop to pupae and then adults. After the mating take place, the female moths lay eggs on the paper stripes (Fig. 2b). The deposited eggs on the paper strips could be collected daily or every 48 h and placed in new rearing containers with a proper amount of the prepared diet (Fig. 2c). The eggs hatch and the emerged larvae feed on the diet and develop through six larval instars before pupations (Fig. 2d). The last instar larvae could be collected to be used in producing the nematodes in vivo. Some larvae should be left in the rearing containers to maintain the colony.

3.1.2 Culture Technique

In the first of many studies of this approach [22–26], they described systems for culturing EPNs based on the White trap that acquire advantage of the infective juveniles (IJs) usual migration away from the cadaver upon emergence (Fig. 3). These methods consist of inoculation, harvest, concentration and (if necessary) decontamination. Insects are inoculated with nematodes on dishes or trays lined with an absorbent

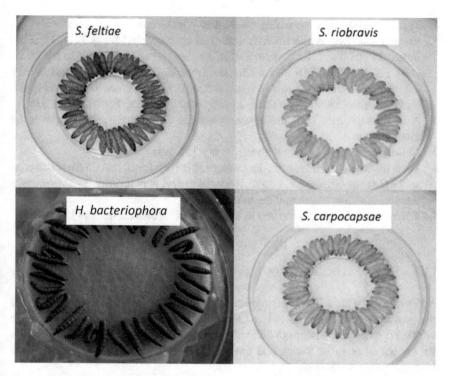

Fig. 3 White-trap: different EPN species (Nematode IJ) emerge from cadavers of *G. mellonella* larvae into water. Modified http://entnemdept.ufl.edu/creatures/nematode/entomopathogenic_nematode.htm

substrate. After 2–5 days, infected insects are transferred to the White traps i.e. harvest dishes. Following harvest, concentration of nematodes can be accomplished by gravity settling and/or vacuum filtration.

Based on the simple White trap methods indicated earlier, the process can be optimized and scaled-up to suit the needs of small field trials or cottage scale commercial ventures. Aspects that can be optimized and scaled-up include nematode species or strain as well as host species, inoculation rate and approach, host density and tray size, harvest, storage, and environmental conditions.

EPNs in vivo production protocol:

EPNs could be maintained as in vivo cultures in *G. mellonella* larvae by following production protocol:

1. Prepare nematode suspension in a concentration of 2000 IJs/mL water.
2. Apply 5 mL water to a Petri-dish (20 cm diameter) padded with two filter paper discs or paper tissue. The volume of water applied should make the discs moistened but not wetted.
3. Place about 200 *G. mellonella* last instar larvae in the previous Petri-dish contaminated with nematodes.

4. Transfer the dead larvae within 48–72 h to White traps.
5. Prepare modified White traps as described by Kaya and Stock [22]. Place an inverted Petri-dish (10 cm diameter) in another bigger Petri-dish (20 cm diameter). Fill the outer Petri-dish with 100 mL water. Drape a piece of muslin over the inverted Petri-dish with its edges touching the water. Place the cadavers were on the muslin (35–40 cadavers/dish) and cover it with the Petri-lid (Fig. 3).
6. After 10 days, new IJs could be emerged from the cadavers migrating over the muslin to the water.
7. Collect the migrated IJs in the water every 1–2 days.
8. Wash the collected EPNs by adding water and leave it for 10 min till the nematodes settle and remove the excess water. Repeat the previous step three times to remove the non-infective stages and the host tissues.
9. The washed IJs could be stored in water or special formulations at 12–14 °C.

3.1.3 Factors Affecting Yields

Production of EPNs differed among insect hosts [27]. Host diet that is improved for insect host production translates into improved efficiency in the overall process. The nutritional quality of insect host diet also impacts the quality and fitness of EPNs that are reared on those insects and thus improved efficiency and lowered costs for the entire process [28–30].

Nematode species is critical and can make a vast difference in IJs yields. Nevertheless, the nematode choice depends greatly on which insect pest one may be targeting (as efficiency will differ by species and strain). Nematode yield is generally proportional to insect host size, yet IJ yield per mg insect (within host species), and susceptibility to infection, is usually inversely proportional to host size or age. In addition to yield, ease of insect culturing and susceptibility to IJs are important factors when choosing a host. Finally, the choice of host species and nematode for in vivo production should depends on nematode yield per cost of insect, and the suitability of the nematode to the target pest [31, 32].

The technique of inoculation can be significant and may be optimized depending on nematode species and insect host [33]. Choices include pipetting, applying nematodes to insect food, or immersion the hosts in a nematode suspension. The inoculation rate (concentration of IJs and amount applied) should be optimized for each particular host and nematode species. In vivo production yields depend on nematode dosage. A dosage that is too low results in low host mortality and a dosage that is too high may result in failed infections due to competition with secondary invaders [33, 34].

Throughout the process, environmental conditions should be optimized such as for temperature, aeration, and relative humidity. Optimum production temperatures lie between 18 and 28 °C for different species [35–37]. It is also crucial to maintain adequate aeration and humidity throughout the production process [12]. To minimize overcrowding effects leading to oxygen deprivation, a pass-through HEPA filter system is implemented. Advances in mechanization and production geared toward application of nematodes through infected host cadavers can improve efficiency and economy of scale [38].

Gaugler et al. [39] and Brown et al. [40] tried a scalable system for in vivo nematode mass production. Unlike the White trap, the LOTEK system of tools and procedures provides process technology for low-cost, high-efficiency mass production. The system consists of: (1) perforated holding trays to secure insect hosts during inoculation, conditioning (synchronizing nematode emergence), and harvesting, (2) an automated, self-cleaning harvester with misting nozzles that trigger IJ emergence and rinse the nematodes through the holding trays to a central bulk storage tank, and (3) a continuous deflection separator for washing and concentrating nematodes. The separator removes 97.5% of the wastewater in three passes, while nematode concentration increased 81-fold. The rearing system offers an increase in efficiency relative to the conventional White trap method with reduced laboratory and space. Morales et al. [41, 42] and Shapiro-Ilan et al. [43] improved the process of in vivo production of EPNs by automated separation of insect from media as well as automated inoculation and harvest (Fig. 4).

3.2   In Vitro Production of EPNs

In vitro culturing of EPNs is based on introducing nematodes to a pure culture of their symbiont in a nutritive medium. Such media must use sterile ingredients to avoid unwanted bacterial contamination, retain the nematode's specific symbiotic bacterium and provide all the necessary nutrients.

3.2.1   Solid Culture

EPNs were reared in vitro for the first time on a solid medium axenically [44, 45]. Thereafter it was realized that growth increased with the presence of bacteria. Then, the importance of the natural bacterial symbiont, and monoxenic culture was recognized [46] and has been the basis for in vitro culture. House et al. [47] formulate a dog food based medium to produce *Neoaplectana carpocapsae* on a commercial scale. Hara et al. [48] and Wouts [49] who stressed on monoxenicity, reported that solid culture was first accomplished in two-dimensional arenas e.g., Petri dishes, containing various media based on dog food, pork kidney, cattle blood, and other animal products at a cost of $0.28 per million.

Bedding [50, 51] reported practical solid culture technology that was a seminal step in nematode production because it leapt from two- to three- dimensional substrates. Bedding flask cultures involved thinly coating crumbed polyurethane foam sponge with poultry offal homogenate. Sterilizing the medium in large autoclavable bags and adding the appropriate bacterium and nematode and was able to produce about 50,000 million IJs of in a week. In Pakistan, different species of EPNs were

Fig. 4 Larval tray system. **A** Stacks of modified type 3 trays sitting on top of one unmodified type 3 tray and a dolly. **B** Open system showing larvae with food on a modified tray (**a**) and frass collected in the unmodified tray at the bottom (**b**). Cited from Shapiro-Ilan et al. [38]

mass produced using chicken offal media [52]. Nematodes can be harvested within 2–5 weeks by placing the foam onto sieves, which are immersed in water. IJs migrate out of the foam, settle downward, and are pumped to a collection tank; the product is cleaned through repeated washing with water. Media for this approach were later improved (for cost and consistency) and may include various ingredients as peptone, yeast extract, eggs, soy flour, and lard. Solid culture method is economically feasible up to a production level of approximately 10×10^{12} nematodes/month [53, 54].

The produced yield of in vitro culture of EPNs depends on different factors quality [55–58].

(1) Nematode inoculum size that affect yield in some strains but not others.
(2) Culture time is inversely related to temperature and should be optimized for maximum yield on a species or strain basis.
(3) Media composition can have a substantial effect on nematode yield. Increasing the quantity of lipids will increase nematode yield and quality. The greatest accumulation of lipids per dry weight was achieved by growing nematodes in *Popillia japonica* and solid culture [59]. It is cleared that level of polar lipids was higher in nematodes produced in artificial media. So, artificial media composition should be adjusted to meet the nutritional composition of a natural host. The optimum physical and chemical components of the medium for maximum production of EPNs were studied by Dunphy and Webster [55]. Tryptic soy broth and yeast extract and D-glucose enhanced the growth and yields of EPNs. Jewell and Dunphy [60] suggested that changes in medium total lipids, neutral and polar lipids, phosphatidylcholine and total protein did not affect nematode development however; changes in total medium carbohydrate did affect IJ yields. Tangchitsomkid et al. [61], Somwong and Petcharat [62] compared the growth of different EPNson different modified artificial media The lipid agar medium in the number of harvested nematodes was 60.8×10^6 IJs per one liter. The cost of nematode production by these modified media is 4.7 times less than that of the lipid agar medium. Yoo et al. [63], Abu Hatab and Gaugler [64] produced *H. bacteriophora* in media containing various lipid sources. They revealed that lipid source significantly affected lipid quantity and quality in *H. bacteriophora*. Media supplemented with extractable insect lipids produced yields 1.9 times higher than did beef fat- or lard-supplemented media. The modified dog biscuit medium recorded the positive result with respect to successful mass production of *H. indicus*. Ehlers [65] described the biology of the nematode-bacterium complex and advised that mass production of EPNs must be directed towards media development and cost reduction, as the bacteria are able to metabolize a variety of protein sources to provide optimum conditions for nematode reproduction. An inoculum level of 2000 IJs per flask yielded highest nematode of 60.11×10^6 IJs per flask which was significantly superior over other inoculum levels [66].

The inoculum size and the time are important for optimizing the final yields of IJs. The highest yield for *H. bacteriophora* was found with an inoculum of 10 IJs per flask, which was tenfold of the optimal inoculum for *S. carpocapsae*. Relationship between inoculum sizes, population development and the final IJ populations of these nematodes should improve the efficiency of commercial nematode production [58].

Bedding et al. [67] developed a culture vessel comprising a tray with side walls and overlapping lids that allowed gas exchange through a layer of polyether–polyurethane foam. These trays are particularly well suited for developing countries as forced aeration is not necessary, making this system independent from cuts in the power supply. Nematodes can be extracted from solid media with centrifugal sifters, or by washing nematodes out of the sponge in simple washing machines and then separating the IJs by sedimentation or migration.

Gauglar and Han [68] stated that the approach was expanded to autoclavable bags with filtered air being pumped in Ehlers [69] reported that mass production relies on the scaling-up of culture volumes from flask cultures to volumes of several cubic meters. Stability of beneficial traits is a prerequisite for production of high quality insect control nematodes. Beneficial traits of nematodes are reproduction potential, longevity of the IJs, their host seeking ability and infectivity and tolerance to environmental stress factors.

In this context, Metwally [70] investigated the possibility of in vitro solid mass production of two Egyptian nematodes *S. carpocapsae* BA2 and *H. bacteriophora* BA1 on different agar media in comparison to two worldwide nematodes *S. riobrave* and *H. marilatus*. The Egyptian isolate *S. carpocapsae* BA2 can be successfully culture on all tested media. This was the first trial to propagate the Egyptian isolates *S. carpocapsae* BA2 and *H. bacteriophora* BA1 on an artificial medium (Fig. 5). The worldwide species *S. riobrave* and *H. marelatus* can be successfully culture on all tested media. Also, we determined the impact of inoculum sizes (bacteria and nematodes) on the growth and yield of *S. carpocapsae* BA2 and which inoculum size result in the greatest yield.

Also, the appropriateness of mass production of local and worldwide EPNs using Bedding flasks has been clarified in two solid media, modified Wouts medium and modified dog food medium. Neither *S. carpocapsae* BA2 nor *H. bacteriophora* BA1 could continue growth in the studied media in Bedding-flasks. However, the world wide species, *S. riobrave*, *S. scaptrisci*, and *H. marelatus* succeeded in propagation in Bedding-flasks (Fig. 6). Worldwide species *S. riobrave* achieved the highest offspring production scoring 45.54 × 10^6 IJs/flask.

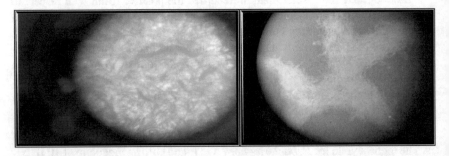

Fig. 5 Microscopic examination of *Steinernema carpocapsae* BA2 propagated on agar plates (x = 40). Reprinted from Metwally [70]

(a)   **(b)**

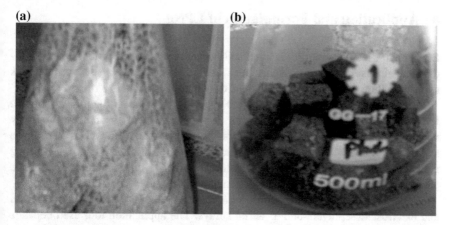

Fig. 6 Infective juveniles (IJs) started in emerging on the walls of the conical flasks. **a** *Steinernema riobrave*, **b** *Heterorhabditis marilatus*. Cited from Metwally [70]

3.2.2 Liquid Culture

Liquid rearing for EPNs was aimed first by Stoll [71]. Symbiotic bacteria are cultured in the shaker followed by the nematodes [72, 73]. The yield reaches approximately 400 IJs/mL at 21–25 °C and the yield was more in the dark.

Factors affecting yield for in vitro liquid culture:

1. The requirements of adequate aeration (without shearing).
2. The life cycles and reproductive biology of both genera. Males and females in Steinernematids are capable of mating in liquid culture [74] thus; maximization of mating is supreme and can be accomplished through regulation of ventilation [75]. However, mating in heterorhabditid is not applicable in liquid culture because the first generation is exclusively hermaphrodites and, although subsequent generations contain amphimictic forms, they cannot mate in liquid culture [74]. Thus, maximizing heterorhabditid yields in liquid culture depends on the degree of recovery. While levels of heterorhabditids recovery in vivo tend to be 100% [76], recovery in liquid culture may range from 0 to 85%.
3. Recovery can be affected by nutritional factors, aeration, CO_2, lipid content, and temperature [63, 76, 77].
4. Also, yield of EPNs from liquid culture may be affected by additional factors including nematode inoculum, species and media [77, 78]. The vital constituent of the liquid culture media is lipid source [63, 79]; glucose [80] and yeast extract content [81].
5. EPNs yield is inversely proportional to the size of the species [77]. Maximum yields reported include 300,000 and 320,000 IJs/mL for *H. bacteriophora* and *S. carpocapsae*, respectively [78].

4 Application and Formulation of EPNs

EPNs have used successfully against soil borne insect pests. But also foliar application against some insect pests has been studied in some cases. Also, formulations and application methods were investigated by many researchers.

Application technique is vital for the success of (EPNs) in biological control. Improved efficiency in applications can be supported through improved delivery means (e.g., optimization of spray equipment). Recently, essential advancement has been made in formulations of EPNs, particularly for foliar applications, e.g., mixing nematodes with surfactants or polymers. Insect host cadavers can improve nematode persistence and decrease the extent of nematodes desired per unit area.

EPNs can be applied with almost all agriculture ground tools, e.g., pressurized and electrostatic sprayers and as aerial sprays. The application tool used depends on the cropping system. The volume, agitation, nozzle type, pressure and recycling time must be taken into consideration [82–84]. Applicators may also be using other techniques such as through baits or subsurface injection. A range of formulations may be used for applying EPNs in aqueous suspension including water dispersible granules (WDG), activated charcoal, alginate and polyacrylamide gels, clay, peat, polyurethane sponge and vermiculite.

4.1 Factors Influencing Application Success

Mainly, the suitable nematode must be matched with the exacting target pest. The virulence of nematodes, persistence and environmental tolerance must be taken into consideration [85]. Furthermore of paramount significance, to be effective, entomopathogenic nematodes usually must be used to soil at minimum rates of 2.5×10^9 IJs/ha or higher [32]. Seasonal re-application is frequently essential where nematode populations will remain high enough to provide effective pest control for 2–8 weeks after application. Biotic agents can affect on nematode applications. EPNs have been announced to act synergistically with some entomopathogens such as *Bacillus thuringiensis* [86] and *Metarhizium anisopliae* Sorokin [87–89] however, other studies indicate antagonism, e.g., *Beauveria bassiana* [90].

Application of EPNs depends on some significant factors including temperature, protection from ultraviolet radiation and sufficient soil moisture [83, 91]. Certainly, entomopathogenic nematode applications for aboveground pests have been strictly limited due to environmental obstructions (e.g., UV radiation or desiccation) that reduce survival and efficacy [83, 92]. Also, Soil parameters can be vital for belowground or surface applications. Soil texture affects nematode movement and survival [91, 93]. Soil pH can affect entomopathogenic nematode distributions [94]. Chemical pesticides and fertilizers may be affected on entomopathogenic nematodes [95]. The relationship between chemical pesticides and EPNs fluctuates based on the nematode species or strain, dosages, and specific chemical and timing of application [96, 97].

Improved efficacy in nematode applications can be supported through enhanced formulation. Efficacy may also be accomplished on foliage with the addition of surfactants to increase leaf exposure [98, 99]. Nematode applications for control of the lesser peach tree borer were significantly enhanced by application of a sprayable gel [100]. Cadaver application technique increase nematode dispersal [101], infectivity [102], and survival [103]. This application may be facilitated through formulations that have been developed to protect cadavers from rupture and enhance ease of handling [104, 105] and progress of mechanized equipment for field application [106]. Also, advanced applications with EPNs can be accomplished through strain improvement or enhanced levels of various beneficial traits of EPN such as environmental tolerance, pathogenicity and progeny production capacity. Species discovery or genetic enhancement via selection, hybridization or molecular manipulation enhances entomopathogenic nematode application [9, 107].

4.2 Nematode Applications

There are many studies on nematodes utilization against insect pests in different crops, vegetables and fruit orchards. Here we present some studies performed using EPNs against some insect pests in different crop systems.

4.2.1 Apple

Sammour and Saleh [96] conducted a laboratory and field experiments to study the compatibility between two Egyptian entomopathogenic nematodes, *Heterorhabditis bacteriophora* (S1) and *Steinernema carpocapsae* (S2) and two well known organophosphate insecticides (Cidial 50% EC and Basudine 60% EC) to control *Zeuzera pyrina* larvae in fasting apple trees.

4.2.2 Sugarbeet

Saleh et al. [108] applied EPNs on sugar beet crop infested with larvae, pupae, and adults of the sugar beet beetle *Cassida vittata*. They applied *S. carpocapsae* at a concentration of 1000 IJs/mL. Within one week, the results were 65, 92, and 57.3% mortality in larvae, pupae, and the adults, respectively. Also, Saleh et al. [109] used entomopathogenic nematodes (EPNs) for the biological control of the sugar beet fly *Pegomyia mixta* in Egypt. They applied *S. feltiae* or *H. bacteriophora* against *P. mixta* on Sugar beet plants in field bioassay. The results were respectively 81.3 or 75.9% reduction in the larval population.

4.2.3 Peach

Saleh et al. [110] studied the efficacy of the new isolate *Heterorhabditis marelatus* D1 from Egypt against the peach fruit fly, *Bactrocera zonata* (Saunders). When the new isolate applied to the soil, resulted over 77% of *B. zonata* adults emerged from their pupae within 48 h after emergence.

4.2.4 Potato

Moawad et al. [111] studied treatments of EPNs *S. carpocapsae* and *H. bacteriophora* for the biological control of the potato tuber moth, *Phthorimaea operculella* (Zell.) infesting potato tubers in the soil. Their work included protective (against insects outside the tubers) and curative (against insects inside the tubers). In the protective treatments the nematodes decreased up to 100% of the pest population.

4.2.5 Maize

Saleh and El-Kifl [112] applied *H. bacteriophora* on the hibernating larvae of corn borer, *Ostrinia nubilalis* inside their tunnels in stored corn stalks. They found that the nematodes were able to kill the hibernating larvae inside their tunnels.

Saleh et al. [113] treated *H. taysearae* and *H. bacteriophora* on corn plants infested with *Sesamia cretica* larvae. Within one week, the applied nematode species resulted 67.8 and 40.6% larval mortality, respectively. El-Wakeil and Hussein [114] evaluated the efficacy of two EPN species against *Sesamia cretica* larvae infesting corn hearts. One week post spraying *H. bacteriophora* and *S. carpocapsae*, the results were 97% and 100% larval mortality, respectively. After two weeks, the larval mortality was 100% due to both EPN species.

4.2.6 Date Palm

Saleh et al. [115] applied the nematodes on soil around the palm trunks targeting the adults of red palm weevil, *Rhynchophorus ferrugineus* in soil or the pupae inside their cocoons aggregated in the palm leaf petioles (Fig. 7). The applications of *H. bacteriophora*, *H. indica*, and *S. carpocapsae* caused mortality in pre-pupae and pupae inside their cocoons reached 98.3%, 90.4%, and 60.3%, respectively. *Steinernema* or *Heterorhabditis* on soil resulted up to 90% mortality in the insect adults.

5 Conclusion

Entomopathogenic nematodes are important biological control agents against soil pests as well as plant-boring pests. Nematodes have been commercially developed by several companies in North America, Europe and Australia and for the control of

Fig. 7 Nematodes application on soil around the palm trunks targeting the adults of red palm weevil, *Rhynchophorus ferrugineus*

a vast range of pests. Progress in developing large-scale production and application technology has led to the expanded use of EPNs. In vivo nematode production is the suitable technique for niche markets and small-scale field-testing. This method requires the least capital outlay and the least amount of technological proficiency, but is blocked by the costs of insect host media. Thus, producing the insect hosts "in-house" in low-cost culture and mechanizing the process is a must for large scale production of EPNs especially in a developing country like Egypt. Also, technical enhancement and reducing the cost of the ingredients of the artificial media will develop efficiency and large-scale production of in vitro solid culture. Also, EPNs have been commercially produced by numerous companies in large liquid fermentation tanks in different industrialized countries. However, this technique requires greater funds investment and an advanced level of technical proficiency. There are different biotic and abiotic factors can affect efficiency of EPN application. Nematodes can be suppressed a diversity of economically significant insect pests.

Our case study deals with the biological control of beet fly *Pegomyia mixta*, the most significant pest of strategic crop of sugar industry in Egypt, using different species of entomopathogenic nematodes. All tested nematodes killed the larvae inside their mines in the sugar beet leaves and pupae in the soil and developed in their bodies. In Egypt, as in many other parts of the world, the peach fruit fly, *Bactrocera zonata* has ever been considered a major orchard pest attacking several fruits. A particularly inviting approach to selective control employs EPNs that enter into the host through its mouth opening, spiracles and anus and kill their hosts through the association with their bacteria when auspiciously applied on both guava fruits infested with *B. zonata* eggs and the soil under these fruits or spraying soil containing old pupae of this pest. The potato tuber moth (PTM), *Phthorimaea operculella* is one of the more serious potato pests in the world. Our work indicated that an integrated control program including EPNs would be valuable against the PTM immature stages outside or inside the potato tubers and may reduce the reliance on chemical insecticides in controlling the PTM. Formulation of nematodes is destined to develop absorption, storage stability, activity, delivery and ease-of-use. Application of nematodes for above ground pests is an advance in the field of bio-insecticides. Our work adds evidence that some formulation adjuvants are valuable in improving field efficiency of EPNs in controlling the larvae of the cotton leaf worm *Spodoptera littoralis* infesting leaves of corn seedlings.

6   Future Prospects

Additional technological advancements are desired to expand and develop the market potential of the nematode-based biopesticides. Recent progress in mass production involves using cheap materials for the nutrient diets of the insect host of in vivo technique and also economic solid media for in vitro production. Isolation of additional species and selective breeding are required for proper classification, for biodiversity studies. This will also contribute to enhance the economic value of EPNs in biological control. Improved efficiency in nematode applications can be supported through improved formulation. Recently, the trend towards improving the aboveground application of nematodes, e.g., mixing EPNs with a polymer and surfactants to increase leaf coverage. Also, efficacy of EPNs can be increased through improved application equipments, e.g., optimizing spray systems (e.g., nozzles, pumps, spray distribution) for increased nematode dispersion and survival. Advanced EPN applications can also be accomplished through strain enhancement. Enhanced strains may possess different valuable traits such as reproductive capacity and environmental tolerance. Genetic enhancement via selection, hybridization or molecular manipulation is a way to improve applications of nematodes.

References

1. Poinar GO Jr (1990) Taxonomy and biology of *Steinernematidae* and *Heterorhabditidae*. In: Gaugle R, Kaya HK (eds) Entomopathogenic nematodes in biological control. CRC Press. Boca Raton, pp 23–61
2. Stock SP, Hunt DJ (2005) Nematode morphology and systematics. In: Grewal PS, Ehlers RU, Shapiro-Ilan DI (eds) Nematodes as biological control agents. CAB International Publishing, Wallingford, UK, pp 3–43
3. Akhurst RJ, Boemare NE (1990) Biology and taxonomy of *Xenorhabdus*. In: Gaugler R, Kaya HK (eds) Entomopathogenic nematodes in biological control. CRC Press, Boca Raton, Florida, pp 75–90
4. Peters A, Ehlers RU (1994) Susceptibility of leatherjackets (*Tipula paludosa* and *T. oleracae*, Tipulidae: Nematocera) to the entomopathogenic nematode *Steinernema feltiae*. J Invertebr Pathol 63:163–171
5. Gaugler R, Boush GM (1979) Nonsusceptibility of rats to the entomogenous nematode *Neoaplectana carpocapsae*. Environ Entomol 8:658–660

6. Lacey LA, Georgis R (2012) Entomopathogenic nematodes for control of insect pests above and below ground with comments on commercial production. J Nematol 44:218–225
7. Lacey LA, Grzywacz D, Shapiro-Ilan DI, Frutos R, Brownbridge M, Goettel MS (2015) Insect pathogens as biological control agents: back to the future. J Invertebr Pathol 132:1–41
8. McMullen II JG, Stock SP (2014) In vivo and in vitro rearing of entomopathogenic nematodes (*Steinernematidae* and *Heterorhabditidae*). J Vis Exp 91:e52096
9. Grewal PS, Ehlers RU, Shapiro-Ilan DI (2005) Nematodes as biological control agents. CABI Publishing, Wallingford, UK
10. Jenkins DA, Shapiro-Ilan D, Goenaga R (2007) Virulence of entomopathogenic nematodes against *Diaprepes abbreviatus* in an oxisol. Florida Entomol 90:401–403
11. Bedding RA, Akhurst RJ (1975) A simple technique for the detection of insect parasitic rhabditid nematodes in soil. Nematol 21:109–110
12. Shapiro-Ilan DI, Gaugler R (2002) Production technology for entomopathogenic nematodes and their bacterial symbionts. J Ind Microbiol Biotechnol 28:137–146
13. Ehlers RU, Shapiro-Ilan DI (2005) Mass production. In: Grewal P, Ehlers RU, Shapiro-Ilan D (eds) Nematodes biological control agents. CABI Publishing, Wallingford, UK, pp 65–79
14. Shapiro-Ilan DI, Gaugler R, Tedders WL, Brown I, Lewis EE (2002) Optimization of inoculation for in vivo production of entomopathogenic nematodes. J Nematol 34:343–350
15. Jorjão AL, Oliveira LD, Scorzoni L, Figueiredo-Godoi LMA, Prata MCA, Jorge AOC, Junqueira JC (2018) From moths to caterpillars: ideal conditions for *Galleria mellonella* rearing for in vivo microbiological studies. Virulence 9:383–389
16. Seung Wook L, DongWoon L, HoYul C (2007) Development of economical artificial diets for greater wax moth, *Galleria mellonella* (L.). Korean J Appl Entomol 46:385–392
17. Birah A, Chilana P, Shukla UK, Gupta GP (2008) Mass rearing of greater wax moth (*Galleria mellonella* L.) on artificial diet. Indian J Entomol 70:389–392
18. ChengHua H, XueHong P, DongFa H, BoHui W, JiLi W (2010) Screening of artificial feed formula for *Galleria mellonella* (L). Guangxi Agric Sci 41:672–674
19. Yan WY, Ting Z, Qiang W, PingLi D, XingXin G, QiHun L, Feng L, HuaiLei S (2010) Optimization of artificial diets for *Galleria mellonella*. Chin Bull Entomol 47:409–413
20. Eischen FA, Dietz A (1990) Improved culture techniques for mass rearing *Galleria mellonella* L. (Lepidoptera: Pyralidae). Entomol News 101:123–128
21. Metwally HM, Hafez GA, Hussein MA, Hussein MA, Salem HA, Saleh MME (2012) Low cost artificial diet for rearing the greater wax moth, *Galleria mellonella* L. (Lepidoptera: Pyralidae) as a host for entomopathogenic nematodes. Egypt J Biol Pest Control 22:15–17
22. Kaya HK, Stock SP (1997) Techniques in insect nematology. In: Lacy I (ed) Manual of techniques in insect pathology. Academic Press, San Diego, pp 281–393
23. White CF (1927) A method for obtaining infective larvae from culture. Science 66:302–303
24. Dutky SR, Thompson JV, Cantwell GE (1964) A technique for the mass rearing the wax moth (Lepidoptera: Galleriadae). Proc Entomol Soc Wash 64:56–58
25. Poinar GO (1979) Nematodes for biological control of insects. CRC Press, Boca Raton, FL
26. Woodering JL, Kaya HK (1988) Steinernematid and heterorhabditid nematodes: a handbook of techniques. Ark Agric Exp Stat South Coop Bull 331:430
27. Salma J, Shahina F (2012) Mass production of eight Pakistani strains of entomopathogenic nematodes (*Steinernematidae* and *Heterorhabditidae*). Pak J Nematol 30:1–20
28. Shapiro-Ilan DI, Guadalupe RM, Morales-Ramos JA, Lewis EE, Tedders WL (2008) Effects of host nutrition on virulence and fitness of entomopathogenic nematodes: lipid and protein based supplements in *Tenebrio molitor* diets. J Nematol 140:13–19
29. Shapiro-Ilan DI, Rojas G, Morales-Ramos JA, Tedders WL (2012) Optimization of a host diet for in vivo production of entomopathogenic nematodes. J Nematol 44:264–273
30. Morales-Ramos JA, Rojas MG, Shapiro-Ilan DI, Tedders WL (2011) Nutrient regulation in *Tenebrio molitor* (Coleoptera: Tenebrionidae): SELF-selection of two diet components by larvae and impact on fitness. Environ Entomol 40:1285–1294
31. Dolinski C, Del-Valle EE, Burla RS, Machado IR (2007) Biological traits of two native Brazilian entomopathogenic nematodes (Heterorhabditidae: Rhabditida). Nematol Bras 31:180–185

32. Dias PVC, Dolinski C, Molina JPA (2008) Influence of infective juvenile doses and *Galleria mellonella* (Lepidoptera: Pyralidae) larvae weight in the in vivo production of *Heterorhabditis baujardi* LPP7 (Rhabditida: Heterorhabditidae). Nematol Bras 32:317–321
33. Shapiro-Ilan DI, Gouge DH, Koppenhöfer AM (2002) Factors affecting commercial success: case studies in cotton, turf and citrus. In: Gaugler R (ed) Entomopathogenic nematology. CABI Publishing, Wallingford, UK, pp 333–356
34. Boff M, Wiegers GL, Gerritsen LJM, Smits PH (2000) Development of the entomopathogenic nematode *Heterorhabditis megidis* strain NLH-E 87.3 in *Galleria mellonella*. Nematology 2:303–308
35. Hazir S, Stock SP, Kaya HK, Koppenhofer AM, Kestin N (2001) Developmental temperature effects on five geographic isolates of *Steinernema feltiae* (Nematoda: Steinernematidae). J Invertebr Pathol 75:81–92
36. Karagoz M, Gulcu B, Hazir S, Kaya HK (2009) Laboratory evaluation of Turkis entomopathogenic nematodes for suppression of the chestnut pests, *Curculio elephas* (Coleoptera: Curculionidae) and *Cydia splendana* (Lepidoptera: Tortricidae). Biocontrol Sci Technol 19:755–768
37. Morton A, Gracía-del-Pino F (2009) Ecological characterization of entomopathogenic nematodes isolated in stone fruit orchard soils of Mediterranean areas. J Invertebr Pathol 102:203–213
38. Shapiro-Ilan DI, Morales-Ramos JA, Rojas MG (2016) In vivo production of entomopathogenic nematodes. In: Glare T, Moran-Diez M (eds) Microbial-based biopesticides. Methods in molecular biology, vol 1477. Humana Press, New York, NY, pp 137–158
39. Gaugler R, Brown I, Shapiro-Ilan D, Atwa A (2002) Automated technology for in vivo mass production of entomopathogenic nematodes. Biol Control 24:199–206
40. Brown I, Gaugler R, Shapiro-Ilan D (2004) LOTEK: an improved method for in vivo production of entomopathogenic nematodes. Int J Nematol 14:9–12
41. Morales-Ramos JA, Rojas MG, Shapiro-Ilan DI, Tedders WL (2009) Automated insect separation system. US patent 8:025-027 B1
42. Morales-Ramos JA, Rojas MG, Shapiro-Ilan DI, Tedders WL (2011) Automated insect separation system. Application no. 12/536,221. Patent number 8025027
43. Shapiro-Ilan DI, Tedders WL, Morales-Ramos JA, Rojas MG (2009) Insect inoculation system and method. Patent disclosure, docket no. 0107.04. Submitted to the US patent and trademark office on 12/11/2009. Application no. 12636245
44. Glaser RW (1932) Studies on *Neoaplectana glaseri*, a nematode parasite of the Japanese beetle (*Popillia japonica*). N J Agric 211:34
45. McCoy EE, Glaser RW (1936) Nematode culture for Japanese beetle control, p 10. New Jersey Department of Agric circular no. 265
46. Poinar GO Jr, Thomas GM (1966) Significance of *Achromobacter nematophilus* Poinar and Thomas (Achromobacteraceae: Eubacteriales) in the development of the nematode DD-136. Parasitology 56:385–390
47. House HL, Welch HE, Cleugh TR (1965) Food medium of prepared dog biscuit for the mass production of the nematode DD-136 (Nematoda: Steinernematidae). Nature 206:8–17
48. Hara AH, Lindegren JE, Kaya HK (1981) Monoxenic mass production of the entomopathogenic nematode, *Neoaplectana carpocapsae* Weiser, on dog food/agar medium. Adv Agric Technol AAT-W 16:1–8
49. Wouts WM (1981) Mass production of the entomogenous nematode, *Heterorhabditis heliothidis* (Nematoda: Heterorhabditidae) on artificial media. J Nematol 13:467–469
50. Bedding RA (1981) Low cost in vitro mass production of *Neoaplectana* and *Heterorhabditis* species (nematoda) for field control of insect pests. Nematology 27:109–114
51. Bedding RA (1984) Large scale production, storage, and transport of the insect-parasitic nematode *Neoaplectana* spp. and *Heterorhabditis* spp. Ann Appl Biol 104:117–120
52. Tabassum KA, Shahina F (2004) In vitro mass rearing of different species of entomopathogenic nematodes in monoxenic solid culture. Pak J Nematol 22:167–175

53. Friedman MJ, Langston SE, Pollitt S (1989) Mass production in liquid culture of insect-killing nematodes. International Patent Appl. WO 89/04602
54. Ramakuwela T, Hatting J, Mark DL, Hazir S, Thiebaut N (2016) In vitro solid-state production of *Steinernema innovationi* with cost analysis. Biocontrol Sci Technol 26:792–808
55. Dunphy GB, Webster JM (1989) The monoxenic culture of *Neoaplectana carpocapsae* DD136 and *Heterorhabditis heliothidis*. Rev Nematol 12:113–123
56. Han R, Cao L, Liu X (1992) Relationship between medium composition, inoculums size, temperature and culture time in the yields of *Steinernema* and *Heterorhabditis* nematodes. Fund Appl Nematol 15:223–229
57. Han R, Cao L, Liu X (1993) Effects of inoculum size, temperature and time on in vitro production of *Steinernema carpocapsae* agriotos. Nematology 39:366–375
58. Wang JX, Bedding RA (1998) Population dynamics of *Heterorhabditis bacteriophora* and *Steinernema carpocapsae* in in vitro solid culture. Fund Appl Nematol 21:165–171
59. Abu Hatab M, Gaugler R (1999) Lipids of in vivo and in vitro cultured *Heterorhabditis bacteriophora*. Biol Control 15:113–118
60. Jewell JB, Dunphy GB (1995) Altered growth and development of *Steinernema carpocapsae* DD136 by *Xenorhabdus nematophilus* mutants. Fund Appl Nematol 18:295–301
61. Tangchitsomkid N, Sontirat S, Chanpaisaeng J (1999) Monoxenic culture of a New Thai strain of entomopathogenic nematodes (*Steinernema* sp. KB strain) on artificial media. Thammasta Int J Sci Technol 4:49–53
62. Somwong P, Petcharat J (2012) Culture of the entomopathogenic nematode *Steinernema carpocapsae* (WEISER) on artificial media. ARPN J Agric Biol Sci 7:229–232
63. Yoo SK, Brown I, Gaugler R (2000) Liquid media development for *Heterorhabditis bacteriophora*: lipid source and concentration. Appl Microbiol Biotechnol 54:759–763
64. AbuHatab M, Gaugler R (2001) Diet composition and lipids of in vitro-produced *Heterorhabditis bacteriophora*. Biol Control 20:1–7
65. Ehlers RU (2001) Mass production of entomopathogenic nematodes for plant protection. Appl Microbiol Biotechnol 56:623–633
66. Anjum S, Prabhuraj A (2007) Mass production of *Heterorhabditis indicus* in different artificial media. Karnataka J Agric Sci 20:859–860
67. Bedding RA, Stanfield MS, Compton GW (1991) Apparatus and method for rearing nematodes, fungi, tissue culture and the like, and for harvesting nematodes. International Patent Appl no. PCT/91/00136, 48p
68. Gaugler R, Han R (2002) Production technology. In: Gaugler R (ed) Entomopathogenic nematology. CABI, Wallingford, UK, pp 289–310
69. Ehlers RU (2008) Stability of beneficial traits during liquid culture of entomopathogenic nematodes. IOBC/wprs Bull 31:4–7
70. Metwally HM (2013) Improving production and potency of bio-insecticides based on entomopathogenic nematodes. Ph.D. thesis, Entomology Department, Ain Shams University, Egypt
71. Stoll NR (1952) Axenic cultivation of the parasitic nematode, *Neoaplectana glaseri*, in a fluid medium containing raw liver extract. J Parasitol 39:422–444
72. Surrey MR, Davies RJ (1996) Pilot scale liquid culture and harvesting of an entomopathogenic nematode, *Heterorhabditis*. Fund Appl Nematol 17:575–582
73. Strauch O, Ehlers RU (2000) Influence of the aeration rate on yields of the biocontrol nematodes *Heterorhabditis megidis* in monoxenic liquid cultures. Appl Microbiol Biotechnol 54:9–13
74. Strauch O, Stoessel S, Ehlers RU (1994) Culture conditions define automictic or amphimictic reproduction in entomopathogenic rhabditid nematodes of the genus *Heterorhabditis*. Fund Appl Nematol 17:575–582
75. Neves JM, Teixeira JA, Simoes N, Mota M (2001) Effect of airflow rate on yield of *Steinernema carpocapsae* Az 20 in liquid culture in an external-loop airlift bioreactor. Biotechnol Bioeng 72:369–373

76. Strauch O, Ehlers RU (1998) Food signal production of *Photorhabdus luminescens* inducing the recovery of entomopathogenic nematodes *Heterorhabditis* spp. in liquid culture. Appl Microbiol Biotechnol 50:369–374

77. Ehlers RU, Niemann I, Hollmer S, Strauch O, Jende D, Shanmugasundaram M, Mehta UK, Easwaramoorthy SK, Burnell A (2000) Mass production potential of the bacto-helminthic bio-control complex *Heterorhabditis indica—Photorhabdus luminescens*. Biocont Sci Technol 10:607–616

78. Han RC (1996) The effects of inoculum size on yield of *Steinernema carpocapsae* and *Heterorhabditis bacteriophora* in liquid culture. Nematology 42:546–553

79. Abu-Hatab M, Gaugler R, Ehlers RU (1998) Influence of culture method on *Steinernema glaseri* lipids. J Parasitol 84(2):215–221

80. Jeffke T, Jende D, Matje C, Ehlers RU, Berthe-Corti L (2000) Growth of *Photorhabdus luminescens* in batch and glucose fedbatch culture. J Appl Microbiol Biotechnol 54:326–330

81. Chavarria-Hernandez N, de la Torre M (2001) Population growth kinetics of the nematode, *Steinernema feltiae*, in submerged monoxenic culture. Biotechnol Lett 23:311–315

82. Wright DJ, Peters A, Schroer S, Fife JP (2005) Application technology. In: Grewal PS, Ehlers RU, Shapiro-Ilan DI (eds) Nematodes as biocontrol agents. CABI, New York, NY, pp 91–106

83. Shapiro-Ilan DI, Gouge DH, Piggott SJ, Patterson Fife J (2006) Application technology and environmental considerations for use of entomopathogenic nematodes in biological control. Biol Control 38:124–133

84. Lara JC, Dolinski C, Fernandes de Sousa E, Figueiredo Daher E (2008) Effect of mini-sprinkler irrigation system on *Heterorhabditis baujardi* LPP7 (Nematoda: Heterorhabditidae) infective juvenile. Sci Agric 65:433–437

85. Shapiro-Ilan DI, Stuart RJ, McCoy CW (2006) A comparison of entomopathogenic nematode longevity in soil under laboratory conditions. J Nematol 38:119–129

86. Koppenhöfer AM, Kaya HK (1997) Additive and synergistic interactions between ento-mopathogenic nematodes and *Bacillus thuringiensis* for scarab grub control. Biol Control 8:131–137

87. Ansari MA, Tirry L, Moens M (2004) Interaction between *Metarhizium anisopliae* CLO 53 and entomopathogenic nematodes for the control of *Hoplia philanthus*. Biol Control 31:172–180

88. Ansari MA, Shah FA, Moens TM (2006) Field trials against *Hoplia philanthus* (Coleoptera: Scarabaeidae) with a combination of an entomopathogenic nematode and the fungus *Metarhizium anisopliae* CLO 53. Biol Control 39:453–459

89. Anbesse SA, Adge BJ, Gebru WM (2008) Laboratory screening for virulent ento-mopathogenic nematodes (*Heterorhabditidis bacteriophora* and *Steinernema yirgalemense*) and fungi (*Metarhizium anisopliae* and *Beauveria bassiana*) and assessment of possible syn-ergistic effects of combined use against grubs of the barley chafer *Coptognathus curtipennis*. Nematology 10:701–709

90. Brinkman MA, Gardner WA (2000) Possible antagonistic activity of two entomopathogens infecting workers of the red imported fire ant (Hymenoptera: Formicidae). J Entomol Sci 35:205–207

91. Kaya HK (1990) Soil ecology. In: Gaugler R, Kaya HK (eds) Entomopathogenic nematodes in biological control. CRC Press, Boca Raton, FL, pp 93–116

92. Arthurs S, Heinz KM, Prasifka JR (2004) An analysis of using entomopathogenic nematodes against above-ground pests. Bull Entomol Res 94:297–306

93. Barbercheck ME (1992) Effect of soil physical factors on biological control agents of soil insect pests. Florida Entomol 75:539–548

94. Kanga FN, Waeyenberge L, Hauser S, Moens M (2012) Distribution of entomopathogenic nematodes in Southern Cameroon. J Invertebr Pathol 109:41–51

95. Bednarek A, Gaugler R (1997) Compatibility of soil amendments with entomopathogenic nematodes. J Nematol 29:220–227

96. Sammour EA, Saleh MME (1996) Combination of entomopathogenic nematodes and insecti-cides for controlling of apple borer, *Zeuzera pyrina* L. (Lepidoptera: Cossidae). J Union Arab Biol (A) Zool 5:369–380

97. Koppenhöfer AM, Grewal PS (2005) Compatibility and interactions with agrochemicals and other biocontrol agents. In: Grewal PS, Ehlers RU, Shapiro-Ilan DI (eds) Nematodes as biological control agents. CABI Publishing, Wallingford, UK, pp 363–381
98. Head J, Lawrence AJ, Walters KFA (2004) Efficacy of the entomopathogenic nematode, *Steinernema feltiae*, against *Bemisia tabaci* in relation to plant species. J Appl Entomol 128:543–547
99. Saleh MME, Hussein MA, Hafez GA, Hussein MA, Salem HA, Metwally HM (2015) Foliar application of entomopathogenic nematodes for controlling *Spodoptera littoralis* and *Agrotis ipsilon* (Lepidoptera: Noctuidae) on corn plants. Adv Appl Agric Sci 3:51–61
100. Shapiro-Ilan DI, Cottrell TE, Mizell RF, Horton DL, Behle B, Dunlap C (2010) Efficacy of *Steinernema carpocapsae* for control of the lesser peach tree borer, *Synanthedon pictipes*: improved aboveground suppression with a novel gel application. Biol Control 54:23–28
101. Shapiro DI, Glazer I (1996) Comparison of entomopathogenic nematode dispersal from infected hosts versus aqueous suspension. Environ Entomol 25:1455–1461
102. Shapiro DI, Lewis EE (1999) Comparison of entomopathogenic nematode infectivity from infected hosts versus aqueous suspension. Environ Entomol 28:907–911
103. Perez EE, Lewis EE, Shapiro-Ilan DI (2003) Impact of host cadaver on survival and infectivity of entomopathogenic nematodes (Rhabditida: *Steinernematidae* and *Heterorhabditidae*) under desiccating conditions. J Invertebr Pathol 82:111–118
104. Shapiro-Ilan DI, Lewis EE, Behle RW, McGuire MR (2001) Formulation of entomopathogenic nematode-infected-cadavers. J Invertebr Pathol 78:17–23
105. Shapiro-Ilan DI, Morales-Ramos JA, Rojas MG, Tedders WL (2010) Effects of a novel entomopathogenic nematode-infected host formulation on cadaver integrity, nematode yield, and suppression of *Diaprepes abbreviatus* and *Aethina tumida* under controlled conditions. J Invertebr Pathol 103:103–108
106. Zhu H, Grewal PS, Reding ME (2011) Development of a desiccated cadaver delivery system to apply entomopathogenic nematodes for control of soil pests. Appl Eng Agric 27:317–324
107. Burnell A (2002) Genetics and genetic improvement. In: Gaugler R (ed) Entomopathogenic nematology. CABI, New York, NY, pp 333–356
108. Saleh MME, Draz KAA, Mansour MA, Hussein MA, Zawrah MFM (2009) Controlling the sugar beet beetle *Cassida vittata* with entomopathogenic nematodes. J Pest Sci 82:289–294
109. Saleh MME, Draz KA, Mansour MA, Hussein MA, Zawrah MF (2011) Controlling the sugar beet fly *Pegomyia mixta* Vill. with entomopathogenic nematodes. Commun Agric Appl Biol Sci 76:297–305
110. Saleh MEE, Metwally HM, Mahmoud YA (2018) Potential of the entomopathogenic nematode, *Heterorhabditis marelatus*, isolate in controlling the peach fruit fly, *Bactrocera zonata* (Saunders) (Tiphritidae). Egypt J Biol Pest Control 28(22). https://doi.org/10.1186/s41938-018-0029-0
111. Moawad SS, Saleh MME, Metwally HM, Ebadah IM, Mahmoud YA (2018) Protective and curative treatments of entomopathogenic nematodes against the potato tuber moth, *Phthorimaea operculella* (Zell.). Biosci Res 15:2602–2610
112. Saleh MME, Elkifl TAH (1994) Virulence of certain entomopathogenic nematodes to the European corn borer, *Ostrinia nubilalis*. Egypt J Biol Pest Control 4:125–131
113. Saleh MME, Matter MM, Hussein MA (2000) Efficiency of entomopathogenic nematodes in controlling *Sesamia cretica* (Lepidoptera: Noctuidae) in Egypt. Bull NRC Egypt 25:181–188
114. El-Wakeil NE, Hussein MA (2009) Field performance of entomopathogenic nematodes and an egg parasitoid for suppression of corn borers in Egypt. Arch Phytopathol Plant Prot 42:228–237
115. Saleh MME, Kassab AS, Abdelwahed MS, Alkhazal MH (2010) Semi-field and field evaluation of the role of entomopathogenic nematodes in the biological control of the red palm weevil *Rhynchophorus ferrugineus*. Acta Hortic 882:407–412

Insect Viruses as Biocontrol Agents: Challenges and Opportunities

Adly M. M. Abd-Alla, Irene K. Meki and Güler Demirbas-Uzel

Abstract Insect viruses were isolated from many insect pests from different families to represent a potential alternative for chemical pesticides. Viruses from families baculoviruses, cypoviruses, and densoviruses have been registered as biological control agents. Insect viruses are considered effective and environmental-friendly which may contribute to the achievement of sustainable agriculture goals through providing a suitable alternative to the chemical insecticides which have negative impacts on the environment and to the non-target organisms. However, the application of insect viruses as bio-control agents also have certain limitations. These include their slow action to their target, narrow host range, problems associated with the large-scale production and the development of insect host resistance against certain viruses. This chapter will discuss the challenges and the prospective use of insect viruses as biological control agents.

Keywords Insect viruses · Field application · Challenges and opportunities

1 Introduction

Insect viruses as biological control agent represent an important component in the integrated pest management (IPM) programs, mainly because it is specific and safe for the environment and compatible with the other integrated pest management components. Under certain situations Insect viruses cause epizootics in the field which provide added control of pests in nature. As insect viruses, i.e. baculoviruses are active in controlling early larval stage of lepidopteran insects, other IPM components such as chemical pesticides can be used to control late larval instars and entomopathogenic nematodes can target pupal stages under ground while mating disruption pheromone can be used to reduce mating in adult stages. Insect viruses were isolated from insect

A. M. M. Abd-Alla (✉) · I. K. Meki · G. Demirbas-Uzel
Insect Pest Control Laboratory, Joint FAO/IAEA Division of Nuclear Techniques in Food and
Agriculture, Vienna International Centre, P.O. Box 100, 1400 Vienna, Austria
e-mail: a.m.m.abd-alla@iaea.org

A. M. M. Abd-Alla
Pests and Plant Protection Department, National Research Centre, Cairo, Egypt

© Springer Nature Switzerland AG 2020
N. El-Wakeil et al. (eds.), *Cottage Industry of Biocontrol Agents and Their
Applications*, https://doi.org/10.1007/978-3-030-33161-0_9

from different insect order such as Lepidoptera, Diptera, Orthoptera and Coleoptera [1–3].

This chapter will cover the following topics; the first part will focuss on the history of insect viruses as biocontrol agents for economic pests, followed by a discussion on the large scale production and commercialization of insect viruses with some examples. Thereafter, the emergence of resistance development and strain composition as challenges for insect viruses will be discussed with highlights on the recent reports in this field. Later the future and perspectives for the use of insect viruses as biocontrol agent will be summarized.

2   History for Insect Viruses as Biocontrol for Economic Pests

Pest management is an important part of both agricultural and forestry production system. In the past, as standard control measure chemicals were used. However, currently due to the negative impact of the chemical pesticides on the environment and the non-target organism and the increase in public concern to these effects, other methods, such as biological control approaches are becoming a viable option for the use of chemicals [4]. Biological control agents are living organisms that interfere with the productivity of other living organisms. In terms of biotechnology, biological control agents are used by human beings for the protection of resources that they want [5]. Biological control agents are composed of a wide range of organisms from vertebrates, insects, mites, plants, fungi, bacteria and viruses. Biological control programs have been successfully used to control noxious weeds, plant pathogens, invertebrate, and vertebrate pests [6]. The production, deployment, and the establishment of biological control agents are crucial parameters in determining the success of these agents for pest management. In addition, the latest improvement in genetic engineering has incorporated some biological control agents into the genomes of crops [7–9].

2.1   Biological Control of Insect, Mites and Nematode Pests

The reason why biological control has been used is to develop new, alternative, and environmentally friendly control agents. For instance, some insect viruses are promising candidates due to their inherent characteristics that appear among the promising candidates. Today, five groups of pathogens containing bacteria, viruses, protozoa, and nematodes are being used for control of insect pests. The microbial agents like viruses, bacteria, fungi and entomopathogenic nematodes were developed as biological control agents for a wide variety of arthropod pests [6, 10].

A variety of insect, mite and nematode pest control agents have been used as bio-control. Reported examples are predators that feed on pests, parasitoids that lay eggs that grow in the pest and later kill it, parasites that weaken the pest and pathogens that infect and kill their pest host [11]. The most common form of the biological control method is the classical one whereby exotic natural enemies are introduced to control the exotic pests [12]. Three hundred million hectares of land (8% of agricultural land) is used for traditional biological control. During the last 120 years, nearly 2000 species of exotic arthropod agents have been announced as arthropod pests in 196 countries or islands. More than 170 species are commercially available for pest control [13]. For instance, *Rodolia cardinalis*, the vedalia ladybird beetle is used as a biological control agent against the insect pest cottony cushion scale, on commercial citrus in California which was one of the first large-scale successes. Since then, the *Vedalia* beetle has now been controlling cotton cushion scale for over 100 years in more than 50 countries. Insect-specific viruses such as baculoviruses, have been extracted, assessed and mass-produced for use against codling moth in apple orchards. So far, the best outcome for management of codling moth occurs when virus applications are combined with mating disruption. Parasitic wasp (*Hymenoptera: Encyrtidae*) is also used as a biocontrol agent in orchards [14].

2.2 Biocontrol of Vertebrate Pests

Biological control technology has also been used against vertebrate pests (rabbits, rats, etc.) and veterinary and medical pests (nuisances, parasites, and diseases). For instance, the myxoma virus (*Leporipoxvirus, Poxviridae*) which causes myxomatosis, was introduced into Australia and spread rapidly throughout and reduced the rabbit population around 75–95% in 1950. In Australia, mosquitos were the main carriers for spreading the virus among rabbits. Finally, the rabbits developed immunity to the virus with virulence decreasing to 50%. [15]. In 1995, a new virus, rabbit hemorrhagic disease (RHD) was established on the mainland of Australia, and rat mortality again decreased to 50–90% specifically in the dry regions. Flies, mosquitos and rabbit fleas were the main vectors. In the temperate areas, rabbit populations returned to pre-RHD levels [15]. The use of viral pathogens led to a dramatic decline at the beginning, however vertebrate pests eventually developed varying degrees of genetic resistance. In addition, the use of microbes has been directed against human beings in times of war through direct application of the agents or using intermediary vectors such as insects to spread disease [16].

2.3 Viruses as Biological Control Agents

Insect viruses are obligate disease-causing organisms that could solely reproduce within a host insect. In the frame of integrated pest management program, viruses are the agents that provide the most safe, efficient and sustainable control method of insect pests. Some of the insect viruses are registered and produced as commercial products. However other viruses are naturally occurring, and they might even initiate an outbreak without any trigger or by unknown triggers especially for fruit fly pests [17]. Virus diseases were reported from more than 800 species of insects and mites. The family baculoviridae composes the most common and widely studied group of pathogenic viruses of insects. Baculoviruses are invertebrate-specific pathogens that in some cases were developed as biopesticides for the control of insects particularly species of the Lepidoptera [18].

Baculoviruses and cypoviruses (cytoplasmic polyhedrosis virus groups) have some biological properties which lead to their successful use as microbial control agents in integrated pest management programs such as, having protective occlusion bodies (polyhedra) which protect and increase the sustainability of these viruses. Baculoviruses generate two different phenotypes in their replicative cycle: the occlusion-derived virus (ODV) needed to spread the infection between larvae, and the budded virus (BV) needed for the dissemination of the infection within the host [19]. They are restricted to a single type of insect and extremely unambiguous for their host range systems. Numerous virus types have been determined among the members of arthropod species.

Research on insect virus in China recorded that more than 200 insect viruses isolates from several virus families like Baculoviridae, Reoviridae, Densovirinae, and Entomopoxvirinae can cause epizootics in natural populations of insects [7]. Some viruses are registered and commercially available, but their selectivity and small potential market might restrict their industrial interest. However, improvements in virus production, formulation and a better understanding of virus epizootiology should shed light to their enhanced role for this group of insect pathogens to become largely used as biological control. As microbial control agents, baculoviruses (nucleopolyhedroviruses and granuloviruses) are the most widely preferred and utilized [2, 20, 21]. The viruses are usually transmitted *per os* and gain access to host tissues via the midgut where the OBs that surrounded the virus rods are softened. In addition to baculoviruses, viruses from crypoviruses and densoviruses are registered as biological control agents in China. Recently nudiviruses such as the virus of *Oryctes rhinoceros* L. has been reported as the most successfully used non-occluded virus. Viruses comprise most of the host-specific entomopathogens, but their main drawbacks are the requirements for invivo production and their sensitivity to ultra-violet degradation.

2.4 Virus Discovery and Detection

Invertebrate pathology is a recently organized discipline, but its roots can be traced to ancient history concerning solutions for preventing disease in honey bees and silkworms [22, 23]. The use of microorganisms for control of insect pests was suggested by Basi, Louis Pasteur, and Elie Metchnikoff [22, 24]. The use of fungi as microbial control agents was suggested by several researchers in the late 19th century. It was not until the development of the bacterium *Bacillus thuringiensis Berliner* that the use of microbes for the control of insects became extensive. Today, plenty of entomopathogens are used for the control of invertebrate pests in a glass house, row crops orchards, ornamentals, range turf, lawn, stored products, forestry, for the abatement of pest and vector insects of veterinary and medical importance [25–27]. Similar to other natural enemies, insect pathogens could be used as a natural control way of the targeted populations. Most of the epizootics are naturally occurring due to viral and fungal pathogens that are responsible for spectacular crashes of insect pest populations [28].

Virus infection begins in the insect's digestive system but spreads throughout the whole body of the host in fatal infections. The body tissues of virus-killed insects are almost completely converted into virus particles. The digestive system is among the last internal organ system to be destroyed, and therefore the infected insects usually continue to feed until they die. Infected insects appear normal until just before their death when they tend to darken in color and move slowly. They often develop slow than uninfected individuals [29–31]. Most virus-infected insects die while attached to the plant on which they feed. Virus-killed insects break open and disseminate virus particles into the environment. These virus particles can infect new insect hosts. Because of the damage of the inner tissues, dead insects often have a "melted" look. The contents of a dead insect can range from milky-white to dark brown or black. While natural virus outbreaks tend to be localized, virus particles can be spread by the movement of infected insects, or the predators such as other insects or birds that come into contact with the infected insects, or non-biological factors like water run-off, rain-splash or air-borne soil particles. Many virus-infected insects also climb to higher positions on their host plant before they die, which maximizes the spread of virus particles after the insect dies and disintegrates. The number of virus infection cycles within a growing season depends heavily on the insect's life cycle. Insect pests with multiple generations per season or longer life cycles can be more heavily impacted by virus outbreaks since there is a greater opportunity for multiple virus infection cycles within a growing season [32].

2.5 Virus Group Used as Biological Control

The major groups of insect viruses that might be used as biocontrol agents for economic pests are presented in the table below (Table 1).

At least 11 groups of viruses, including the Baculoviridae, Reoviridae, Poxviridae, and the Iridoviridae, are known to cause diseases in insects [33]. Baculoviridae are confined to arthropods; replicating in Lepidoptera (butterflies and moths), Hymenoptera (wasp), Diptera (flies), Coleoptera (beetles), Neuroptera (lacewings), Arachnida (spiders) and Crustacea (prawns and shrimps) [34]. In other groups of viruses, some members are also pathogenic to vertebrates and/or plants [35]. Only

Table 1 Major groups of entomopathogenic viruses

Virus groups	Nucleic acids	Morphology of virus particle	Presence of inclusion body	Host range in arthropods	Related virus groups
Baculoviridae (a) Nuclear polyhedrosis virus (b) Granulosis virus	Circular dsDNA	Bacilliform	+	Lepidoptera, diptera, hymenoptera, coleoptera, neuroptera, crustacea, mites	No viruses with similar morphology found in vertebrates or plants
Nudiviridae	Circular dsDNA	Bacilliform	–	Coleoptera	No viruses with similar morphology found in vertebrates or plants
Reoviridae (cytoplasmic polyhedrosis viruses, cypoviruses)	Segmented dsRNA	Isometric	+	Diptera, lepidoptera, hymenoptera, crustacea	Other members of Reoviridae infect vertebrate and plant hosts
Poxviridae – Entomopoxviruses	Linear dsDNA	Brick-shaped	+	Lepidoptera, diptera, coleoptera, orthoptera	Other members of Poxviridae infect vertebrate hosts
Iridoviridae (Iridoviruses)	Linear dsDNA	Isometric	+	Diptera, lepidoptera, coleoptera, hymenoptera, hemiptera	Other proposed members of Iridoviridae infect vertebrate hosts
Parvoviridae (densoviruses)	Linear dsDNA	Isometric	–	Diptera, lepidoptera, orthoptera	Other members of parvoviridae infect vertebrate hosts
Picornaviridae Small RNA viruses (many groups)	ssRNA	Isometric	–	Diptera, lepidoptera, hymenoptera, hemiptera orthoptera, coleoptera hymenoptera, hemiptera	Viruses with similar morphological and bio-chemical characteristics infect vertebrate and plant hosts

Adapted from [33]

one insect-pathogenic virus Nodamura virus (family: Nodaviridae) is known to infect a vertebrate and an insect and is transmissible to suckling mice by *Aedes aegypti* [36]. Most of the research and developments of viral bioinsecticides activities are focused on Baculoviridae due to its safety and the large number of viruses isolated from economic pests. Baculoviruses include over 1690 viruses that have been recorded from more than 1100 species of insects and mites [37, 38]. Of these, three families (Baculoviridae, Polydnaviridae, Ascoviridae) are specific to insects and related arthropods. The baculoviruses are the most widely exploited virus group for biocontrol: they are very different from viruses that infect vertebrates and are considered safe to be used as biopesticides.

The mode of pathogenesis and replication of entomopathogenic viruses vary according to the family, but the infection nearly always occurs through ingestion by the host. Virions then bind to receptors in the gut and penetrate epithelial cells and initiate infection. In baculoviruses, the infection often spreads to the hemocoel and then to essential organs and tissues, particularly fat bodies. Acute infections lead to host death in 5–14 days. There are two genera of baculoviruses: nucleopolyhedroviruses (NPV) and granuloviruses (GV). The host range of baculoviruses is restricted to the order, and usually the family of origin of the host. The commercial production for baculoviruses as biopesticides are considered to present a minimum risk to people and wildlife. Mass production of baculoviruses can only be done in vivo but it is economically viable only for larger hosts such as Lepidoptera, and the formulation and application are straightforward [37]. Currently, there are approximately 16 biopesticides based on baculoviruses presented for use or under development. The majority of these products are targeted against *Lepidoptera*. For example, codling moth granulovirus, CpGV (*Cydia pomonella* Granulovirus) is an effective biopesticide of codling moth caterpillar, pests of apples [39].

The baculovirus virions are enveloped rod-shaped nucleocapsids comprising circular, supercoiled, double-stranded DNA. The virions of GVs are individually occluded in a protein matrix (granulin). In the NPVs single enveloped (SNPV) or multiple enveloped (MNPV) virions are occluded in a protein matrix (polyhedrin). The occlusion bodies or polyhedra are dissolved in the alkaline environment of the host's insect midgut after being absorbed by the host. The free virions enter the gut epithelial cells and replicate in the nuclei. Non-occluded virus particles that are budded from the gut cells into the hemocoel attack other tissues (fat, tracheal matrix, hypodermis, etc.) within the host. Virus particles that are occluded within polyhedra are generally the infective inoculum for subsequent hosts. Part of baculovirus virions transmission may be facilitated by predators and ovipositing parasitoids via mechanical transmission [28, 35].

As with other biocontrol agents, there are three basic strategies for the use of entomopathogenic viruses as microbial agents, which include inoculation, augmentation, and conservation. In most crops the use of viral pathogens of insects is intentive and does not use their full epizootic potential, but take the advantage of their virulence and specificity [33]. Baculoviruses registered for use or under development for insect pest control are presented in Table 2.

Table 2 Baculoviruses registered for use or under development for control of insect pests in agroecosystems, stored products, and forestry

Virus	Target insect	Crop	Commercial product	Country	Reference
Anticarsia gemmatalis NPV	The velvet bean caterpillar	Soybeans		Brazil, Paraguay	[52]
Cydia pomonella (CpGV)	The codling moth	Pear and apple	Cyd-X and VirosoftCP4 in North America and in Europe include Carpovirusine™ (France), Madex™ and Granupom™ (Switzerland), Granusal™ (Germany), and Virin-CyAP	North America and Europe	[53]
Helicoverpa armigera NPV	The cotton bollworm, corn earworm, tobacco budworm budworms, corn earworm	Cotton, *Row Crops*	Gemstar LC (Certis USA)	China, USA	[54]
Mamestra brassicae NPV	Cabbage moth	Vegetables		China	
Autographa californica NPV	Beet armyworm Alfalfa looper and several other lepidopteran species	Vegetables		China	[55]
Spodoptera exigua NPV	Beet armyworm	Vegetables	Spod-X LC (Certis USA)	China	
Spodoptera litura NPV	Cotton leafworm	Vegetables		China	

(continued)

Table 2 (continued)

Virus	Target insect	Crop	Commercial product	Country	Reference
Plutella xylostella GV	Diamondback moth	Vegetables		China	
Trichoplusia ni NPV	Cabbage looper	Vegetables			[56]

Adapted from [7]

3 Large Scale Production and Commercialization of Insect Viruses

Baculoviruses have been isolated from more than 500 host species, with most of them from nucleopolyhedroviruses; 456 in Lepidoptera, 30 in Hymenoptera and 27 viruses in Diptera. Granuloviruses are specific to Lepidoptera with 148 reported cases [39]. The use of insect viruses as biological control agents for insect pests was mainly focused on baculoviruses, one cypovirus (CPV) and one densovirus which have been registered as commercial bioinsecticides in China [7, 8].

There have been many reviews on the development of baculoviruses that are currently registered and used to control insects on large scale. Some of these viruses are listed in Table 1. Based on the treated area, three viruses are the most used ones and are reviewed by Rohrman [40]. In brief, the NPV of the velvet bean caterpillar *Anticarsia gemmatalis* isolated in 1977 in Brazil became the most successful example of a virus used as a biological pesticide with more than 2 million ha treated area per year in 2006, providing effective and safe control of larvae of the key crop defoliator of soybean fields [21, 41, 42]. The second most widely used viral pesticide is the granulovirus of the codling moth, *Cydia pomonella* (CpGV) in many countries in North America and Europe since 2000 for the control of the insect on pear and apple crops. The virus was isolated in Mexico in 1963 [43] and it is currently produced under many commercial products in different countries (see Table 1). The third most widely used viral biopesticide is the NPV of the cotton bollworm, *Helicoverpa armigera* isolated in china in 1976, and was commercially produced as biopesticide in 1993 to control this pest. In 2005, the production of this biopesticide was around 1600 tons of infected insect [7, 8, 40]. In addition to these examples many other viral products are currently being used as biological control, some of them are list in Table 1.

3.1 Method for the Large-Scale Production of Viral Pesticides

Insect viruses are obligate pathogens and so far, cannot be produced outside the host cells. Therefore, the most common method for the virus production is in vivo through infecting the host insect in the production facility or in field by using the host insect cell line for the virus production.

3.1.1 Virus Production by Infecting Insect Host in Production Facilities

Most of the viral biopesticides are produced through infection the insect host in production facilities such as the production of CpGV and HearMNPV. This method requires the establishment of large-scale rearing facility of the insect host and infection of the susceptible stage (larvae) at relatively later stage with optimized virus concentration and ensuring the production of the highest viral production. The maintenance of mass production facility of the insect host requires the implementation strict hygiene measures to avoid accidental pathogen infection in the insect colony. In most cases, two separate production lines should be maintained, healthy colony and viral production line. This method is mainly suitable for insect host with biology enabling the economic mass-rearing.

3.1.2 Virus Production by Collecting Infected Larvae from Treated Filed

In some cases where the mass rearing of the host is not economically feasible or the production of the virus through infecting the host is not enough; collecting the infected larvae from the field remain the suitable option. This method was used in Brazil to produce the NPV of the velvet bean caterpillar, *Anticarsia gemmatalis* when the laboratory production was not found to be economically viable. In this case the virus production was carried out in farmers' fields. Plots of soybeans that were naturally infested with *A. gemmatalis* were sprayed with virus and then the dead larvae were collected 8–10 days after virus application. In this method, the major problem is that the virus production will depend on the natural host, prevalence and consequently the cost of the collection. In Brazil, individuals were able to collect about 1.8 kg of larvae/day at a cost in the mid-1990s of about $15. However, the viral production varied in the 1990s from enough virus to treat 650,000 to 1.7 million ha/year. By 1999, the production of virus was not sufficient to meet the demand [40].

3.1.3 Virus Production in Insect Host Cell Line

Due to the various challenges associated with the production of baculoviruses as biopesticides in infected larvae system such as the high cost and the unexpected failure in the healthy colony, the production of the viruses under fully controlled system such as insect cell line became an option to meet the market requirement. The cell culture growing in suspension that display a doubling time of 24 h or less and can be scaled up to 10,000 L in airlift or stirred tank bioreactors represent a feasible option for the virus large scale production. In some cases, the viral host is not feasible for mass-rearing and there is no cell line available for production of the virus and therefore an alternative host cell line such as the production of the nonoccluded *Oryctes* nudivirus, OrNV, using an adherent coleopteran cell line are used [44].

There are several cell culture systems that have received attention for their ability to produce insect viruses in large scale such as the *Heliothes zea* (HzAMI) cell line to produce *Helicoverpa armigera* nucleopolyhedrovirus (HearNPV), the *Spodoptera frugiperda* SF9 cell line to produce *S. frugiperda* multicapsid nucleopolyhedrovirus SfMNPV and the *Anticarsia gemmatalis* cell line to produce *Anticarsia gemmatalis* nucleopolyhedrovirus, AgMNPV [45–51]. However, managing problems related to stability of the virus strains in culture, enhancing virus yields per cell through an understanding of how the host cell responds to the infecting virus, and the development of chemically defined media and feeds for the desired production systems remains the main challenges for cell culture-based production of insecticidal viruses [44]. In the above-mentioned examples, the virus production remains an option to overcome the production cost problem. However, in the case of the production of the non-occluded *Oryctes* nudivirus (OrNV) in the slow growing adherent coleopteran cell line DSIR-HA-1179, it is mandatory to overcome limitations associated with production of the virus in infected larvae. This system is well described by Reid et al. [44].

4 Resistance Development and Strain Composition as Challenges for Insect Viruses

As stated previously in this chapter, baculoviruses are among the insect viruses that are regarded safe for use as biological control agents due to their host-specificity and therefore, in this chapter they have been used as examples. Their efficient use as biological control agents is determined by the interaction of the virus and the insect-host in the field. The most important requirement for a successful control is the correct virus dosage to mortality ratio of an insect pest. This ratio changes with time due to circumstances in the field that lead to insect resistance to baculoviruses thus affecting the efficiency of the virus as a control measure. Different insect populations present variable responses to particular virus infections as well as a considerable variability in individual insect responses to different virus dosage. There are three main factors that determine the response of an insect to virus infection which contribute to the resistance development. These are categorised into the developmental, environmental and genetic, with each of them influencing the expression of insect resistance to baculoviruses. The developmental and environmental factors mostly lead to short-term resistance while the genetic factors mainly affect the long-term expression of resistance. This section of the chapter will discuss the mechanisms of resistance development in insects to baculoviruses and how the interplay between these factors influence the application of baculoviruses as biocontrol agents.

4.1   Mechanisms of Resistance Development

4.1.1   Developmental Factors

These are factors related to the growth and maturation of an insect through differ-ent stages. The relationship between the age of an insect and its response to virus infections has been reported in many insects such as *Lymantria dispar* and *Helio-this virescens* resistance to baculovirus [57–59]. It has been demonstrated that the resistance of an insect increases with larval age mainly due to the ability of an insect host to renew midgut cells during larval development which allows the elimination of the infected cells [60]. Further, as larval weight is related to developmental stage, the increase in resistance is reported to be directly proportional to the weight of the larvae. The increase in the larval weight reduces the surface-volume ratio of the midgut of the larvae and therefore increase the probability of the virus particles to pass through the midgut without attaching to the susceptible epithelial cells [60, 61]. Levy et al. [62] demonstrated increased resistant of *Anticarsia gemmatalis* larvae to AgMNPV which was correlated to increasing thickness of the peritrophic membrane (PM) which would protect the infection of the midgut cells. However, the presence of stilbene-derived optical brighteners which interfere with the synthesis of chitin, (a component of the PM), is known to reduce developmental resistance [63, 64]. Other physiological changes related to larval age/weight such as increase in gut PH with age are also known to affect an insect's resistance to virus infection. Therefore, the highest level of resistance is observed in the final-instar larvae just prior to pupation which confirms the age-related resistance. Moreover, this has also been associated with the shift in the balance of juvenile and molting hormones at the late instar stage [65]. This resistance justifies the importance of early treatments when using baculoviruses for pest control.

4.1.2   Environmental Factors

This involves the external influences that may trigger the insect defense mechanisms and therefore lead to the expression of resistance. Environmental factors affect the relationship between the virus and its insect-host, by acting either directly to the virus and affect its prevalence in the field or directly on the insect and alter its response to virus infection. This was initially evidenced in silk moth *Bombyx mori*, whereby high viral infections were observed in *B. mori* larvae in autumn than in spring [66]. These differences were associated with the quality of the mulberry leaves fed by the larvae due to changes in the sucrose, protein and cellulose levels in the leaves. For instance, protein levels directly affect the antiviral and protease activity of the larval digestive fluids while sucrose deficiency increases the uptake to the virus by the midgut epithelial cells and hence increase their susceptibility to virus infection. Physical factors such as temperature and light have also been shown to influence insect resistance [61]. For example, high temperature treatments increase resistance levels of late-instar larvae to virus infections. In the case of light, the level of resistance increases with the duration of larvae exposure to light, as reported for *B. mori* larvae that were reared in constant darkness were found to be more susceptible to virus infection than those reared in light [66]. This suggests that insects exposed to abnormal environmental conditions are more susceptible to virus infections.

4.1.3 Genetic Factors

These are factors that influence gene expressions, therefore regulating the immune defense mechanisms of an insect to virus infection. There are several genetic factors that contribute to the variability in resistance development in insects. First, most of the baculoviruses used as biological control agents are endemic since they have coevolved with their hosts. Due to this virus-host coevolution, particular host genes may become fixed in a certain population to offer resistance to virus infections. Differences in susceptibility to baculovirus infections have been reported in geographically distinct populations of the same species. Other studies investigating different baculovirus strains concluded that the resistance maybe influenced by a complex genetic mechanism or by single autosomal genes (either dominant or recessive alleles). For instance, resistance of *Cydia pomonella* (codling moth) to *C. pomonella* granulovirus (CpGV) is restricted specifically to the CpGV-M genotype and not to other isolates such as CpGV-I12 or CpGV-R5 [67, 68]. This resistance difference between the isolates has been associated with a viral gene, pe38, which for instance allows CpGV-R5 to replicate and not CpGV-M in resistant larvae [69]. In addition, the heredity of CpGV-M resistance is described as monogenic and sex-linked, since heterozygous and homozygous males showed different levels of resistance which could be as a result of a gene dosage effect [67, 70, 71]. Second, studies on selection of resistance genes within a population have shown that some individual insects possess a resistance gene, however, this requires several generations under selection pressure from the virus. For example, after selection against AgMNPV, *Anticarsia gemmatalis* reverted to the original levels after a few generations without virus treatment [72, 73]. On the other hand, *C. pomonella* resistance to CpGV-M remained stable for over 30 generations without virus treatment [74]. The existence of viruses in latent forms has also been shown to contribute to resistance, since they can be activated by several stressors including other viruses or virus strains which are genetically variable and subject to selection for changes in virulence.

4.2 Interplay Between Developmental, Environmental and Genetic Factors to Resistance Development

The interaction between these three major factors contribute to resistance development by affecting the insect's defense mechanisms to viral attack. First, resistance to virus infection can develop at any stage of virus infection following the initial viral attack. The most common route of baculoviruses entry is *per os* through the midgut lumen followed by the attachment, entry and establishment of infection in the midgut columnar cells and later the entry of virus particles into hemocoel to initiate secondary infections. Although there are several defense mechanisms that can act at any stage of the infection process, majority of the defences are activated during the initial viral invasion which prevents successful entry of the virus to the susceptible midgut cells. This contributes to resistance development to peroral infection and can greatly be affected by the developmental factors. For example, one of the insect's defense mechanism against viral invasion involves discharge of infected midgut cells into the gut lumen at each larval molting stage and their replacement with new cells. In addition, this type of defense (discharge and regeneration of columnar cells) can be influenced by genetic factors since in some insect populations, such as *B. mori* infected with cytoplasmic polyhedrosis virus (CPV), the regenerated columnar cells became re-infected while, *Hyphantria cunea* larvae infected with the BmCPV, the regenerated cells became immune to subsequent infection [60, 75].

On the other hand, environmental factors can interact with both genetic and developmental factors and influence the initial defense mechanism; light can influence the cell composition of the midgut epithelium, temperature can induce cellular discharge while both light and nutrient levels can affect the synthesis of antiviral agents. Temperature and photoperiod can affect the metabolic rates of an insect via hormone production and affect the rate of virus infection. For example, in two populations of *Pieris brassicae* larvae subjected to the same nutritional stress presented difference in response to GV infection; one population showed increase in susceptibility while the other showed resistance. This shows a correlation between the three major factors affecting the expression of resistance in that, as an insect develops a high virus dose is required to initiate a lethal infection, however, there are defense mechanisms that exist to counterattack the infection which are genetically (host or viral) influenced and may be subject to environmental stimuli [76].

The effects of the environmental factors in the field application appear to cause small changes in response since they involve aspects of nutrition or climate that are uncontrollable. Genetic factors are associated with the developmental factors which mainly cause age-related resistance. The genetic factors are subject to selection which can affect long term procedures for baculovirus application since a relatively small change in response could alter the cost effectiveness of viral control. For instance, frequent application of baculoviruses at high dose can increase the risk of resistance build-up by destabilizing any coevolved host-virus balance and therefore promote spread of resistance. As stated earlier, the age-related (developmental) resistance likely develops into resistant individuals than into susceptible ones. While the resistance may develop in an early instar larva, the selection pressure is usually stronger if late instar larvae were exposed to virus. Hence, in many cases the most cost-effective method of baculoviruses application in the field is against the early larval instars of the pests which reduces the risk of selection for increased resistance as well as early protection to the crops [60, 65, 77].

5 Future Perspectives for the Use of Insect Viruses as Biocontrol Agent

The specificity and the production of secondary inoculum make baculoviruses and other insect viruses attractive alternatives to chemicals insecticides and ideal components of Insect Pest Management systems due to their lack of unwanted negative effects on nontargeted beneficial insects including other biological control organisms or any negative impact on the environment and the ecosystems [78–80]. In addition, the use of insect viruses as bioinsecticides is compatible with many other components of biological control agents such as insect predators and parasitoids or other insect pathogens such as entomopathogenic bacteria or fungi in the frame of integrated pest-management. In addition, the fact that insect viruses are unable to infect mammals, including humans, makes them very safe to handle and attractive candidates as alterative biopesticides to avoid the use of the harmful chemical pesticides. However, despite the above-mentioned advantages, insect virus biopesticides products still represent a small fraction of the insect pesticides market, mainly due to certain limitations such as the narrow host range, the slow killing and loss of effect due to the exposure to UV light in the sun and recently the development of resistance in the host insect against the used viruses. Therefore, the future for the continuous use of the insect viruses will depend on the success to overcome these limitations.

The narrow host range can be faced through the use of biopesticide composed of virus mixture to increase the range of effectiveness of one product that can be used against several pests which will increase the market value of such product. The development of formulation which include protectant materials against UV could increase the sustainability of the viral product that can tolerate the UV effect and therefore increase the virus persistence [81, 82]. The use of recombinant bocaviruses that include the deletion of virus genes that delay the virus killing (e.g. the deletion of the ecdysteroid UDP-glucosyltransferase (*egt*) gene) or the expression of toxins that accelerate the killing effect has been developed for some viruses. However there are several challenges facing the large-scale production of these viruses as fast killing of the host affect negatively the amount of produced virus from infected host [7–9]. Finally, the use of the correct virus (or a mixture of virus) strains in the biopesticides to overcome the development of resistant against the virus in the host population might help to face the resistance challenges [68]. The success in facing the abomination limitation will shape the future use of insect viruses as biopesticides to control the major insect pests.

References

1. Beas-Catena A, Sánchez-Mirón A, García-Camacho F, Contreras-Gómez A, Molina-Grima E (2014) Baculovirus biopesticides: an overview. J Anim Plant Sci 24:362–373
2. Hunter-Fujita FR, Entwistle PF, Evans HF, Crook NE (1998) Insect viruses and pest management. In: Insect viruses and pest management [cited 2019 Apr 12]. Available from: https://www.cabdirect.org/cabdirect/abstract1105344

3. Lacey LA, Grzywacz D, Shapiro-Ilan DI, Frutos R, Brownbridge M, Goettel MS (2015) Insect pathogens as biological control agents: back to the future. J Invertebr Pathol 132:1–41
4. Herzfeld D, Kay S (2011) Integrated pest management. In: Private pesticide applicator study manual, 19th edn. University of Minnesota Extension, Minnesota. Available from: http://apps. extension.umn.edu/agriculture/pesticide-safety/ppat_manual/Intro.pdf
5. Thézé J, Lopez-Vaamonde C, Cory JS, Herniou EA (2018) Biodiversity, evolution and ecological specialization of baculoviruses: a treasure trove for future. Appl Res Viruses 10
6. Lacey LA, Frutos R, Kaya HK, Vail P (2001) Insect pathogens as biological control agents: do they have a future? Biol Cont 21:230–248
7. Sun X (2015) History and current status of development and use of viral insecticides in China. Viruses 7:306–319
8. Sun X, Peng H (2007) Recent advances in biological control of pest insects by using viruses in China. Virol Sin 22:158–162
9. Sun X, Wang H, Sun X, Chen X, Peng C, Pan D et al (2004) Biological activity and field efficacy of a genetically modified *Helicoverpa armigera* single-nucleocapsid nucleopolyhedrovirus expressing an insect-selective toxin from a chimeric promoter. Biol Control 29:124–137
10. Kaya HK, Lacey LA (2007) Introduction to microbial control. In: Lacey LA, Kaya HK (eds) Field manual of techniques in invertebrate pathology: application and evaluation of pathogens for control of insects and other invertebrate pests. Springer Netherlands, Dordrecht [cited 2019 Apr 12], pp 3–7. Available from: https://doi.org/10.1007/978-1-4020-5933-9_1
11. Bale JS, van Lenteren JC, Bigler F (2008) Biological control and sustainable food production. Philosoph Trans Royal Soc B Biol Sci 363:761–776
12. Carruthers RI, Onsager JA (1993) Perspective on the use of exotic natural enemies for biological control of pest grasshoppers (Orthoptera: Acrididae). Environ Entomol 2:885–903
13. van Lenteren JC (2019) IOBC internet book of biological control—IOBC-Global, International Organisation for Biological Control [cited 2019 Apr 16]. http://www.iobc-global.org/publications_iobc_internet_book_of_biological_control.html
14. Vincent C, Andermatt M, Valéro J (2007) Madex® and VirosoftCP4®, viral biopesticides for codling moth control. In: Vincent C, Goethel MS, Lazarovits G (eds) Biological control: a global perspective, pp 336–343
15. Robinson AJ, Holland MK (1995) Testing the concept of virally vectored immunosterilisation for the control of wild rabbit and fox populations in Australia. Aust Veter J 72:65–68
16. Lockwood JA (2008) Six-legged soldiers: using insects as weapons of war. Oxford University Press
17. Dyck VA, Gardiner GT (1992) Sterile-insect release program to control the codling moth *Cydia pomonella* (L.) (Lepidoptera: Olethreuridae) in British Columbia, Canada. Acta Phytopathol Entomol Hung 27:219–222
18. Black BC, Brennan LA, Dierks PM, Gard IE (1997) Commercialization of baculoviral insecticides. In: The baculoviruses. Springer, Berlin, pp 341–387
19. Toprak U, Bayram Ş, Gürkan MO (2005) Gross pathology of SpliNPVs and alterations in Spodoptera littoralis Boisd. (Lepidoptera: Noctuidae) morphology due to baculoviral infection. Tarim Bilimleri Dergisi 11:65–71
20. Cory JS, Evans HF (2007) Viruses. In: Lacey LA, Kaya HK (eds) Field manual of techniques in invertebrate pathology: application and evaluation of pathogens for control of insects and other invertebrate pests. Springer Netherlands, Dordrecht [cited 2019 Apr 12], pp 149–74. Available from: https://doi.org/10.1007/978-1-4020-5933-9_7
21. Moscardi F (1999) Assessment of the application of baculoviruses for control of lepidoptera. Annu Rev Entomol 44:257–289
22. Steinhaus EA (2019) Disease in a minor chord: being a semihistorical and semibiographical account of a period in science when one could be happily yet seriously concerned with the diseases of lowly animals without backbones, especially the insects. The Ohio State University Press [cited 2019 Apr 12]. Available from: https://kb.osu.edu/handle/1811/29317

23. Steinhaus EA (1956) Potentialities for microbial control of insects. J Agric Food Chem 4:676–680
24. Steinhaus EA (1957) Microbial diseases of insects. Ann Rev Microbiol 11:165–182
25. Burges HD, Horace D (1981) Microbial control of pests and plant diseases 1970–1980. Academic Press [cited 2019 Apr 12]. Available from: http://agris.fao.org/agris-search/search.do?recordID=US201300331594
26. Kaya HK, Lacey LA (2000) Introduction to Microbial Control. In: Lacey LA, Kaya HK (eds) Field manual of techniques in invertebrate pathology: application and evaluation of pathogens for control of insects and other invertebrate pests. Springer Netherlands, Dordrecht [cited 2019 Apr 12]. pp 1–4. Available from: https://doi.org/10.1007/978-94-017-1547-8_1
27. Tanada Y, Kaya HK (1993) Insect pathology. Academic Press
28. Evans HF (1986) Ecology and epizootiology of baculoviruses [cited 2019 Apr 12]. Available from: http://agris.fao.org/agris-search/search.do?recordID=US881247488
29. Toprak U, Harris S, Baldwin D, Theilmann D, Gillott C, Hegedus DD et al (2012) Role of enhancin in *Mamestra configurata* nucleopolyhedrovirus virulence: selective degradation of host peritrophic matrix proteins. J Gen Virol 93:744–753
30. Volkman LE, Summers MD, Hsieh CH (1976) Occluded and nonoccluded nuclear polyhedrosis virus grown in *Trichoplusia ni*: comparative neutralization comparative infectivity, and in vitro growth studies. J Virol 19:820–832
31. Wang P, Granados RR (1997) An intestinal mucin is the target substrate for a baculovirus enhancin. PNAS 94:6977–6982
32. Braunagel SC, Summers MD (2007) Molecular biology of the baculovirus occlusion-derived virus envelope [cited 2019 Apr 16]. Available from: https://www.ingentaconnect.com/content/ben/cdt/2007/00000008/00000010/art00006
33. Payne CC (1982) Insect viruses as control agents. Parasitology 84:35–77
34. Martignoni ME, Iwai PJ (1986) A catalog of viral diseases of insects, mites, and ticks. Gen Tech Rep PNW-GTR-195 Portland, OR: US Department of Agriculture, Forest Service, Pacific Northwest Research Station, 57 p [cited 2019 Apr 12]; 195. Available from: https://www.fs.usda.gov/treesearch/pubs/26278
35. Groner A (1986) Specificity and safety of baculoviruses [cited 2019 Apr 12]. Available from: http://agris.fao.org/agris-search/search.do?recordID=US881246788
36. Matthews R (1982) Classification and nomenclature of viruses. Fourth report of the International Committee on Taxonomy of Viruses. Intervirology 17:1–199
37. Grzywacz D, Moore S (2017) Production, formulation, and bioassay of baculoviruses for pest control. In: Lacey LA (ed) Microbial control of insect and mite pests (Chap. 7). Academic Press [cited 2019 Apr 16], pp 109–24. Available from: http://www.sciencedirect.com/science/article/pii/B978012803527600007X
38. Onstad D (1998) Two databases provide information about insect pathogens. ASM NEWS 64:9
39. Rohrmann GF (2013) Baculovirus molecular biology [Internet], 3rd edn. Bethesda (MD), National Center for Biotechnology Information (US) [cited 2019 Apr 6]. Available from: http://www.ncbi.nlm.nih.gov/books/NBK114593/
40. Rohrmann GF (2013) Baculoviruses as insecticides: three examples. National Center for Biotechnology Information (US) [cited 2019 Apr 6]. Available from: https://www.ncbi.nlm.nih.gov/books/NBK138299/
41. Allen GE, Knell JD (1977) A nuclear polyhedrosis virus of *Anticarsia gemmatalis*: I. Ultrastructure, replication, and pathogenicity. Florida Entomologist 60:233–240
42. Carner GR, Turnipseed SG (1977) Potential of a nuclear polyhedrosis virus for control of the velvetbean caterpillar in soybean. J Econ Entomol 70:608–610
43. Tanada Y (1964) A granulosis virus of the codling moth, *Carpocapsa pomonella* (Linnaeus) (Olethreutinae, Lepidoptera). J Insect Pathol 6:78–80
44. Reid S, Chan LCL, Matindoost L, Pushparajan C, Visnovsky G (2016) Cell culture for production of insecticidal viruses. Methods Mol Biol 1477:95–117
45. de Almeida AF, de Macedo GR, Chan LCL, da Pedrini MRS (2010) Kinetic analysis of in vitro production of wild-type *Spodoptera frugiperda* nucleopolyhedrovirus. Braz Arch Biol Technol 53:285–291

46. Lua LHL, Reid S (2000) Virus morphogenesis of *Helicoverpa armigera* nucleopolyhedrovirus in *Helicoverpa zea* serum-free suspension culture. J Gen Virol 81:2531–2543

47. Mena JA, Kamen AA (2011) Insect cell technology is a versatile and robust vaccine manufacturing platform. Expert Rev Vaccines 10:1063–1081

48. Micheloud GA, Gioria VV, Eberhardt I, Visnovsky G, Claus JD (2011) Production of the *Anticarsia gemmatalis* multiple nucleopolyhedrovirus in serum-free suspension cultures of the saUFL-AG-286 cell line in stirred reactor and airlift reactor. J Virol Methods 178:106–116

49. Micheloud GA, Gioria VV, Pérez G, Claus JD (2009) Production of occlusion bodies of *Anticarsia gemmatalis* multiple nucleopolyhedrovirus in serum-free suspension cultures of the saUFL-AG-286 cell line: Influence of infection conditions and statistical optimization. J Virol Methods 162:258–266

50. Nguyen Q, Qi YM, Wu Y, Chan LCL, Nielsen LK, Reid S (2011) In vitro production of *Helicoverpa baculovirus* biopesticides—automated selection of insect cell clones for manufacturing and systems biology studies. J Virol Methods 175:197–205

51. Pedrini MRS, Reid S, Nielsen LK, Chan LCL (2011) Kinetic characterization of the group II *Helicoverpa armigera* nucleopolyhedrovirus propagated in suspension cell cultures: Implications for development of a biopesticides production process. Biotechnol Prog 27:614–624

52. Ignoffo CM (1964) Production and virulence of a nuclear polyhedrosis virus from larvae of *Trichoplusia ni* (Hubner) reared on a semisynthetic diet. J Insect Pathol 6:318–329

53. Stern V, Federici B (1990) Granulosis virus: biological control for western grape leaf skeletonizer. Calif Agric 44:21–22

54. Ignoffo CM, Garcia C (1985) Host spectrum and relative virulence of an Ecuadoran and a Mississippian biotype of *Nomuraea rileyi*. J Invertebr Pathol 45:346–352

55. Vail PV, Jay DL, Hink WF (1973) Replication and infectivity of the nuclear polyhedrosis virus of the alfalfa looper, *Autographa californica*, produced in cells grown in vitro. J Inverteb Pathol 22:231–237

56. Hostetter DL, Puttler B (1991) A new broad host spectrum nuclear polyhedrosis virus isolated from a celery looper, *Anagrapha falcifera* (Kirby), (Lepidoptera: Noctuidae). Environ Entomol 20:1480–1488

57. Kirkpatrick BA, Washburn JO, Volkman LE (1998) AcMNPV pathogenesis and developmental resistance in fifth instar *Heliothis virescens*. J Inverteb Pathol 72:63–72

58. McNeil J, Cox-Foster D, Gardner M, Slavicek J, Thiem S, Hoover K (2010) Pathogenesis of *Lymantria dispar* multiple nucleopolyhedrovirus in *L. dispar* and mechanisms of developmental resistance. J Gen Virol 91:1590–1600

59. McNeil J, Cox-Foster D, Slavicek J, Hoover K (2010) Contributions of immune responses to developmental resistance in *Lymantria dispar* challenged with baculovirus. J Insect Physiol 56:1167–1177

60. Engelhard EK, Volkman LE (1995) Developmental resistance in fourth instar *Trichoplusia ni* orally inoculated with *Autographa californica* M nuclear polyhedrosis virus. Virology 209:384–389

61. Briese D (1986) Host resistance to microbial control agents. Fortschritte der Zoologie 32:233–256

62. Levy SM, Falleiros ÂMF, Moscardi F, Gregório EA (2011) The role of peritrophic membrane in the resistance of *Anticarsia gemmatalis* larvae (Lepidoptera: Noctuidae) during the infection by its nucleopolyhedrovirus (AgMNPV). Arthropod Struct Dev 40:429–434

63. Murillo R, Lasa R, Goulson D, Williams T, Muñoz D, Caballero P (2003) Effect of Tinopal LPW on the insecticidal properties and genetic stability of the nucleopolyhedrovirus of *Spodoptera exigua* (Lepidoptera: Noctuidae). J Econ Entomol 96:1668–1674

64. Wang P, Granados RR (2001) Molecular structure of the peritrophic membrane (PM): identification of potential PM target sites for insect control. Arch Insect Biochem Physiol 47:110–118

65. Grove MJ, Hoover K (2007) Intrastadial developmental resistance of third instar gypsy moths (*Lymantria dispar* L.) *to L. dispar* nucleopolyhedrovirus. Biol Cont 40:355–361

66. Zafar B, Wani SA, Malik MA, Ganai MA (2013) A review: disease resistance in silkworm. Bombyx Mori 4:157–166

67. Berling M, Blachere-Lopez C, Soubabere O, Lery X, Bonhomme A, Sauphanor B et al (2009) *Cydia pomonella* granulovirus genotypes overcome virus resistance in the codling moth and improve virus efficiency by selection against resistant hosts. Appl Environ Microbiol 75:925–930

68. Eberle KE, Asser-Kaiser S, Sayed SM, Nguyen HT, Jehle JA (2008) Overcoming the resistance of codling moth against conventional *Cydia pomonella* granulovirus (CpGV-M) by a new isolate CpGV-I12. J Inverteb Pathol 98:293–298

69. Gebhardt MM, Eberle KE, Radtke P, Jehle JA (2014) Baculovirus resistance in codling moth is virus isolate-dependent and the consequence of a mutation in viral gene pe38. Proc Natl Acad Sci USA 111:15711–15716

70. Asser-Kaiser S, Heckel DG, Jehle JA (2010) Sex linkage of CpGV resistance in a heterogeneous field strain of the codling moth *Cydia pomonella* (L.). J Inverteb Pathol 103:59–64

71. Asser-Kaiser S, Fritsch E, Undorf-Spahn K, Kienzle J, Eberle KE, Gund NA et al (2007) Rapid emergence of baculovirus resistance in codling moth due to dominant, sex-linked inheritance. Science 317:1916–1918

72. Fuxa JR, Richter AR (1998) Repeated reversion of resistance to nucleopolyhedrovirus by *Anticarsia gemmatalis*. J Inverteb Pathol 71:159–164

73. Fuxa JR, Richter AR (1989) Reversion of resistance by *Spodoptera frugiperda* to nuclear polyhedrosis virus. J Inverteb Pathol 53:52–56

74. Undorf-Spahn K, Fritsch E, Huber J, Kienzle J, Zebitz CPW, Jehle JA (2012) High stability and no fitness costs of the resistance of codling moth to *Cydia pomonella* granulovirus (CpGV-M). J Inverteb Pathol 111:136–142

75. Yamaguchi K (1977) Regeneration of the midgut epithelial cells in the silkworm, *Bombyx mori*, infected with the cytoplasmic polyhedrosis virus. J Sericult Sci Jpn 46:170–180

76. David WAL, Gardiner BOC (1965) Resistance of *Pieris brassicae* (Linnaeus) to granulosis virus and the virulence of the virus from different host races. J Inverteb Pathol 7:285–290

77. Abot AR, Moscardi F, Fuxa JR, Sosa-Gómez DR, Richter AR (1996) Development of resistance by *Anticarsia gemmatalis* from Brazil and the United States to a nuclear polyhedrosis virus under laboratory selection pressure. Biol Cont 7:126–130

78. Cunningham JC (1995) Baculoviruses as microbial insecticides [cited 2019 Apr 12]. Available from: https://www.cfs.nrcan.gc.ca/publications?id=21623

79. Groner A (1990) Safety to nontarget invertebrates of baculoviruses. In: Safety of microbial insecticides, pp 135–47

80. Huber J (1986) Use of baculoviruses in pest management programs [cited 2019 Apr 12]. http://agris.fao.org/agris-search/search.do?recordID=US19880053890

81. Dougherty EM, Guthrie KP, Shapiro M (1996) Optical brighteners provide baculovirus activity enhancement and UV radiation protection. Biol Cont 7:71–74

82. Petrik DT, Iseli A, Montelone BA, Van Etten JL, Clem RJ (2003) Improving baculovirus resistance to UV inactivation: increased virulence resulting from expression of a DNA repair enzyme. J Invertebrate Pathol 82:50–56

Biocontrol Products for Suppressing Plant Diseases

Biological Control of Phyto-pathogenic Bacteria

Hassan Abd-El-Khair

Abstract The pathogenic bacteria can attack many plants causing different symptoms include necrosis, tissue maceration, wilting and hyperplasia and resulting diseases and damage to crops. The bacteria enter the host plant through natural openings or wounds and then it colonized locally intercellular spaces and systematically the vascular system of host. Virulence of bacterial pathogen was increased by increase of bacterial metabolites production viz. enzymes, toxins and/or plant hormones often under control of quorum sensing mechanisms. Application of effective chemicals or resistance sources against bacterial plant diseases are limited because of copper compounds may cause phytotoxic or rusting to plants as well as antibiotics application has not enough disease control. Therefore, the biological control can be successfully applied for crop protection against bacterial pathogens, where the biological control depended on the use of natural enemies viz. bacteria, fungi and viruses which they was common in any agricultural system.

Keywords Agricultural system · Pathogenic bacteria · Fungi · Bacteria · Biological control

1 Introduction

Phytopathogenic bacteria can be controlled by different methods such as cultural practices, crops rotation, resistant varieties, soil sterilization, seed-disinfection, foliar spray of copper and zinc compounds and hot water treatment [1, 2]. Application of chemicals is generally much less successful for controlling bacterial diseases than the fungal diseases. Applications of traditional bactericides such as copper compounds are limit for controlling bacterial plant diseases, where their application may cause phytotoxic or rusting. The accumulation of copper in the soil causes decline in plants vigor; inhibited of others microflora and strains of bacteria may be become resistant [2, 3]. The antibiotics, such as streptomycin sulphate, was applied for controlling bacterial diseases, e.g. fire blight disease in pome fruits, black rot disease in cabbages,

H. Abd-El-Khair (✉)
Plant Pathology Department, National Research Centre, Dokki, Giza, Egypt
e-mail: khairhafz@yahoo.com

© Springer Nature Switzerland AG 2020
N. El-Wakeil et al. (eds.), *Cottage Industry of Biocontrol Agents and Their Applications*, https://doi.org/10.1007/978-3-030-33161-0_10

bacterial spot disease in peach, wildfire disease in tobacco, canker disease in citrus and some diseases in ornamental plants, but not against crown gall by *A. tumefaciens*. Application of antibiotics has not enough to control plant bacterial diseases may be due to non-persistence, side phytotoxic effect, the highest costs and development of resistance in bacterial populations as well as after few days of spraying, the antibiotic in the content plants of antibiotic gradually decreased and necessary spray weekly repeated [4, 5].

Therefore, the biological control may be one of the crop protection methods which are relatively new for controlling bacterial plant pathology in the field. For good biological controlling of the bacterial disease should knew the disease cycle, where bacterial diseases develop rapidly when the environmental conditions were conducive for disease development. Seed dressing or coating with a biological control agent as well as dipping the seedling in a bioagent suspension are the visible methods to deliver a agent onto plant surface [6]. The biological control also can be alternative tool for controlling of bacterial plant diseases; especially chemicals couldn't control the crown gall disease (*Agrobacterium tumefaciens*) and fire blight disease (*Erwinia amylovora*). The biological control also occurs in nature, where bio-control agents can be reduced the inoculums density of many plant pathogens such as fungi, bacteria, virus, viroid and nematode. The natural products by biological control agents also have led to the development of "biorational" pesticides, where the biological agents have different mechanisms included; competition, antibiosis, parasitism, plant growth stimulation and induce systemic resistance as well as it considered as environmentally friendly measure [7, 8].

In this book chapter will indicate the common phytopathogenic bacterial species able to cause plant diseases and focuses on new biological control approach by showing the mechanisms of bio-control agents; the beneficial effects of common bio-control agents and field application of biological control agents.

2 Phytopathogenic Bacteria

The phytopathogenic bacteria cause many several plant diseases throughout the world. The common bacterial symptoms i.e. blights; spots on leaves or fruits, deadening of tissue on leaves, stems or trunk of trees; soften on roots or tubers and galls stems or crown roots [9, 10]. Survival of plant pathogenic bacteria can be occurred in nature on plant debris, in or on seeds as well as in soils, where infected seeds or any plant part can be sources of bacterial inoculums. Bacteria can be sucked into the plants through stomata, hydathodes, lenticels, wounds formed on roots, stems or leaves and through specific feeding placements of insects. Diagnosis of bacterial diseases depends on characteristic symptomatology, infectious agents' isolation and physiological, biochemical or molecular characters [11, 12]. The phytopathogenic bacteria are single-cell microorganism, in range of 1–2 μm in sizes and couldn't see with un-aided eye. All plants bacterial pathogenic are Gram-negative, short rod in shape, singly or in pairs, except *Streptomyces* are filamentous and *Corynebacterium* formed rod shape as Y or V shape. The pathogenic bacteria moved by the flagella as well as it characterized by produce clear colonies on nutrient agar medium. The bacterial colonies differed in their size, shape, form of edges, elevation and color. Most of pathogenic bacteria are aerobic, while someone is anaerobic. Some bacteria produced a fluorescent pigments on the surface of king B agar medium, viz. fluorescent pseudomonads, while other bacteria can produce characteristic volatile compounds, as odor, in rotten potatoes such as soft rot *Erwinia* [13, 14]. The most common plant bacterial pathogens will be presented in the following paragraphs.

2.1 Agrobacterium tumefaciens

Agrobacterium tumefaciens causes the crown galls disease [15] more than 140 plant species of eudicots, especially viz. almond, apples, cherry, pears, peach, raspberry and roses. It is rod-shaped and Gram-negative bacterium. The disease gains its name from the larges galls which typically occur at the crown of the plants. The bacterial cellular of *A. tumefaciens* can be saprophytically live up to two years in soils. The galls bacteria chemotactical move into the sites of wounds and through host cells, when a nearby is the wounds caused by insect feeding near of soil line, transplanting injury and/or any other wounds [16, 17]. Insertion of small segments of DNA (as T-DNA plasmid) into the plant cells, where it randomly incorporated at location into the plants genome causes galls symptom. The resulting tissues are white and/or creamy color and the plant cell may be having one or more nuclei. The tumors are formed on the roots and/or stems of the plants by tissues continue to enlarge, depended on site of original wounds. The bacterial cells are occupied in the intercellular spaces around the periphery of galls, while it not found in center of enlarging tumors. Degradation of the tumors release bacterial cells back into soil, where it carries away with soil and/or water, or remains in soil for next season of growth [18–20].

2.2 Dickeya (dadantii *and* solani)

Dickeya genus is Gram-negative bacterium in the family Enterobacteriaceae. *Dickeya* is the resulted from re-classificay of 75 *Pectobacterium* (*Erwinia*) *chrysanthemi* strains as a newly genus. Many species of *Dickeya* spp., viz. *Dickeya dadantii* are plant bacterial pathogens. The families members are facultative anaerobic, sugars fermentation to lactic acid, reductase nitrates, while not formed oxidases. *D. dadantii* moved by a motile by peritrichous flagella, nonsporing, straight rod-shaped cell with rounded ends. *D. dadantii* could cause plant disease symptoms viz. blights and soften, where the bacteria produces a lot of pectinase enzymes could macerate and break down the plant materials of cell walls. *Dickeya solani* causes diseases of blacklegs and soften to potatoes. The symptoms caused by bacterium are often indistinguishable about those caused by *Pectobacterium*. But the symptoms are the most virulent and caused disease at low inoculum level [21–23].

2.3 Erwinia amylovora

A severe fire blight could affect apples, pears and some members of the family Rosaceae [24]. It is Gram-negative bacterium in the family Enterobacteriaceae, rod shaped and motile with peritrichous flagella. Pears are the most susceptible, than other rosaceous plants [25, 26]. The name of fire blight disease because of the bacterium affects the pear branches have persistent the blackened leaves and the tree appears as scorched by fire. *E. amylovora* also is greatly historical important to phyto-bacteriologists, where it was the first bacterium demonstrated that the bacteria could cause plant diseases. The fire blight Symptoms were firstly recorded in close orchard in New York City, and then the bacterium spreads westward and/or across continents, during the 20th century. The bacterium is often found in a watery polysaccharide matrix, called ooze as well as it produces characteristic colonies (small, mucoid, domed, round and glistening). *E. amylovora* strains are having a fermentative metabolism, oxidase negative and catalase positive [3, 5, 27–29].

2.4 Pectobacterium carotovorum *and* Pectobacterium atrosepticum

Pectobacterium carotovorum (*Pcc*) is a Gram-negative bacterium in the family Entero-bacteriaceae as a member of the genus *Erwinia* [30]. The species is a plant pathogen with a diverse host range, including many agricultural plant species. It produces pectolytic enzymes that hydrolyze pectin between individual plant cells. *Pectobacterium atrosepticum* (*Pca*) is a species of Gram-negative bacterium of the family Enterobacteriaceae. It is a plant pathogen causing blackleg of potato. *Pcc* caused soften disease to several plant crops, while *Pca* caused the blackleg disease to potatoes. The soft rot disease affects many cruciferous and other crops causing a watery, soft, foul-smelling rot. *Pca* is found in temperate countries and in highland areas in the tropics [29, 31–33].

2.5 Pseudomonas syringae *Pathovars*

Pseudomonas syringae is a member of the genus *Pseudomonas*. It is a rod-shape and Gram-negative moved with polar flagella. The bacterium could infect wide ranges of plant species and includes more than 50 bacterial pathovars. The *syringae* group gave negative reaction for oxidase and/or arginine dihydrolase and positive reaction for levan production. Many bacterial strains could secrete the plant toxin syringomycin (lipodepsinonapeptide) and produce fluorescene pigments on King's B medium due to product the siderophore pyoverdin [34]. Several pathovars of *P. syringae*, i.e. *P. syringae* pv. *aceris* infect maple *Acer* species; *P. syringae* pv. *actinidiae* infect kiwifruit; *P. syringae* pv. *aptata* infect beets; *P. syringae* pv. *atrofaciens* and *P. syringae* pv. *lapsa* infect wheat; *P. syringae* pv. *dysoxylis* infect the kohekohe tree; *P. syringae* pv. *japonica* infect barley; *P. syringae* pv. *panici* infect *Panicum* grass species; *P. syringae* pv. *papulans* infect crabapple; *P. syringae* pv. *phaseolicola* causes halo blight of beans; *P. syringae* pv. *pisi* infect peas; *P. syringae* pv. *syringae* infect *Syringa*, *Prunus* and *Phaseolus* species, *P. syringae* pv. *glycinea* infect soybean and *P. syringae* pv. *delphinii* [35–37].

2.6 Ralstonia solanacearum

R. solanacearum is soil borne bacterium, Gram-negative, an aerobic, nonspore forming and moves by polar flagella. The bacterium causes bacterial wilts by colonizing the xylem of ranges of plant species such as tomato, pepper, eggplant and Irish potato. The bacterium usually enters the plant through wounds by swimming with flagella and chemotaxically attracted toward exudates of roots. The pathogen also gets into the xylems through natural openings. The bacterium is able to systematically move and multiply within susceptible hosts, after invading, and then the wilts symptoms was occurred. After extensive colonization by bacterium; the wilts symptom usually occurs as the most visible pathogenic effect [38–40].

2.7 Xanthomonas oryzae

The genus *Xanthomonas* is a member of the family Xanthomonadaceae. *X. oryzae* pv. *oryzae* is Gram-negative and rod-shape. The bacterium could produce extracellular polysaccharide and yellow soluble pigments (xanthomonadin). *X. oryzae* pv. *oryzae* could cause blight disease to of rice, as the important rice disease, in many countries such as Asia, Australia, Africa, Latin America, the Caribbean and the USA. The disease symptoms firstly appear as pale to grey green on the leaves of young plants and as water-soaked streaks near the leave tips or margins. The lesions coalesce and then become yellowish-white with wavy edges. Then, the whole leaves may infect and become whitish or grayish in color and then die. Systemic infection resulted as wilts, desiccate of leaves and death in young rice transplants. The leaves of older rice plants become yellowish and then die. The plants of rice plants may infect with bacterium by rice seeds, stems or roots which left behind after harvest or by alternative host weeds [41–43].

2.8 Xanthomonas campestris *Pathovars*

Xanthomonas campestris pathovars are members in the genus *Xanthomonas* of the family Xanthomonadaceaeis. The bacterium is a rod-shape and Gram-negative. The bacterial species can cause varieties of plant diseases, includes black rot disease in cruciferous and wilts disease in turfgrasses [44]. The pathovars of bacterium can be classified based on the host plants into: *Xanthomonas campestris* pv. *begoniae*; *X. campestris* pv. *campestris*; *X. campestris* pv. *cannabis*; *X. campestris* pv. *carota*; *X. campestris* pv. *corylina*; *X. campestris* pv. *graminis*; *X. campestris* pv. *glycines*; *X. campestris* pv. *malvacearum*; *X. campestris* pv. *mori*; *X. campestris* pv. *pelargonii*; *X. campestris* pv. *phaseoli*; *X. campestris* pv. *prunii*; *X. campestris* pv. *sesame*, etc. [45–47].

2.9 Xanthomonas axonopodis

The *Xanthomonas* genus currently contains about 20 of bacterial species includes *axonopodis*. The *axonopodis* pathovars causes economic important diseases on significance plant hosts. *Xanthomonas axonopodis* pv. *manihotis* could induce combination symptoms such as angular leaves lesions, blights, wilts, stems exudate and stems canker [48]. The bacterial species indicate high specific degrees and some species could split into multiple pathovars based on specific host. *Xanthomonas axonopodis* subsp. *citri* causes important canker disease on citrus trees viz. lime, oranges, lemons and pamelo. *Xanthomonas axonopodis* (syn. *campestris*) pv. *vesicatoria* (*Xanthomonas euvesicatoria* or *Xanthomonas perforans*) causes the bacterial leaf spot disease, where the infection with pathogen begin after transplanting and this disease may lost the citrus yields. *Xanthomonas axonopodis* pv. *punicae* also causes bacterial blights of pomegranate [47, 49].

2.10 Xylella fastidiosa

Xylella fastidiosa is an aerobic, Gram negative which it is transmitted exclusively by xylem fluid feeding sap insects. The bacterium could cause economically loss in many agricultural important plants, includes almond, blueberry, citrus, coffee, grape, oleander, peach, plum causing economic serious diseases such as bacterial leaf scorch (oleander and coffee), peach phony disease, grape Pierce's disease and variegated chlorosis disease of citrus. *X. fastidiosa* cause disease symptoms on plants includes margins necrosis, leaves abscission, dieback, growth delayed in the spring and decline of vigor become lead to death. For example; grapevine Pierce's disease causes leaves yellowing and then the leaves die. The disease leaves concentric zone series of discolored tissues and fruit clusters which may be wilting and drying up.

Citrus variegated chlorosis also could infect the commercial oranges. On the upper side of leaves, the disease causes chlorotic areas with corresponding brown, while on the lower side formed gummy lesions [50–52].

3 Mechanism of Biocontrol Agents

The biological control may be resulted from different interaction types among microorganisms (Table 1). Understanding the mechanisms of the interactions between biocontrol agents and pathogens may indicate the create conditions conducive for successful or improvement of biological control strategies [53, 54]. Many of microorganism viz. *Trichoderma* spp., *Bacillus subtilis*, *Pseudomonas flourescens*, *Erwinia herbicola*, *Bacillus thuringiensis* and *Agrobacterium radiobacter* can be widely used as bio-control agents against bacterial plant diseases. *A. tumefaciens* biologically controlled when plant materials dipg in cells inculums of *A. radiobacter*

Table 1 Types of interspecies antagonisms leading to biological control of plant pathogens [58]

Type	Mechanism	Examples
Direct antagonism	Hyperparasitism/predation	Lytic/some non-lytic mycoviruses *Ampelomyces quisqualis* *Lysobacter enzymogenes* *Pasteuria penetrans* *Trichoderma virens*
Mixed-path antagonism	Antibiotics	2,4-diacetylphloroglucinol Phenazines Cyclic lipopeptides
	Lytic enzymes	Chitinases Glucanases Proteases
	Unregulated waste products	Ammonia Carbon dioxide Hydrogen cyanide
	Physical/chemical interference	Blockage of soil pores Germination signals consumption Molecular cross-talk confused
Indirect antagonism	Competition	Exudates/leachates consumption Siderophore scavenging Physical niche occupation
	Induction of host resistance	Contact with fungal cell walls Detection of pathogen-associated, molecular patterns Phytohormone-mediated induction

(K84 strain), while *E. amylovora* can be controlled by using *E. herbicola*, *P. fluorescens* and avirulent strains *P. syringae* [24]. *P. fluorescens* gave promising inhibitory effects against *X. vesicatoria* and *Clavibacter michiganensis* subsp. *michiganensis* (bacterial cankers and wilts diseases of tomatoes, respectively) [55, 56]. *P. fluorescens* also can be considered as a biocontrol agent against bacterial wilt disease in potato and tobacco and fire blight disease on peach and apple [57].

3.1 Antibiosis

Many microbes produce and secrete one or more compounds, such as low-molecular weight compounds or antibiotics, which may have direction activities against other microorganisms (Table 2). Antibiosis may play an important role for suppressing the plant pathogenic bacteria. The biocontrol agents are able to produce volatile

Table 2 Some antibiotics produced by bio-control agents and their applied for controlling plant diseases

Sources	Antibiotics	Plant disease	References
Agrobacterium radiobacter	Agrocin 84	Crown gall	Kerr [67]
Bacillus amyloliquefaciens FZB42	Bacillomycin, fengycin	Wilt	Koumoutsi et al. [68]
Bacillus cereus UW85	Zwittermicin A	Damping-off	Smith et al. [69]
Bacillus subtilis AU195	Bacillomycin D	Aflatoxin contamination	Moyne et al. [70]
B. subtilis QST713	Iturin A	Damping-off	Kloepper et al. [71]
Bacillus subtilis BBG100	Mycosubtilin	Damping-off	Leclere et al. [72]
Pantoea agglomerans C9-1	Herbicolin	Fire blight	Sandra et al. [73]
Pseudomonas fluorescens F113	2, 4-diacetylphloroglucinol	Damping-off	Shanahan et al. [74]
P. fluorescens 2-79 and 30-84	Phenazines	Take-all	Thomashow et al. [75]
P. fluorescens Pf-5	Pyoluteorin, pyrrolnitrin	Damping-off	Howell and Stipanovic [76]
Lysobacter sp strain SB-K88	Xanthobaccin A	Damping-off	Islam et al. [77]
Trichoderma virens	Gliotoxin	Root rot	Wilhite et al. [78]
Burkholderia cepacia	Pyrrolnitrin, pseudane	Damping-off and rice blast	Homma et al. [79]

antibiotics such as aldehydes, alcohols and ketones and hydrogen cyanide (HCN) as well as nonvolatile antibiotics such as polyketides i.e. 2,4-diacetylphloroglucinol (DAPG) and mupirocin; heterocyclic nitrogenous compounds, i.e. phenazine derivatives: pyocyanin, phenazine-1-carboxylic acid (PCA) and hydroxyphenazines and pyrrolnitrin (Prn) as phenylpyrrole antibiotic [59, 60]. *Bacillus* spp. produces varieties of antibiotic lipopeptides e.g. bacillomycin, iturins, surfactin and Zwittermicin A [61], *Bacillus cereus* (UW85 strain) was able to produce the antibiotic zwittermycin and kanosamine [62]. The genetically engineering of WCS358r strain (*Pseudomonas putida*), produces DAPG and phenazine, highly suppresses the wheat take-all disease in the fields [63]. *P. fluorescens* (2–79 strain) has antagonistic affect against take-all disease in wheat by producing a phenazine antibiotic. *P. fluorescens* (CHA0 strain) produces antibiotic substances including DAPG, hydrogen cyanide and pyloluteorin that able to suppress take-all disease in wheat. Among other bacteria, by *A. radiobacter* (K84 strain) produce antibiotic agrocin K84 for bio-controlling of crown gall caused by *A. tumefaciens* strains. *Chaetomium globosum*; *Trichoderma harzianum* and *Trichoderma* spp. produce chaetomin; peptaibols and pyrones [64–66].

3.2 Competition

The microbes could successfully colonize the phyllospheres and rhizospheres by effective competition on the available nutrients, where nutrient competition has an important role in disease suppression. Infection with plant pathogens occurs only after stimulants by plant hosts, where the specific stimulant for germinating of pathogen spores may be came from seeds germination or grown roots. The stimulantion factors may be including volatile components such as ethanol and acetaldehyde or fatty acids. The competition on nutrients, in the rhizpshere, may occur when the biocontrol agent could decrease the substances availability limiting the pathogen growth. The competition on iron in alkaline soils, for example, becomes limiting factor for growth microbes in this soil [80]. The fluorescent pseudomonads viz. *P. fluorescens* and *P. putida* could produce special siderophores, which have very high affinity to iron, becoming limited resource by others. *P. putida* when colonize the roots system was able produce agglutinin (glycoprotein). *P. fluorescens* strains produced iron-chelating salicylic acid, at low iron available, may induce systemic resistance against *Fusarium* wilt disease in radish. The plant surfaces of plant hosts supply nutrients include plant exudates, leachates or senesced tissues. The nutrients also may obtain from the waste products of insects or from the soil [81–83].

3.3 Hyperparasitism

The hyperparasitism means that the pathogens are directly attack with specific biocontrol agent which could kill it or it propagule. The first stage of mycoparasitism is the chemotropic growth, where the biocontrol agent could grow toward the target fungus. The second stage is recognition; include specific interaction between lectin of pathogen or carbohydrate receptors on biocontrol agent surface. The third stage is attachment by cell wall degradation such as chitinases and b-1, 3-glucanase [84]. The final stage is penetration, where the biocontrol agent could produce structures like appressoria for penetrating the cell wall of pathogenic fungus [85]. Many mycoparasites occur on a wide range of fungi which play an important role in disease control. *Trichoderma lignorum* (*Trichoderma viride*) able to parasitize hyphae of *Rhizoctonia solani* to control damping off of citrus seedling as well as other *Trichoderma* species for controlling *Rhizoctonia bataticola* and *Armillaria mellea*. Recently, *T. harzianum* and *T. hamatum* marketed as wound dressings as decay inhibitors in ornamental and forest trees *Coniothyrium minitans* and *Sporidesm*.

3.4 Cell-Wall Degrading Enzymes

Many bio-control agents produce extracellular hydrolytic enzymes that can interfere with pathogen growth and activities. The biocontrol agents were able to produce the lytic enzymes, which hydrolyze wide varieties of polymeric compounds, includes chitins, celluloses, hemicelluloses and proteins. These enzymes secrete by microbes resulted as suppression of activities of plant pathogens. *Serratia marcescens* produces chitinase, while *Lysobacter* and *Myxobacteria* produce lytic enzymes, where Chitins and b-1, 3-glucan the major constituent of cellular wall of fungi. *P. fluorescens* (CHA0 strain) produces antibiotics, siderophores and hydrogen cyanide. A chitinase (ChiA strain) deficient mutant of *S. marcescens* inhibited elongation of fungal germ tube and reduced *Fusarium* wilts of pea seedlings in greenhouse conditions. The transgenic bacterium, when ChiA strain inserted into *Escherichia coli*, could reduce incidence of Southern blights disease (*Sclerotium rolfsii*) in beans [86, 87].

3.5 Induction of Systemic Resistance

The inducible systemic acquired resistance (ISR) in plants can induce by plants inoculation with necrogenic pathogen, non-pathogen, certain natural compounds and synthetic chemicals. The defenses induction in plant hosts may be locally and/or systemically in the nature, depend on the types, sources and amounts of stimulation. The responses of defense may be include the physical thickens of cells wall by lignification, callose deposition, low-molecular-weight substances accumulation (e.g., phytoalexins) or various proteins synthesis (e.g., chitinase, glucanase, peroxidase and others pathogenesis-related (PR) proteins) [88, 89]. The first ISR pathway, mediated by salicylic acid compound, can be produce by pathogens infection and leads to the PR proteins expression. The second phenotype of ISR, mediated by jasmonic acid and/or ethylene, can be produce by application of some non-pathogenic rhizobacteria. This defense system can be resulted by plant growth-promoting rhizobacteria (PGPR) which colonized the plants and effectively controlled plant diseases by inducing ISR. These strains could colonize the plant roots resulted the control against foliar plant diseases [90]. PGPR inoculation was effective for controlling angular leaves spots disease (*Pseudomonas syringae* pv. *lachrymans*) and bacterial wilts disease (*Erwinia tracheiphila*). The phytoalexins amounts were increased in inoculated plants, when inoculated with *P. fluorescens* (WCS417r strain), compared to non-bacterized plants [91]. *P. fluorescens* (CHA96 strain) could induce PR-proteins (e.g., endo-chitinase and b-1, 3-glucanases) in the intercellular fluid of plant leaves [92]. The lipopolysaccharide with the Oantigenic side chain produced by *P. fluorescens* (WCS374 strain strain) could induce systemic resistance in radishes against *Fusarium* wilt disease [83]. *P. fluorescens* (CHA0 strain) effectively controlled wheat take-all disease (*Gaeumannomyces graminis* var. *tritici*). The CHA0 strain can produce metabolites that may stress the plant when the metabolites are delivered into the plant cells [92].

4 The Beneficial Effect of Common Biocontrol Agents

4.1 Bacillus *Species*

Bacillus is a genus of spore-forming, Gram-positive and rod-shape. The genus *Bacillus* was able to produce phytostimulation viz. cytokinins in the rhizosphere. *B. subtilis* strain synthesise zeatin riboside-type cytokinins, dihydrozeatin-riboside and isopentenyl-adenosine. The tissues of roots and shoots of lettuce plants, when inoculated with *Bacillus* (*Bacillus amyloliquefaciens* and *B. subtilis*), contained a greater amount of cytokinin. It has high levels of other plant hormones such as indolyl-3-acetic acid and abscisic acid, than untreated plants. *Bacillus megaterium* was able to reduce iron and make it available to the plant. The PGPR produced volatile organic compounds including aldehydes, ketones and alcohols that responsible for inducing root development. Other volatiles, such as 1-octen-3-ol and butyrolactone, might also participate in this plant–bacteria interaction. The beneficial effect of PGPR is due to a single compound or the synergistic activity of several compounds [72, 93, 94]. Strains of *Bacillus* were able to synthesise lipopeptide-type compounds viz. lipopeptides from the fengycin, surfactin and iturin families which have effectively suppressive against pathogens [95, 96]. Production of lipopetide antibiotic, by mutated *B. subtilis* (M40 strain), has highly antagonistic effect than *B. subtillus* (wild strain) against *Erwinia carotovora* var. *carotovora* [97]. *Bacillus* spp. viz. *B. amyloliquefaciens*, *B. cereus*, *B. subtilis*, *Bacillus pasteurii*, *Bacillus pumilus*, *Bacillus mycoides* or *Bacillus sphaericus* were able to induce ISR mechanisms, plant defence systems, aganist various pathogens such as fungi, bacteria, viruses and nematodes [98]. *B. subtilis* strain FB17 produces acetoin (3-hydroxy-2-butanone) that it is responsible for triggering ISR in plants. *Bacillus* volatile compounds that trigger ISR in plants are dependent on ethylene and jasmonate pathways, but independent of salicylic acid [71, 99].

4.2 Pseudomonas *Species*

Pseudomonas is a genus of non-spore-forming, Gram-negative and rod-shape. *Pseudomonas* is natural bio-control agent living in disease-suppressive soils and it's able to rapidly grow and good colonise, where the rhizosphere is a highly competitive ecosystem for spaces and foods. Bacterium is able to produce various compounds such as antibiotics, polysaccharides and siderophores. *P. fluorescens* (WCS417Rr strain) plays an important role for root colonisation of tomato plants, or tissue penetration and as well as living as endosymbionts. *Pseudomonas* synthesis the siderophores, as important factor for the colonization of plant roots, especially under iron-limiting conditions. The siderophores can prevent soil pathogens from iron uses, where it is an essential element for growth of many micro-organisms. *Pseudomonas* protected the potato plants by a good production of pseudobactin-type siderophores [8, 100, 101].

Pseudomonas strains are capable to synthesis a wide spectrum of antibiotics that characterised by their suppressive activities against plant pathogens. *Pseudomonas* synthesis several compounds such as Plt; PCA; DAPG; Prn; HCN and bacteriocin [102]. The fluorescent *Pseudomonas* (CHAO strain) produces more than 10 compounds with pathogen biocontrol activity and plant-growth promotion, such as DAPG, Plt, Prn, HCN, indoleacetic acid, salicylic acid, pyochelin and siderophores [103]. *Pseudomonas* also induces systemic resistance in plants, where *P. fluorescens* (EP1strain), *P. putida* (5-48 strain) and *P. fluorescens* can protect sugarcane, oak and tomato plants against soil borne pathogens. *Pseudomonas chlororaphis* synthesis the 2R, 3R-butanediol compound is responsible for triggering ISR in tobacco plants. *P. chlororaphis* (O6 strain), similar to *B. subtilis* (GB03strain) induced systemic resistance in plants by 2R, 3R-butanediol, where it played an important role in plant–bacteria communication [104].

4.3 Trichoderma *Species*

Trichoderma spp. is worldwide occurrence and easily isolated from soil. *Trichoderma* are rapid growth and production of numerous conidia that play an important role in characterization this genus. *Trichoderma* spp. was able to parasitize on other fungi and produce toxin. The mycoparasitism processes includes the coiling of pathogens hyphae, penetrate and dissolute the cytoplasm of host. *T. lignorum* when compete with *R. solani* on nutrient medium, the mycoparasitism mechanism is favored. *T. virens*, produce a new antibiotic, highly inhibits *Phytophthora* species and *Pythium ultimum* [105]. *Trichoderma* species can compete for space and nutrients to grow in the rhizosphere, where *Trichoderma koningii* strains are excellent root colonizers against *R. solani* on cotton seedlings. *Trichoderma* species could inhibit the fungi growth in vitro tests. The rootsegments treated with *T. virens*, soil heavily infestation with *Macrophomina phaseolina*, only *T. virens* grows from the roots on agar medium at room temperature [106].

Trichoderma species produce chitinases and/or glucanases enzymes which suppress the plant pathogen, where these enzymes breaking down the polysaccharides and chitin of fungal cell walls. *T. harzianum* also was produced proteases, where these enzymes break down hydrolytic enzymes into peptides or their constituents of amino acids and reduced the act capacity on cells of plants [107]. *Trichoderma* species can induce systemic resistance in inoculating roots of cucumber seedlings, where spores of *T. harzianum* began the defense responses of plants in roots and leaves of cucumber plants. The plant response was marked by an increase the peroxidase and chitinase activities as well as increased the callose deposition in the inner surfaces of cells wall. *T. harzianum* also induced pathogenesis-related proteins in cucumber roots [108].

5 Field Application of Biological Control Agents

The successful applied biological control of different phytobacterial species will be presented in the following subsections.

5.1 Agrobacterium tumefaciens

Control of pathogenic *A. tumefaciens* strains began thorough nursery stock, where the pathogen infested soils must be planted by non-susceptible varieties of monocotyledonous crops, such as corn or wheat, for many years. Application of *A. radiobacter* is inexpensive and effective means for controlling the crown gall disease development due to produce antibiotic agrocin 84 by *A. radiobacter* (K84 strain). The bacteriocin-sensitive pathogenic strains were effectively controlled, but insensitive pathogenic strains are not biological controlled [109]. Garrett [110] also revealed that crown gall can be biological controlled by a bacteriocin of non-pathogen *A. radiobacter* (84 strain) in the rhizosphere of susceptible plants. In vitro tests, the live cells of strain 84 have inhibition effect aganist *A. tumefaciens* and inhibited the galls formation in cherry leaves scars. The dipping of the new rootstock in strain 84 inhibited the crown gall in a field experiment. On the other hand, the heat-killed bacterial cells had no effect. Application of *A. radiobacter* (K84 strain) also could protect young *Chrysanthemum* plants against *Agrobacterium* pathogen. The trials indicated that the efficiency of cuttings treatment by immerged in a suspension of strain K84, for controlling pathogen strains [111]. Non-pathogenic *A. tumefaciens* was successfully applied as biological control agent against crown gall in grapevines [112].

The effectiveness of *A. radiobacter* strains i.e. K84, 0341 and non-agrocin-producing mutant (K84 Agr) was applied for controlling of crown galls in stone fruit rootstocks. Strains of K84 and 0341 could control crown galls on plum trees when two *A. tumefaciens* strains (resistant to agrocin 84) inoculated in the soils. Strain K84 could control the disease on peach trees when strains of *A. tumefaciens* (sensitive or resistant to agrocin 84) inoculated in the soils. The effectiveness of K84 strain was higher against the sensitive strain. Strains of K84 and K84 Agr could control crown galls on plum or peach tres in soil treated by *A. tumefaciens* (sensitive or resistant to agrocin 84) strains [113]. Both strains K84 and K1026 of *A. radiobacter* controlled the sensitive strains in peach seedlings. The K1026 strain-treated plants showed no galls with resistant strains, while some galls notice in plants treated with K84 strain. Agrobacteria, recovery of from galls, revealed that all bacterial isolates of controls and K1026 strain or the most isolates of K84 strain-treated plants, showed the same characteristics of inoculated strains in experiments with sensitive and resistant strains [114, 115].

A. radiobacter (K84 strain), its genetically modified (GEM strain) and K1026 strain were tested for their effectiveness against local Tunisian strains (C58 and B6 strains). Strain K84 was effective against all crown gall isolates in tomato and tagetes, except of B6 strain. The strains GEM and K1026 were very effective against all isolates. The antagonists more effectively controlled crown gall disease on bitter almond-tree rootstocks, than quince BA29 and peach_almond GF677 rootstocks under field conditions [116]. The bacterial strains of grapevine or K84 strain (*A. radiobacter* biovar 2) suppressed the galls in stems of tomato seedlings, when inoculated with *A. tumefaciens* biovar 3. All *A. radiobacter* strains reduced incidence and size of gall than pathogen only. Strain VAR03-1 reduced the galls formation and size on roots of grapevines [117]. A nonpathogenic *Agrobacterium vitis* (VAR03-1 strain) significantly reduced the tumors numbers and the severity of disease in grapevine, rose and tomato when planted in soils infested with both *A. vitis*, *Agrobacterium rhizogenes* and *A. tumefaciens*. Strain VAR03-1 and nonpathogenic *A. rhizogenes* (K84 strain) had almost identical inhibitory effects on crown gall of rose and tomato. The inhibitory effect of VAR03-1strain on grapevine was superior to K84strain as well as VAR03-1strain greatly controlled crown galls due to *A. vitis* in grapevine fields [118].

Isolates of *A. radiobacter*, i.e. UHFBA-8, UHFBA-11 and UHFBA-12, completely inhibited gall formation onto tomato stems by Agrocin production. Root dip treatment of peach rootstocks with UHFBA-11 isolate has more reduction aganist crown gall incidence than untreated. Isolate UHFBA-8, as roots dipping treatment, reduced the crown galls incidence in cherry rootstock Colt, compared to untreated plants. *A. radiobacter* isolate viz. UHFBA-8 and 11, as Rifampicin resistant mutants, efficiently colonized the root system of peach and Colt throughout growing season. *A. radiobacter* produced many mechanisms of biological control such as efficient colonization of roots, binding and physical blockages of infection sites and agrocin production [119]. *B. megaterium*, *Pseudomonas asplenii*, *Pseudomonas fragi* (two isolates), *Pseudomonas viridilivd, Paenibacillus polymyxa, Curtobacterum* sp., *Curtobacterum* sp. and *Curtobacterium flaccumfaciens* had inhibitory activity against *A. tumefaciens* in vitro tests. *C. flaccumfaciens* highly reduced the crown gall incidence in shoots of rose and leaves of kalanchoe, than galling in fruits of squash. *P. asplenii, P. viridilivd* and *P. polymyxa* could reduce the crown galls incidence of in leaves of kalanchoe and/or fruits of squash than rosees shoot. *P. fragi* isolates highly reduced the galling in squash fruits than shoots of rose or leaves of kalanchoe. Isolate *B. megaterium* could completely suppress the galls development in shoots of rose [120].

5.2 Dickeya *(dadantii and* solani*)*

Dickeya spp. is necrotrophic bacteria that infect large numbers of economic plant species [121]. Out of 1165 rhizobacteria, 18 antagonistic bacteria (*Pseudomonas* or *Bacillus*) could inhibit the growth of *Dickeya* spp. and *Pectobacterium* spp. in vitro tests. The most isolates suppressed maceration of potato tuber slices causing by *Pectobacterium* spp. strains. Despite the poor efficacy was recorded against *Dickeya* spp. on potato slices [122]. The bacteriophages, as Myoviridae or Siphoviridae family's members, inhibited the growth of *D. dadantii* strains (antibiotic resistant) in vitro. No disease progression was detected in *Dickeya* infected plants treated with bacteriophage. These results suggested that *Dickeya* strains can be the biological controlled [123].

From successive screenings of 10,000 bacterial isolates, 58 strains (*Pseudomonas* spp. and *Bacillus* spp.) could inhibit the growth of *Dickeya* spp. and/or *Pectobacterium* spp. The biocontrol agents decreased the growth of *Dickeya* sp. and *Pectobacterium* sp. pathogens in vitro tests. In greenhouse assays, the antagonstics also reduced the soft rot incidence on potato plants when artificially inoculated with *Dickeya dianthicola*. *P. putida* (PA14H7 strain) or *P. fluorescens* (PA3G8 and PA4C2 strains) repeatedly decreased the blackleg severity prevent *D. dianthicola* transmission to daughter tubers [124]. *Serratia plymuthica* A30, potato endophyte, could protect potato plants against blackleg caused by *Dickeya* or *Pecto-bacterium* in the storage or the filed conditions. These results suggest that potato tubers treatment with *S. plymuthica* after harvest can reduce the soften severity in storage and prevent the soften bacteria transmission to daghter tubers in field application [125].

5.3 Erwinia amylovora

Fire blight disease is difficult to control by chemicals, where the strategies of control combination among different measures, to eliminate the disease sources, to prevent the plant infection. *E. herbicola* (Eh252 strain) a nonpathogenic epiphytic bacterium, when sprays onto blossoms of apple before *E. amylovora* inoculation, reduced incidence fire blight [126] as well as *E. herbicola, P. fluorescens* and avirulent strains *P. syringae* can biological control of fire bight disease [5]. *P. fluorescens* (A506 strain) significantly reduced the *E. amylovora* colonization in pear under greenhouse conditions [127]. *Pantoea agglomerans* (E325 and C9-1strains) and *P. fluorescens* (A506 strain) suppressed the growth of *E. amylovora* and reduced the disease incidence, as single treatment [128]. *P. agglomerans* (Pa21889 strain), *B. subtilis* (BsBD170 strain) and *Rahnella aquatilis* (Ra39 strain) significantly reduced blossom blight of apple [129]. Talc-based formulation of *P. agglomerans* (Eh-24 strain) reduced the blighted blossoms percentage on pears infected with *E. amylovora* [130]. *P. fluorescens* (A506 strain), *P. agglomerans* (C9-1 and E325 strains) and *B. subtilis* (QST strain) reduced the blossom infection in Eastern United States [131].

Of 120 antagonistic epiphytic bacterial isolates, four representative strains *P. fluorescens, P. agglomerans, P. putida* and *S. marcescens* showed the maximum growth inhibition against *E. amylovora* under laboratory. *P. agglomerans* and *P. fluorescens* were more promising than *P. putida* and *S. marcescens*. The selected bacteria highly reduced the disease in field experiment, where *P. agglomerans* was the most effective antagonist, while the least effective was *S. marcescens* [132]. *Streptomyces* sp. (C1-4 strain) suppressed fire blight disease symptoms in the leaf tissues of apple and pear shoots [133]. *P. agglomerans* (P10c strain), when applied a twice significantly, reduced fire blight incidence and showed a positive evolution of survival ability in the stigma in field application [134]. *Pseudomonas graminis* (49M strain) protected apple blossoms and shoots against fire blight, than *P. fluorescens* (A506 strain) and *Pantoea vagans* (C9-1strain) in greenhouse. Therefore, it can be applied a new biopesticide against fire blight [135]. From 114 bacterial isolates; nine bacterial strains (2328B-5, 2328B-3, 2025-1, Ach1-1, 2074-1, 2321-5, Ach2-1, 2066-7 and 2025-11) gave antibacterial activity against *E. amylovora* in vitro tests [136]. *P. agglomerans*, fluorescent *Pseudomonas* sp., *Enterobacter* sp. and *Serratia* sp. were able to reduce the disease severity of *E. amylovora* on immature fruits or flowers [137].

5.4 Pectobacterium carotovorum *and* Pectobacterium atrosepticum

Pseudomonas fluorescens (F113 strain), which produces DAPG, inhibited *Erwinia carotovora* subsp. *atroseptica* in vitro strain. The F113 strain (wild-type) or F113Rif strain (the spontaneous rifampicin-resistant mutant) could inhibit the *E. carotovora* subsp. *atroseptica* growth in vitro and prevented soften on wounds of potato tubers under in vivo conditions. The strain F113Rif could reduce the *E. carotovora* subsp. *atroseptica* population of in soil and on potato tubers in unplanted and planted soil, respectively. The results indicated that *P. fluorescens* (F113strain) is a promising biocontrol agent against *E. carotovora* subsp. *atroseptica* [138]. Pretreatment of the melon cotyledons with *P. fluorescens* has antagonistic effect aganist *Erwinia carotovora* subsp. *carotovora* infection and reduced their deleterious effect on fresh and dry weights of melon seedlings [139]. *Sterptomyces plicatus* (101 strain) and *Streptomyces* sp. (OE7 strain) controlled soften on potatoes, where it can apply in programs of integrated controls against soften pathogens of potatoes [140, 141].

B. *subtilis* and *T. harzianum* reduced the activity of pectolytic enzymes (viz. PG and PME) produced by *E. carotovora* subsp. *carotovora* in vitro tests [142]. The field experiment was designed by Abd El-Khair and Haggag [142], in a randomized complete block, for testing the bactericidal activity of *B. subtilis* and *T. harzianum* against soft rot erwiniae. Application of *T. harizanum and B. subtilis* could protect the daughter potato tuber against soften when assayed at harvest time, where the percentage of softening tubers were zero, than untreated plants (Table 3). *T. harzianum* improved the plant height, number of leaves and potato yield, followed *B. subtilis*. When the harvest tubers stored at 3 months of storage, *T. harzianum* could protect stored potato, where the lowest potato decay was recorded, followed by *B. ubtilis*. These results demonstrated that *T. harzianum and B. subtilis* treatments can be efficient method for disinfected potato tubers by field applied to produce the healthy potato tubers stored for long time (Table 3).

B. *subtilis* has a positive reaction against *E. carotovora* and decreased the soft rot disease severity [143]. *Pseudomonas aeruginosa* (pY11T-3-1 strain) reduced the occurrence of soften on tubers or fruits potatoes, peppers, tomatoes, cucumbers and eggplants. The strain may be provide a more environment and economic alternatives

Table 3 Effects of *Trichoderma harzianum* and *Bacillus subtilis* on percentages of soften potato tubers at harvest and after storage and on growth and yield parameters under field application [142]

Parameters	Treatments		
	Trichoderma harzianum	*Bacillus subtilis*	Control
Soften % at harvest	0.0b	0.0b	20.0a
Soften % after storage (3 months)	10.0c	22.2b	62.5a
Tubers weight losses % after storage (3 months)	26.9c	47.7b	72.9a
Growth parameters			
Stem height (cm)	62.7b	59.7c	54.3d
Increase %	10.0	16.0	–
Leaves no./pit	43.0b	37.3c	22.7d
Increase %	89	65	–
Tuber yield parameters			
Tuber weight (g)	89.0a	83.0b	79.8c
Increase %	11.5	4.0	–
Tuber yield (kg)	0.89a	0.75b	0.64c
Increase %	39.5	17.1	–

for controlling soil borne pathogens [144].Yeasts viz. *Rhodotorula* spp. (Rh$_1$ and Rh$_2$) and *Saccharomyces cerevisae* (Sc1) reduced the soft rot incedince caused by *P. carotovorum* subsp. *carotovorum* in Chinese cabbage [145]. *P. putida* suppressed *E. carotovora* infection, where N-Acyl homoserine lactones serve as the vital quorum-sensing signal that regulate the virulence of the pathogenic bacterium [146].

Mean followed by the same letter in each row are not significantly differed using L.S.D. teat ($P = 0.05$).

Bacillus spp.; *B. thuringiensis*, *B. cereus*, *B. subtilis*, *B. megaterium* and *B. pumilus* also showed a good activity against growth of *P. carotovorum* subsp. *carotovorum* using disk method [147]. *Pseudomonas* isolates significantly inhibited the growth of *P. carotovorum* in vitro experiments as well as good prevention of soft rot disease on Valerian rhizome under greenhouse conditions [148]. *Lactobacillus* sp. (E-45strain) and *Bacillus* (E-65 strain) significantly inhibited the growth of *E. carotovora* subsp. *carotovora* in vitro. Strain E-65 has the stronger antagonistic activity against pathogen in vitro and stored potatoes [149]. *P. putida* controlled *E. carotovora* in carrots, where *P. putida* mixed with neem cake gave the best reduction of *E. carotovora* with a significantly increase in the carrot yield under field conditions [150]. *P. fluorescens* and *B. subtilis* had significant anatgonstic effect on *E. carotovora* subsp *carotovora* isolates infected Sponta potato tubers, than *B. thuringiensis* in pot experiments [151]. *Gliocladium* sp. (T.N.C73 strain) also could inhibit the bacterial growth in disc diffusion tests [152]. *Bacillus amyloliquefaciens* subsp. *plantarum* controlled soften bacteria in green peppers or Chinese cabbages in vivo assays [153].

Trichoderma pseudokoningii (SMF2 strain) induced systemic resistance, against soft rot disease in Chinese cabbage, by producing Trichokonins through the activation of salicylic acid signaling pathway [154]. *T. virdi*, *P. fluorescens* and *B. subtilis* showed the stronger antagonistic activity against *E. carotovora*, respectively [155]. Sandipan et al. [156] revealed that *P. fluorescens* showed the maximum growth inhibition against *E. carotovora* subsp. *carotovora*, than *T. viride* in potatoes. *Trichoderma asperellum* reduced the negative effect caused by *E. carotovora* on the young orka seedlings [157]. *Streptomyces diastatochromogenes* sk-6 successfully reduced soft rot disease in stored potatoes, as pretreatment of potato tubers, and it protects the potato tubers against phytopathogenic bacteria in the early period of their reproduction [158]. *P. carotovorum*, causing potato tuber soft rot, uses *N*-acyl-L-homoserine lactones (AHLs) to control the production of virulence factors via quorum sensing (QS). Some bacteria produce enzymes to inactivate the AHLs signals of pathogenic bacteria. *Bacillus* sp., *Variovorax* sp., *Variovorax paradoxus* and *A. tumefaciens* showed putative AHLs activity for controlling the production of virulence factors by *P. carotovorum*, where AHLs-degrading endophytic bacteria can be utilized as a novel biocontrol agent of potato tuber soft rot in Vietnam [159]. *Streptomyces* spp. showed the strongest effect in vitro and in pots experiment, followed by *P. fluorescence*, *B. subtilis* or *P. aeruginosa*, respectively. The severity of soften disease also was the lowest disease severity values with *Streptomyces* spp., *P. fluorescence*, *B. subtilis* or *P. aeruginosa*, respectively [160].

5.5 Pseudomonas syringae *Pathovars*

P. agglomerans could suppress the basal kernel blight development (*P. syringae* pv. *syringae*) in barley fields, where the efficacy of biocontrol strain was affected by time and rate of application [161]. Six strains of *P. syringae* pathovars were effective against *P. syringae* pv. *glycinea* in vitro and in planta. The epiphytic behaviors of the antagonistic *P. syringae* (22d/93 strain) and its two antibiotic-resistant mutants were not significantly different in soybean fields, where the pathogen development was significantly reduced during the whole growing season [162]. *B. subtilis* controlled *P. syringae* pv. *tomato* in infecting *Arabidopsis* roots in vitro and in soil application by producing surfactin as lipopeptides, where surfactin role in biological control was tested with a *B. subtilis* mutant (M1strain) against *P. syringae* infectivity in *Arabidopsis. B. subtilis*, when colonize the roots produces lipopeptide level could sufficient kill *P. syringae* [163]. The antagonistic activities of 206 bacterial isolates, inculdes 62 and 35 genus obtained from phyllosphere of pome trees, were tested against *P. syringae* pv. *syringae* (tip leaf necrosis disease) in vitro and in vivo tests. Strains of RK 84, 85, 113 or 154 (*P.agglomerans*); RK164 strain (*Leclercia adecar-boxylata*); RK142 strain (*P. putida*); RK114strain (*Curtobacterium flaccumfaciens*); RK135strain (*Erwinia rhapontici*); RK137strain (*Alcaligenes piechaudii*); RK102 strain (*Serratia liquefaciens*) and their combinations significantly reduced diseases development [164].

The phages namely Ph1, Ph2 and ph1+2 reduced the disease severity of halo blight disease (*P. syringae* pv. *phaseolicola*), than *P. fluorescence* and *P. putida*. These phages may be useful tool for controlling of halo blight pathogen in greenhouse, where mixed phages was more effective than single treatment [165]. In vitro tests, 21 strains of *Bacillus* spp. could display antagonistic effect on *P. syringae* pv. *syringae* in citrus. Antagonistic bacteria confirmed their antimicrobial effect on *P. syringae* under greenhouse conditions, where it able to reduce the stem necrosis at 10 weeks after *P. syringae* inoculation [166]. The endophytic bacteria, which recovered from *Leptospermum scoparium*, medicinal plant produces essential oils showing antimicrobial activity, showed antagonistic effect against *P. syringae* pv. *actinidiae* (bacterial canker disease) in kiwifruit in vitro. The endophytic bacteria were able to produce multiple antibiotics, such as phenazine, DAPG and hydrogen cyanide. Three of endophytic bacteria transmissmated by wounds inoculation to kiwifruit could inhibit P. *syringae* pv. *actinidiae* colonization and reduced disease severity in two different commercial cultivars [167]. *B. amyloliquefaciens* (SS-12.6 or SS-38.4 strains) and *B. pumilus* (SS-10.7 strain) produced crude extracts of lipopeptides able to biological control of *P. syringae* pv. *aptata* in sugar beet [168].

5.6 Ralstonia solanacearium

Among of fluorescent psedumonads (125 strains) and non-fluorescent bacteria (52 strains); *P. fluorescens* (Pfcp strain) suppressed the bacterial wilt pathogen in treated banana, eggplant and tomato plants, than *Bacillus* spp. (B33 and B36 strains) [169]. All the tested bio-control agents also reduced the bacterial wilt disease to various degrees, where the degree of disease suppression by other microbes varied with the time of application [170]. Of 118 rhizobacteria, six strains, i.e. RP87, B2G, APF1, APF2, APF3 and APF4 had well inhibitory against bacterial wilt pathogen in vitro screening. Treatments of soil or seedlings of tomato with strains of APF1 or B2G significantly reduced disease incidence and increased the fresh and dry weight of tomato plants under greenhouse conditions as well as the two antagonstic strains are promising in field applications [171]. Out of the 50 fluorescent *Pseudomonas* isolates; Pf S2, Pf Wt3 and PfW1 isolates only showed inhibition against the growth of *R. solanacearum* in potato in vitro tests and under greenhouse conditions. The isolates increased the plant growth viz. plant height and dry weight [172].

Antagonistic microbes like Mycorrhizal fungi, *Streptomyces* sp. and *Trichoderma* sp. play an important role in biological control of *R. solanacearum* [173]. Of 73 isolated antagonists, eight were controlled bacterial wilt in tomatoes and peppers in vitro and in vivo tests, where all treatments significantly reduced disease symptoms. *B. megaterium, Candida ethanolica, Enterobacter cloacae* and *Pichia guillermondii* highly suppressed the disease and increased plants height and fruits weight [174]. Of six selected isolates had antagonistic or suppress effect against *R. solanacearum* in dual culture assays, *Staphylococcus epidermidis* and *B. amyloliquefaciens* showed significantly lower disease incidence than control in greenhouse [175]. *B. subtilis, Enterobacter aerogenes, P. fluorescens* and *P. putida*, were isolated from rhizosphere of tomato, increased seed germination over untreated control except *E. aerogenes*. *P. fluorescens* highly reduced bacterial wilts of tomatoes in greenhouse conditions, followed by *P. putida, B. subtilis* and *E. aerogenes*, respectively. In field trails, *P. fluorescens* caused the highest wilt disease reduction, while *P. putida* exhibited the lowest [176].

Field application of *B. subtilis, Trichoderma hamatum* and *Trichoderma album* were made for controlling the incidence or severity of bacterial wilts caused by *Ralstonia solanacearium* by Abd El-Khair and Seif El-Nasr [40]. *B. subtilis, T. album* and *T. hamatum* were applied separately, as potato tuber-pieces or soil treatment in potato cv. Nicola, during the summer growing season (February/May of 2007). After of 30 days of planting, the treatments could significantly reduce the bacterial wilts disease severity (WS) and protect the potato plants against the disease (Table 4). As tuber pieces treatment; *B. subtilis* and *T. hamatum* reduced the WS to 4% and disease control (DC) reached to 80%, than WS (8%) and DC (60%) with *T. album*. In soil treatment, *T. hamatum* could prevent completely bacterial wilt disease, where the disease incidence was zero, than WS (2 and 6%) and DC (90.0 and 70.0%) with *B. subtilis* and *T. album*, respectively. After sixty days of planting, both *T. hamatum* and *T. album* reduced the WS to 20.0% and DC was 47.4% as tuber pieces treatment,

Table 4 Effects of bio-control agents application on wilt disease severity caused by *Ralstonia solanacearum* in potato under field conditions [40]

Bio-control agents	Percents of bacterial wilt severity (WS) and disease control (DC)/days							
	30				60			
	BWS		PDC		BWS		PDC	
	Tuber pieces	Soil	Tuber pieces	Soil	Tuber pieces	Soil	Tuber pieces	Soil
February–May of 2007 season—potato cv. Nicola								
Bacillus subtilis	4.0c	2.0c	80.0	90.0	22.0b	20.0b	42.1	47.4
Trichoderma album	8.0b	6.0b	60.0	70.0	20.0c	18.0c	47.4	52.6
Trichoderma hamatum	4.0c	0.0d	80.0	100.0	20.0c	20.0b	47.4	47.4
Control	20.0a	20.0a	–	–	38.0a	38.0a	–	–
October–January of the 2007/2008—potato cv. Diamante								
Bacillus subtilis	2.0b	0.0b	85.7	100.0	8.0b	8.0b	50.0	50.0
Trichoderma album	0.0c	0.0b	100.0	100.0	8.0b	0.0c	50.0	100.0
Trichoderma hamatum	0.0c	0.0b	100.0	100.0	0.0c	0.0c	100.0	100.0
Control	14.0a	14.0a	–	–	16.0a	16.0a	–	–

Mean followed by the same letter in each column, in each expermint, are not significantly differed using L.S.D. teat ($P = 0.05$)

than WS (22.0%) and DC (42.1%) with *B. subtilis*, respectively. In soil treatment, *T. album* reduced the WS to 18.0% (DC 52.6%), than the WS 20.0% (DC 47.4%) with both *B. subtilis* and *T. hamatum*. The treatments also improved the growth parameters of potato plants cv. Nicola such as plant height, stem number/pit and leaves number (Table 5). *T. album* highly improved the plant height, followed by *B. subtilis* and *T. hamatum*, whereas *T. hamatum* highly increased the stem number/pit and number of leaves, than *B. subtilis* or *T. album*. The treatments also increased the tested tuber yield parameters (Table 2).

In another experiment, the same bio-control agents applied as whole tuber dressing or soil treatment with whole potato tubers cv. Diamante during winter growing season (October/January of 2007/2008). The treatments could protect the potato plants and improve growth and yield parameters (Tables 4 and 5). After 30 days of planting, *T. hamatum* and *T. album* protected completely the potato plants against wilt disease infection, when applied as whole tuber dressing or soil treatment (Table 4). *B. subtilis* could protect completely the as soil treatment, while the bacterium reduced BWS (2.0%) and PDC (85.7%) as whole tuber treatment. After 60 days of planting, *T. hamatum* protected the potato plants with a rate of 100% against the disease when

Table 5 Effects of bio-control agents on growth and tuber yield parameters of potato under field conditions [40]

Bio control agents	Growth parameters (60 days after planting)						Yield parameters at harvest time			
	Plant height (cm)		Stems no./pit		Leaves no./pit		Tuber weight		Yield (g)/pit	
	Tuber pieces	Soil	Tuber pieces	Soil	Tuber pieces	Soil	Tuber pieces	Soil	Tuber pieces	Soil
February–May of 2007 season—potato cv. Nicola										
B. subtilis	43.1b	44.6b	1.5b	1.7b	39.1b	49.9b	32.9a	22.0a	527.6a	494.4b
T. album	45.4a	45.4a	1.3c	1.3c	39.0b	46.6c	20.4c	20.0b	448.8b	533.3a
T. hamatum	39.3c	41.4c	1.6a	1.8a	43.4a	50.7a	25.8b	17.1c	463.6b	376.2c
Control	39.1d	39.1d	1.2d	1.2d	37.8c	37.8d	16.8d	16.8c	319.9c	319.9
October–January of the 2007/2008—potato cv. Diamante										
B. subtilis	28.0a	32.5a	3.0b	3.6a	24.8c	40.6a	58.8c	101.6c	323.4c	387.3
T. album	26.2b	27.3c	3.7a	3.6a	39.7a	31.7c	68.7a	109.1b	381.9a	542.6a
T. hamatum	26.0b	31.1b	2.9b	3.5a	31.2b	38.1b	61.5	110.7a	369.0b	508.0b
Control	21.7c	21.7d	2.9b	2.9b	23.2c	23.2d	56.9d	56.9d	229.5d	229.5d

Mean followed by the same letter in each column, in each expermint, are not significantly differed using L.S.D. teat ($P = 0.05$)

applied as whole tuber dressing, followed by DC 50 with *B. subtilis* and *T. album*. In soil treatment, *T. album* and *T. hamatum* were resulted 100% disease control, while *B. subtilis* gave DC 50%. *B. subtilis* significantly increased the potato height, than *T. album* or *T. hamatum* with significant differences were recorded among biocontrol agents and the control except between *T. hamatum* and *T. album* in tuber dressing. *T. album* significantly improved the stems number, while both *T. album* and *B. subtilis* gave the best leaves number on potato plants in both tuber and soil treatments, respectively. *T. album* and *T. hamatum* gave the highest yield in tuber and soil treatments, that *B. subtilis* treatment (Table 5).

P. aeruginosa (T1 strain), *Pseudomonas* sp. (AM12 strain), *Pseudomonas* sp. (AM13 strain), *Pseudomonas* sp. (BH25 strain) and *P. putida* (R6 strain) had antagonistic effect against *R. solanacearum* (Tom5 strain) isolated from a wilted tomato plants. In bioassays; BH25 strain reduced the pathogenic effects caused by Tom5. BH25 strain also could improve the fresh and dry weights and seedlings vigor index [177]. From 298 rhizobacteria isolates were common in rhizospheres or rhizoplanes of tomatoes or eucalyptus; nine isolates (UFV-11, 32, 40, 56, 62, 101, 170, 229 and 270) suppressed bacterial wilt disease in vitro or in vivo experiments. The UFV-56 isolate (*B. thuringiensis*) or UFV-62 isolate (*B. cereus*) could suppress wilts development in eucalyptus at the early stage [178]. Application endophytic bacteria can induce biological control aganist plant diseases, where the bacteria associated with plant roots could exert atagonstic activities which directly or indirectly prevent the plant development by solubilize minerals in soil, phytohormones synthesises, soil borne pathogens suppressive and/or induce systmic resistance [179]. *B. amyloliquefaciens* (DSBA-11 and DSBA-12 strains), isolated from rhizospheric of wilted tomato plants, had the best antagonistic effect and plant growth promoting ability.

These strains DSBA-11 and DSBA-12 showed the highly antagonistic activity, than *B. subtilis* (DTBS-5 strain), *B. cereus* (JHTBS-7strain) and *B. pumilus* (MTCC-7092strain), against *R. solanacearum* in vitro conditions. *B. amyloliquefaciens* (DSBA-11strain) showed the maximum growth inhibition of *R. solanacearum*, followed by DSBA-12 and *B. subtilis* strains. Strain of DSBA-11 has better phosphorus solubilizing ability and indole acetic acid production, than other strains of *Bacillus* spp. in vitro conditions. The minimum bacterial wilt disease incidence in cultivar Pusa Ruby was recorded with DSBA-11 strain, followed by DSBA-12 strain after 30 days of inoculation. The bio-control efficacy was higher in DSBA-12 strain treated plants, followed by treated with MTCC- 7092 strain, under glasshouse conditions [180].

5.7 Xanthomonas orzyae

Methods of controlling rice bacterial blight are limited in effectiveness, where usage the chemical pesticides causing toxic effects on the main consumers, kills non-target organisms and environment contamination. Application of bio-control agents has efficiency and safety for humans and other non-target organisms as well as they leave no toxic residues in foods. Especially, we need to apply environment- safe tools to reduce the grain yield losses in rice resulted from *X. oryzae* pv. *oryzae* infection [43, 181]. *Aspergillus ochraceus, Aspergillus flavus, Aspergillus niger, Fusarium pallid-oroseum, Fusarium chlamydosporum, Micrococcus* sp., *Penicillium janthinellum, Pseudomonas acidovorus* and *Streptomyces* sp., isolated from rice leaves, could inhibit *X. oryzae* pv. *oryzae* in vitro [182]. *In planta* experiments, the effect of endophytic *Streptomyces* spp. against bacterial leaf blight disease was non-significantly differed about those in controls. But, the treated rice plants could significantly produce taller or higher tiller numbers, than control plants. The biocontrol mechanisms of *Streptomyces* spp. may due to produce chitinase, phosphatase and siderophore [183]. *Streptomyces toxytricini* significantly reduced *X. oryzae* pv. *oryzae*-related yield loss in infected rice cultivars in field, than the rice yield in healthy rice cultivars [184]. The endophytic strains as *B. amyloliquefaciens* (A1, A3 and A13 strains), *B. subtilis* (A15 strain), *Bacillus methylotrophicus* (A2 strain) and two rhizospherial *Bacilli* (D29 and H8 strains) showed a high antagonistic activity against *X. oryzae* pv. *oryzae* in vitro tests. Four bacterial strains significantly increased the inhibition rate against the pathogen in greenhouse as well as the fresh and dry weight in treated plants of rice. The tested strains produce siderophores, indole-3-acetic acid, and able to solubilize of phosphate [185].

A total eight isolates of non-pathogenic phyllosphere (actinomycetes) were positively controlled *X. oryzae* pv. *oryzae* in vitro as well as its significantly reduced the disease severity [181]. *P. fluorescens* (RR8 strain) has strong antibacterial activity against *X. oryzae* pv. oryzae in vitro studies, where the bacterium can be serve as potential biocontrol agent against bacterial blight disease [43]. Out of 512 bacteria, isolated from the rice rhizosphere, *P. aeruginosa* showed antagonism against *X. oryzae* pv. *oryzae* as well as it able to solubilize phosphorus and produce phytohormone indole acetic acid and siderophores in vitro tests. Strain BRp3 suppressed different strains of bacterial pathogen in rice. The crude extract of BRp3 contains siderophores i.e.1-hydroxy-phenazine, pyocyanin or pyochellin; rhamnolipids; 4-hydroxy-2-alkylquinolines or novel 2, 3, 4-trihydroxy-2-alkylquinolines might be responsible for antagositic effects on bacterial pathogen [186].

5.8 Xanthomanas campestris *Pathovars*

Twenty isolates of cabbage phylloplane yeasts showed an antibacterial effect on *X. campestris* pv. *campestris* in the fields, where the disease severity was highly reduced with isolates LR32, LR42 and LR19 strain, respectively [187]. Both R14 strain (*B. subtilis*), C116 strain (*B. pumilus*), RAB7 strain (*Bacillus megaterium* pv. *cerealis*) and C210 strain (*B. cereus*) had antagonistic effects against *X. campestris* pv. *campestris* (LFR-3 strain) in vitro [188]. The endophytic *B. subtilis* (BB strain) was tested for controlling black rot disease (*X. campestris* pv. *campestris*) in cabbages, cauliflowers, rapes and broccolis during three growth seasons. Strain BB controlled the black rot disease in all crops of *Brassica* in the seasons of short rainy or dry.

Application of biological control was effectively reduced the disease in broccoli, in main rainy season in clay loam soil and sandy loam soil, than cabbage or rape [189]. *Bacillus* spp. was significantly reduced the incidence or severity of black rots in cabbages under field conditions, when the bacteria applied through the roots, than through the seeds or foliage application. The promising antagonists include *B. cereus*, *Bacillus lentimorbus* and *B. pumilus* strains [190]. Application of *R. aquatilis* (two strains), as soil, seeds, roots or leaves treatments, reduced susceptibility of tomato toward *X. campestris* pv. *vesicatoria* which produced irregular yellow-necrotic areas on plants. *R. aquatilis*, foliar application, effectively reduced the pathogen. The seed treatment also produced the highest fresh and dry weight, than foliar, soil or root treatments [191].

Four *Bacillus* isolates had as positive effect of antibiosis and hemolysis on all strains of *X. campestris* pv. *campestris* causing crucifers black rot disease. The antimicrobial and hemolytic activities correlation revealed that the lipopeptides cause the mechanisms of antibiosis, where *Bacillus* isolates can produce bioactive or surfactant compounds at late growth phase [192]. Three *Bacillus* strains, isolated from the peppers rhizosphere grews in greenhouse or field, suppressed the populations of *X. axonopodis* pv. *vesicatoria*, as well as decreased the disease development, when applied single and mixed, in experiments of greenhouse and field, respectively. The treatments increased the growth parameters such as stem diameter, root elongation, root and shoot dry weight and yield responded to the treatments in the field experiments as successful biological control of bacterial spot disease [193]. The compost's suppressive effect on *X. campestris* pv. *vesicatoria* growth mainly due to biotic factors, where sterilised compost samples were not effective. The suppressive characteristics of agroindustrial-based compost were mixed with a strain of *B. pumilus* (MSW231strain) was effective against *X. campestris* pv. *Vesicatoria* [194].

5.9 Xanthomonas axonopodis

Citrus canker incited by *X. axonopodis* pv. *citri* is a serious disease of acid lime all over the world. The citrus canker disease was controlled by application of antibiotics and some agrochemicals as spraying treatments, but little work applied the biological management. Management of the bacterial angular leaf spot disease caused by *X. axonopodis* pv. *malvacearumm* using of synthetic chemicals was recently discouraged because of nature hazardous or environment pollution. Among seven strains; MMP and Pf1 strains (*P. fluorescens*) highly inhibited of *X. axonopodis* pv. *malvacearum* (bacterial blight of cotton) in vitro tests. Strains of MMP and Pf1 strains, as talc formulation, has the best inhibiting of *X. axonopodis* pv. *malvacearum* growth. The Pf1 strain, as wet seeds treatment, significantly improved the seeds germination and vigour of cotton seedlings. Pf1 strain, as seeds treatment, significantly suppressed the blights incidence, than foliar application of. Application of MMP or Pf1 strains induced a new peroxidase [195]. Of 453 isolates of bacterial flora of citrus leaves in Iran; 26 strains inhibited growth of *X. axonopodis* pv. *citri* in vitro conditions. The bacterial strains were identified as *P. fluorescens*, *Pseudomonas viridiflava* and *P. syringae*, while the last group as *Bacillus* spp. based on physiological and biochemical characteristics [196]. *Pseudomonas* sp. could protect eucalyptus seedlings when applied before or after *X. axonopodis* inoculation in greenhouse experiments, where the numbers of lesions on leaves was lower than in untreated control plants. High antibiotic activity against *X. axonopodis* and leaf blight suggests that it has potential to control the disease in eucalyptus seedlings [197]. Single spray of aqueous suspension of *B. subtilis* resulted in a satisfactory decline of the disease [198].

The secondary metabolites of *Pseudomonas* sp. (LN strain) were effective on citrus canker disease (Xac 306 strain). The free-cells supernatant of LN strain was firstly treated with methanol (AMF), followed by ethyl acetate (AEF) and then it fractionated by vacuum liquid chromate-graphy (VLC). The activity of all fractions tested on bacterial growth by using Xac 306 well agar diffusion test or minimum inhibition concentration. Cytotoxicity effect studies demonstrated that the tested concentrations of EAF, VLC2 and VLC3 the fractions had non genotoxic effect. The fraction of VLC3 significantly reduced the incidence of canker lesions in citrus only [199]. The formulated isolate phages (CF and SM) were reduced the disease severity of halo blight (*X. axonopodis*) in pepper in greenhouse and field conditions respectively, compared with unformulated phages.

Application of skim milk and corn flour increased the phage longevity and phage population in greenhouse and open field respectively. These formulated phages may be useful as a tool to biocontrol of halo blight disease [200]. From forty actinomycete strains, isolated from natural sources, eight strains had high antagonistic activity against *X. axonopodis* pv. *punicae* causing oily spot disease in pomegranate. The extracted compounds belonged to amino glycosides from *Streptomyces violaceusnige* produces the maximum inhibition on isolates of pathogens. The extracted compounds or *S. violaceusnige* could effectively prevent the *Xanthomonas* growth on inoculated fruits of pomegranate in vivo [201]. Of 14 strains of the genus *Paenibacillus*; only two strains only had inhibitory effects against *X. axonopodis* pv. *malvacearumm* (The bacterial angular leaf spot disease in cotton), compared to the control [202]. A total of 53 bacteria isolated from pomegranate phylloplane were identified as *Bacillus*, *Alcaligenes*, *Myroides*, *Brevibacterium*, *Pantoea*, *Proteus*, *Enterobacter*, *Lysinibacillus*, *Paenibacillus* and *Stenotrophomonas* species. About 27 phylloplane bacteria revealed potential bio-control ability against pathogen in vitro evaluation. Assay of secondary metabolites from 14 phylloplane bacteria, *B. subtilis* (P49) and Bacterium fjat (P31) exhibited maximum inhibition of pathogen growth. Field application of *B. subtilis* and Bacterium fjat recorded maximum disease reduction on leaves, followed by on fruits, respectively [203]. *Bacillus* spp., *Pseudomonas* spp., *Rhodococcus* and the combinations controlled *X. axonopodis* pv. *phaseoli*. It is clear that the use of combinations of these organisms increased the efficacy of the biocontrol of several strains [204].

5.10 Xylella fastidiosa

A current control strategy for *X. fastidiosa* includes exclusion of bacterium or vector, Vector control, systemic insecticides where secondary spread is important [205]. The population of *X. fastidiosa* occurred as cultural of endophytic bacterium in branches or leaves of healthy plants sweet oranges or in tangerine (*Citrus reticulata* cv. *Blanco*) plants were assessed by Lacava et al. [206]. Endophytic bacteria identified as belonging to the genus *Methylobacterium*. The *X. fastidiosa* growth of stimulated by *Methylobacterium extorquens*, while it inhibited by *Curtobacterium flaccumfaciens* in vitro. The endophytic viz. *B. pumilus*, *C. flaccumfaciens*, *E. cloacae*, *Methylobacterium* spp. (i.e. *M. extorquens*, *M. fujisawaense*, *M. mesophilicum*, *M. radiotolerans* and *M. zatmanii*), *Nocardia* sp., *P. agglomerans* and *X. campestris* were tested against *X. fastidiosa,* the causal Citrus variegated chlorosis (CVC). A relationship between CVC symptoms and isolation frequency *Methylobacterium* spp. revealed that the genus was frequently isolated from symptomatic plants.

The frequent of *C. flaccumfaciens* significantly frequently isolated from asymptomatic plants than CVC-symptoms plants, while *P. agglomerans* was frequently isolated from tangerine or sweet oranges were asymptomatic, symptomatic or showing CVC symptoms [52, 207]. *X. fastidiosa* (six strains) was applied for biological controlling of natural Pierce's disease progression. Strain EB92-1 only (obtained from elderberry) provides good controlling on disease in flame seedless or Cabernet sauvignon, while Syc86-1 strain did not ineffect in the vineyard tests. Strain PD95-6 grape was lower severity of disease in flame seedless, than nontreated vines. Strain PD91-2 of grape could delay symptoms on Cabernet sauvignon for 12–18 months. Benign inoculation of susceptible grapevines with *X. fastidiosa* strains, especially strain EB92-1 can biological control of Pierce's disease in commercial vineyards in Florida and where the disease occurs [208].

6 Conclusion and Future Prospects

For good control of bacterial plant diseases must correct identify the plant disease and their pathogens by known the pathogen's life cycle and how it relates to the cycle of disease development. This information is needed to develop a suitable management program that attacks the pathogen at the weakest point in it life cycle. The development of alternative approaches to control pathogens of crops utilizing bio-control agents is necessary to reduce risk pesticides. Application of biological control for control phytopathogenic bacteria depended on the use of natural enemies. To success of biological control, the bio-control agent must grow very fast, the environment is favorable for their growth and development and it must be applied at pre-planting or prior to the onset of disease.

The bio-control also agents had different mechanisms such as antibiosis, competition, hyperparasitism, cell-wall degrading enzymes and induction of systemic resistance which play an important role for controlling plant diseases, where inducing systemic resistance protects plants not only against the attacking pathogen, but against other types of pathogens. Therefore, the biological control successfully applied for controlling many bacterial plant diseases as safe alternative tools replace the chemical bactericides which had mammalian toxicity and environmental pollution. Especially, some commercial bio-pesticides products, I.e. BlightBan A506 (*Pseudomonas fluorescens* A506, Nufarm, Inc.; Kodiak (*Bacillus subtilis strain GB03,* Gustafson LLC) and Plant Shield (*Trichoderma harzianum strain KRL-AG2*, BioWorks, Inc) were occurred.

References

1. Gadoury DM, McHardy WE, Rosenberger DA (1989) Integration of pesticide application schedules for disease and insect control in apple orchards of the northern United States. Plant Dis 73:98–105
2. Agrios G (1997) Plant pathology, 4th edn. Academic Press, pp 1–635
3. Van der Zwet T, Beer SV (1995) Fire blight—its nature, prevention and control. A practical guide to integrated disease management. Agr Inf Bull 631:91
4. Jones AL, McManus PS, Chiou CS (1996) Epidemiology and genetic diversity of streptomycin resistance in *E. amylovora* in Michigan. Acta Hort 338:333–340
5. Vanneste JL (2000) Fire blight. The disease and causative agent, *Erwinia amylovora*. CABI Publications, pp. 1–370
6. Arwiyanto T (2014) Biological control of plant diseases caused by bacteria. J Perlindungan Tanaman Indonesia 18:1–12
7. Andrews JH (1992) Biological control in the phyllosphere. Annu Rev Phytopathol 30:603–635
8. Raaijmakers JM, Weller DM (2001) Exploiting genotypic diversity of 2,4-diacetylphloroglucinol-producing *Pseudomonas* spp.: characterization of superior root-colonizing *P. fluorescens* strain Q8r1-96. Appl Environ Microbiol 67:2545–2554
9. Abd El- Kahir H (2004) Efficacy of starner in controlling the bacterial soft rot pathogen in onion. Ann Agric Sci Ain Shams Univ Cairo 49:721–731
10. Abd El- Khair H (2004) Variation and control of *Erwinia carotovora* subsp. *carotovora* isolates the causal agent of potato soft rot disease. Ann Agric Sci Ain Shams Univ Cairo 49:377–388
11. Coplin DL, Rowan RG, Chisholm DA, Whitmoyer RE (1981) Characterization of plasmids in *Erwinia stewartii*. Appl. Env. Microbiol. 42:599–604
12. Schaad NW, Jones JB, Chun W (2001) Laboratory guide for identification of plant pathogenic bacteria, 3rd edn. The American Phytopathological Society Press, St. Paul
13. Lelliott RA, Stead D (1987) Methods for the diagnosis of bacterial diseases of plants. Blackwell Scientific Publications, Oxford, UK, pp 1–216
14. Vidaver AK, Lambrecht PA (2004) Bacteria as plant pathogens. The Plant Health Instructor. https://doi.org/10.1094/PHI-I-2004-0809-01
15. Smith EF, Townsend CO (1907) A plant tumor of bacterial origin. Science 25:671–673
16. Young JM, Kuykendall LD, Martínez-Romero E, Kerr A, Sawada H et al (2001) A revision of *Rhizobium* Frank 1889, with an emended description of the genus, and the inclusion of all species of *Agrobacterium* Conn 1942 and *Allorhizobium undicola* de Lajudie *et al.* 1998 as new combinations: *Rhizobium radiobacter, R. rhizogenes, R. rubi.* Int J Sys Evol Microbiol 51:89–103

17. Pitzschke A, Hirt H (2010) New insights into an old story: *Agrobacterium*-induced tumor formation in plants by plant transformation. EMBO J 29:1021–1032

18. Schell J, Van Montagu M (1977) The Ti-plasmid of *Agrobacterium tumefaciens*, a natural vector for the introduction of NIF genes in plants? Basic Life Sci 9:159–179

19. Goodner B, Hinkle G, Gattung S, Miller N et al (2001) Genome sequence of the plant pathogen and biotechnology agent *Agrobacterium tumefaciens* C58. Science 294(5550):2323–2328

20. Gelvin SB (2010) Plant proteins involved in *Agrobacterium*-mediated genetic transformation. Annu Rev Phytopathol 48:45–68

21. Samson R, Legendre JB, Christen R, Saux MFL, Achouak W, Gardan L (2005) Transfer of *Pectobacterium chrysanthemi* (Burkholder et al. 1953) Brenner et al. 1973 and *Brenneria paradisiaca* to the genus *Dickeya* gen. nov. as *Dickeya chrysanthemi* comb. nov. and *Dickeya paradisiaca* comb. nov. and delineation of four novel species, *Dickeya dadantii* sp. *nov.*, *Dickeya dianthicola* sp. *nov.*, *Dickeya dieffenbachiae* sp. *nov.* and *Dickeya zeae* sp. *nov.* Int J Syst Evol Microbiol 55:1415–1427

22. Ma B, Hibbing ME, Kim HS, Reedy RM, Yedidia I, Breuer J, Breuer J, Glasner JD, Perna NT, Kelman A, Charkowski AO (2007) Host range and molecular phylogenies of the soft rot enterobacterial genera *Pectobacterium* and *Dickeya*. Phytopathol 97:1150–11639

23. Zhang Y Fan Q, Loria R (2016) A re-evaluation of the taxonomy of phytopathogenic genera Dickeya and Pectobacterium using whole-genome sequencing data. Syst Appl Microbiol 39:252–259

24. Van der Zwet T, Keil HL (1979) Fire blight—a bacterial disease of Rosaceous plants. Agric handbook, vol 510. Department of Agriculture, Washington D.C, U.S

25. Abd El-Khair H, Seif El-Nasr HI (2002) Epidemiology and control of fire blight disease in pears. Arab Univ J Agric Sci, Ain Shams Univ, Cairo 10:1059–1069

26. Barakat FM, Seif El-Nasr HI, Mikhail MS, Abd El-Khair H (2002) Effect of some fungicides and bactericides on the growth *Erwinia amylovora*, the causal of fire blight of pear. In: The First conference of the Central Agric Pesticide Lab, vol 1, pp 328–337, 3–5 Sep 2002

27. Sands DC (1990) Physiological criteria-determinative tests. In: Klement Z, Rudolph K, Sands DC (eds) Methods in phyto-bacteriology. Akadémiai Kiadó, Budapest, pp 133–143

28. Abd El-Khair H, Barakat FM, Mikhail MS, Seif El-Nasr HI (2003) Differentiation between Egyptian isolates of *Erwinia amylovora*, based on cellular protein patterns. In: Proceedings of 10th Congress of Phytopathology, 9–10.12.2003, Giza, Egypt, pp 339–353

29. Haggag KHE, Abd El-Khair H (2006) Antibacterial activity of some Egyptian medicinal plants against *Erwinia carotovora* subsp. *carotovora* isolates in potato. Egypt J Appl Sci 21:428–441

30. Faquihi H, Mhand RA, Ennaji M, Benbouaza A, Achbani E (2014) *Aureobasidium pullulans* (De Bary) G. Arnaud, a biological control against soft rot disease in potato caused by *Pectobacterium carotovorum*. Int J Sci Res 3:1779–1786

31. Perombelon M, Kelman A (1980) Ecology of the soft rot erwinias. Annu Rev Phytopathol 18:361–387

32. Perombelon MCM (2002) Potato diseases caused by soft rot erwinias: An overview of pathogenesis. Plant Pathol 51:1–12

33. Toth Ian K, Bell Kenneth S, Holeva Maria C, Birch PRJ (2003) Soft rot erwiniae: from genes to genomes. Mol Plant Pathol 4:17–30

34. Arnold DL, Gibbon MJ, Jackson RW, Wood JR, Brown J, Mansfield JW et al (2001) Molecular characterization of avrPphD, a widely-distributed gene from *Pseudomonas syringae* pv. *phaseolicola* involved in non-host recognition by pea (*Pisum sativum*). Physiol Mol Plant Pathol 58:55–62

35. Green S, Studholme DJ, Laue BJ, Dorati F, Lovell H, Arnold D, Cottrell JE, Bridgett S, Blaxter M, Huitema E, Thwaites R, Sharp PM, Jackson RW, Kamoun S (2010) Comparative genome analysis provides insights into the evolution and adaptation of *Pseudomonas syringae* pv. *aesculi* on *Aesculus hippocastanum*. PLoS ONE, 5:e10224

36. Zhang J, Li W, Xiang T, Liu Z, Laluk K, Ding X, Zou Y, Gao M, Zhang X, Chen S, Mengiste T, Zhang Y, Zhou JM (2010) Receptor-like cytoplasmic kinases integrate signaling from

multiple plant immune receptors and are targeted by a *Pseudomonas syringae* effector. Cell Host Microbe 7:290–301

37. Abd El-Khair H, Nofal MA (2001) Flowers bacterial soft rot of bird of paradise (*Strelitzia reginae,* Banks) in Egypt and its control. Arab Univ J Agric Sci, Ain Shams Univ, Cairo 9:397–410

38. Denny TP (2006) Plant-pathogenic *Ralstonia* species. In: Gnanamanickam SS (ed) Plant-associated bacteria. Springer, Dordrecht, pp 573–644

39. Genin S (2010) Molecular traits controlling host range and adaptation to plants in *Ralstonia solanacearum.* New Phytol 187:920–928

40. Abd El-Khair H, Seif El-Nasr HI (2012) Applications of *Bacillus subtilis* and *Trichoderma* spp. for controlling the potato brown rot in field. Arch Phytopathol Plant Prot 45:1–15

41. Mew T, Alvarez A, Leach J, Swings J (1993) Focus on bacterial blight of rice. Plant Dis 77:5–12

42. Verdier V, Vera Cruz C, Leach JE (2011) Controlling rice bacterial blight in Africa: Needs and prospects. J Biotechnol 159:320–328

43. Ramanamma C, Santoshkumari M (2017) Biological control of blight of rice using RR8 rhizosphere bacteria. Int J Current Microbiol Appl Sci 5(Special Issue):124–128

44. Johnson BJ (1994) Biological control of annual bluegrass with *Xanthomonas campestris* pv. *poannua* in bermudagrass. Hort Sci 29:659–662

45. Bora LC, Gangopadhyay S, Chand JN (1994) Biological control of bacterial leaf spot (*Xanathomans campestris* pv. *vignaeradiatae* Dye) of mung bean with phylloplane antagonists. AGRISsince 23:162–168

46. Jalali I, Parashar RD (1995) Biocontrol of *Xanthomonas campestris* pv. *campestris* in *Brassica juncea* with phylloplane antagonist. Plant Disease Res10:145–147

47. Vauterin L, Rademaker J, Swings J (2000) Synopsis on the taxonomy of the genus *Xanthomonas.* Phytopathology 7:677–682

48. Babu AGC, Thind BS (2005) Potential use of combinations of *Pantoea agglomerans, Pseudomonas fluorescens* and *Bacillus subtilis* for the control of bacterial blight of rice. Ann the Sri Lanka Dept Agric 7:23–37

49. Young JM, Park DC, Shearman HM, Fargier E (2008) A multilocus sequence analysis of the genus *Xanthomonas.* Syst Appl Microbiol 5:366–377

50. Hopkins DL (1989) *Xylella fastidiosa*: xylem-limited bacterial pathogen of plants. Ann Rev Phytopathol 27:271–290

51. Araújo WL, Marcon J, Maccheroni W Jr, Van Elsas JD, Van Vuurde JWL, Azevedo JL (2002) Diversity of endophytic bacterial populations and their interaction with *Xylella fastidiosa* in citrus plants. Appl Enviro Microbiol 68:4906–4914

52. Zhang S, Cruz ZF, Kumar D, Hopkins DL, Gabriel DW (2011) The *Xylella fastidiosa* biocontrol strain EB92-1 genome is very similar and syntenic to Pierce's disease strains. J Bacteriol 193:5576–5577

53. Lo CT (1998) General mechanisms of action of microbial biocontrol agents. Plant Pathol Bull 7:155–166

54. Ahanger RA, Bhatand HA, Dar NA (2014) Biocontrol agents and their mechanism in plant disease management. Sciencia Acta Xaveriana, An Int Sci J 5:47–58

55. Tzeng KC, Lin YC, Hsu ST (1994) Foliar fluorescent pseudomonads from crops in Taiwan and their antagonism to phytopathogenic bacteria. Plant Pathol Bull 3:24–33

56. Nishioka MF, Nakashima N, Matsuyama N (1997) Antibacterial activities of metabolites produced by *Erwinia* spp. against various phytopathogenic bacteria. Ann Phytopathol Soc Japan 63:99–102

57. Défago G, Haas D (1990) Pseudomonads as antagonists of soilborne plant pathogens: modes of action and genetic analysis. In: Bollag JM, Stotsky G (eds) Soil biochemistry. Marcel Dekker Inc., New York

58. Pal KK, Gardener BM (2006) Biological control of plant pathogens. The Plant Health Instructor. https://doi.org/10.1094/PHI-A-2006-1117-02

59. Ahmad F, Ahmad I, Khan MS (2008) Screening of free-living rhizospheric bacteria for their multiple plant growth promoting activities. Microbiol Res 163:173–181
60. DeSouza JTA, Arnould C, Deulvot C, Lemanceau P, Gianinazzi-PearsonV Raaijmakers JM (2003) Effect of 2,4-diacetylphloroglucinol on *Pythium*: cellular responses and variation in sensitivity among propagules and species. Phytopatholol 93:966–975
61. Notz R, Maurhofer M, Schnider-Keel U Duffy B, Haas D, Defago G (2001) Biotic factors affecting expression of the 2,4-diacetylphloroglucinol biosynthesis gene phlA in *Pseudomonas fluorescens* biocontrol strain CHA0 in the rhizosphere. Phytopathol 91:873–881
62. Nalini S, Parthasarathi R, PrabudossV (2016) Production and characterization of lipo-peptide from *Bacillus* SNAU01 under solid state fermentation and its potential application as anti-biofilm agent. Biocatal Agric Biotechnol 5:123–132
63. Bakker PA, Glandorf DC, Viebahn M, Ouwens TW, Smit E, Leeflang P, Wernars K, Thomashow LS, Thomas-Oates JE, Van Loon LC (2002) Effects of *Pseudomonas putida* modified to produce phenazine-1-carboxylic acid and 2, 4-diacetyl-phloroglucinol on the microflora of field grown wheat. Antonie Van Leeuwenhoek 81:617–624
64. Kerr A (1989) Commercial release of a genetically engineered bacterium for the control of crown gall. Agric Sci 2:41–48
65. Ghisalberti EL, Sivasithamparam K (1991) Antifungal antibiotics produced by *Trichoderma* spp. Soil Biol Biochem 23:1011–1020
66. Maurhofer M, Keel C, Haas D, Defago G (1995) Influence of plant species on disease suppression by *Pseudomonas fluorescens* strain CHA0 with enhanced antibiotic production. Plant Pathol 44:40–50
67. Kerr A (1980) Biological control of crown gall through production of agrocin 84. Plant Dis 64:25–50
68. Koumoutsi A, Chen XH, Hene A, Liesegang H, Gabrielle H, Frnke P, Vater J, Borris H (2004) Structural and functional characterization of gene clusters directing nonribosomal synthesis of bioactive lipopeptides in *Bacillus amyloliquefaciens* strain FZB42. J Bact 186:1084–1096
69. Smith KP, Havy MJ, Handelsman J (1993) Suppression of cottony leak of cucumber with *Bacillus cereus* UW85. Plant Dis 77:139–142
70. Moyne AL, Shelby R, Cleveland TE, Tuzun S (2001) Bacillomycin D: an iturin with antifungal activity against *Aspergillus flavus*. J Appl Microbiol 90:622–629
71. Kloepper JW, Ryu CM, Zhang S (2004) Induced systemic resistance and promotion of plant growth by *Bacillus* spp. Phytopathology 94:1259–1266
72. Leclère V, Béchet M, Adam A, Guez JS, Wathelet B, Ongena M, Thonart P, Gancel F, Chollet-Imbert M, Jacques P (2005) Mycosubtilin overproduction by Bacillus subtilis BBG100 enhances the organism's antagonistic and biocontrol activities. Appl Enviro Microbiol 71:4577–4584
73. Sandra AI, Wright CH, Zumoff LS, Steven VB (2001) *Pantoea agglomerans* strain EH318 produces two antibiotics that inhibit *Erwinia amylovora* in vitro. Appl Enviro Microbiol 67:282–292
74. Shanahan P, O'Sullivan DJ, Simpson P, Glennon JD, O'Gara F (1992) Isolation of 2,4-Diacetphloroglucinal from a fluorescent pseudomonad and investigation of physic-ological parameters influencing its production. Appl Enviro Microbiol 17:107–113
75. Thomashow LS, Weller DM, Bonsall RF, Pierson LS (1990) Production of the antibiotic phenazine-1-carboxylic acid by fluorescent *Pseudomonas* in rhizosphere of wheat. Appl Enivron Microbiol 56:908–912
76. Howell CR, Stipanovic RD (1980) Suppression of *Pythium ultimum* induced damping-off of cotton seedlings by *Pseudomonas fluorescens* and its antibiotic, pyoluterin. Phytopathol 70:712–715
77. Islam TM, Hashidoko Y, Deora A, Ito T, Tahara S (2005) Suppression of damping-off disease in host plants by the rhizoplane bacterium *Lysobacter* sp. strain SB-K88 in linked to plant colonization and antibiosis against soilborne Peronosporomycetes. Appl Eviron Microbiol 71:3786–3796

78. Wilhite SE, Lunsden RD. Strancy DC (2001) Peptide synthetase gene in Trichod-erma virens. Appl Environ Microbiol 65:5055–5062
79. Homma Y, Kato Z, Hirayman F, Konno K, Shirahama H, Suzui T (1989) Production of antibiotics by *Pseudomonas capacia* as an agent for biological control of soilbrone plant pathogens. Soil Biochem 21:723–728
80. Leong SA, Expert D (1989) Siderophores in plantpathogen interactions. In: Kosuge T, Nester EW (eds) Plant-microbe interactions, molecular and genetic perspectives, vol. 3. McGraw-Hill, New York, pp 62–83
81. Loper JE, Buyer JS (1991) Siderophores in microbial interactions on plant surfaces. Molec Plant Microbe Interact 4:5–13
82. Hamdan H, Weller DM, Thomashow LS (1991) Relative importance of fluorescent siderophores and other factors in biological control of *Gaeumannomyces graminis* var. *tritici* by *Pseudomonas fluorescens* 2-7 9 and M4-80R. Appl Environ Microbiol 57:3270–3277
83. Leeman M, Den Ouden FM, Van Pelt JA, Dirkx FPM, Steijl H, Bakker PAHM, Schippers B (1996) Iron availability affects induction of systemic resistance to *Fusarium* wilt of radish by *Pseudomonas fluorescens*. Phytopathol 86:149–155
84. Di Pietro A (1993) Chitinolytic enzymes produced by *Trichoderma harzianum*: antifungal activity of purified endochitinase and chitobiosidase. Phytopathol 83:302–307
85. Chet I (1987) *Trichoderma* application, mode of action and potential as biocontrol agent of soil-borne pathogenic fungi. In: Chet I (ed) Innovative approaches to plant disease control. Wiley, New York, pp 137–160
86. Haran S, Schickler H, Peer S, Logeman S, Oppenheim A, Chet I (1993) Increased constitutive chitinase activity in transformed *Trichoderma harzianum*. Biol Control 3:101–108
87. Shapira R, Ordentlich A, Chet I, Oppenheim AB (1989) Control of plant diseases by chitinase expressed from cloned DNA in *Escherichia coli*. Phytopathol 79:1124–1249
88. Sequeira L (1983) Mechanisms of induced resistance in plants. Ann Rev Microbiol 37:51–79
89. Hammerschmidt R, Lamport DTA, Muldoon EP (1984) Cell wall hydroxyproline enhancement and lignin deposition as an early event in the resistance of cucumber to *Cladosporium cucumerinum*. Physiol. Plant Pathol 24:43–47
90. Alstrom S (1995) Evidence of disease resistance induced by rhizosphere pseudomonads against *Pseudomonas syringae* pv. *phaseolicola*. J Gen Appl Microbiol 41:315–325
91. Van Peer RG, Niemann GJ, Schippers B (1991) Induced resistance and phytoalexin accumulation in biological control of *Fusarium* wilt of carnation by *Pseudomonas* sp. strain WCS417r. Phytopathol 81:728–734
92. Maurhofer M, Hase C, Meuwly P, Metraux JP, Defago G (1994) Induction of systemic resistance to tobacco necrosis virus. Phytopathol 84:139–146
93. Glick BR (1995) The enhancement of plant growth by free-living bacteria. Can J Microbiol 41:109–117
94. Valencia-Cantero E, Hernandez-Calderón E, Velázquez-Becerra C, López-Meza JE, Alfaro-Cuevas R, Lopez-Bucio J (2007) Role of dissimilatory fermentative iron-reducing bacteria in Fe uptake by common bean (*Phaseolus vulgaris* L.) plants grown in alkaline soil. Plant Soil 291:263–273
95. Tsuge K, Akiyama T, Shoda MJ (2001) Cloning, sequencing, and characterization of the iturin A operon. J Bacteriol 183:6265–6273
96. Blom D, Fabbri C, Connor EC, Schiestl FP, Klauser DR, Boller T, Eberl L, Weisskopf L (2011) Production of plant growth modulating volatiles is widespread among rhizosphere bacteria and strongly depends on culture conditions. Environ Microbiol 13:3047–3058
97. Masih H, Singh AK, Kumar Y, Srivastava A, Singh RK, Mishra SK, Shivam K (2011) Isolation and optimization of metabolite production from mutant strain of *Bacillus* sp. with antibiotic activity against plant pathogenic agents. J Pharmac Biomed Sci 11:1–4
98. Ryu CM, Farag MA, Hu CH, Reddy MS, Wei HX, Paré PW, Kloepper JW (2003) Bacterial volatiles promote growth in *Arabidopsis*. Proc National Acad Sci USA 100:4927–4932
99. Rudrappa T, Biedrzycki ML, Kunjeti SG, Donofrio NM, Czymmek KJ, Paré PW, Bais HP (2010) The rhizobacterial elicitor acetoin induces systemic resistance in Arabidopsis thaliana. Commun Integrat Biol 3:130–138

100. Weller DM (2007) Pseudomonas biocontrol agents of soilborne pathogens: looking back over 30 years. Phytopathol 97:250–256
101. Cornelis P (2010) Iron uptake and metabolism in pseudomonads. Appl Microbiol Biotechnol 86:1637–1645
102. Voisard C, Keel C, Haas D, Défago G (1989) Cyanide production by Pseudomonas fluorescens helps suppress black root rot of tobacco under gnotobiotic conditions. EMBO J 8:351–358
103. Haas D, Défago G (2005) Biological control of soil-borne pathogens by fluorescent pseudomonads. Nature Rev Microbiol 3:307–319
104. Compant S, Duffy B, Nowak J, Clément C, Barka EA (2005) Use of plant growth-promoting bacteria for biocontrol of plant diseases: principles, mechanisms of action and future prospects. Appl Environ Microbiol 71:4951–4959
105. Weindling R (1934) Studies on a lethal principle effective in the parasitic action of *Trichoderma lignorum* on *Rhizoctonia solani* and other soil fungi. Phytopathol 24:1153–1179
106. Howell CR (2003) Mechanisms employed by *Trichoderma* species in the biological control of plant diseases: the history and evolution of current concepts. Plant Dis 87:4–10
107. Kapat A, Zimand G, Elad Y (1998) Effect of two isolates of *Trichoderma harzianum* on the activity of hydrolytic enzymes produced by *Botrytis cinerea*. Physiol Mol Plant Pathol 52:127–137
108. Yedidia I, Srivastva AK, Kapulnik Y, Chet I (2001) Effect of *Trichoderma harzianum* on microelement concentrations and increased growth of cucumber plants. Plant Soil 235:235–242
109. Kerr A, Htay K (1974) Biological control of crown gall through bacteriocin production. Physiol Plant Pathol 41:37–40
110. Garrett CME (1978) Biological control of crown gall, *Agrobacterium tumefaciens*. Ann Appl Biol 89:96–97
111. Amiot AF, Róux J, Faivre M (1982) Biological control of *Agrobacterium tumefaciens* (Schmit *et* Townsend) Conn on *Chrysanthemum* with K84 *Agrobacterium radiobacter* (Beijerinck *et* Var Delder) Conn strain. In: ISHS Acta Horticulturae, 125: Symposium on Chrysanthemum. https://doi.org/10.17660/actaHortic.125.30
112. Thomson JA (1986) The potential for biological control of crown gall disease on grapevines. Trends of Biotechnol 4:219–224
113. López MM, Gorris MT, Salcedo CI, Montojo AM, Miró M (1989) Evidence of biological control of *Agrobacterium tumefaciens* strains sensitive and resistant to agrocin 84 by different *Agrobacterium radiobacter* strains on stone fruit trees. Appl Environ Microbiol 55:741–746
114. Vicedo B, Penalver R, Asins MJ, Lopez MM (1993) Biological control of *Agrobacterium tumefaciens*, colonization, and pAgK84 transfer with *Agrobacterium radiobacter* K84 and the Tra-mutant strain K1026. Appl Environ Microbiol 59(1):309–315
115. Ryder MH, Jones DA (1991) Biological control of crown gall using *Agrobacterium* strains K84 and K1026. Austr J Plant Physiol 18:571–579
116. Rhouma A, Ferchichi A, Hafsa M, Boubaker A (2004) Efficacy of the non-pathogenic *Agrobacterium* strains K84 and K1026 against crown gall in Tunisia. Phytopathol Mediterr 43:167–176
117. Nalini S, Parthasarathi R, PrabudossV (2016) Production and characterization of lipo-peptide from Bacillus SNAU01 under solid state fermentation and its potential application as anti-biofilm agent. Biocatal Agric Biotechnol 5:123–132. -->
118. Kawaguchi A, Inoue K, Ichinose Y (2008) Biological control of crown gall of grapevine, rose and tomato by nonpathogenic *Agrobacterium vitis* strain VAR03-1. Phytopathol 98:1218–1225
119. Gupta AK, Khosla K, Bhardwaj SS, Thakur A, Devi S, Jarial RS, Sharma C, Singh KP, Srivastava DK, Lal R (2010) Biological control of crown gall on peach and cherry rootstock colt by native *Agrobacterium radiobacter* isolates. Open Horticul J 3:1–10
120. Tolba IH, Soliman MA (2013) Efficacy of native antagonistic bacterial isolates in biological control of crown gall disease in Egypt. Ann Agric Sci 58:43–49

121. Czajkowski R (2016) Bacteriophages of soft rot Enterobacteriaceae—a minireview. FEMS Microbiol Let 363:230
122. Krzyzanowska DM, Potrykus M, Golanowska M, Polonis K, Gwizdek-Wisniewska A Lojkowska E, Jafra S (2012) Rhizosphere bacteria as potential biocontrol agents against soft rot caused by various *Pectobacterium* and *Dickeya* spp. strains. J Plant Pathol 94:367–378
123. Delfan AS, Etemadifar Z, Emtiazi G, Bouzari M (2015) Isolation of *Dickeya dadantii* strains from potato disease and biocontrol by their bacteriophages. Braz J Microbiol 46:791–797
124. Essarts YR, Cigna J, Laurent Q, Caron A, Munier E, Cirou AB, Hélias V, Faure D (2015) Biocontrol of the potato blackleg and soft rot diseases caused by *Dickeya dianthicola*. Appl Environ Microbiol 82:268–278
125. Hadizadeh I, Peivastegan B, Hannukkala A, Van der Wolf JM, Nissinen R, Pirhonen M (2019) Biological control of potato soft rot caused by *Dickeya solani* and the survival of bacterial antagonists under cold storage conditions. Plant Pathol 68:297–311
126. Vanneste JLYUJ, Beer SV (1992) Role of antibiotic production by *Erwinia herbicola* Eh252 in biological control of *Erwinia amylovora*. J Bacteriol 174:2785–2796
127. Wilson M, Lindow SE (1993) Interactions between the biological control agent *Pseudomonas fluorescens* A506 and *Erwinia amylovora* in pear blossoms. Phytopathol 83:117–123
128. Pusey PL (2002) Biological control agents for fire blight of apple compared under conditions limiting natural dispersal. Plant Dis 86:639–644
129. Laux P, Wesche J, Zeller W (2003) Field experiments on biological control of fire blight by bacterial antagonists. J Plant Dis Prot 110:401–407
130. Özaktan H, Bora T (2004) Biological control of fire blight in pear orchards with a formulation of Pantoea agglomerans strain Eh 24. Braz J Microbiol 35:224–229. http://dx.doi.org/10.1590/S1517-83822004000200010
131. Sundin GW, Yoder KS, Aldwinckle HS (2009) Field evaluation of biological control of fire blight in the Eastern United States. Plant Dis 93:386–394
132. Gerami E, Hassanzadeh N, Abdollahi H, Ghasemi A, Heydari A (2013) Evaluation of some bacterial antagonists for biological control of fire blight disease. J Plant Pathol 95:127–134
133. Doolotkeldieva T, Bobusheva S (2016) Fire blight disease caused by *Erwinia amylovora* on rosaceae plants in Kyrgyzstan and biological agents to control this disease. Adv Microbiol 6:831–851
134. Smail AB, Abderrahman O, Abdessalem T (2016) Evaluation of biological control agent *Pantoea agglomerans P10c* against fire blight in Morocco. Afr J Agric Res 11:1661–1667
135. Mikiciński A, Sobiczewski P, Puławska J, Maciorowski R (2016) Control of fire blight *Erwinia amylovora* by a novel strain 49M of *Pseudomonas graminis* from the phyllosphere of apple (*Malus* spp.). Europ J Plant Pathol 145:265–276
136. Ameur A, Rhallabi N, Doussomo ME, Benbouazza A, Ennaji MM, Achbani E (2017) Selection and efficacy biocontrol agents in vitro against fire blight (*Erwinia amylovora*) of the rosacea. Int Res J Eng Technol 4:539–545
137. Sharifazizi M, Harighi B. Sadeghi A (2017) Evaluation of biological control of *Erwinia amylovora*, causal agent of fire blight disease of pear by antagonistic bacteria. Biol Cont 104:28–34
138. Cronin D, Loccoz YM, Fenton A, Dunne C, Dowling DN O'Gara F (1997) Ecological interaction of a biocontrol *Pseudomonas fluorescens* strain producing 2,4-diacetyl-phloroglucinol with the soft rot potato pathogen *Erwinia carotovora* subsp. *atroseptica*. FEMS Microbiol Ecol 23:195–106
139. El-Hendawy HH, Zeid IM, Mohamed ZK (1998) The biological control of soft rot disease in melon caused by *Erwinia carotovora* subsp. *carotovora* using *Pseudomonas fluorescens*. Microbial Res 153:55–60
140. Zamanian S, Shahidi BGH, Saadoun H (2005) First report of antibacterial properties of a new of *Sterptomyces plicatus* (strain 101) against *Erwinia carotovra* subsp. *carotovra* from Iran. Biotechnol 4:114–120
141. Baz M, Lahbabi D, Samri S, Val F, Hamelin G, Madore I, Bouarab K, Beaulieu C, Ennaji MM, Barakate M (2012) Control of potato soft rot caused by *Pectobacterium carotovorum*

and *Pectobacterium atrosepticum* by Moroccan actinobacteria isolates. World J Microbiol Biotechnol 28:303–311

142. Abd El-Khair H, Haggag KHE (2007) Application of some bactericides and bioagents for controlling the soft rot disease in potato. Res J Agric Biol Sci 3:463–473

143. Juan NM, Jessica CS, Luigi CP, Marcia CL, Ricardo FP, Renate ST (2008) Biocontrol of *Erwinia carotovora* on Calla (*Zantedeschia* sp.). Agro Sur 36:59–70

144. Dong F, Zhang XH, Li YH, Wang JF, Zhang SS, Hu XF, Chen JS (2010) Characterization of the endophytic antagonist pY11T-3-1 against bacterial soft rot of Pinellia ternate. Let Appl Microbiol 50:611–617

145. Mello MRF, Silveira EB, Viana IO, Guerra ML, Mariano RLR (2011) Use of antibiotics and yeasts for controlling Chinese cabbage soft rot. Hortic Bras 29:78–83

146. Qianqian L, Ni H, Meng S, He Y, Yu Z, Li L (2011) Suppressing *Erwinia carotovora* pathogenicity by projecting N-acyl homoserine lactonase onto the surface of *Pseudomonas putida* cells. J Microbiol Biotechnol 21:1330–1335

147. Issazadeh K, Rad SK, Zarrabi S, Rahimibashar MR (2012) Antagonism of *Bacillus* species against *Xanthomonas campestris* pv. *campestris* and *Pectobacterium carotovorum* subsp. *carotovorum*. Afr J Microbiol Res 6:1615–1620

148. Ghods-Alavi BS, Ahmadzadeh M, Behboudi K, Jamali S (2012) Biocontrol of rhizome soft rot (*Pectobacterium carotovorum*) on valerian by *Pseudomonas* spp. under in vitro and greenhouse conditions. J Agric Technol 8:1913–1923

149. Rahman MM, Ali ME, Khan AA, Akanda AM, Kamal Uddin MD, Hashim U, AbdHamid SB (2012) Isolation, characterization and identification of biological control agent for potato soft rot in Bangladesh. Sci World J Article ID 723293, 6 p. https://doi.org/10.1100/2012/723293

150. Algeblawi A, Adam F (2013) Biological control of *Erwinia carotovora* subsp *carotovora* by *Pseudomonas fluorescens*, *Bacillus subtilis* and *Bacillus thurin-giensis*. Int J Chem Environ Biol Sci 1:771–774

151. Sowmya DS, Rao MS, Kumar RM, Gavaskar J, Priti K (2012) Biomana-gement of *Meloidogyne incognita* and *Erwinia carotovora* in carrot (*Daucus carota* L.) using *Pseudomonas putida* and *Paecilomyces lilacinus*. Nematol medit 40:189–194

152. Saputra H, Puspita F, Nugroho TT (2013) Production of an antibacterial compound against the plant pathogen *Erwinia carotovora* subs. *carotovora* by the biocontrol strain *Gliocladium* sp. T.N.C73. J Agric Technol 9:1157–1165

153. Zhao Y, Li P, Huang K, Wang Y, Hu H, Sun Y (2013) Control of postharvest soft rot caused by *Erwinia carotovora* of vegetables by a strain of *Bacillus amyloliquefaciens* and its potential modes of action. World J Microbiol Biotechnol 29:411–420

154. Li HY, Luo Y, Zhang XS, Shi WL, Gong ZT, Shi M, Chen LL, Chen XL, Zhang YZ, Song XY (2014) Trichokonins from *Trichoderma pseudokoningii* SMF2 induce resistance against Gram-negative *Pectobacterium carotovorum* subsp. *carotovorum* in Chinese cabbage. FEMS Microbiol Lett 354:75–82

155. Makhlouf, Abeer, H. and Abdeen, Rehab (2014) Investigation on the effect of chemical and biological control of bacterial soft root disease of potato in storage. J Biol Agric Healthcare 4:31–44

156. Sandipan PB, Chaudhary RF, Shanadre CM, Rathod NK (2015) Appraisal of diverse bioagents against soft rot bacteria of potato (*Solanum tuberosum* L.) caused by *Erwinia carotovora* subsp. *carotovora* under in vitro test. Europ J Pharmac Medical Res 2:495–500

157. Idowu OO, Olawole OI, Idumu OO, Salami AO (2016) Bio-control effect of *Trichoderma asperellum* (Samuels) Lieckf. and *Glomus intraradices* Schenk on okra seedlings infected with *Pythium aphanidermatum* (Edson) Fitzp and *Erwinia carotovora* (Jones). American J Exp Agric 10:1–12

158. Doolotkeldieva T, Bobusheva S, Suleymankisi A (2016) Biological control of *Erwinia carotovora* ssp. *carotovora* by *Streptomyces* species. Adv Microbiol 6:104–114

159. Ha NT, Minh TQ, Hoi PX, Thuy NTT, Furuya N, Long HH (2018) Biological control of potato tuber soft rot using *N*-acyl-L-homoserine lactone-degrading endophytic bacteria. Current Sci 115:1921–1927

160. Salem EA, Abd El-Shafea YM (2018) Biological control of potato soft rot caused by *Erwinia carotovora* subsp. *carotovora*. Egypt J Biol Pest Cont 19:28:94

161. Kiewnick AB, Jacobsen BJ, Sands DC (2000) Biological control of *Pseudomonas syringae* pv. *syringae*, the causal agent of basal kernel blight of Barley, by antagonistic *Pantoea agglomerans*. Phytopathol 90:368–375

162. Völksch B, May R (2001) Biological control of *Pseudomonas syringae* pv. *glycinea* by epiphytic bacteria under field conditions. Microb Ecol 41:132–139

163. Bais HP, Fall R, Vivanco JM (2004) Biocontrol of *Bacillus subtilis* against infection of *Arabidopsis* roots by *Pseudomonas syringae* is facilitated by biofilm formation and surfactin production. Plant Physiol. https://doi.org/10.1104/pp.103

164. Kotan R, Sahin F (2006) Biological control of *Pseudomonas syringae* pv. *syringae* and nutritional similarity in carbon source utilization of pathogen and its potential biocontrol agents. J Turk Phytopath 35:1–13

165. Hassan EO, El-Meneisy AZA (2014) Biocontrol of halo blight of bean caused by *Pseudomonas phaseolicola*. Int J Virol 10:235–242

166. Mougou I, M'hamdi NB (2018) Biocontrol of *Pseudomonas syringae* pv. *syringae* affecting citrus orchards in Tunisia by using indigenous *Bacillus* spp. and garlic extract. Egypt J Biol Pest Cont 19:28:60

167. Wicaksono WA, Jones EE, Casonato S, Monk J, Ridgway HJ (2018) Biological control of Pseudomonas syringae pv. actinidiae (Psa), the causal agent of bacterial canker of kiwifruit, using endophytic bacteria recovered from a medicinal plant. Biol Cont 116:103–112

168. Nikolić I, Berić T, Stankovic SS (2019) Biological control of *Pseudomonas syringae* pv. *aptata* on sugar beet with *Bacillus pumilus* SS-10.7 and *Bacillus amyloliquefaciens* (SS-12.6 and SS-38.4) strains. J Appl Microbiol https://doi.org/10.1111/jam.14070

169. Anuratha CS, Gnanamanickam SS (1990) Biological control of bacterial wilt caused by *Pseudomonas solanacearum* in India with antagonistic bacteria. Plant Soil 124:109–116

170. Lwin M, Ranamukhaarachchi SL (2006) Development of biological control of *Ralstonia solanacearum* through antagonistic microbial populations. Int J Agric Biol 8:1560–8530

171. Lemessa F, Zeller W (2007) Screening rhizobacteria for biological control of *Ralstonia solanacearum* in Ethiopia. Biol Cont 42:336–344

172. Kuarabachew H, Assefa F, Hiskias Y (2007) Evaluation of Ethiopian isolates of *Pseudomonas fluorescens* as biocontrol agent against potato bacterial wilt caused by *Ralstonia (Pseudomonas) solanacearum*. Acta Agric Slov 90:125–135

173. Tahat MM, Sijam K (2010) *Ralstonia solanacearum*: the bacterial wilt causal agent. Asian J Plant Sci 9:385–393

174. Nguyen MT, Ranamukhaarachchi SL (2010) Soil-borne antagonists for biological control of bacterial wilt disease caused by *Ralstonia solanacearum* in tomato and pepper. J Plant Pathol 92:395–406

175. Nawangsih AA, Damayanti I, Wiyono S, Kartika JG (2011) Selection and characterization of endophytic bacteria as biocontrol agents of tomato bacterial wilt disease. HAYATI J Biosci 18:66–70

176. Seleim MAA, Saead FA, Abd-El-Moneem KMH, Abo-ELyousr KAM (2011) Biological control of bacterial wilt of tomato by plant growth promoting rhizobacteria. Plant Pathol J 10:146–153

177. Maji S, Chakrabartty PK (2014) Biocontrol of bacterial wilt of tomato caused by *Ralstonia solanacearum* by isolates of plant growth promoting rhizobacteria. Astu J crop Sci 8:208–214

178. Santiago TR, Grabowski C, Rossato M, Romeiro RS, Mizubuti ESG (2015) Biological control of Eucalyptus bacterial wilt with rhizobacteria. Biol Cont 80:14–22

179. Aino M (2016) Studies on biological control of bacterial wilt caused by Ralstonia solanacearum using endophytic bacteria. J Gen Plant Pathol 82:323–325

180. Singh D, Yadav DK, Chaudhary G, Rana VS, Sharma RK (2016) Potential of *Bacillus amyloliquefaciens* for biocontrol of bacterial wilt of tomato incited by *Ralstonia solanacearum*. J Plant Pathol Microbiol 7:327

181. Ilsan NA, Nawangsih AA, Wahyudi AT (2016) Rice phyllosphere actinomycetes as biocontrol agent of bacterial leaf blight disease on rice. Asian J Plant Pathol 10:1–8

182. Sindhan GS, Parashar RD, Indra H (1997) Biological control of bacterial leaf of rice caused by *Xanthomonas oryzae* pv. *oryzae*. Plant Dis Res 12:29–32

183. Hastuti RD, Estari Y, Suwanto A, Saraswati R (2012) Endophytic *Streptomyces* spp. as biocontrol agents of rice bacterial leaf blight pathogen (*Xanthomonas oryzae* pv. *oryzae*). HAYATI J Biosci 19:155–162

184. Van Hop D, Phuong HPT, Quang ND, Ton PH, Ha TH, Van Hung N, Van NT, Van Hai T, Kim Quy NT, Anh Dao NT, Thi-Thom V (2014) Biological control of *Xanthomonas oryzae* pv. *oryzae* causing rice bacterial blight disease by *Streptomyces toxytricini* VN08-A-12, isolated from soil and leaf-litter samples in Vietnam. Biocontrol Sci 1:103–111

185. El-Shakh ASA, Kakar KU, Wang X, Almoneafy AA, Ojaghian MR, Li B (2015) Controlling bacterial leaf blight of rice and enhancing the plant growth with endophytic and rhizobacterial *Bacillus* strains. Toxicol Environ Chem 97:766–785

186. Yasmin S, Hafeez FY, Mirza MS, Rasul M, Arshad HMI, Zubair M, Iqbal M (2017) Biocontrol of bacterial leaf blight of rice and profiling of secondary metabolites produced by rhizospheric *Pseudomonas aeruginosa* BRp3. Front Microbiol. https://doi.org/10.3389/fmicb.2017.01895

187. Assis SMP, Mariano RLR, Michereff SJ, Silva G, Maranhão EAA (1999) Antagonism of yeasts to *Xanthomonas campestris* pv. *campestris* on cabbage phylloplane in field. Revista de Microbiol 30:191–195

188. Luna CL, Mariano RLR, Souto-Maior AM (2002) Production of a biocontrol agent for crucifers black rot diseaseproduction of a biocontrol agent for crucifers black rot disease. Braz J Chem Eng 19:133–140

189. Wulff EG, Mguni CM, Mortensen CN, Keswani CL, Hockenhull J (2002) Biological control of black rot (*Xanthomonas campestris* pv. *campestris*) of brassicas with an antagonistic strain of *Bacillus subtilis* in Zimbabwe. Europ J Plant Pathol 108:317–325

190. Massomo SMS, Mortensen CN, Mabagala RB, Newman MA, Hockenhull J (2004) Biological control of black rot (*Xanthomonas campestris* pv. *campestris*) of cabbage in Tanzania with *Bacillus* strains. J Phytopathol 152:98–105

191. El-Hendawy HH, Osman ME, Sorour NM (2005) Biological control of bacterial spot of tomato caused by *Xanthomonas campestris* pv. *vesicatoria* by *Rahnella aquatilis*. Microbiol Res 160:343–352

192. Monteiro L, Mariano RLR, Souto AMM (2005) Antagonism of *Bacillus* spp. against *Xanthomonas campestris* pv. *campestris*. Braz Arch Biol Technol 48:23–29

193. Mirik M, Aysan Y, Çinar Ö (2008) Biological control of bacterial spot disease of pepper with Bacillus strains. Turkish J Agric For 32(5):381–390

194. Suárez-Estrella F, Ros M, Vargas-García MC, López MJ, Moreno J (2014) Control of *Xanthomonas campestris* pv. *vesicatoria* using agroindustrial waste-based compost. J Plant Pathol 96:243–248

195. Salah Eddin K, Marimuthu T, Ladhalakshmi D, Velazhahan R (2007) Biological control of bacterial blight of cotton caused by *Xanthomonas axonopodis* pv. *malvacearum* with *Pseudomonas fluorescens*. Arch Phytopathol Plant Prot 40:291–300

196. Montakhabi MK, Rahimian H, Falahati RM, Jafarpour B (2011) In vitro investigation on biocontrol of *Xanthomonas axonopodis* pv. *citri* cause of citrus bacterial canker by citrus antagonistic bacteria. J Plant Prot (Agric Sci Technol) 24:368–376

197. Lopes LP, Oliveira Jr AG, Beranger JPO, Góis CG. Vasconcellos FCS, San Martin JA. Andrade CGTJ, Mello JCP, Andrade G (2012) Activity of extracellular compounds of *Pseudomonas* sp. against *Xanthomonas axonopodis* in vitro and bacterial leaf blight in eucalyptus. Trop Plant Pathol 37. https://doi.org/10.1590/s1982-56762012000400001

198. Das R, Mondal B, Mondal P, Khatua DC, Mukherjee N (2014) Biological management of citrus canker on acid lime through *Bacillus subtilis* (S-12) in West Bengal, India. J Biopest 7(supp):38–41

199. Murate LS, de Oliveira AG, Higashi AY, Barazetti AR Simionato AS, da Silva CS, Simões GC, dos Santos IMO, Ferreira MR, Cely MVT, Navarro MOP, de Freitas VF, Nogueira MA,

de Mello JCP, Leite Jr RP, Andrade G (2015) Activity of secondary bacterial metabolites in the control of citrus canker. Agric Sci 6:295–303

200. Tewfike TA, Desoky SM (2015) Biocontrol of *Xanthomonas axonopodis* causing bacterial spot by application of formulated phage. Ann Agric Sci Moshtohor 53:615–624

201. Chavan NP, Pandey R, Nawani N, Nanda RK, Tandon GD, Khetmalas MB (2016) Biocontrol potential of actinomycetes against *Xanthomonas axonopodis* pv. *punicae*, a causative agent for oily spot disease of pomegranate. Biocontrol Sci Technol 26:351–372

202. Osman TMT, Algam SAE, Ali ME, Osman EHB, Mahdi AA (2016) In vitro screening of some biocontrol agents against *Xanthomonas axonopodis* pv. *malvacearum* isolated from infected cotton plants. Int J Agric, For Plantat 2:270–278

203. Puneeth ME (2016) Biocontrol of bacterial blight of pomegranate caused by *Xanthomonas axonopodis* pv. *punicae* (Hingorani and Singh) Vauterin et al. MSc. Thesis, University of Agricultural Sciences, Plant Pathology, Bengaluru (Abstract)

204. Corrêa BO, Soares VN, Sangiogo M, de Oliveira JR, Andréa BMAB (2017) Interaction between bacterial biocontrol-agents and strains of *Xanthomonas axonopodis* pv. *phaseoli* effects on biocontrol efficacy of common blight in beans. Afr J Microbiol Res 11:1294–1302

205. Oliver R, Owens W, Hopkins DL (2008) Interaction of a biological control strain and a pathogenic strain of Xylella fastidiosa in grapevine. J Plant Pathol 90S:195

206. Lacava PT, Arau´jo WL, Marcon J, Maccheroni W Jr, Azevedo JL (2004) Interaction between endophytic bacteria from citrus plants and the phytopathogenic bacteria *Xylella fastidiosa*, causal agent of citrus-variegated chlorosis. Let Appl Microbiol 39:55–59

207. Hopkins DL, Thompson CM (2008) Biological control of Pierce's disease in the vineyard with a benign strain of *Xylella fastidiosa*. J Plant Pathol 90S:115

208. Hopkins DL (2005) Biological control of Pierce's disease in the vineyard with strains of *Xylella fastidiosa* Benign to Grapevine. Plant Dis 89:1348–1352

Biocontrol Agents for Fungal Plant Diseases Management

Younes M. Rashad and Tarek A. A. Moussa

Abstract Plant fungal diseases are the most destructive diseases where the fungal pathogens attack many economic crops causing yield losses, which affect directly many countries' economy. The great Irish Famine in 19th century was due to potato (a great portion of Irish diets) was attacked by an oomycete pathogen *Phytophthora infestans* causing late blight disease which destroyed the potato crop for several years (1845–1852). Since this date the plant fungal diseases have a great attention from the researchers. Control of fungal diseases using different fungicides has dangerous effects on human beings as well as animals by precipitating in the plant tissues and then transfer to human and animals causing many health complications. Hence, the biological control of plant pathogenic fungi became the most important issue, due to the chemical risk to control the fungal diseases. From 1990's the importance of using microorganisms was increased as biocontrol agents to decrease the chemical uses and their hazardous for human and animal health topics. In this chapter, using of different microorganism as biological control agents of plant fungal diseases were reviewed, as well as using chemicals in controlling fungal diseases and their effects on plants, environment and common health impacts.

Keywords Fungal pathogens · Biological control · Chemical control · Biocontrol agents

1 Introduction

Fungi are non-chlorophytic, spore-forming, eukaryotic organisms. Most of the fungal species are saprophytes. So, about 20,000 species out of more than 100,000 fungal species are parasites causing diseases in crops [1–4]. Most of plants may be attacked

Y. M. Rashad
Plant Protection and Biomolecular Diagnosis Department, Arid Lands Cultivation Research Institute, City of Scientific Research and Technological Applications, Alexandria, Egypt

T. A. A. Moussa (✉)
Botany and Microbiology Department, Faculty of Science, Cairo University, Giza 12613, Egypt
e-mail: tarekmoussa@cu.edu.eg

© Springer Nature Switzerland AG 2020
N. El-Wakeil et al. (eds.), *Cottage Industry of Biocontrol Agents and Their Applications*, https://doi.org/10.1007/978-3-030-33161-0_11

by one or more species of fungal pathogens. On the other hand, the fungal species can attack only one plant species (Specialist) or many plant species (Generalist).

In the last century, most of diagnostic characters used in the identification of the phytopathogenic fungi were not evidently accurate, so any identifying character such as type of fruiting body, spores can scope the search for a particular phylum. Most diagnosis depends on visual signs and symptoms for diagnosis of fungal diseases [5]; therefore, there were many problems and difficulties in combating these pathogens. It is very important to identify the plant fungal pathogens to know their taxonomic groups, which affects significantly for managing these pathogenic fungi.

This chapter is concerned with the use of biological control agents instead of chemical control against the fungal plant pathogens. The biological control has many advantages in relation to soil fertility, plant, animal and human health.

2 Fungal Pathogenesis

Fungal pathogenesis is the stage of disease in which the pathogenic fungus is in close association with the tissue of host. There are three stages:

1. Inoculation: the transfer of pathogenic fungus to the infection area, in which the plant is invaded (the infection area may be natural openings such as stomata, hydathodes, or lenticels), wounds or unbroken plant surface.
2. Incubation: the period between the invasion of the pathogenic fungus and the symptoms appearance.
3. Infection: the appearance of symptoms associated with the establishment and pathogen spread.

Fungal pathogens cause symptoms which may be general or localized. In most cases, necrosis of host tissue, stunting, distortions and plant tissue abnormality and organs changes as a result of fungal infections [6].

One of the important pathogenic fungi characteristics, is virulence (infection ability). There are many properties of a fungal pathogen that contribute the ability to spread and destroy the tissue. Most of the virulence factors are enzymes to destruct plant cell walls [7–9], toxins which are cell killers, exopolysaccharides to block the path of cell fluid [10, 11], and many substances which interfere cell growth. The pathogenic species differ in virulence and hence the substances which involved in the invasion and destruction of host tissue.

3 Control of Fungal Diseases

The fungal plant diseases control is critical to the safe food production, and it cause serious problems in the use of land for agricultural, water, and other inputs. Plants carry inherent disease resistance in both natural and cultivated systems, so control of fungal diseases is successful for many crops [12].

3.1 Chemical Control

Along the years, many chemicals have been used to control fungal plant pathogens. Some of these have been substituted as cheaper, effective, or less hazardous substances [13]. Pruning cuts, stumps and wounds can be protected against fungal pathogens by painting with special chemicals on the surfaces exposed to environment. Plant structures such as tubers, cuttings, rhizomes, bulbs and corms which used in vegetative propagation, are often immersed in chemicals before planting. In case of trees fungal infections, fungicide was injected inside trees or by pouring into a hole made into the tissues.

Most of chemicals have been used as fungicides, where they interfere with many metabolic processes in fungal cells. The biological activity of a fungicide is restricted to its metabolism in the fungal cell and the chemicals that are transported within the plant was affected by metabolism of the plant cell. Many fungicides have low toxicity to mammals [14].

Antibiotics are chemical substances produced by microorganisms which are capable of injuring or destroying living organisms. They have been used worldwide to control bacterial and fungal diseases where many ordinary plant protection methods have failed. On the contrary, there are few antibiotics are used to control plant fungal diseases [15].

The development of resistant strains of fungi to chemicals was discussed in the 1970s and the community became aware with health and environmental impact of these chemicals in l980s and 1990s. The use of agricultural chemicals causes significant public health problems [16]. The worry about the risk of humans and domestic animals poisoning, livestock products contaminations, their impact on the beneficial insects, hazardous residue in food products, ecological imbalances at the level of microorganism and the possibility of contamination of water with subsequent fish loss and buildup of residues in groundwater. For that reasons, fungicides should be avoided and be used only in the heavy infection situations [17].

El-Abyad et al. [7] concluded that under pyradure stress, the virulence of sugar beet pathogens *Rhizoctonia solani* and *Sclerotium rolfsii* was reduced in vivo and in vitro. The reduction in the virulence of *R. solani* and *S. rolfsii* was due to decreased inoculum potential of the two pathogens under pyradure stress in situ and production of cell wall degrading enzymes in vitro. Under salinity stress, the resistance shown by the sugar beet cultivars against infection by *R. solani* and *S. rolfsii* was to be due to the maturation of cell wall composition of these cultivars with age [8].

3.2 Biological Control

Owing to the hazardous effects inflicted by chemical fungicides on non-target organisms and the surrounding environment, many researchers have focused during the last few decades on finding an alternative option for control of fungal plant diseases, that is, biological control. The broad definition of biological control is "suppression of pathogenic organisms and reducing their effects on hosts as well as favoring the crops beneficial organisms using wild or modified organisms, genes, gene products, or biological induction of systemic resistance" [18]. Biological control agents include many antagonistic microorganisms such as fungi, bacteria, or viruses [19].

3.2.1 Bacteria as Biocontrol Agents

Numerous bacterial species are extensively utilized as biological control agents to control of several phytopathogenic fungi. In addition, these bioagents have many beneficial effects on the treated plants. Members of many bacterial genera, epiphytic and/or endophytic, are used in this concern. The most common bacteria utilized as bio-control agents include some species of the genera *Bacillus, Pseudomonas, Streptomyces, Rhizobium, Burkholderia, Gluconobacter, Azoarcus, Herbaspirillum,* and *Klebsiella* [20, 21].

Bacillus spp.

Bacillus Cohn (Firmicutes, Bacillales, Bacillaceae) is a genus of gram-positive, aerobic, rods (bacilli) bacteria, which can form spores, and comprises 377 species and 8 subspecies [22]. Members of this genus have a wide distribution and found in soil, decaying matter, water, air, in/on living plants and animals, and in some severe habitats [23]. *Bacillus* spp. have a great importance and been involved in many uses in agricultural, industrial, and pharmaceutical applications such as production of diverse antibiotics, lipopeptides, enzymes, and bioactive secondary metabolites [24, 25]. Several antibiotics are known to be produced by *Bacillus* spp. such as fengycin, sublichenin, subtilosin A, gramicidin, sublancin, bacillomycin, tochicin, bacitracin, polymyxin, bacilysocin and neotrehalosadiamine [26, 27]. A broad set of hydrolytic enzymes are produced also by *Bacillus* spp. like chitinases, β-1,3(4)-glucanase, proteases, and lipases [28, 29]. The high capability of *Bacillus* spp. for production of these diverse of structurally and functionally different antagonistic substances make them pioneers in the field of the bio-fungicides. Moreover, most of *Bacillus* spp. utilized as biocontrol agents possess a growth enhancing activity on the host plant. Of the world biopesticides market, commercial *B. thuringiensis*-based products share about 90% [30].

Several studies have elucidated the use of *Bacillus* spp. in the biological control of different pathogenic fungi [28, 31–33]. The most common *Bacillus* spp. utilized in biocontrol of plant diseases include *B. subtilis, B. thuringiensis, B. fortis, B. amyloliquefaciens, B. vallismortis, B. pumilus, B. sphaericus, B. cereus, B. licheniformis, B. polymyxa, B. megaterium, B. mycoides, B. mojavensis,* and *B. pasteurii* [25, 34]. Chen et al. [35] investigated the antifungal activity of the potent strain *B. velezensis* LM2303 which achieved a control efficiency of 72.3% against wheat *Fusarium* head blight caused by *F. graminearum,* in the field. Moreover, this strain showed antagonistic potency in vitro against different pathogenic fungi. Genomic mining of *B. velezensis* LM2303 results in identification of 13 biosynthetic gene clusters encoding for antimicrobial substances (fengycin B, iturin A, surfactin A, butirosin), as well as siderophores (bacillibactin and teichuronic acid). Furthermore, encoding-genes responsible for root colonization, growth enhancement, and immune system induction were identified. Generally, the direct biocontrol mechanisms exerted by *Bacillus* spp. against the phytopathogenic fungi include antibiosis via biosynthesis of various antifungal substances (antibiotics, lipopeptides, enzymes), competition for space and/or nutrients by colonizing the plant surface or production of various siderophores, while, the indirect mechanisms include induction of the plant systemic resistance leading to triggering many fungitoxic substances such as phenolic compounds and defense-related enzymes, as well as plant growth promotion via inducing the biosynthesis of plant growth regulators [34].

Pseudomonas Spp

Pseudomonas Migula (Gammaproteobacteria, Pseudomonadales, Pseudomonadaceae) is a genus of aerobic, gram-negative, rods, motile bacteria, which cannot form spores, and contains 254 species and 18 subspecies [22, 36]. *Pseudomonas* spp. can resist diverse biotic and abiotic extreme conditions, use numerous organic substances, and exhibit high metabolic and physiological diversity. Owing to their elevated resistances, they can inhabit a wide range of habitats such as soil, aquatic environments, and air, in/on plants or animals [37]. This distribution is ascribed to the capability to synthesize a long list of antagonistic substances enabling them to compete with the surrounding microbiota such as phenazines, pyochelin, rhizoxins, pyrrolnitrine, hydrogen cyanide, 2,4-diacetylphloroglucinol, and pyoluteorin [38]. Although some members of the genus *Pseudomonas* are phytopathogenic, many are of great benefit providing the plant with protection against the attacking pathogens.

The biocontrol mechanisms utilized by *Pseudomonas* spp. include rivalry for nutrients and space, biosynthesis of antagonistic substances and enzymes, or by triggering plant immune system against various pathogenic fungi [39]. Furthermore, some *Pseudomonas* spp. promote the plant growth, and inhibit soil-borne pathogens [40]. Roles of *Pseudomonas* spp. in enhancing the plant growth include biosynthesis of growth regulators, nitrogen fixation, phosphate mineralization, as well as sequestering iron by secretion of siderophores [41]. Many *Pseudomonas* spp. are widely utilized as bioagents against many fungal diseases and commercially represent a big sector in the biopesticides market. Aielloa et al. [42] studied the biocontrol ability of the endophyte *P. synxantha* DLS65 against the postharvest brown rot of stone fruit in vitro and in vivo. A considerable growth suppression of both fungi was achieved by using *P. synxantha* in vitro. In addition, a significant reduction in the disease symptoms was also reported in the storage even after 20 days at 0 °C. The rivalry for nutrients or space, secretion of fungitoxic substances or volatile organic compounds were named to be a projected as biocontrol mechanisms by *P. synxantha*.

Streptomyces spp.

Streptomyces Waksman and Henrici (Actinobacteria, Actinomycetales, Actinomycetaceae) is a bacterial genus which include aerobic, filamentous, gram-positive species that produce fungus-like mycelia and aerial hyphae with branches that carry chains of spherical to ellipsoidal spores [43]. Currently, this genus comprises 848 species and 38 subspecies with annual increase in the species number [22]. Members of genus *Streptomyces* have wide distribution and found in various habitats such as soil, water, decaying vegetation, endophytic, epiphytic, even in extreme habitats such as deep-sea sediments, volcanic soils, frozen soils, and desert soils [44, 45]. *Streptomyces* spp. are highly recognized as antibiotics, enzymes, and bioactive secondary metabolites producers [46, 47]. Indeed, antibiotics produced by *Streptomyces* genus represent the largest share, approximately two-thirds, of the known antibiotics so far, and their number has exponentially increased every year [48, 49]. The most common antibiotics identified from *Streptomyces* spp. are streptomycin, pimaricin, neomycin, phenalinolactones A-D, cypemycin, warkmycin, and grisemycin [50, 51]. Various enzymes are also reported to be produced by *Streptomyces* spp. like chitinases, proteases, peroxidases, β-1,3 glucanases, laccases, and tyrosinases [46, 52, 53]. Furthermore, a large set, around 7600, of bioactive compounds synthesized by *Streptomyces* spp. like anticancer, antiviral, antihypertensive, immunosuppressive, and antioxidant were also reported [54].

Biocotrol of phytopathogenic fungi using members of genus *Streptomyces* has been extensively investigated by various researchers [55–57]. Different species are common in this concern such as *S. lydicus, S. vinaceusdrappus, S. griseoviridis, S. griseorubens, S. tsusimaensis, S. griseofuscus, S. spororaveus, S. tendae, S. humidus, S. hygroscopicus, S. caviscabies, S. philanthi, S. sindeneusis,* and *S. flavotricini* [58–61]. Of sixteen endophytic actinobacteria screened for their fungitoxic effect against pathogenic mycoflora, *S. asterosporus* SNL2exhibited the strongest antifungal activity in vitro, especially against *F. oxysporum* f. sp. *radicis lycopersici*, the causal agent of tomato root rot [62]. Moreover, application of this isolate led to a considerable reduction the severity of tomato root rot by 88.5%. In another study, the fungitoxic activity of the cultural secondary metabolites produced by *S. griseorubens* E44G was evaluated in vitro on the growth and ultrastructure of mycelial cells of *F. oxysporum* f. sp. *lycopersici* [63]. Investigations using the transmission electron microscope showed many noxious effects in the fungal mycelia after treatment with the culture filtrate at 400 μL.

The ultra-cytochemical study revealed the digestion of chitin of the cell wall after the exposure to the bacterial filtrate, indicating the production of the lytic enzyme chitinase by *S. griseorubens* E44G as a biocontrol mechanism. The biocontrol modes of action utilized by *Streptomyces* spp. include physical contact (hyperparasitism), rivalry for space/nutrients, antibiosis via biosynthesis of hydrolytic enzymes, antibiotics and fungitoxic substances [56]. Indirect mechanisms via triggering plant resistance, and/or improving the plant growth may be involved also [57]. However, the biocontrol mechanisms used by a biocontrol agent are affected by the other conditions like soil type, temperature, pH, humidity, and existence of surrounding microorganisms [61].The *S. aureofaciens* filtrate was inhibited the germination of *F. solani* and in vivo seed coating was the most efficient method for controlling the pathogenicity of *F. solani* by *S. aureofaciens* [64].

Rhizobium spp.

Members of *Rhizobium* Frank (Alphaproteobacteria, Rhizobiales, Rhizobiaceae) are aerobic, rod-shaped, gram-negative, motile, non-spore producing, nitrogen-fixing bacteria, which comprises 112 species. *Rhizobium* spp. are widely distributed and found as free-living in soil or colonize legumes roots forming nodules, nitrogen-fixing symbioses [22, 65]. Members of genus *Rhizobium* are categorized according to their associated leguminous plant, and growth rate. The most known species include *R. leguminosarum, R. phaseoli, R. trifolii, R. lentis, R. japonicum, R. aggregatum,* and *R. sullae*. In addition to nitrogen fixating and growth enhancing effects (phytohormones biosynthesis), *Rhizobium* spp. are well known as biological control agents against numerous pathogenic mycoflora like *Rhizoctonia solani, F. solani, F. oxysporum, Macrophomina phaseolina, Sclerotinia sclerotiorum, Pythium* sp. and *Sclerotium rolfsii* [66–68].

The antagonistic modes of action utilized by *Rhizobium* spp. include rivalry for space and nutrients by secretion of siderophores, in addition to antibiosis via production of antibiotics such as bacteriocins and trifolitoxin, lytic enzymes, and fungitoxic substances such as hydrogen cyanide. Furthermore, triggering of plant immune system against attacking pathogens is widely reported for many species of *Rhizobium* via induction of hypersensitivity responses, defense-related genes, and production of antifungal compounds and molecules [69]. Volpiano et al. [70] investigated the antagonistic activity of different *Rhizobium* strains toward *S. rolfsii* in vitro and in vivo. A mycelial growth inhibition up to 84% in vitro and a significant decrease in the incidence of collar rot of common bean by 18.3 and 14.5% in the pot and field experiments were reported by strains SEMIA 439 and 4088. In addition, the antagonistic mechanism through volatile compounds by strain SEMIA 460 was also reported. Hemissi et al. [71] investigated the antifungal potential of some *Rhizobium* strains against *R. solani* in vitro and the incidence of *Rhizoctonia* root rot of chickpea under greenhouse conditions. Among the 42 tested *Rhizobium* strains, 24 isolates exhibited varied extent of antifungal activity against *R. solani* in vitro. Biosynthesis of fungitoxic substances and phosphorous solubilization were recognized as biocontrol mechanisms by some tested *Rhizobium* strains. In addition, a considerable disease reduction was recorded by applying these strains.

Others

Other genera including *Burkholderia, Gluconobacter, Azoarcus, Herbaspirillum,* and *Klebsiella* are known also as antifungal agents against phytopathogenic fungi, and plant growth-promoting rhizobacteria [72, 73]. Many *Burkholderia* species are known to produce antifungal substances like phenazine iodinin, and hydrolytic enzymes. Rivalry for space and/or nutrients with other microorganisms and triggering plant immunity against pathogens were also reported. Anti-spore germination activity by *Burkholderia* spp. was recorded against spores of *Penicillium digitatum, S. sclerotiorum, Aspergillus flavus, A. niger, Phytophthora cactorum,* and *Botrytis cinereal* [74]. Detoxification and degradation of the virulence factor of a pathogen is another biocontrol mechanism utilized by some bacterial biocontrol agents. Some strains of *B. cepacia* and *B. ambifaria*have the ability to hydrolyze the mycotoxin fusaric acid, responsible for root rot and wilt diseases, which produced by some pathogenic *Fusarium* spp., as well as inhibit their mycelial growth [75]. Detoxification of fusaric acid by *K.oxytoca* was reported also via biosynthesis of detoxificating proteins that attach to the toxins [76].

The biocontrol activity of *B. gladioli* pv. *agaricicola* was studied against *Verticillium dahliae,* in vitro and in situ on tomato [77]. A significant fungitoxic effect was recorded by the bacterial strain ICMP12322 in vitro against the pathogenic fungus. In addition, a considerable disease reduction was achieved by application of this strain in the pot experiment. In another study, Bevardi et al. [78] reported a potent antagonistic activity by *G. oxydans* against the blue mold fungus *P. expansum.* A pronounced inhibition in the fungal growth up to 95% was achieved in vitro test. In vitro biocontrol activity of three growth-promoting rhizobacteria *Azospirillum brasilense* SBR, *Azotobacter chroococcum* ZCR, and *K. pneumoneae* KPR was investigated against the pathogenic mycoflora *F. oxysporum, S. sclerotiorum,* and *Pythium* sp. and in pots on cucumber [79]. A significant inhibition in fungal growth up to 100% in vitro and 56% decrease in the damping-off incidence were recorded by applying the tested bacterial biocontrol agents.

3.2.2   Fungi as Biocontrol Agents

Many antagonistic fungi have been extensively utilized as bio-fungicides against various phytopathogenic fungi. Owing to their widespread occurrence, persistence, multifunctional antifungal activities against plenty of pathogenic mycoflora, and relative ease of culturing and maintenance in vitro, they have attained a broad approbation in this concern. The most common fungi used as bio-control agents include members of the genera *Trichoderma, Gliocladium, Clonostachys, Penicillium, Chaetomium, Myrothecium, Laetisaria, Coniothyrium,* and arbuscular mycorrhizal fungi.

Trichoderma spp.

Trichoderma Pers. (Ascomycota, Sordariomycetes, Hypocreales) is a prevalent fungal genus of increasing interest due to their diverse bioactivities, global distribution, varied metabolites production, and competitive and reproductive potentiality. Members of *Trichoderma* found mostly in all types of ecosystems as soil-borne, on decaying plant materials, endophytic, epiphytic, on other fungi, and/or in aquatic habitats [80–83].

Many species of *Trichoderma* genus are geographically limited, some are widely distributed, while, few have a cosmopolitan distribution [84]. According to Bissett et al. [85], more than 250 of *Trichoderma* spp. have been listed. However, in the recent few years, more than 45 new species have been described [86–93]. Species of genus *Trichoderma* can synthesis several hydrolytic enzymes and antimicrobial substances which provide them with ecological dominance under varied environmental conditions and the ability to perform many biological functions. One of the most important characteristics of *Trichoderma* spp. is the high and numerous potentialities to antagonize a broad spectrum of fungal phytopathogens which qualify them as the most common bio-control agents. Indeed, commercial *Trichoderma*-based products represent more than 50% of fungal bio-fungicides market.

During the last years, use of *Trichoderma* spp. as bio-fungicides against various phytopathogenic fungi has attracted high scientific attention [94–96]. For example, El-Sharkawy et al. [97] studied foliar application of two isolates of *T. harzianum* and *T. viride* as bio-fungicides against wheat rust under greenhouse conditions. A significant anti-spore germination of *Puccinia graminis* uredospores was recorded in vitro. Under greenhouse conditions, a considerable reduction in the disease measures and improvement of wheat growth and yield parameters were reported. The antifungal activity was attributed to their production of some antifungal secondary metabolites. The antifungal potentiality of *T. harzianum* WKY1 against *Colletotrichum sublineolum*, causative of sorghum anthracnose, was studies by Saber et al. [98]. In vitro, a pronounced growth inhibition in the mycelia of *C. sublineolum* was recorded as well as a decrease in the disease severity under greenhouse conditions.

Both direct and indirect biocontrol mechanisms evolved by *Trichoderma* species have been discussed including rivalry for space or nutrients, antibiosis, and mycoparasitism. In addition, triggering of plant immune responses and enhancement of their growth were also reported [99]. However, predominance of one mechanism does not mean that the others are not contributed to the antagonistic behavior of the bioagent. Production of a large set of enzymes like cellulases, amylases, lipases and pectinases, as well as secondary metabolites such as siderophores, in addition to their high reproductive capacity provides *Trichoderma* spp. with antagonistic ability to compete the fungal pathogens for space and/or nutrients [100].

Biosynthesis of numerous antifungal lytic enzymes [101], as well as various antibiotic, secondary metabolites, volatile, and nonvolatile antifungal compounds by *Trichoderma* species are well known and recognized. In addition to phenolic compounds, production of various antibiotics like, trichodermol, viridian, gliovirin, harzianolide, harzianum A, trichodermin and koninginins has been also reported

[102]. However, it is difficult to differentiate between competition and antibiosis in agar plate. The inhibition zones result from antibiosis are indistinguishable from those produced by the nutrients shortage.

Mycoparasitism (obtaining nutrients from the fungal pathogen) may be contributed to the antagonistic behavior of some *Trichoderma* spp. [103–105]. However, the ability to parasitize pathogenic fungi is not a simple process; it involves specificity between both fungi. It depends primarily on the chemical attraction by the pathogenic fungus and the cell signaling in *Trichoderma* which includes recognition (sensing their prey), as well as capability for production of lytic enzymes [106]. A successful mycoparasitic process involves chemical recognition by *Trichoderma* sp. to their prey fungus, chemical attraction, connection, coiling around their fungal prey and penetrating them mechanically through sending appressoria into the prey mycelium or chemically through secretion of cell-wall hydrolytic enzymes, and sometimes secretion of some antifungal secondary metabolites [107].

Moreover, some *Trichoderma* spp. are identified as endophytes [108–111] that can trigger the plant systemic acquired resistance against attaching pathogens [109]. Moreover, they induce plant tolerance against drought and salinity [112]. Upregulation of different defense-related genes are also reported as a response to the endophytic *Trichoderma*, in addition to some phytochemicals [113]. In this regard, Park et al. [110] recorded a markedly inhibition in the disease development in ginseng, caused by *B. cinerea* and *Cylindrocarpon destructans*, as a response to application of the endophytic *T. citrinoviride*.

Gliocladium spp.

Gliocladium spp. (Ascomycota, Sordariomycetes, Hypocreales) are frequently found as soil-borne, endophytes, epiphytes, on other fungi, on plant debris, freshwater, and coastal soils [59, 114, 115]. *Gliocladium* spp. have a worldwide distribution and exceptional ecological versatility. They inhabit numerous ecosystems like tropical, temperate, subarctic, and desert areas [116]. Species of this genus are reported as producers of a vast range of secondary metabolites which exhibit different bioactivities such as antifungal, antibacterial, nematicidal, anti-tumour activities, as well as hydrocarbons and their derivatives (myco-diesel), and ligninolytic enzymes [117–120]. Taxonomically, many *Gliocladium* spp. were reclassified and moved to the genus *Clonostachys* due to significant molecular and morphological differences from the type form of *Gliocladium* spp. For instance, *G. catenulatum* is renamed to *C. rosea* f. *catenulata*, and *G. roseum* is renamed to *C. rosea* f. *rosea* [121, 122]. Furthermore, other species were transferred to the genus *Trichoderma* such as *G. virens* which is now classified as *T. virens*.

Species of the genus *Gliocladium* are widely known as bio-fungicides for many pathogenic mycoflora. The most common species used as biocontrol agents are *C. rosea* f. *rosea* (syn. *G. roseum*), *C. rosea* f. *catenulata* (syn. *G. catenulatum*), and *T. virens* (syn. *G. virens*). *Gliocladium* spp. have a potent antagonistic activity against various fungal mycopathogens like *P. ultimum, B. cinerea, F. graminearum, F. udum,*

Phytophthora cinnamomi, P. citricola, Alternaria alternata, Verticillium spp. and *Chaetomium* spp. [123–125]. Borges et al. [126] recorded significant biocontrol efficiency for *C. rosea* against tomato gray mold. Application of *C. rosea* recorded 100% biocontrol efficiency in stem and ≥90% in the entire tomato plant. Tesfagiorgis et al. [127] recorded a disease reduction (90%) in powdery mildew of zucchini when treated with *C. rosea* under greenhouse conditions.

Production of different antagonistic metabolites by *Gliocladium* spp. has been reported such as gliotoxin and viridin by *G. flavofuscum* [128]. According to the type of the antibiotic produced by strains of *T. virens* they can be differentiated into two groups (P and Q). Members of group P synthesis gliovirin which poses narrow antifungal spectrum activity, primarily, against oomycetes [129], while, members of group Q synthesis gliotoxin which poses a broad range of antifungal as well as antibacterial activities [130]. Another species of *Gliocladium* has been reported as a producer of a set of volatile antifungal substances against *P. ultimum* and *V. dahliae*. Of them, the antifungal antibiotic annulene was identified [131]. Mycoparasitism against different fungal pathogens was also reported as a proposed biocontrol mode of action of *Gliocladium* spp. [132, 133]. In a recent study, 199 candidate mycoparasites isolated from agricultural soils in southwestern Greece, of them, the isolate *Gliocladium* sp. G21-3 was the most aggressive mycoparasite and a competent antagonist against sclerotia of *S. sclerotiorum* [134].

Penicillium spp.

Penicillium Link (Ascomycota, Eurotiomycetes, Eurotiales) is a diverse genus which contain more than 400 species with a cosmopolitan distribution. *Penicillium* spp. are found as soil-borne, on decaying crops, on wood, fresh and dry fruits, water, and in indoor air. They are well known as organic materials decomposers, causative of food spoilage, producers of mycotoxins and enzymes, air allergens, and/or causative of postharvest decay of some crops [135]. Members of genus *Penicillium* are widely recognized as synthesizers of diverse bioactive substances such as antibiotics, antitumor agents, nephrotoxin, and ergot alkaloids [136].

Some *Penicillium* species are known as bio-fungicides against fungal diseases. The endophytic *P. oxalicum* T 3.3 exhibited an aggressive antifungal activity against anthracnose of dragon fruit, caused by *Colletotrichum gloeosporioides*. Production of β-glucanase and chitinase was reported for this biocontrol agent [137]. Sreevidya et al. [138] reported a remarked biocontrol activity of *P. citrinum* against botrytis gray mold of chickpea in the greenhouse and field. The antifungal activity was attributed to their production of mycotoxin citrinin. In addition, production of lytic enzymes like protease and glucanases were also reported. The biocontrol activity (75%) of *P. citrinum* was reported on charcoal rot of sorghum under greenhouse condition [139]. De Cal et al. [140] reported a markedly decrease in the powdery mildew of strawberry in vitro and in vivo via application of *P. oxalicum*.

Chaetomium spp.

Chaetomium spp. Kunze (Ascomycota, Sordariomycetes, Sordariales) are filamentous fungi which exist as soil-borne, air-borne, endophytic, epiphytic, on any cellulose containing materials, and on plant debris. It comprises more than 160 described species with a cosmopolitan distribution [141]. Some of these fungi act as biofungicides to control numerous pathogenic mycobiota like *A. raphani*, *A. brassicicola*, and *P. ultimum*. Zhao et al. [142] reported a potent antagonistic activity by the endophytic *C. globosum* CDW7 against rape sclerotinia rot, caused by *S. sclerotiorum*. Seven secondary metabolites were identified from their culture filtrate including the antifungal metabolites flavipin, chaetoglobosin A-E and V_b, for which their antagonistic potential was attributed. Hung et al. [143] reported also an in vitro mycelial growth inhibition of *P. nicotianae* by 50 ~ 56% when grew against the antagonists *C. globosum*, or *C. cupreum* in biculture tests and against their crude extracts. Furthermore, *C. cupreum* parasitized *P. nicotianae* and degraded their mycelia after 30 days of incubation. In pot experiment, use of *Chaetomium* spp. lowered the disease severity of citrus root rot by 66–71%. *Chaetomium* species have been reported as producers of lytic enzymes which involved in the mycoparasitism [144, 145]. In addition, numerous antifungal secondary metabolites were reported from the culture filtrates of *Chaetomium* spp. like flavipin, chaetoviridins, chaetoglobosins, and rubrorotiorin [142, 146, 147].

Myrothecium spp., *Laetisaria* spp., and *Coniothyrium Minitans*

Myrothecium spp. Tode (Ascomycota, Sordariomycetes, Hypocreales) are filamentous fungi that poses a universal distribution and found as soil-borne or on plants. It comprises more than 35 described species [148]. *Myrothecium* spp. are recognized as producers of various bioactive substances such as trichothecenes mycotoxins (roridin A, verrucarin A, and 8beta-acetoxy-roridin H) [149, 150], as well as lytic enzymes like proteinases and lipases [151]. Some of *Myrothecium* spp. have a potential antagonistic behavior against several fungal phytopathogens, weeds, insects, and nematodes [152, 153]. Barros et al. [154] reported a biocontrol activity of *Myrothecium* sp. against *S. sclerotiorum* in vitro and in vivo experiments. A considerable decrease in the soybean mold disease up to 70% was recorded by application of the biocontrol agent.

   Laetisaria Burds. (Basidiomycota, Agaricomycetes, Corticiales) is a genus of 4 species with widespread distribution. The soil-borne fungus *L. arvalis* is well recognized as a bio-fungicide against some pathogenic mycoflora. Among the 28 biocontrol agents tested by Brewer and Larkin [155], the isolate *L. arvalis* ZH-1 significantly reduced the disease incidence of potato black scurf by 60%. In another study, soil treatment with *L. arvalis* led to a markedly decrease in tomato damping-off, caused by *P. indicum*, recording 72% seed germination [156]. Furthermore, Bobba and Conway [157] reported the competition for nutrients as an antagonistic mechanism by *L. arvalis* against the pathogenic fungus *S. rolfsii* in the competitive colonization experiment.

Coniothyrium minitans W. A. Campb. (Ascomycota, Dothideomycetes, Pleosporales) is a worldwide distributed fungus. It is a naturally obligate mycoparasite on sclerotia of the fungal pathogens *S. sclerotiorum*, *S. minor*, *S. trifoliorum*, and *S. rolfsii* [158, 159]. In this regard, Chitrampalam et al. [160] studied the antifungal activity of *C. minitans* on *S. minor*, the causal of the lettuce drops, in vitro and in vivo. A total sclerotial mortality was recorded in the culture plates. In the field experiment, a significant reduction in the lettuce drop was achieved; this reduction was correlated with a reduction in the existence levels of the sclerotia. During the mycoparasitic process by *C. minitans*, the outer pigmented layer of the sclerotia has been mechanically penetrated and enzymatically using lytic enzymes [161]. However, the antibiosis mechanism via production of the antifungal secondary metabolite macrosphelide A was also reported [162].

Arbuscular Mycorrhizal Fungi (AMF)

AMF are soil fungi (Mucoromycota, Glomeromycotina) which comprise about 300 species in 3 classes, 5 orders, 15 families and 38 genera [163, 164]. They are obligate endophytes that live in mutualism with roots of 80% of the vascular plants [165]. AMF are found in all terrestrial ecosystems with varied extent of pH, salinity, organic matter, and environmental conditions. They have a cosmopolitan distribution, where they have been reported from all continents [166]. In the arbuscular mycorrhizal association, the fungus attains carbon from the photosynthesis of the plant, while the plant takes many advantages from the fungus. AMF supply the mycorrhizal host with water, and minerals via their extra radical hyphal network. Moreover, AMF improve the plant growth and metabolic processes, increase their resistance to drought, salinity, heavy metals, as well as enhance their immunity against various pathogenic mycobiota [167].

Many researchers have extensively studied the biocontrol activity of AMF to control different types of phytopathogenic fungi like *A. solani, Aphanomyces euteiches, Cercospora arachidicola, Cercosporidium personatum, Erysiphe graminis, F. solani, F. verticillioides, Gaeumannomyces graminis, M. phaseolina, P. cactorum, P. aphanidermatum, R. solani, S. cepivorum,* and *V. dahliae* [168–172]. Olowe et al. [173] investigated biocontrol activity of *Glomus clarum* and *G. deserticola* against maize ear rot. A considerable reduction in the disease effects on the plant growth parameters was recorded by application of AMF. El-Sharkawy et al. [97] investigated the biocontrol of wheat stem rust by using AMF and *Trichoderma* spp. under greenhouse conditions. A markedly decrease in the disease measures as well as enhancement in the growth and yield parameters were recorded. Moreover, an induction in the activities of some defensive enzymes and total phenol content were also recorded. The likely biocontrol mechanisms exerted by AMF comprise direct rivalry with other soil-borne pathogenic fungi for nutrients, space, and colonization sites, changing of the soil microbial composition in the rhizosphere area [174, 175].

Furthermore, AMF may indirectly decrease the losses resulting from the disease by damage compensation, growth improvement and triggering the plant immunity against the phytopathogens attack [170, 172]. In this regard, Abdel-Fattah et al. [176] reported triggering multiple defense-related reactions in bean plants against infection with *Rhizoctonia* root rot as a result of application of AMF. Some ultrastructural and biochemical responses were recorded including cell-wall thickening, cytoplasmic granulation, increase in the cell organelles number, nuclear hypertrophy, and accumulation of fungitoxic compounds (phenolics) and triggering of defensive enzymes activity. However, achieving a genetic polymorphism (86.8%) as well as triggering of the transcriptional expression level of defense-related genes were also reported [177].

3.3   Induction of Systemic Resistance and Defense-Related Genes in Plant

Plants have a strategy against fungal infection by evolving multiple immune mechanisms [178, 179]. The first immune response is started by the recognition of pathogen-associated molecular patterns conserved (PAMPs), like lipopolysaccharides, flagellin, chitin and glycoproteins by what is called Pattern-Recognition Receptors (PRRs) which located on the surface of cell [180]. The understanding of PAMP stimulates PAMP-triggered immunity (PTI), including oxidative burst, MAPK (mitogen-activated protein kinase) activation, deposition of callose, defense-related genes induction, and antimicrobial compounds accumulation [181–183]. The pathogens can successfully suppress PTI by secreting different effectors, like small RNAs and proteins to suppress host PTI in the host cells [184–186]. On the other hand, plants have secreted resistant proteins to recognize the specific effectors of pathogen, leading to an effector-triggered immunity (ETI), whereas ETI is more rapid and powerful than PTI and stimulates comparable defense responses set as in PTI but in an accelerated and powerful way [178, 179, 183, 187].

The starting of PTI or ETI from the infected loci often stimulates resistance induced in tissues that give resistance against a wide range of pathogens [39]. This systemic acquired resistance (SAR) is often correlated with level of salicylic acid (SA) increased and regulate the activation of pathogenesis related (PR) genes and comprises one or more long-distance signals that increase the capacity to enhanced defensive in intact parts of plant [188]. Also, beneficial microbes in the rhizosphere can induce systemic resistance (ISR). In most cases, ISR is SA-independent and develops without accumulation of PR proteins. *P. fluorescens* is still able to induce ISR that does not synchronize with enhanced SA levels.

4 Case Study

In Egypt, many researchers concerned with the biological control of fungal diseases, my research group studied many bioagents for control of many plant fungal diseases such as *Streptomyces* spp. [64, 189], *Pseudomonas* spp. and *Bacillus* spp [190–192] and some fungal species such as *Gliocladium* spp., *Paecilomyces* spp., *Penicillium* spp. and *Trichoderma* spp. [189]. The *Trichoderma harzianum* was used widely as a bioagent, which observed the most potent organisms among bacterial and fungal species used against sugarbeet pathogen *R. solani* in the study carried out by Moussa [189] and shown in Table 1. The mechanism of *T. harzianum* to control the fungal pathogens was by mycoparasitism on the pathogen hyphae and observed using scanning electron microscope (SEM) (Figs. 1, 2 and 3).

Hyphal interactions between *T. harzianum* and *R. solani* were observed by scanning electron microscopy. *T. harzianum* attached to the host by hyphal coils (Figs. 1, 2 and 3).

Table 1 Control of sugar beet root rot disease caused by *R. solani* with different antagonists

Antagonist	Disease incidence (%)		
	Seed coating	Seed soaking	Soil pre-inoculation
Control	42.53a	71.43	75.68a[a]
Bacteria			
Bacillus cereus	14.85defg	66.4bc	48.18a
B. subtilis	10.67efgh[a]	81.2a	52.91a
Fungi			
Gliocladium deliquescens	15.51def	51.9c	20.68b
Paecilomyces marquandii	8.8fgh[b]	52.3c	50.93a
Penicillium vermiculatum	10.12efgh	65.5bc[a]	50.93a
Trichoderma harzianum	6.48 h	62.9bc	17.16b
T. koningii	13.53efg	69.5ab	48.18a
T. pseu-dokoningii	25.94b	63.6bc	52.9a
T. viride	25.63bc	60.5bc	48.18a

[a]Values within a row followed by the same letter are not significantly different at 5% level according to Duncan's multiple range test (DMRT)
[b]Values within the column followed by the same letter are not significantly different at 5% level (based on DMRT)

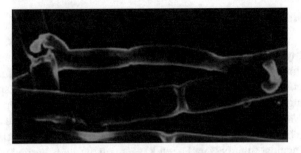

Fig. 1 Scanning electron micrographs of *Trichoderma harzianum* hyphae interacting with those of *Rhizoctonia solani* in which hypha of *T. harzianum* coiling around and penetrating one of *R. solani*. Partial degradation of host cell wall can be observed (X 8500) [190]

Fig. 2 Scanning electron micrographs of *Trichoderma harzianum* hyphae interacting with those of *Rhizoctonia solani* in which hooks of *T. hrzianum* attached to hyphae of *R. Solani* (X 2000) [190]

Fig. 3 Scanning electron micrographs of *Trichoderma harzianum* hyphae interacting with those of *Rhizoctonia solani* in which appressorium-like structure formed by *T. harzianum*, attached to a hyphae of *R. solani* with partial degradation of host cell wall (X 8500) [190]

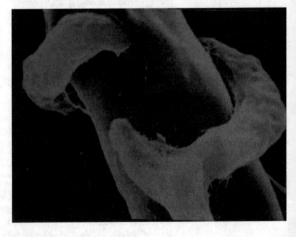

In another case study, the research was developed to study the effect of bioagent on the host plant as well as fungal pathogens. Some bacterial species were known as plant growth promoting rhizobacteria (PGPR) which secret some compounds to enhance plant growth, it was found that all growth parameters of *Cucumi ssativus* L. cv. Market were increased in absence and presence of the fungal pathogen *P. aphanidermatum* in greenhouse experiment as shown in Table 2. On the other hand, the use of *P. aeruginosa* and *B. amyloliquefaciens* separately inhibit the fungal pathogen *P. aphanidermatum* [191]. Another study on the biocontrol of *F. graminearum* which attacks wheat, in which it was concluded that the use of *B. subtilis* and *Pseudomonas fluorescens* increased the growth parameters of wheat and suppress the growth of *F. graminearum*, also *P. fluorescens* was the most efficient than *B. subtilis* or in mixture [192].

In a recent study conducted by the authors, the biocontrol activity of a mixture of arbuscular mycorrhizal fungi was investigated against *Rhizoctonia* root rot of common bean, caused by *Rhizoctonia solani* Kühn, under natural conditions. The obtained results exhibited a considerable reduction in the disease severity and incidence by the mycorrhizal colonization. In addition, a significant enhancement of the shoot and root lengths and dry weights, and the leaf area was observed in the colonized plants when compared with the control plants. Moreover, the mineral nutrient concentrations and yield parameters were also improved. Transmission electron microscope observations showed some defense-related ultrastructural changes including cell wall thickening and cytoplasmic granulation. The biochemical analysis of the colonized plants showed an accumulation of the phenolic compounds, which have a fungitoxic activity, and induction of the defense-related enzymes phenylalanine ammonia lyase, peroxidase and polyphenoloxidase [176]. Furthermore, the molecular examination indicated an induction of the transcriptional expression level of the defense-related genes chitinase and β-1,3-glucanase as a response to the mycorrhizal colonization [177].

5 Conclusion and Future Prospects

In this chapter, the authors tried to highlight the most important biological control practices all over the world and focused on Egypt as a home country, it is found that through the past century, the attention to biological control of economic crops has increased from both the government and the researchers starting from the ordinary application of biocontrol agents in contact directly to the soil and in form of gelatin capsules to insertion of the resistance genes in the plant and produce what we know today GM plants (genetically modified plants). In Egypt, the biological control of different diseases becomes common due to the awareness of farmers about the benefits of biocontrol applications.

Table 2 The efficacy of three isolated species from PGPR on growth parameters of *Cucumis sativus* L. cv. Market more in the presence or absence of pathogenic *Pythium aphanidermatum* under greenhouse condition

Plant treaded with	Plant growth parameters					Chlorophyll content (unit)
	Length (cm)			Weight (g)		
	Plant	Stem	Root	Fresh	Dry	
Untreated	17.5 ± 0.8 [b]	7.17 ± 0.72	10.33 ± 1.59	0.25 ± 0.06	0.027 ± 0.005	12.45 ± 3.38
Pythium aphanidermatum (*Pa*)	P	P	P	P	P	P
Bacillus subtilis (*Bs*)	23.7 ± 1.04[a]	13.5 ± 0.87[a]	10.37 ± 0.29	0.62 ± 0.02[a]	0.033 ± 0.007[a]	16.05 ± 1.5[a]
Bacillus amyloliquefaciens (*Ba*)	17.0 ± 1.30	8.97 ± 0.90	8.030 ± 2.00	0.53 ± 0.01[a]	0.029 ± 0.004[a]	12.43 ± 0.50
Pseduomonas aeruginosa (*Psa*)	19.3 ± 1.15[a]	9.00 ± 1.00	10.13 ± 2.08	0.61 ± 0.07[a]	0.031 ± 0.002	16.00 ± 5.15[a]
(*Bs*) + (*Ba*)	22.0 ± 5.20[a]	10.5 ± 1.32[a]	11.50 ± 5.70	0.65 ± 0.01[a]	0.037 ± 0.003	16.03 ± 3.03[a]
(*Bs*) + (*Psa*)	21.83 ± 0.7[a]	9.00 ± 2.00	12.8 ± 2.70[a]	0.59 ± 0.03[a]	0.029 ± 0.001	12.6 ± 1.90
(*Ba*) + (*Psa*)	15.7 ± 0.20	10.0 ± 0.87	5.67 ± 0.58	0.19 ± 0.03	0.020 ± 0.002	11.27 ± 0.8
(*Bs*) + (*Ba*) + (*Psa*)	11.83 ± 1.6	5.00 ± 1.00	6.83 ± 1.76	0.19 ± 0.03	0.022 ± 0.002	5.6 ± 2.030
(*Bs*) + (*Pa*)	P	P	P	P	P	P
(*Ba*) + (*Pa*)	15.0 ± 0.21	6.00 ± 0.90	9.00 ± 0.16	0.18 ± 0.32	0.22 ± 0.010	8.33 ± 0.13
(*Psa*) + (*Pa*)	12.33 ± 0.6	5.67 ± 0.58	6.67 ± 0.58	0.17 ± 0.03	0.17 ± 0.010	5.70 ± 0.13
(*Bs*) + (*Ba*) + (*Pa*)	P	P	P	P	P	P
(*Bs*) + (*Psa*) + (*Pa*)	P	P	P	P	P	P
(*Ba*) + (*Psa*) + (*Pa*)	P	P	P	P	P	P
(*Bs*) + (*Ba*) + (*Psa*) + (*Pa*)	P	P	P	P	P	P

Mean of three replicates ± SD

[a] significant at level 5%

[b] P: Plants can't survival under fungus infection rate (1.3×10^7 propagules g^{-1} soil)

References

1. Hawksworth L (1991) The fungal dimension of biodiversity: magnitude, significance, and conservation. Mycol Res 95:641–655
2. US EPA (2005) Human health risk assessment protocol for hazardous waste combustion facilities. EPA530-R-05-006
3. Gonzalez-Fernández R, Prats E, Jorrín-Novo JV (2010) Proteomics of plant pathogenic fungi. J Biomed Biotechnol 2010:932527
4. Vadlapudi V, Naidu KC (2011) Fungal pathogenicity of plants: molecular approach. Eur J Exp Bio 1:38–42
5. Crous PW, Hawksworth DL, Wingfield MJ (2015) Identifying and naming plant-pathogenic fungi: past, present, and future. Annu Rev Phytopathol 53:247–267
6. Jibril SM, Jakada BH, Kutama AS, Umar HY (2016) Plant and pathogens: pathogen recognition, Invasion and plant defense mechanism. Int J Curr Microbiol App Sci 5:247–257
7. El-Abyad MS, Abu-Taleb AM, Abdel-Mawgoud T (1996) Effect of the herbicide pyradur on host cell wall-degradation by the sugarbeet pathogens *Rhizoctonia solani* Kühn and *Sclerotium rolfsii* Sacc. Can J Bot 74:1407–1415
8. El-Abyad MS, Abu-Taleb AM, Abdel-Mawgoud T (1997) Response of host cultivar to cell wall-degrading enzymes of the sugarbeet pathogens *Rhizoctonia solani* Kühn and *Sclerotium rolfsii* Sacc. under salinity stress. Microbiol Res 152:9–17
9. Moussa TAA, Tharwat NA (2007) Optimization of cellulase and β-glucosidase induction by sugarbeet pathogen *Sclerotium rolfsii*. Afr J Biotechnol 6:1048–1054
10. Moussa TAA, Shanab SMM (2001) Impact of cyanobacterial toxicity stress on the growth activities of some phytopathogenic *Fusarium* spp. Az J Microbiol 53:267–282
11. Moussa TAA, Ali DMI (2008) Isolation and identification of novel disaccharide of α-L-Rhamnose from *Penicillium chrysogenum*. World Appl Sci J 3:476–486
12. Martinelli F, Scalenghe R, Davino S, Panno S, Scuderi G, Ruisi P, Villa P, Stroppiana D, Boschetti M, Goulart LR, Davis CE, Dandekar AM (2014) Advanced methods of plant disease detection. A Rev Agron Sustain Dev 35:1–25
13. Shuping DSS, Eloff JN (2017) The use of plants to protect plants and food against fungal pathogens: a review. Afr J Tradit Complement Altern Med 14:120–127
14. Patel N, Desai P, Patel N, Jha A, Gautam HK (2014) Agro-nanotechnology for plant fungal disease management: a review. Int J Curr Microbiol App Sci 3:71–84
15. Al-Agamy MHM (2011) Tools of biological warfare. Res J Microbiol 6:193–245
16. Nicolopoulou-Stamati P, Maipas S, Kotampasi C, Stamatis P, Hens L (2016) Chemical pesticides and human health: the urgent need for a new concept in agriculture. Front Public Health 4:148
17. Bale JS, van Lenteren JC, Bigler F (2008) Biological control and sustainable food production. Philos Trans R Soc Lond B Biol Sci 363(1492):761–776
18. Gnanamanickam SS (2002) Biological control of crop diseases. Marcel Dekker Inc, New York, USA
19. Compant S, Duffy B, Nowak J et al (2005) Use of plant growth promoting bacteria for biocontrol of plant diseases: Principles, mechanisms of action and future prospects. Appl Environ Microbiol 71:4951–4959
20. Hong CE, Park JM (2016) Endophytic bacteria as biocontrol agents against plant pathogens: current state-of-the-art. Plant Biotechnol Rep 10:353
21. Carmona-Hernandez S, Reyes-Pérez JJ, Chiquito-Contreras RG et al (2019) Biocontrol of postharvest fruit fungal diseases by bacterial antagonists: a review. Agronomy 9:121
22. Parte AC (2018) LPSN—List of prokaryotic names with standing in nomenclature (bacterio.net), 20 years on. Int J Syst Evol Microbiol 68:1825–1829
23. Connor N, Sikorski J, Rooney AP et al (2010) Ecology of speciation in the genus *Bacillus*. Appl Environ Microbiol 76:1349–1358
24. Todar K (2012) Bacterial resistance to antibiotics. The microbial world. Lectures in microbiology, University of Wisconsin-Madison

25. Fira D, Dimkić I, Berić T et al (2018) Biological control of plant pathogens by *Bacillus* species. J Biotechnol 285:44–55
26. Tojo S, Tanaka Y, Ochi K (2015) Activation of antibiotic production in bacillus spp. by cumulative drug resistance mutations. Antimicrob Agents Chemother 59(12):7799–7804
27. Halami PM (2019) Sublichenin, a new subtilin-like lantibiotics of probiotic bacterium *Bacillus licheniformis* MCC 2512^T with antibacterial activity. Microb Pathog 128:139–146
28. Saber WIA, Ghoneem KM, Al-Askar AA et al (2015) Chitinase production by *Bacillus subtilis* ATCC 11774 and its effect on biocontrol of *Rhizoctonia* diseases of potato. Acta Biol Hung 66(4):436–448
29. Contesini FJ, Melo RR, Sato HH (2018) An overview of *Bacillus* proteases: from production to application. Crit Rev Biotechnol 38(3):321–334
30. Kumar S, Singh A (2015) Biopesticides: present status and the future prospects. J Fertil Pestic 6:e129
31. Dimkić I, Živković S, Berić T et al (2013) Characterization and evaluation of two *Bacillus* strains, SS-12.6 and SS-13.1, as potential agents for the control of phytopathogenic bacteria and fungi. Biol Control 65(3):312–321
32. Guo Q, Dong W, Li S et al (2014) Fengycin produced by *Bacillus subtilis* NCD-2 plays a major role in biocontrol of cotton seedling damping-off disease. Microbiol Res 169(7):533–540
33. Zhang X, Zhou Y, Li Y et al (2017) Screening and characterization of endophytic *Bacillus* for biocontrol of grapevine downy mildew. Crop Prot 96:173–179
34. Shafi J, Tian H, Ji M (2017) *Bacillus* species as versatile weapons for plant pathogens: a review. Biotechnol Biotechnol Equip 31(3):446–459
35. Chen L, Heng J, Qin S et al (2018) A comprehensive understanding of the biocontrol potential of *Bacillus velezensis* LM2303 against *Fusarium* head blight. PLoS ONE 13(6):e0198560
36. Tchagang CF, Xu R, Overy D et al (2018) Diversity of bacteria associated with corn roots inoculated with Canadian woodland soils, and description of *Pseudomonas aylmerense* sp. nov. Heliyon 4(8):e00761
37. Gomila M, Peña A, Mulet M et al (2015) Phylogenomics and systematics in *Pseudomonas*. Front Microbiol 18(6):214
38. Bosire EM, Rosenbaum MA (2017) Electrochemical potential influences phenazine production, electron transfer and consequently electric current generation by *Pseudomonas aeruginosa*. Front Microbiol 8:892
39. Pieterse CMJ, Zamioudis C, Berendsen RL et al (2014) Induced systemic resistance by beneficial microbes. Ann Rev Phytopathol 52:347–375
40. Kumar P, Dubey RC, Maheshwari DK et al (2016) Isolation of plant growth-promoting *Pseudomonas* sp. PPR8 from the rhizosphere of *Phaseolus vulgaris* L. Arch Biol Sci 68(2):363–374
41. Panpatte DG, Jhala YK, Shelat HN et al (2016) *Pseudomonas fluorescens*: a promising biocontrol agent and PGPR for sustainable agriculture. In: Singh D, Singh H, Prabha R (eds) Microbial inoculants in sustainable agricultural productivity. Springer, New Delhi
42. Aielloa D, Restucciaa C, Stefani E et al (2019) Postharvest biocontrol ability of *Pseudomonas synxantha* against *Monilinia fructicola* and *Monilinia fructigena* on stone fruit. Postharvest Biol Tech 149:83–89
43. Chater KF (2016) Recent advances in understanding *Streptomyces*. F1000Res 5:2795
44. Santhanam R, Okoro CK, Rong X et al (2012) *Streptomyces deserti* sp. nov., isolated from hyper-arid Atacama Desert soil. Antonie Van Leeuwenhoek 101(3):575–581
45. Zhang L, Ruan C, Peng F et al (2016) *Streptomyces arcticus* sp. nov., isolated from frozen soil. Int J Syst Evol Microbiol 66(3):1482–1487
46. Al-Askar AA, Rashad YM, Hafez EE et al (2015) Characterization of Alkaline protease produced by *Streptomyces griseorubens* E44G and its possibility for controlling *Rhizoctonia* root rot disease of corn. Biotechnol Biotechnol Equip 29(3):457–462
47. Le Roes-Hill M, Prins A, Meyers PR (2018) *Streptomyces swartbergensis* sp. nov., a novel tyrosinase and antibiotic producing actinobacterium. Antonie Van Leeuwenhoek 111(4):589–600

48. Romero-Rodríguez A, Maldonado-Carmona N, Ruiz-Villafán B et al (2018) Interplay between carbon, nitrogen and phosphate utilization in the control of secondary metabolite production in Streptomyces. Antonie Van Leeuwenhoek 111:761–781
49. Barreiro C, Martínez-Castro M (2019) Regulation of the phosphate metabolism in *Streptomyces* genus: impact on the secondary metabolites
50. Gebhardt K, Meyer SW, Schinko J et al (2011) Phenalinolactones A-D, terpenoglycoside antibiotics from Streptomyces sp. Tü 6071. J Antibiot (Tokyo) 64:229–232
51. Helaly SE, Goodfellow M, Zinecker H et al (2013) Warkmycin, a novel angucycline antibiotic produced by *Streptomyces* sp. Acta 2930*. J Antibiot (Tokyo) 66(11):669–674
52. Rashad YM, Al-Askar AA, Ghoneem KM et al (2017) Chitinolytic *Streptomyces griseorubens* E44G enhances the biocontrol efficacy against *Fusarium* wilt disease of tomato. Phytoparasitica 45(2):227–237
53. Hafez EE, Rashad YM, Abdulkhair WM et al (2019) Improving the chitinolytic activity of *Streptomyces griseorubens* E44G by mutagenesis. J Microbiol Biotechnol Food Sci 8(5):1156–1160
54. Ara I, Bukhari NA, Aref N et al (2014) Antiviral activities of streptomycetes against tobacco mosaic virus (TMV) in *Datura* plant: evaluation of different organic compounds in their metabolites. Afr J Biotechnol 11:2130–2138
55. Wang SM, Liang Y, Shen T et al (2016) Biological characteristics of *Streptomyces albospinus* CT205 and its biocontrol potential against cucumber Fusarium wilt. Biocontrol Sci Techn 26(7):951–963
56. Jung SJ, Kim NK, Lee DH et al (2018) Screening and evaluation of *Streptomyces* species as a potential biocontrol agent against a wood decay fungus. Gloeophyllum Trabeum Mycobiol 46(2):138–146
57. Gowdar SB, Deepa H, Amaresh YS (2018) A brief review on biocontrol potential and PGPR traits of *Streptomyces* sp. for the management of plant diseases. J Pharmacogn Phytochem 7(5):03–07
58. Al-Askar AA, Abdulkhair WM, Rashad YM (2011) In vitro antifungal activity of *Streptomyces spororaveus* RDS28 against some phytopathogenic fungi. Afr J Agric Res 6(12):2835–2842
59. Al-Askar AA, Abdulkhair WM, Rashad YM et al (2014a) *Streptomyces griseorubens* E44G: a potent antagonist isolated from soil in Saudi Arabia. J Pure Appl Microbiol 8:221–230
60. Law JW, Ser HL, Khan TM et al (2017) The potential of *Streptomyces* as biocontrol agents against the rice blast fungus, *Magnaportheoryzae* (*Pyricularia oryzae*). Front Microbiol 8:3
61. Bubici G (2018) *Streptomyces* spp. as biocontrol agents against *Fusarium* species. CAB Rev 13,50
62. Goudjal Y, Zamoum M, Sabaou N et al (2016) Potential of endophytic streptomyces spp. for biocontrol of fusarium root rot disease and growth promotion of tomato seedlings. Biocontrol Sci Technol 26(12):1691–1705
63. Al-Askar AA, Baka ZA, Rashad YM et al (2015) Evaluation of *Streptomyces griseorubens* E44G for the biocontrol of *Fusarium oxysporum* f. sp. *lycopersici*: ultrastructural and cytochemical investigations. Ann Microbiol 65:1815–1824
64. Moussa TAA, Rizk MA (2002) Biocontrol of sugarbeet pathogen *Fusarium solani* (Mart.) Sacc. by *Streptomyces aureofaciens*. Pak J Biol Sci 5:556–559
65. Poole P, Ramachandran V, Terpolilli J (2018) Rhizobia: from saprophytes to endosymbionts. Nat Rev Microbiol 16(5):291–303
66. Al-Ani RA, Adhab MA, Mahdi MH et al (2012) *Rhizobium japonicum* as a biocontrol agent of soybean root rot disease caused by *Fusarium solani* and *Macrophomina phaseolina*. Plant Protect Sci 48:149–155
67. Tamiru G, Muleta D (2018) The effect of rhizobia isolates against black root rot disease of Faba Bean (*Vicia faba* L) caused by *Fusarium solani*. Open Agr J 12:131–147
68. Jacka CN, Wozniaka KJ, Porter SS et al (2019) Rhizobia protect their legume hosts against soil-borne microbial antagonists in a host-genotype-dependent manner. Rhizosphere 9:47–55

69. Das K, Prasanna R, Saxena AK (2017) Rhizobia: a potential biocontrol agent for soilborne fungal pathogens. Folia Microbiol 62(5):425–435
70. Volpiano CG, Lisboa BB, São José JFB et al (2018) *Rhizobium* strains in the biological control of the phytopathogenic fungi *Sclerotium* (*Athelia*) *rolfsii* on the common bean. Plant Soil 432:229–243
71. Hemissi I, Mabrouk Y, Abdi N et al (2011) Effects of some *Rhizobium* strains on chickpea growth and biological control of *Rhizoctonia solani*. Afr J Microbiol Res 5(24):4080–4090
72. Ahemad M, Kibret M (2014) Mechanisms and applications of plant growth promoting rhizobacteria: current perspective. J King Saud Univ Sci 26(1):1–20
73. Katiyar D, Hemantaranjan A, Singh B (2016) Plant growth promoting Rhizobacteria-an efficient tool for agriculture promotion. Adv Plants Agric Res 4(6):426–434
74. Elshafie HS, Camele I, Ventrella E et al (2013) Use of plant growth promoting bacteria (PGPB) for promoting tomato growth and its evaluation as biological control agent. Int J Microbiol Res 5:452–457
75. Simonetti E, Roberts IN, Montecchia MS et al (2018) A novel *Burkholderia ambifaria* strain able to degrade the mycotoxin fusaric acid and to inhibit *Fusarium* spp. growth. Microbiol Res 206:50–59
76. Toyoda H, Katsuragi KT, Tamai T et al (1991) DNA sequence of genes for detoxification of fusaric acid, a wilt-inducing agent produced by *Fusarium* species. J Phytopathol 133:265–277
77. Elshafie HS, Sakr S, Bufo SA et al (2017) An attempt of biocontrol the tomato-wilt disease caused by *Verticillium dahliae* using *Burkholderia gladioli* pv. *agaricicola* and its bioactive secondary metabolites. Int J Plant Biol 8(1):57–60
78. Bevardi M, Frece J, Mesarek D et al (2013) Antifungal and antipatulin activity of *Gluconobacter oxydans* isolated from apple surface. Arh Hig Rada Toksikol 64(2):279–284
79. Hassouna MG, El-Saedy MA, Saleh HM (1998) Biocontrol of soil-borne plant pathogens attacking cucumber (Cucumis sativus) by Rhizobacteria in a semiarid environment. Arid Land Res Manage 12(4):345–357
80. Al-Askar AA, Ghoneem KM, Rashad YM (2012) Seed-borne mycoflora of alfalfa (*Medicago sativa* L.) in the Riyadh Region of Saudi Arabia. Ann Microbiol 62(1):273–281
81. Al-Askar AA, Ghoneem KM, Rashad YM et al (2014) Occurrence and distribution of tomato seed-borne mycoflora in Saudi Arabia and its correlation with the climatic variables. Microb Biotechnol 7(6):556–569
82. Jaklitsch WM, Voglmayr H (2015) Biodiversity of *Trichoderma* (Hypocreaceae) in Southern Europe and Macaronesia. Stud Mycol 80:1–87
83. Samuels GJ (2006) *Trichoderma*: systematics, the sexual state, and ecology. Phytopathol 96(2):195–206
84. Goh J, Nam B, Lee JS et al (2018) First report of six *Trichoderma* species isolated from freshwater environment in Korea. Korean J Mycol 46(3):213–225
85. Bissett J, Gams W, Jaklitsch W et al (2015) Accepted *Trichoderma* names in the year 2015. IMA Fungus 6(2):263–295
86. Qin WT, Zhuang WY (2016) Two new hyaline-ascospored species of *Trichoderma* and their phylogenetic positions. Mycologia 108:205–214
87. Qin WT, Zhuang WY (2016) Seven wood-inhabiting new species of the genus *Trichoderma* (Fungi, Ascomycota) in Viride clade. Sci Rep 6:27074
88. Qin WT, Zhuang WY (2016) Four new species of *Trichoderma* with hyaline ascospores from central China. Mycol Prog 15:811–825
89. Qin WT, Zhuang WY (2017) Seven new species of *Trichoderma* (Hypocreales) in the Harzianum and Strictipile clades. Phytotaxa 305:121–139
90. Chen K, Zhuang WY (2017) Discovery from a large-scaled survey of *Trichoderma* in soil of China. Sci Rep 7(1):9090
91. Chen K, Zhuang WY (2017) Seven soil-inhabiting new species of the genus *Trichoderma* in the Viride clade. Phytotaxa 312:28–46
92. Zhang YB, Zhuang WY (2017) Four new species of *Trichoderma* with hyaline ascospores from southwest China. Mycosphere 8(10):1914–1929

93. Zhang YB, Zhuang WY (2018) New species of *Trichoderma* in the Harzianum, Longibrachia-tum and Viride clades. Phytotaxa 379(2):131–142

94. Abdel-Fattah GM, Shabana YM, Ismail AE et al (2007) *Trichoderma harzianum*: a biocontrol agent against *Bipolaris oryzae*. Mycopathologia 164(2):81–89

95. Malmierca MG, Cardoza RE, Alexander NJ et al (2012) Involvement of *Trichoderma* tri-chothecenes in the biocontrol activity and induction of plant defense-related genes. Appl Environ Microbiol 78:4856–4868

96. Ganuza M, Pastor N, Boccolini M et al (2018) Evaluating the impact of the biocontrol agent *Trichoderma harzianum* ITEM 3636 on indigenous microbial communities from field soils. J Appl Microbiol 126:608–623

97. El-Sharkawy HH, Rashad YM, Ibrahim SA (2018) Biocontrol of stem rust disease of wheat using arbuscular mycorrhizal fungi and *Trichoderma* spp. Physiol Mol Plant Pathol 103:84–91

98. Saber WIA, Ghoneem KM, Rashad YM et al (2017) *Trichoderma harzianum* WKY1: an indole acetic acid producer for growth improvement and anthracnose disease control in sorghum. Biocontrol Sci Technol 27(5):654–676

99. Srivastava M, Pandey S, Shahid M et al (2015) Biocontrol mechanisms evolved by *Tricho-derma* sp. against phytopathogens: a review. Bioscan 10:1713–1719

100. Strakowska J, Blaszczyk L, Chelkowski J (2014) The significance of cellulolytic enzymes produced by *Trichoderma* in opportunistic lifestyle of this fungus. J Basic Microbiol 54:S2–S13

101. Gajera HP, Bambharolia RP, Patel SV et al (2012) Antagonism of *Trichoderma* spp. against *Macrophomina phaseolina*: evaluation of coiling and cell wall degrading enzymatic activities. Plant Pathol Microbiol 3:2157–7471

102. Vinale F, Sivasithamparam K, Ghisalberti EL et al (2014) *Trichoderma* secondary metabolites active on plants and fungal pathogens. Open Mycol J 8:127–139

103. Ojha S, Chatterjee NC (2011) Mycoparasitism of *Trichoderma* spp. in biocontrol of fusarial wilt of tomato. Arch Phytopathol Plant Protect 44(8): 771–782

104. Qualhato TF, Lopes FA, Steindorff AS et al (2013) Mycoparasitism studies of Trichoderma species against three phytopathogenic fungi: evaluation of antagonism and hydrolytic enzyme production. Biotechnol Lett 35(9):1461–1468

105. Guzmán-Guzmán P, Alemán-Duarte MI, Delaye L et al (2017) Identification of effector-like proteins in *Trichoderma* spp. and role of a hydrophobin in the plant-fungus interaction and mycoparasitism. BMC Genetics 18:16

106. Omann MR, Lehner S, Escobar Rodríguez C et al (2012) The seven-transmembrane receptor Gpr1 governs processes relevant for the antagonistic interaction of *Trichoderma atroviride* with its host. Microbiol 158(Pt 1):107–118

107. Mukherjee M, Mukherjee PK, Horwitz BA et al (2012) *Trichoderma*-plant-pathogen interac-tions: advances in genetics of biological control. Indian J Microbiol 52(4):522–529

108. Taribuka J, Wibowo A, Widyastuti SM et al (2017) Potency of six isolates of biocontrol agents endophytic *Trichoderma* against fusarium wilt on banana. J Degrade Min Land Manage 4(2):723–731

109. Park Y-H, Kim Y, Mishra RC et al (2017) Fungal endophytes inhabiting mountain-cultivated ginseng (*Panax ginseng* Meyer): diversity and biocontrol activity against ginseng pathogens. Sci Rep 7:16221

110. Park Y-H, Mishra RC, Yoon S et al (2019) Endophytic *Trichoderma citrinoviride* isolated from mountain-cultivated ginseng (*Panax ginseng*) has great potential as a biocontrol agent against ginseng pathogens. J Ginseng Res. (in press) https://doi.org/10.1016/j.jgr.2018.03.002

111. Chen J-L, Sun S-Z, Miao C-P et al (2016) Endophytic *Trichoderma gamsii* YIM PH30019: a promising biocontrol agent with hyperosmolar, mycoparasitism, and antagonistic activities of induced volatile organic compounds on root-rot pathogenic fungi of *Panax notoginseng*. J Ginseng Res 40(4):315–324

112. Ek-Ramos MJ, Zhou W, Valencia CU et al (2013) Spatial and temporal variation in fungal endophyte communities isolated from cultivated cotton (*Gossypium hirsutum*). PLoS ONE 8:1–13

113. Martínez-Medina A, Fernández I, Sánchez-Guzmán MJ et al (2013) Deciphering the hormonal signalling network behind the systemic resistance induced by *Trichoderma harzianum* in tomato. Front Plant Sci 4:206

114. Kim JY, Yun YH, Hyun MW (2010) Identification and characterization of gliocladium viride isolated from mushroom fly infested oak log beds used for shiitake cultivation. Mycobiology 38(1):7–12

115. Nur A, Salam M, Junaid M et al (2014) Isolation and identification of endophytic fungi from cocoa plant resistante VSD M.05 and cocoa plant suscebtible VSD M.01 in South Sulawesi, Indonesia. Int J Curr Microbiol App Sci 3(2):459–467

116. Sutton JC, Li D-W, Peng G et al (1997) *Gliocladium roseum*: a versatile adversary of *Botrytis cinerea* in crops. Plant Dis 81(4):316–328

117. Strobel GA, Knighton B, Kluck K et al (2008) The production of myco-diesel hydrocarbons and their derivatives by the endophytic fungus *Gliocladium roseum* (NRRL 50072). Microbiology 154:3319–3328

118. Song HC, Shen WY, Dong JY (2016) Nematicidal metabolites from *Gliocladium roseum* YMF1.00133. Appl Biochem Microbiol 52:324–330

119. Zhai MM, Qi FM, Li J et al (2016) Isolation of secondary metabolites from the soil-derived fungus *Clonostachys rosea* YRS-06, a biological control agent, and evaluation of antibacterial activity. J Agric Food Chem 64:2298–2306

120. Rybczyńska-Tkaczyk K, Korniłłowicz-Kowalska T (2018) Activities of versatile peroxidase in cultures of *Clonostachys rosea* f. *catenulata* and *Clonostachys rosea* f. *rosea* during biotransformation of alkali lignin. J AOAC Int 101(5):1415–1421

121. Schroers HJ (2001) A monograph of bionectria (ascomycota, hypocreales, bionectriaceae) and its clonostachys anamorphs. Stud Mycol 46:1–214

122. Schroers HJ, Samuels GJ, Seifert KA et al (1999) Classification of the mycoparasite *Gliocladium roseum* in Clonostachys as *C. rosea*, its relationship to *Bionectria ochroleuca*, and notes on other *Gliocladium*-like fungi. Mycologia 91(2):365–385

123. Jabnoun-Khiareddine H, Daami-Remadi M, Ayed F et al (2009) Biocontrol of tomato verticillium wilt by using indigenous *Gliocladium* spp. and *Penicillium* sp. isolates. Dyn Soil Dyn Plant 3(1):70–79

124. Agarwal T, Malhotra A, Trivedi PC et al (2011) Biocontrol potential of *Gliocladium virens* against fungal pathogens isolated from chickpea, lentil and black gram seeds. J Agric Technol 7(6):1833–1839

125. Hassine M, Jabnoun-Khiareddine H, Aydi Ben Abdallah R et al (2017) In vitro and in vivo antifungal activity of culture filtrates and organic extracts of *Penicillium* sp. and *Gliocladium* spp. against *Botrytis cinerea*. J Plant Pathol Microbiol 8(12):427

126. Borges ÁV, Saraiva RM, Maffia LA (2015) Biocontrol of gray mold in tomato plants by *Clonostachys rosea*. Trop plant pathol 40(2):71–76

127. Tesfagiorgis HB, Laing MD, Annegarn HJ (2014) Evaluation of biocontrol agents and potassium silicate for the management of powdery mildew of zucchini. Biol Control 73:8–15

128. Ayent AG, Hanson JR, Truneh A (1992) Metabolites of *Gliocladium flavofuscum*. Phytochemistry 32(1):197–198

129. Howell CR (2006) Understanding the mechanisms employed by *Trichoderma* virens to effect biological control of cotton diseases. Phytopathology 96(2):178–180

130. Anitha R, Murugesan K (2005) Production of gliotoxin on natural substrates by *Trichoderma virens*. J Basic Microbiol 45(1):12–19

131. Stinson M, Ezra D, Hess WM, Sears J, Strobel G (2003) An endophytic *Gliocladium* sp. of *Eucryphiacordifolia* producing selective volatile antimicrobial compounds. Plant Science 165(4):913–922

132. Sun ZB, Li SD, Zhong ZM et al (2015) A perilipin gene from *Clonostachys rosea* f. *catenulata* HL-1-1 is related to sclerotial parasitism. Int J Mol Sci 16:5347–5362

133. Sun ZB, Sun MH, Li SD (2015) Identification of mycoparasitism-related genes in *Clonostachys rosea* 67-1 active against *Sclerotinia sclerotiorum*. Sci Rep 5:18169

134. Tsapikounis FA (2015) An integrated evaluation of mycoparasites from organic culture soils as biological control agents of sclerotia of *Sclerotinia sclerotiorum* in the laboratory. BAOJ Microbio 1(1):001

135. Yin G, Zhang Y, Pennerman KK et al (2017) Characterization of blue mold *Penicillium* species isolated from stored fruits using multiple highly conserved loci. J Fungi (Basel) 3(1):E12

136. Kozlovsky AG, Zhelifonova VP, Antipova TV (2013) Biologically active metabolites of *Penicillium* fungi. J Org Biomol Chem 1:11–21

137. Mamat S, Md Shah UK, Remli NAM et al (2018) Characterization of antifungal activity of endophytic *Penicillium oxalicum* T 3.3 for anthracnose biocontrol in dragon fruit (*Hylocereus* sp). Int J Agric Environ Res 4(1):65–76

138. Sreevidya M, Gopalakrishnan S, Melø TM (2015) Biological control of *Botrytis cinerea* and plant growth-promotion potential by *Penicillium citrinum* in chickpea (*Cicer arietinum* L.) Biocont Sci Technol 25:739–755

139. Sreevidya M, Gopalakrishnan S (2016) *Penicillium citrinum* VFI-51 as bio agent to control charcoal rot of sorghum (*Sorghum bicolor* (L.) Moench). Afr J Microbiol Res 10(19):669–674

140. De Cal A, Redondo C, Sztejnberg A et al (2008) Biocontrol of powdery mildew by *Penicillium oxalicum* in open-field nurseries of strawberries. Biol Control 47(1):103–107

141. Doveri F (2013) An additional update on the genus *Chaetomium* with descriptions of two coprophilous species, new to Italy. Mycosphere 4:820–846

142. Zhao SS, Zhang YY, Yan W et al (2017) *Chaetomium globosum* CDW7, a potential biological control strain and its antifungal metabolites. FEMS Microbiol Lett 364(3):fnw287

143. Hung PM, Wattanachai P, Kasem S et al (2015) Efficacy of *Chaetomium* species as biological control agents against *Phytophthora nicotianae* root rot in citrus. Mycobiology 43(3):288–296

144. Abdel-Azeem AM, Gherbawy YA, Sabry AM (2016) Enzyme profiles and genotyping of *Chaetomium globosum* isolates from various substrates. Plant Biosyst 150(3):420–428

145. Wanmolee W, Sornlake W, Rattanaphan N et al (2016) Biochemical characterization and synergism of cellulolytic enzyme system from *Chaetomium globosum* on rice straw saccharification. BMC Biotechnol 16(1):82

146. Xue M, Zhang Q, Gao JM et al (2012) Chaetoglobosin V_b from endophytic *Chaetomium globosum*: absolute configuration of chaetoglobosins. Chirality 24:668–674

147. Ye Y, Xiao Y, Ma L et al (2013) Flavipin in *Chaetomium globosum* CDW7, an endophytic fungus from Ginkgo biloba, contributes to antioxidant activity. Appl Microbiol Biotechnol 97:7131–7139

148. Seifert K, Morgan-Jones G, Gams W, Kendrick B (2011) The genera of hyphomycetes. CBS biodiversity series no. 9:1–997. CBS-KNAW Fungal Biodiversity Centre, Utrecht, Netherlands

149. Ruma K, Sunil K, Prakash HS (2014) Bioactive potential of endophytic *Myrothecium* sp. isolate M1-CA-102, associated with *Calophyllum apetalum*. Pharm Biol 52(6):665–676

150. Nguyen LTT, Jang JY, Kim TY et al (2018) Nematicidal activity of verrucarin A and roridin A isolated from *Myrothecium verrucaria* against Meloidogyne incognita. Pestic Biochem Physiol 148:133–143

151. Chavan SB, Vidhate RP, Kallure GS et al (2017) Stability studies of cuticle degrading and mycolytic enzymes of *Myrothecium verrucaria* for control of insect pests and fungal phytopathogens. Indian J Biotechnol 16:404–412

152. Lamovšek J, Urek G, Trdan S (2013) Biological control of root-knot nematodes (*Meloidogyne* spp.): microbes against the pests. Acta Agric Slov 101(2):263–275

153. Chen Y, Ran SF, Dai DQ et al (2016) Mycosphere essays 2. *Myrothecium*. Mycosphere 7(1):64–80

154. Barros DCM, Fonseca ICB, Balbi-Peña MIP et al (2015) Biocontrol of *Sclerotinia sclerotiorum* and white mold of soybean using saprobic fungi from semi-arid areas of Northeastern Brazil. Summa Phytopathologica 41(4):251–255

155. Brewer MT, Larkin RP (2005) Efficacy of several potential biocontrol organisms against *Rhizoctonia solani* on potato. Crop Prot 24:939–950

156. Krishnamoorthy AS, Bhaskaran R (1990) Biological control of damping-off disease of tomato caused by *Pythium indicum* Balakrishnan. J Biol Control 4(1):52–54

157. Bobba V, Conway KE (2003) Competitive saprophytic ability of *Laetisaria arvalis* compared with *Sclerotium rolfsii*. Proc Okla Acad Sci 83:17–22

158. Whipps JM, Sreenivasaprasad S, Muthumeenakshi S et al (2008) Use of *Coniothyrium minitans* as a biological control agent and some molecular aspect of sclerotial mycoparasitism. Eur J Plant Pathol 121:323–330

159. Zeng W, Wang D, Kirk W et al (2012) Use of *Coniothyrium minitans* and other microorganisms for reducing *Sclerotinia sclerotiorum*. Biol Control 60(2):225–232

160. Chitrampalam P, Wu BM, Koike ST et al (2011) Interactions between *Coniothyrium minitans* and *Sclerotinia minor* affect biocontrol efficacy of *C. minitans*. Phytopathology 101:358–366

161. Giczey G, Kerenyi Z, Fulop L et al (2001) Expression of cmg1, and exo-beta-1,3-glucanase gene from *Coniothyrium minitans*, increases during sclerotial parasitism. Appl Environ Microbiol 67:865–871

162. Tomprefa N, Hill R, Whipps J et al (2011) Some environmental factors affect growth and antibiotic production by the mycoparasite *Coniothyrium minitans*. Biocontrol Sci Techn 21:721–731

163. Goto BT, Silva GA, Assis D et al (2012) Intraornatosporaceae (Gigasporales), a new family with two new genera and two new species. Mycotaxon 119(1):117–132

164. Spatafora JW, Chang Y, Benny GL et al (2016) A phylumlevel phylogenetic classification of zygomycete fungi based on genome-scale data. Mycologia 108(5):1028–1046

165. Kehri HK, Akhtar O, Zoomi I et al (2018) Arbuscular mycorrhizal fungi: taxonomy and its systematics. Int J Life Sci Res 6(4):58–71

166. Brundrett MC (2009) Mycorrhizal associations and other means of nutrition of vascular plants: understanding the global diversity of host plants by resolving conflicting information and developing reliable means of diagnosis. Plant Soil 320(1–2):37–77

167. Chen M, Arato M, Borghi L et al (2018) Beneficial services of arbuscular mycorrhizal fungi—from ecology to application. Front Plant Sci 9:1270

168. Al-Askar AA, Rashad YM (2010) Arbuscular mycorrhizal fungi: a biocontrol agent against common bean *Fusarium* root rot disease. Plant Pathol J 9(1):31–38

169. Olawuyi OJ, Odebode AC, Oyewole IO et al (2014) Effect of arbuscular mycorrhizal fungi on *Pythium aphanidermatum* causing foot rot disease on pawpaw (*Carica papaya* L.) seedlings. Arch Phytopathol Plant Prot 47(2):185–193

170. Spagnoletti FN, Leiva M, Chiocchio V et al (2018) Phosphorus fertilization reduces the severity of charcoal rot (*Macrophomina phaseolina*) and the arbuscular mycorrhizal protection in soybean. J Plant Nutr Soil Sci 181(6):855–860

171. Zhang Q, Gao X, Ren Y et al (2018) Improvement of verticillium wilt resistance by applying arbuscular mycorrhizal fungi to a cotton variety with high symbiotic efficiency under field conditions. Int J Mol Sci 19(1):241

172. Mohamed I, Eid KE, Abbas MHH et al (2019) Use of plant growth promoting Rhizobacteria (PGPR) and mycorrhizae to improve the growth and nutrient utilization of common bean in a soil infected with white rot fungi. Ecotoxicol Environ Saf 171:539–548

173. Olowe OM, Olawuyi OJ, Sobowale AA et al (2018) Role of arbuscular mycorrhizal fungi as biocontrol agents against *Fusarium verticillioides* causing ear rot of *Zea mays*L. (Maize). Curr Plant Biol 15:30–37

174. Vierheilig H et al (2008) the biocontrol effect of mycorrhization on soilborne fungal pathogens and the autoregulation of the AM symbiosis: one mechanism, two effects? In: Varma A (ed) Mycorrhiza. Springer, Berlin, Heidelberg

175. Vos CM, Yang Y, De Coninck B et al (2014) Fungal (-like) biocontrol organisms in tomato disease control. Biol Control 74:65–81

176. Abdel-Fattah GM, El-Haddad SA, Hafez EE et al (2011) Induction of defense responses in common bean plants by arbuscular mycorrhizal fungi. Microbiol Res 166(4):268–281

177. Hafez EE, Abdel-Fattah GM, El-Haddad SA et al (2013) Molecular defense response of mycorrhizal bean plants infected with *Rhizoctonia solani*. Ann Microbiol 63(3):1195–1203

178. Chisholm ST, Coaker G, Day B, Staskawicz BJ (2006) Host-microbe interactions: shaping the evolution of the plant immune response. Cell 24;124(4):803–14
179. Jones JD, Dangl JL (2006) The plant immune system. Nature 444(7117):323–329
180. Dangl JL, Jones JD (2001) Plant pathogens and integrated defence responses to infection. Nature 411(6839):826–833
181. Altenbach D, Robatzek S (2007) Pattern recognition receptors: from the cell surface to intracellular dynamics. Mol Plant Microbe Interact 20(9):1031–1039
182. Schwessinger B, Zipfel C (2008) News from the frontline: recent insights into PAMP-triggered immunity in plants. Curr Opin Plant Biol 11(4):389–395
183. Niu D, Xia J, Jiang C, Qi B, Ling X, Lin S, Zhang W, Guo J, Jin H, Zhao H (2016) *Bacillus cereus* AR156 primes induced systemic resistance by suppressing miR825/825* and activating defense-related genes in *Arabidopsis*. J Integr Plant Biol 58(4):426–439
184. Speth C, Willing E-M, Rausch S, Schneeberger K, Laubinger S (2013) RACK1 scaffold proteins influence miRNA abundance in Arabidopsis. The plant J 76(3):433–445
185. Katiyar-Agarwal S, Jin H (2010) Role of small RNAs in host-microbe interactions. Annu Rev Phytopathol 48:225–246
186. Weiberg A, Wang M, Lin FM, Zhao H, Zhang Z, Kaloshian I, Huang HD, Jin H (2013) Fungal small RNAs suppress plant immunity by hijacking host RNA interference pathways. Science 342(6154):118–123
187. Göhre V, Robatzek S (2008) Breaking the barriers: microbial effector molecules subvert plant immunity. Annu Rev Phytopathol 46:189–215
188. Fu ZQ, Dong X (2013) Systemic acquired resistance: turning local infection into global defense. Annu Rev Plant Biol 64:839–863
189. Moussa TAA (1999) Towards the biological control of some root-rot fungal pathogens of sugarbeet in Egypt. Ph.D. Thesis, Cairo University
190. Moussa TAA (2002) Studies on biological control of sugarbeet pathogen *Rhizoctonia solani* Kühn. J. Biol. Sci. 2:800–804
191. Elazzazy AM, Almaghrabi OA, Moussa TAA, Abdel-Moneim TS (2012) Evaluation of some plant growth promoting rhizobacteria (PGPR) to control *Pythiumaphanidermatum* in cucumber plants. Life Sci. J. 9(4):3147–3153
192. Moussa TAA, Almaghrabi OA, Abdel-Moneim TS (2013) Biological control of the wheat root rot caused by *Fusarium graminearum* using some PGPR strains in Saudi Arabia. Ann Appl. Biol. 163:72–81

Production, Formulation and Application of Fungi-Antagonistic to Plant Nematodes

Ezzat M. A. Noweer

Abstract Nematodes are invertebrate roundworms that inhabit most environments on earth. They comprise one of the largest and most diverse groups of multicultural organisms in existence. Plant parasitic nematodes spend at least some part of their lives in soil, one of the most complex environments. Their activities are not only influenced by variation in soil physical factors such as temperature, moisture and aeration but also by a vast array of living organisms, including other nematodes, bacteria, fungi, algae, protozoon's, insects, mites and other soil animals. The biological component of the soil ecosystem is particularly important in limiting and more or less stabilizing nematode populations. For biological control to be successful, it must be supported by a backbone of basic ecological research. Nematophagous fungi can be fungal egg-parasites, nematode-trapping fungi that capture nematodes using modified hyphal traps, or endoparasitic that parasitizes the nematode by means of small conidia or zoospores. The history of attempts to use predaceous fungi to control plant- parasitic nematodes had been the subject of several reviews. Most research on microbial agents that attack nematodes in soil has concerned fungi, especially those that form traps to ensnare their prey. There are many attempts to production and formulation of the nematophagous fungi to plant nematodes. Many factors affecting for nematophagous fungi enhancement as Indigenous Nematophagous Fungi Present in the Soil, Time after Nematicide Application and Ecological Habitat. Duration of Cover Crop for Enhancing Nematophagous Fungi in Field Conditions. Biological control agents are generally produced in commercial quantities by one of two fermentation methods. The oldest and perhaps the simplest is solid substrate fermentation, which involves growing the microorganism on the surface of a substrate (e.g. bran) that has been impregnated with nutrients. However, for most other applications, it has been largely superseded by submerged culture fermentation, in which micro-organisms are grown in a liquid medium. A number of types of formulation, including dusts, granules, wet table powders and liquids have been used in biological crop protection products, but granular formulations are generally considered to be most suitable for micro-organisms that are to be applied to soil. The dry nature of these formulations means that an antagonist of nematodes needs to have the capacity

E. M. A. Noweer (✉)
Plant Pathology Department, National Research Centre, Dokki, Cairo, Egypt
e-mail: enoweer@hotmail.com

to survive desiccation, if it is to be seriously considered for commercial development as a biological control agent. Granular biological products have traditionally been produced by blending the organism with a carrier such as clay or ground corn cobs and alien introducing a material to bind the organism to the carrier, but in recent years there has been considerable interest in encapsulating biological control agents in gallants such as sodium alginate. Formulation of microbial products in forms which have extended shelflives and which can be applied to soil using conventional farm equipment is likely to present a major challenge to those interested in commercializing antagonists of nematodes. Organisms which cannot be dried without loss of viability are likely to be difficult to formulate and are unlikely to retain their activity in storage for more than a few months unless expensive storage conditions are employed.

Keywords Plant nematodes · Nematophagous fungi · Formulation · Biological products · Storage and application · Ecosystem · Microbial products

1 Introduction

Nematodes are invertebrate roundworms that inhabit most environments on earth. They comprise one of the largest and most diverse groups of multicultural organisms in existence. Plant parasitic nematodes spend at least some part of their lives in soil, one of the most complex environments. Their activities are not only influenced by variation in soil physical factors such as temperature, moisture and aeration but also by a vast array of living organisms, including other nematodes, bacteria, fungi, algae, protozoon's, insects, mites and other soil animals. This biological component of the soil ecosystem is particularly important in limiting and more or less stabilizing nematode populations. In the last few years' man tended to reduce the use of pesticides to control pests for a safer environment. The uses of these toxic substances accumulate in the food and cause a serious potential hazard for human health. There can be no doubt that the 'Nematicide crisis' has created a situation where an increase in the resources allocated to research on biological control of nematodes is essential. However, it is important that the area in which research is needed is not interpreted too narrowly.

For biological control to be successful, it must be supported by a backbone of basic ecological research. In this book chapter, we aimed to have a sound knowledge of the population dynamics of nematodes, of the threshold levels needed to cause economic damage, of the role that parasites, predators and other soil organisms play in regulating nematode populations, and of the complex interrelationships that occur between nematodes and other components of the soil ecosystem. In the long term, the additional knowledge accumulated from such studies is likely to impact on many areas of Nematology and may in fact prove more valuable than any new biological control measures that might be developed.

2 What Are Nematophagous Fungi?

Nematophagous fungi are fungi that feed on nematodes. These fungi can be fungal egg-parasites, nematode-trapping fungi that capture nematodes using modified hyphal traps, or endoparasites that parasitize the nematode by means of small conidia or zoospores. There are various ways for soil-borne fungi to suppress nematode multiplication. A detailed review of fungi as biocontrol agents against plant-parasitic nematodes has been published by Kerry and Jaffee [1] and others. In summary, there are five mechanisms that fungi use to suppress nematodes. Some of these interactions are direct whereas others are indirect.

The direct mechanism is performed by:

(1) Fungi that feed on nematodes directly, known as nematophagous fungi; fungi interact with nematodes in an indirect manner by several mechanisms including.
(2) Fungi that kill nematodes by mycotoxin [2].
(3) Through the destruction of the feeding sites of sedentary nematodes in roots [3].
(4) Fungi that are nonpathogenic to plants but compete with nematodes in roots and significantly reduce nematode multiplication [4]. Many of these fungi are used as potential nematode biocontrol agents.

The indirect mechanism is performed by:

(5) Mycorrhizal fungi improve the growth of nematode infected plants and may also affect nematode development [5].

According to a survey of nematophagous fungi in Ireland by Gray [6], nematophagous fungi were found in all of the habitats examined, among which, permanent pasture, coniferous leaf litter, and coastal vegetation had the most frequent incidence of nematophagous fungi. Other habitats examined by Gray included coniferous leaf litter, old and partly revegetated dung, permanent grassland pasture, cultivated land, moss cushions, decaying vegetation and compost, and peat land [6]. In addition, many other studies i.e., Barron [7] also supported the hypothesis that nematophagous fungi are widely distributed and have great potential to be explored as biocontrol agents.

However, as stated by Kerry [8]: "The successful introduction of such an agent depends on whether a suitable niche for the microorganism exists or can be created and until we know much more about the factors that affect the activity of nematophagous fungi in soil, their full potential as control agents for nematodes will not be realized".

3   Groups of Nematophagous Fungi According to Their Feeding Habits

3.1   Nematode-Trapping Fungi

Facultative fungi that form trapping structure to trap nematodes, there are 6 types of traps reported by Barron [7]. Traps come in many forms: **adhesive hyphae, networks, knobs, rings, constricting rings** and **non-constricting rings**.

Adhesive hyphae: e.g. Zygomycotina

   Stylopage, Cystopage

Adhesive traps: e.g. Deuteromycota

   Monacrosporium cionopagum (branches)
   M. ellipsosporium (knobs)
   Arthrobotrys oligospora (networks)

Non-adhesivetraps: e.g. Deuteromycota

   Arthrobotrys dactyloides (constricting ring)
   Dactylella leptospora (non-constricting ring)

Fungi that form traps:
   Several species of soil-dwelling fungi produce traps to ensnare nematodes before they infect them. Some trapping fungi can proliferate in soil in the absence of nematodes while others are more dependent on nematodes as a nutrient source for growth (Figs. 1 and 2).

3.2   Facultative Parasitic Fungi Attacking Sedentary Stages of Nematodes

These are facultative fungi that are commonly soil saprophytes, and are opportunistic fungi isolated from the sedentary stages (female and egg stages) of sedentary nematodes such as *Heterodera, Globodera*, and *Meloidogyne* (Fig. 3). They do not form specialized infection structures except appressoria. They can survive and proliferate in soil in the absence of nematodes.
   Eg. Hyphomycotina

Acremonium, Cylindrocarpon, Fusarium, Paecilomyces, Verticilium

   The above images were scanned from the "Parasites and Predators of Plant-Parasitic Nematodes" slide set that was prepared by the Education and Biological Control of Nematodes Committees and issued by the Society of Nematologists in 1990.

Adhesive hyphae   Adhesive network

Adhesive knob   Constricting ring

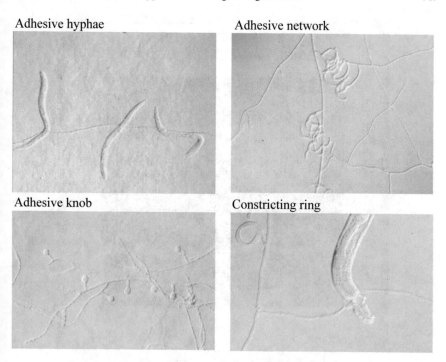

Fig. 1 Some types of traps, Kerry [8]

3.3 Endoparasitic Fungi

These are obligate parasitic fungi that have limited growth in soil outside the colonized nematode cadaver. They can infect vermiform nematodes by producing **adhesive spores attached to cuticle** of passing nematodes (Figs. 4 and 5).

Eg. Hyphomycotina

> *Hirsutella rhossiliensis*
> *Drechmeria coniospora*
> *Verticilium* spp.

Some can infect vermiform nematodes by producing **conidia spores** that can be **ingested** by nematodes (Fig. 6).

Harposporium anguillulae

Some can **infect vermiform nematodes** by producing **motile zoospores that encyst** on the nematode's surface (Fig. 7).

Eg. Oomycota

> *Myzocytium* spp.

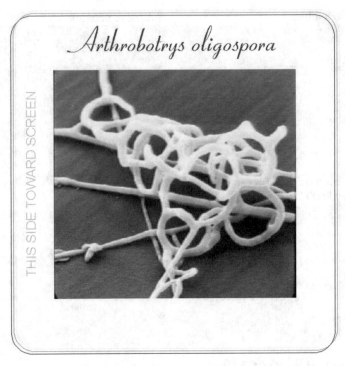

Fig. 2 Adhesive traps networks *Arthrobotrys oligospora*, Kerry [8]

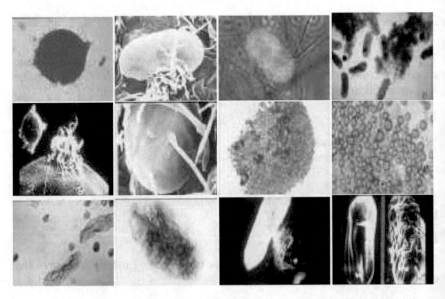

Fig. 3 "Fungal Parasites of Sedentary Females and Eggs" by the Nematologists Society-1990

Fig. 4 Adhesive conidium's of *Hirsutella rhossiliensis* on conidiophores, Timper and Brodie [9]

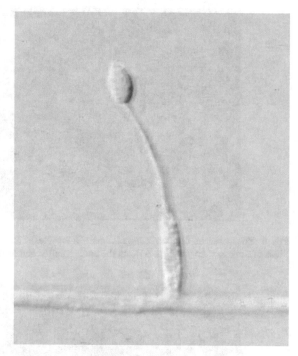

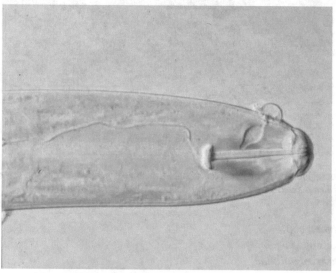

Fig. 5 Conidium's of *H. rhossiliensis* on the cuticle of a nematode with infection bulb inside the body cavity, Jaffee and Muldoon [10]

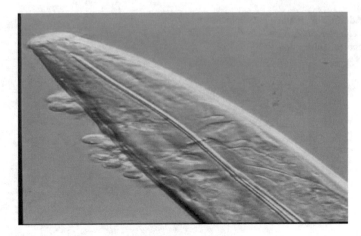

Fig. 6 Zoospores of *Catenaria anguillulae* (Oomycetes) encysted on *Xiphinema americanum*. Signs of infection are apparent beneath the cuticle, Jaffee and Muldoon [10]

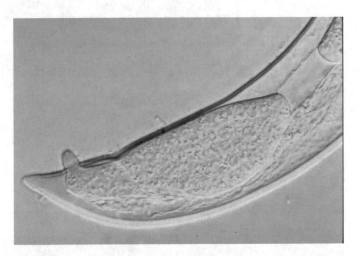

Fig. 7 Sporangium and discharge tube (through which zoospores swim into the soil solution) in tail of *Xiphinema americanum*, Jaffee and Muldoon [10]

Lagenidium spp.
Chytridiomycota
Catenaria anguillulae

Some can **infect sedentary nematodes** when the nematodes were **exposed on the root** surface.

Eg. Oomycota

Nematophthora gynophila

4 Isolating of Fungi-Antagonistic to Plant Nematodes

4.1 Isolation of Fungi Using Nematode Water Agar (NWA) Medium

Twenty grams of agar were added to one liter of distilled water and dissolved if needed on a water bath. The medium was then poured in 10 ml aliquots into a series of glass tubes and sterilization was made by autoclaving for 20 min at 15 lbs pressure. The idea of using such a poor medium is to cut down the growth of other moulds such as Mucorales and the more vigorously growing Hyphomycetes in order to give chance to the more delicate growing nematophagous fungi. Addition of steriled nematodes to culture plates stimulates trap formation in nematophagous fungi [11].

4.2 Isolation of Fungi Using Potato Dextrose Agar (PDA) Medium

Potato dextrose agar (PDA) was prepared from 200 gm peeled and sliced potato + 20 gm D-Glucose + 15 gm agar + 100 ml distilled water. For isolating of fungi from the nematode- infested soil samples, P.D.A medium was used according to Harwing et al. [12]. Pure cultures from different isolates were grown on PDA slants cultured in the dark at 25 °C using the single colon and hyphal tip technique [13]. The Propagating purified fungi cultures were renewed monthly.

4.3 Identification of the Isolated Fungi

Identification of the more vigorous fungi isolated on PDA medium was made by examining inocula 2 days after inoculation as the growth of these fungi was quick. Eight days after incubation, a single conidiophore from each culture was transferred to PDA slants to purificate and incubates at 25 °C. After 15 days, all isolates were stained with methyl blue and examined microscopically. Identification of fungi was made according to Barnett [14] and Alexopoulos [15].

Identification of the more delicate fungi and nematophagous fungi grown on Nematode Water Agar (NWA) medium was accomplished by referring to the detailed descriptions and keys offered by Barnett [14], Alexopoulos [15] and Noweer [16]. Trapping system, septation of the hyphae, spores in clusters or single on conidiophores, shape of spores and number of spore cells were the bases of identification.

4.4 Ecology of Nematophagous Fungi

An ecology study of nematophagous fungi conducted by Gray [17] revealed that different types of nematophagous fungi have different edaphic preferences. The saprophytic NTF (formed adhesive nets) are found in soil with low organic matter and low moisture due to their saprophytic nature. When nutrients or moisture condition improved, the saprophytic NTF are able to compete with other soil organisms by feeding on the expanding nematode population. In contrast, NTF that form rings are more common in soil with high organic matter and moisture. Endoparasitic fungi that produce conidia are strongly influenced by organic matter. While most of the NTF (except those that formed adhesive branches) are not affected by nematode densities, Endoparasitic fungi that form injective spores are nematode-density dependent. In general, the conidia-forming Endoparasitic were isolated from samples with comparatively high soil moisture and low pH. Little is known about edaphic preference of the nematophagous fungi with unmodified adhesive hyphae, except that they are more frequently recovered from soils with higher pH. Table 1 summarizes the soil factors which affect different nematophagous fungi.

Base on their ecological preferences, nematode-trapping fungi (NTF) are separated into two groups: saprophytic and parasitic NTF [18].

Saprophytic NTF—form 3-dimensional-network traps in response to the presence of nematodes. Under low nematode population densities, they remain saprophytic. Therefore, they are regarded as inefficient nematode-trappers.

Parasitic NTF—have low saprophytic ability, but form traps spontaneously. This group consists of NTF that form constricting rings, adhesive branches and are more effective nematode trappers than the saprophytic NTF [19].

Table 1 Effect of soil edaphic factors on distribution of nematophagous fungi, Gray [17]

Nematophagous fungi	Organic matter	pH	Moisture	Nematode densities
Nematode-trapping	NS[z*]	Low	NS	NS
Net	Low[y]	Low	Low	NS
Ring	High[x]	Low	High	NS
Adhesive hyphae	NS	High	NS	NS
Adhesive branch	NS	NS	NS	High
Adhesive knobs	NS	Low	NS	NS
Endoparasitic	High	Low	High	High

[*z]NS = effects of the edaphic factors is not significant
[y]Low = lower value of the edaphic factor is preferred by the group of nematophagous fungi
[x]High = higher value of the edaphic factor is preferred by the group of nematophagous fungi

4.5   Occurrence and the Morphological Identity of Some Antagonistic Fungi Isolated from Soils Infested with Root-Knot Nematode

The study aims towards the studying of microbial agents that attack root-knot nematodes. Soils which famous by its heavy organic manure application riches in the antagonistic fungi. Occurrence of nematode-antagonistic fungi in Abd-Elsamad village, Giza, sandy soils which are well-known by its heavy organic manure application were studying by collecting soil samples from fruit orchards, field crop and vegetables during two consecutive years. Seven species of nematode-trapping fungi *Arthrobotrys conoides*, *A. dactyloides*, *A. oligospora*, *Dactylaria brochopaga*, *D. Thaumasia* var.*longa*, *Dactylella gephyropaga* and *Stylopaga hadra* were isolated from the root-knot nematode-positive samples. All of these fungi were identified, described and photographed. Four nematode-endoparasitic fungi *Catenaria angiulella*, *Cephalosporium balanoides*, *Haptoglosa heterospora* and *Harposporium anguillula* were isolated from nematode bodies. All of these fungi were identified, described and photographedand *Verticilium chlamydosporium* was isolated from egg-masses of *Meloidogyne incognita* (Figs. 8, 9, 10, 11, 12, 13, 14, 15, 16, 17, 18 and 19), Noweer [16].

Fig. 8 (1–3): *Stylopaga hadra,* Noweer [16]

Fig. 9 (4–6): *Arthrobotrys conoides*, Noweer [16]

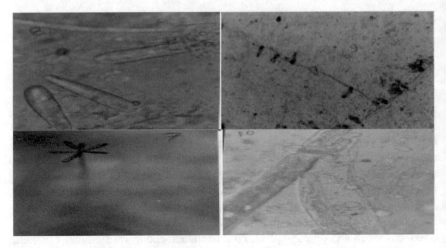

Fig. 10 (7–10): *A.dactyloides,* Noweer [16]

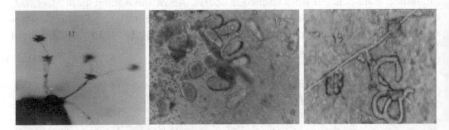

Fig. 11 (11–13): *A. oligospora,* Noweer [16]

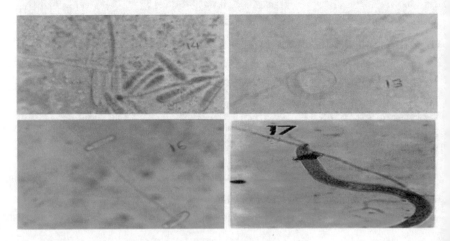

Fig. 12 (14–17): *Dactylaria brochopaga,* Noweer [16]

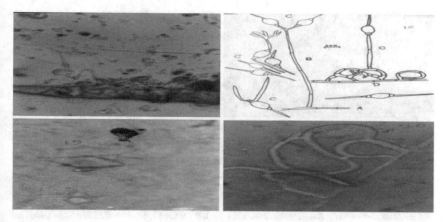

Fig. 13 (18–21): *D. Thaumasia* var.*longa,* Noweer [16]

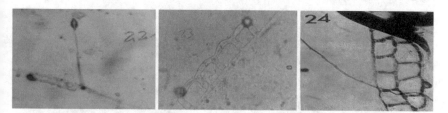

Fig. 14 (22–24): *Dactylella gephyropaga,* Noweer [16]

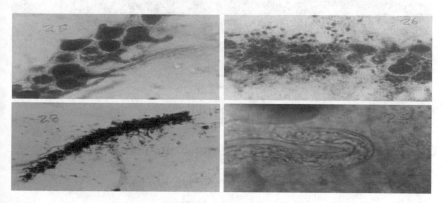

Fig. 15 (25–28): *Catenaria angiulella,* Noweer [16]

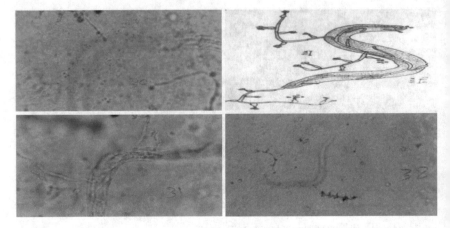

Fig. 16 (29–32): *Cephalosporium balanoides,* Noweer [16]

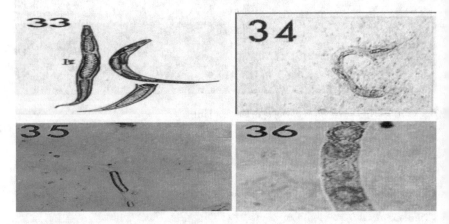

Fig. 17 (33–36):*Haptoglosa heterospora,* Noweer [16]

5 Production, Formulation of Fungi-Antagonistic to Plant Nematodes

5.1 The History of Using Fungi as a Biocontrol Agent to Plant Nematodes

The history of attempts to use predaceous fungi to control plant- parasitic nematodes had been the subject of several reviews. Most research on microbial agents that attack nematodes in soil has concerned fungi, especially those that form traps to ensnare their prey. The nematode-trapping fungi; the first nematode- trapping fungus, the

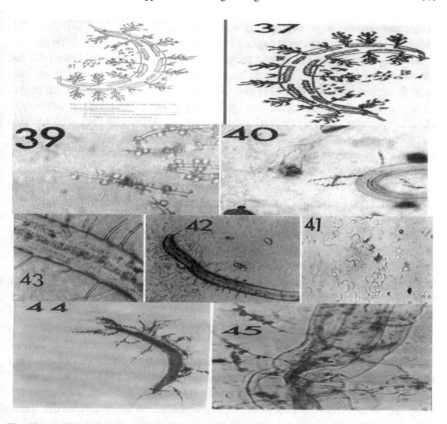

Fig. 18 (38–45): *Harposporium anguillula,* Noweer [16]

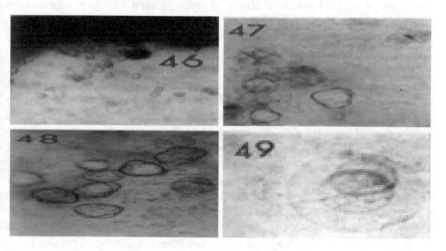

Fig. 19 (46–49):*Verticiliumchlamydosporium,* Noweer [16]

ubiquitous *Arthrobotrys oligospora*, was described by Fersenius [20]. Zopf [21] observed that the fungus could capture motile nematodes and parasitize them.

The major genera of predacious fungi, *Arthrobotrys, Dactylaria, Dactylella* and *Monacrosporium*, probably contain more than 100 species. Many of these have been described by Drechsler [22] and Galuilina [23] have shown that the majority or predatory fungi lies under four speices *Dactylaria, Dactylella, Arthrobotrys, Monacrosporium*. In a survey conducted in Egypt by Aboul-Eid [24], who reported that *Arthrobotrys conoides, A. oligospora, Dactylaria brochophaga* and *D. thaumasia* var, *longa* were found in organic-manure soils. Rao and Malek [25] found that the fungi *Arthrobotrys dactyloides, A. arthrobotryoides* and *Dactylaria thaumasia* slowed the population increase of *Pratylenchus penetrans* on alfa-alfa in the laboratory and greenhouse. Of the three tested fungi, *A. dactyloides* was the most effective antagonist.

Mai and Chen [26] used the nematode trapping fungi *Arthrobotrys superba, A. dactyloides, A. arthrobotryoides,* and *Dactylella doedycoides* to reduce the penetration of alfa- alfa roots by *Pratylenchus penetrans* sterilized soil. Nematode- destroying fungi play a major role in recycling the carbon, nitrogen, and other important elements from the rather substantial of nematodes which browse on microbial primary decomposers.

In a greenhouse studies Godoy et al. [27] indicated that *Peacilomyces lilacinus* and*Verticilium chlamydosporium* were effective in reducing *M. arrenaria* infestations. The classical work in that area was carried out by Linford et al. [28]. They added chopped green pineapple tops to nematode infested soil in pots and estimated nematode populations and the activity of the predaceous fungi. When certain soil fungi are cultured they produce some toxic metabolites in the culture media. These toxic metabolites have been used by different workers for nematode control [29–32]. Rosenzweiget al. [33] tested the ability of seven predator's fungi to capture nine different nematodes which included free living, ecto and endo plant-parasitic nematodes. They found that the fungi displayed no selectivity at all with each fungus being able to trap and consume all the different nematodes tested.

Niblack and Hussey [34] reported that a nematode- trapping fungus, *A. amerospora,* combined in three commercial preparations with *Rhizobium japanicum* in oculum was evaluated for control of *Heterodera glycines* on soybean (Glycine max) in the field and greenhouse. He found that *A. amerospora* was not considered a probiocontrol agent for *Heteroderaglycines* on soybean.

Paecilomyces lilacinus has shown promise as a biological control agent several nematode species [35]. Also, Dube and Smart [36] found that the fungus *P. lilacinus* penetrated the nematode egg and destroyed the embryo; it also attacked and grew inside developing females resulting in their death. Isolates known to be pathogenic to nematode eggs have sometimes failed to provide any nematode control, despite being present at relatively high populations [37]. Murray and Wharton [38] reported that the nematode trapping fungus *A. oligospora* traps and invades all the free living juvenile stages of the trichostrongyle nematode *Trichostrongylms colubriformis*.

In a greenhouse experiment conducted to evaluate the effect of nematophagous fungi *Arthrobotrys oligospora* and pigeon droppings as a soil amendment on the population dynamics of *Meloidogyne incognita* on Muskmelon, Ali [39] indicated that was a trend towards nematode population decrease with greater efficiency when the fungus was introduced into soil 2 weeks prior to planting and nematode inoculation. Inoculums density of *A. oligospora* was positively correlated with number of juveniles and galls per gram of root.

Voss and Wyss [40] investigated the potential of the nematophagous endoparasitic fungus *Cetenaria anguillula* as a control agent against several plant parasitic nematodes, and special emphasis was placed on variability studies in which 19 isolates were compared. For this purpose, an easy and reproducible biotest system was developed in order to quantity the specific virulence of isolates under defined in vitro conditions.

A field inoculated with *Heterodera schachtii* for about 12 year's natural inhabitants and continuously cropped with host plants exhibited a gradual decline in sugar- beet cyst nematode damage. After a decade of cultivation, significant differences could no longer be obtained between control plots and those treated with nematicides. Soil analysis revealed a remarkable variety of fungal antagonists including parasites of females and eggs; and at least 6 species of nematode- trapping fungi. There, *Arthrobotrys* spp. was present at high densities and one or more could be isolated from any 0.5 gm of soil. Averages of 44% of all cysts recovered following a cabbage crop were infected by fungi and 57% of the eggs within these cysts were parasites [41].

The effect of culture filtrates (CF) of *F. oxysporium niveum*, *Macrphomina phaseolina*, *Sclerotium battaticola* and *Trichoderma herzianum*, obtained from infected cantaloupe roots, on juveniles mortality and egg hatch of *M. incognita* was studied by Ali and Barakat [42]. Their results indicated that, the culture filtrates of all the tested fungi demonstrated toxic effect and killed the nematode juveniles and inhibited the egg hatch to a varying degree. Also, numbers of *M. incognita* juveniles, eggmasses and root galls as well as disease severity of the pathogenic fungi were greatly suppressed by the addition of the antagonistic fungus *Trichoderma herzianum* to the soil.

Hertz [43] reported that conidia of *A.oligospora* germinated directly into adhesive traps when applied close to cow faces on water agar plates, the conidial trap is considered a survival structure enabling the fungus to overcome fungi stasis. Traps adhere to the surface of passing nematodes, thus facilitating the spread of the fungus, before penetration of the nematode cuticle and immobilization of the nematode take place.

In a greenhouse experiment, Hoffmann and Sikora [44] tested the influence of organic matter on the efficacy of nematode-trapping fungi in reducing the early penetration of rape roots by *Heterodera schachii*. Field application of egg and larval parasitic fungi and chemicals for controlling root- knot nematodes on some medicinal herb as mentioned by Park et al. [45]; they found that the number of root- gall, egg mass and nematode density of *Paeonia albifora* was suppressed in *P.lilacinus* treated plots. The fungi *A. dactyloides*, *A. oligosora*, *Macrosporium ellipsosporium*, and*M. cionopagum*, killed most of the *Pratylenchus penetrans* adults and juveniles added to the fungus cultures [9].

Ali et al. [46] reported that all antagonistic fungi *A. oligospora, A. conoides* significantly reduced the number of hatching eggs. Anter et al. (1994) found that, in a greenhouse experiment, *A. conoides* and *A. oligospora* showed the highest effect on reducing *M. incognita* numbers during the first 4 weeks. Whereas *Paecilomyces lilacinus, Verticiliumchlamydosporium*and *Trichoderma herzianum* gave their maximum effect, 8 weeks after planting. Reddy and Sharma [47] studied the effect of fungal cultures and nematicides on larval penetration, root gall and egg- masses counts of *M. incognita* on tomato. The results indicated that the fungal cultures (*Paecilomyces lilacinus, Aspergillus niger* and *Fusarium oxysporium*) reduced the penetration of *M. incognita* in all the treated plots in comparison to control. Also the root gall and egg-mass production was low in all the treatments and were not different from each other superior over control. *A.oligospora* and *A. conoides* showed high effect on the activity of the second stage juvenile of *M. incognita* compared to the control treatment [48].

The nematophagous fungus *A. oligospora* which can live saprophytic ally as well as predatorily forms sticky reticulate traps in the presence of linig nematodes to kill and consume them [49]. When the susceptibility of *M. javanica* and *Heterodera schachtii* to the nematode- trapping fungi *Monacrosporium ellipsosporium* and *M. cionopagum* was compared by Jaffee and Muldoon [10], results indicated that *Meloidogyne* spp. in general, were more susceptible than *H. schachtii* to *M. ellipsosporium*and *M. cionopagum.*

Granular formulations of *Dactylella candida* and *Arthrobotrys dactyloides* were prepared by encapsulating different quantities of fungal biomass in alginate, or by subjecting encapsulated biomass to further fermentation. Results of experiments with these formulations showed that the presence of nutrients and the quality and quantity of biomass in granules determined their level of activity against nematodes [50].

Reddy et al. [51] reported that *Trichoderma harzianum* (2 and 4 g dosages) in combination with neem (*Azadirachta indica*), Karanj *(Pongamia pinnota)* and castor oil *(Ricinus communis)* cakes (at 20/40 g dosages) was effective in increasing the growth of acid lime (*Citrus aurantifolia*) seedlings and reducing the population of *Tylenchulus semipenetrans* both in soil and root experiments. The parasitization of citrus nematode females with *T. harzianum* increased in the presence of the oil cakes.

Safullah [52] indicated that *T. harzianum* and *Verticilium chlamydosporium,* previously isolated from *Globodera rostochiensis,* were tested against *G. pallida* and *G. rostochiensis* males the toxic metabolites from the fungi, released into the medium, killed males on agar plates. *V. chlamydosporium* was more effective than *T. harzianum.* Aboul-Eid et al. [53] studied the effects of different fungal and bacterial treatments on *Meloidogyne incognita* infecting Tomato. They found that the nematode trapping fungus*Dactylaria brochopaga* the more effective on reproduction of root-knot nematode *M. incognita.Arthrobotrys oligospora, A. Conaides, Arthrobotrys* sp.*, Dactylaria shelensis, Dactylaria* sp. And *Monacrosporium bembicodes* were tested by Duponnois et al. [54] for their trapping ability against *Meloidogyne mayaguensis.* Most of the *Arthrobotrys* strains and one *Dactylaria* strain decreased the development of the nematodes. The growth of the tobacco plants was consequently improved, but certain fungi have proper phytostimulant effects by acting on the soil structure through commensalism mechanisms with the plant roots.

The effect of different fungal filtrates against *Meloidogyne incognita* and *M. javanica* were studied in vitro by Sankaranarayanan [55]. The culture filtrates of *Trichoderma harzianum* and *T. kaningii* recorded 100% mortality within 24 h of exposure in both the nematode species. Culture filtrates of different fungi under study, except *T. harzianum*(PDBCTH 7 and PDBCTH 8) against *M. incognita* and*T. harzianum* (PDBTH 8) against *M. javanica*, recorded 100% nematode mortality at 96 h of exposure. Abdel-Bariet al. [56]. Reported that the effect of different fungal filtrates on *Meloidogyne incognita* larvae in laboratory bioassay tests. They found that the fungus filtrate of *Trichoderma viridi* was the more effective on mortality of root-knot nematode M. *incognita*. Effect of temperature, PH and nematode starvation on induction of rings of *Dactylaria brochopaga* was studied by Bandyopadhyay and Singh [57]. The results showed that there was no ring formation at 10 and 35 °C. The optimum PH for ring formation in presence of *Hoplolaimus indicus* was 7; the percentage increasing with incubation time. The ring formation of *D. brochopaga* in the presence of *H. indicus* and *M. incognita* was adversely affected by increasing the starvation period of the nematodes.

Al-Shalaby and Noweer [58] tested the effects of Five Plant Extracts on the Reproduction of Root-Knot Nematode *Meloidogyne incognita* Infested Peanut under Field Condition. They found that the neem extract was the more effective on reproduction of root-knot nematode *M. incognita*. Noweer andAl-Shalaby [59] assessed the effects of Some Aromatic and Medicinal Plants as Amendments against *Meloidogyne incognita* on Peanut under Field Condition. They reported that all of the medicinal plants as Amendments were effective on reproduction of root-knot nematode *M. incognita*. Noweer [60] studied the efficacy of the nematode-trapping fungus *Dactylaria brochopaga* and biofertilizers, on controlling the root-knot nematode *Meloidogyne incognita* infecting Tomato. He found that addition of the fungus mixed with biofertilizers were more effective on reproduction of root-knot nematode *M. incognita* than the biofertilizers or the fungus alone. Noweer and Hasabo [61] assessed the effect of different management practices for controlling the root-knot nematode *Meloidogyne incognita* on squash. They reported that all of the management practices were effective on reproduction of root-knot nematode *M. incognita*. Hasabo and Noweer [62] managed the root-knot nematode *Meloidogyne incognita* on eggplant using some plant extracts. They found that all of the plant extracts were effective on reproduction of root-knot nematode *M. incognita* especially the neem extract. Aboul-Eidet al. [63] studied the effect of a nematode trapping fungus *Dactylaria brochopaga* on *M. incognita infesting* olives and coconut palms in Egypt. They found that the fungus were effective on reproduction of root-knot nematode M. *incognita*.

Noweer and El-Wakeil [64] assessed the combination of the entomopathogenic nematode *Heterorhabditis bacteriophora* and the nematode-trapping fungi *Dactylaria brochopaga* and *Arthrobotrys conoides* for controlling *Meloidogyne incognita* in tomato fields. They found that addition of the fungus mixed with the entomopathogenic nematode were more effective on reproduction of root-knot nematode M. *incognita*. Noweer and Dawood [65] evaluated the efficiency of propolis extract on faba bean plants and its role against nematode infection. The data revealed that the propolis extract as soil drench reduced the juvenile-*Meloidogyne sp.*-population density per one kg soil and number of root-galls per one gm roots especially at the higher concentration (1000 mg/L). Noweer [66] investigated using the Nematode Biocide Dbx-1003 for controlling Citrus Nematode Infecting Mandarin, and Interrelationship with Co inhabitant fungi; who found that addition of the Nematode Biocide Dbx-1003 was effective on reproduction of Citrus Nematode. An Egyptian population of *Dactylaria brochopaga* proved to be more effective as nematode-antagonist since it negatively affected nematode population larvae through production of traps which capture the larvae and dissolve nematode outer cuticle and digest the inner content of the victim [16].

Noweer and Aboul-Eid [67] studied the Biological control of root-knot nematode *Meloidogyne incognita* infesting cucumber *Cucumis sativus* L. cvs. Alfa by the nematode-trapping fungus *Dactylaria brochopaga* under field conditions. They found that the nematode-trapping fungus *D. brochopaga* alone or in combination with yeast, molasses and vermiculite reduced the juvenile-*Meloidogyne incognita*-population density per one kg soil and number of root-galls per one gm roots. Noweer and Al-Shalaby [68] evaluated nematophagous fungi *Dactylaria brochopaga* and *Arthrobotrys dactyloides* against *Meloidogyne incognita* infesting peanut plants under field conditions. They found that population densities of *M. incognita* in soil were significantly reduced in all treatments compared with control, as well as gall formation on peanut roots. Noweer [69] studied the effects of some nematode-trapping fungi on the root-knot nematode *Meloidogyne* sp. infesting white bean *Phaseolus vulgaris* and sugar beet *Beta vulgaris sp.vulgaris* under field conditions. He found that the fungus *Dactylaria brochopaga* affected the development and reproduction of *Meloidogyne incognita* on white bean and sugar beet under field conditions.Aboul-Eid et al. [70] evaluated the impact of the nematode-trapping fungus, *Dactylaria brochopaga* as a biocontrol agent against *Meloidogyne incognita* infesting Superior grapevine. They found that all treatments significantly reduced *M. incognita* J_2 in soil and number of root galls compared with the untreated control. Significant yield increases have been observed with all treatments compared with the untreated control. Spores suspension twice applications gave the highest yield production.

Noweer [71] studied in a field trial to use the nematode-trapping fungus *Arthrobotrys dactyloides* to control the root-knot nematode *Meloidogyne incognita* infesting bean plants. He found that the fungus *Arthrobotrys dactyloides were* affected on the development and reproduction of *Meloidogyne incognita* on bean plantsunder field conditions. The fungus, *Verticiliumchlamydosporium* is among the most effective biological control agents against plant-parasitic nematodes. *V. chlamydosporium* parasitizes eggs of root-knot and cyst nematodes [1]. Also, it infects nematode eggs and sedentary females of cyst nematodes by hyphae produced on actively growing mycelium [72]. Survival and spread of the fungus occur through chlamydospores, micro conidia, and mycelium [1]. *V. chlamydosporium* colonizes the rhizosphere, which facilitates the infection of egg masses protruding from female root-knot nematodes on infected roots [73]. Noweer and Al-Shalaby [74] studied the effect of *V. chlamydosporium* combined with some organic manure on *Meloidogyne incognita* and other soil micro-organisms on tomato under field condition they found that decreased the counts of *Meloidogyne incognita* juveniles in soil, as well as, gall formation on roots. the fungus *Dactylaria brochopaga* or the fungus *Verticilium chlamydosporium* was affected on the development and reproduction of Meloidogyne incognita on Eggplant plants variety Balady under field conditions especially for the (F1 + F2AVYM). This was indicated by the lower numbers of juveniles in soil, lower numbers of galls per 5 gm roots, the % reduction in population density of soil larvae, in treatment of the nematophagous *A. dactyloides*.

Noweer [75] investigated the effect of the nematode-trapping fungus *Dactylaria brochopaga* and the nematode egg parasitic fungus *Verticilium chlamydosporium* in Controlling Citrus Nematode Infesting Mandarin, and Interrelationship with the Co inhabitant Fungi. Data revealed that the mixed compound treatment greatly affected the citrus nematode numbers both in soil and roots, in comparing with those of Vydate or induced by mixed compound was 97% and 70%; respectively in soil and roots. Rates of reproduction increase of the citrus nematode also reached 3% and 30% in both soil and roots; respectively. Vydate treatment resulted in a relatively lesser percentages. Growth of the concomitant fungus, *Trichoderma* sp. was increased specially in the last samples of October 2017, however those of fungi; *Aspergillus flavus*, *Fusarium* sp. And *Rhizopus* sp. was reduced, due to mixed compound treatment. *Aspergillus niger* and *Penicillium* sp. were not affected by the presence of the mixed compound. Vydate did not affect the co inhabitant fungi to a great extent.

6 Application and Safety Problems of Fungi-Antagonistic to Plant Nematodes

6.1 Examples of Use of Cover Crops for Enhancement of Nematophagous Fungi

Earlier studies by Linford et al. [28] and Cooke and Godfrey [76] demonstrated that incorporation of cabbage leaves into the soil enhanced nematophagous fungi. Later it was found that legume crops tend to enhance nematophagous fungi better than other crops. Root-knot nematode numbers were suppressed when soil amended with alfalfa was inoculated with *Arthrobotrys conoides* [77]. The efficient nematode trapping fungal species, *A. dactyloides* and *Monocosporium ellipsospora*, appeared only in micro plots amended with alfalfa [78]. When alfalfa meal was incorporated into the soil, suppression of root-knot nematodes by nematode-trapping fungi (NTF) increased. Pea rhizosphere enhanced the densities and species diversity of nematode-trapping fungi better than white mustard or barley [79]. The population density of the NTF and formation of conidia traps, structures that can overcome fungi stasis were much higher in the pea rhizosphere than the root-free soil [79].

6.2 Screening Cover Crops for Nematophagous Fungi Enhancement

Three cover crops were evaluated for their NTF enhancement ability in a greenhouse [80]. Soil from a pineapple field was either amended with chopped leaf tissues of sunn hemp (*Crotalaria juncea*), rapeseed (*Brassica napus*), marigold (*Tagetes erecta*) or pineapple (*Ananas comosus*) at 1% (w/w) and compared with soil treated with 1,3-dichloropropene (1,3-d) or bare soil. Three months after cowpea seedlings were planted into these soils, NTF numbers were higher in treatments receiving leaf amendments as compared to 1, and 3-d treated soil or bare soil (Fig. 20). However, only soil treated with sunn hemp had higher propgules of parasitic NTF (the most efficient nematode trapper) than 1, 3-d and bare soil. Therefore, among the cover crops tested, sunn hemp is recommended cover crop for NTF enhancement.

Other potential legumes tested for NTF enhancement were cowpea and velvet bean (*Mucuna deeringiana*). A soil collected from South West Florida Research and Education Center, University of Florida, Immokalee, FL was amended with chopped leaf tissues of sunn hemp (SH), cowpea (CP) or velvetbean (VB) at 1% w/w for 7 days in plastic pots and assayed for nematophagous fungal population densities. Soil amended with higher C: N ratio crop biomass such as black oat (*Avenasativa*) and soil without amendment (BS) were included as a control. At 21 days after plating (dap), sunn hemp enhanced *Dactylaria eudermata,* a NTF forming three-dimensional nets as compared to the control ($P < 0.05$). Cowpea enhanced the abundance of *Catenaria anguillulae* ($P < 0.05$), which is a zoosporic forming endoparasitic fungus that was

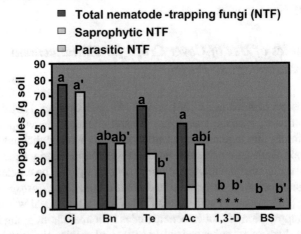

Fig. 20 Nematode-trapping fungal population densities in soil treated with sunn hempNematode-trapping fungal population densities in soil treated with sunn hemp (*Crotalaria juncea*, Cj), rapeseed (*Brassica napus*, Bn) or marigold (*Tagetes erecta*, Te) at 1% (w/w) were compared with soil amended with pineapple leaves (*Ananas comosus*, Ac), treated with 1, 3-dichloropropene (1, 3-d), or remained bare (BS). Columns with same letters were not different according to Waller-Duncan k-ratio t-test ($P \leq 0.05$), Wang et al. [80]

found to attack many species of plant-parasitic nematodes in Florida [81]. In contrast, effect of velvetbean on nematophagous fungi was not significant (Fig. 21). Both sunn hemp and cowpea enhanced *Harposporium anguillulae* ($P < 0.05$), but this fungus mainly only infect free-living nematodes, with a mouth cavity larger enough to ingest the fungal conidia of *H. anguillulae*.

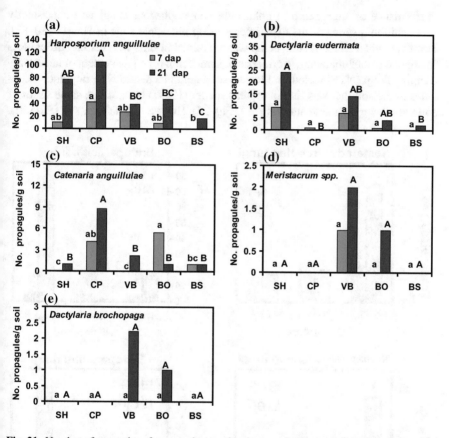

Fig. 21 Number of propgules of nematophagous fungi per g of different soil treatments Number of propagules of nematophagous fungi per g of soil treated with leaf amendment of sunn hemp (SH), cowpea (CP), velvetbean (VB), or black oat (BO) or non-amended bare soil (BS). Means are average of 4 replications at 7 (red bar) and 21 (blue bar) days after plating (dap). Means followed by the same letter were not different among the treatments at each dap according to Waller-Duncan k-ratio ($P \leq 0.05$) test, Wang et al. [82]

7 Factors Affecting Cover Crops for Nematophagous Fungi Enhancement

7.1 Indigenous Nematophagous Fungi Present in the Soil

Performance of sunn hemp in enhancing nematophagous fungi was consistently promising in a series of the experiments using pineapple soils in Hawaii [80–83]. However, when effect of sunn hemp was examined in soils with distinct differences in organic matter content, performance of sunn hemp for enhancement of nematode-trapping fungi (NTF) varied. In Florida, sunn hemp increased NTF population densities in the soil that was rich in organic matter (HYW) but not in a same series of soil with lower organic matter (NYW) (Fig. 22). However, in another soil (Expt. II),

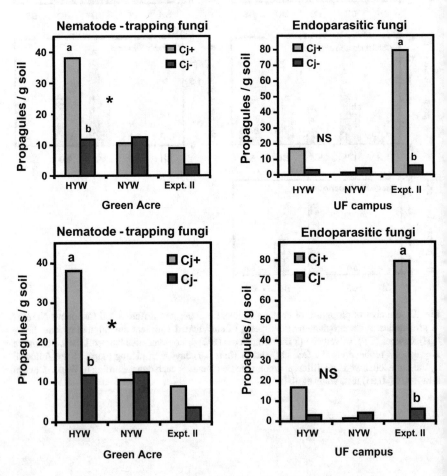

Fig. 22 Population densities of nematophagous fungi in three soils, Wang et al. [82]

Table 2 The initial organic matter content (%)

Soil	Initial organic matter content (%)
HYW	8
NYW	2.5
Expt II	2

(HYW, NYW, and Expt II) with different organic matter contents. Columns with different letters indicate difference between sunn hemp treatments according to analysis of variance ($P < 0.05$). Cj+ and Cj− indicate with and without sunn hemp amendment respectively

sunn hemp increased endoparasitic fungal population densities even though this soil had low organic matter (Fig. 22). These results indicated that performance of cover crop depends on the species of nematophagous fungi present in the soils.

Table 2 shows the initial organic matter content (%) as the followings:

7.2 Time After Nematicide Application

Some nematicide might be suppressive to activity of nematophagous fungi. Population densities of parasitic NTF were reduced in field recently treated with 1, 3-dichloropropene (1,3-D) (soil D205, Fig. 23) [82]. Ability of sun hemp to increase

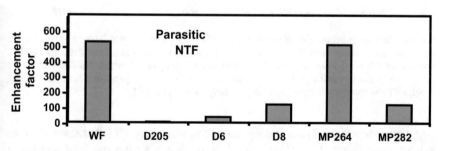

Enhancement factor = [(Cj +)-(Cj -)] / (Cj -)

WF   = fallow >10 yrs
D205 = 2 months post-plant and 1,3-D.
D6   = 4 months post-plant and 1,3-D.
D8   = 4 yrs post-plant, 1,3-D>1 yr.
M264 =15 months post-plant and after 1,3-D.
M282 =2 yrs post-plant and after 1,3-D.

Fig. 23 Ability of sunn hemp to enhance population densities of parasitic nematode-trapping fungi (NTF) as measured by an enhancement factor in six pineapple field soils with different time periods after the last 1,3-dichloropropene (1,3-D) application. Cj+ and Cj− indicate with and without sunn hemp amendment respectively, Wang et al. [82]

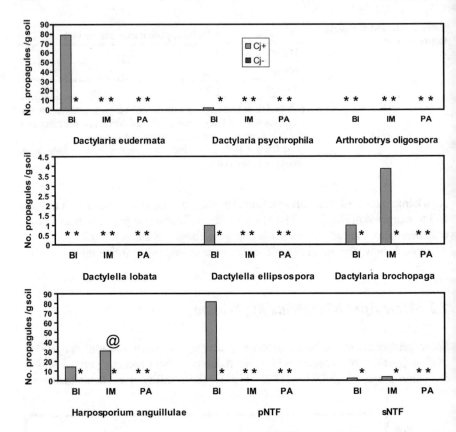

Fig. 24 Number of propagules of nematophagous fungi per g of soil in soil treated or not treated with sunn hemp amendment (Cj+ or Cj− respectively). In soil collected from Buck Island (BI), Immokalee (IM) and Pine Acres (PA). * signified no propagules detected. No difference were detected between Cj+ and Cj− for all the fungi ($P > 0.05$). pNTF = parasitic nematode-trapping fungi, sNTF = saprophytic nematode-trapping fungi, Wang et al. [82]

population densities of parasitic nematode-trapping fungi (NTF) measured by an enhancement factor (Fig. 23) increased as the time after the last 1,3-D treatment occurred, except on MP282 soil which had a different soil texture than the other soils. A high enhancement factor in the 10-years fallow soil indicates that enhancement of NTF is not due to the planting of pineapple (Fig. 24).

7.3 Ecological Habitat

Three soils collected from distinctly different agricultural sites were used to examine the ability of sunn hemp to enhance nematophagous fungi. These soils were collected from Buck Island (BI), a long-term pasture; Immokalee (IM), 10 years of

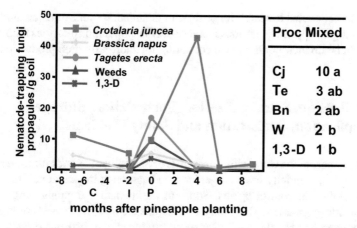

Fig. 25 Nematode-trapping fungal population densities in a pineapple field. Plots were planted with sunn hemp (Cj), rapeseed (Bn), or marigold (Te), or left fallow with weeds (W) or treated with 1,3-dichloropropene (1,3-D). Values in the table are means of 4 replications according to repeated measure analysis over time from covercrop planting to 9 months after pineapple planting. Means followed by the same letters were not different according to Waller-Duncan k-ratio t-test ($P < 0.05$), Wang et al. [83]

vegetable cultivation amended with compost yearly; and Pine Acres (PA), fallow with native weeds. Although sunn hemp can induce various species of nematophagous fungi in BI, only population densities of *Harposporium* and *Dactylaria brochopaga* were stimulated by sunn hemp amendment in IM soil. On the other hand, no nematophagous fungi were enhanced by sunn hemp in PA soil (Fig. 25).

7.4 Duration of Cover Crop for Enhancing Nematophagous Fungi in Field Conditions

Under field conditions, sunn hemp is recommended to be planted for 2–3 months and then plowed under for maximum nitrogen (N) input into the soil. In a pineapple field in Hawaii, sunn hemp was grown for 3 months and incorporated into the soil. One month after incorporation, pineapple crowns were planted. Repeated measure analysis of the population densities of nematode-trapping fungi (NTF) revealed that NTF were higher in sunn hemp treated plots than weed fallow or plots treated with 1, 3-dichloropropene (Fig. 13). Population densities of NTF remained high in sunn hemp treated plot 4 months after pineapple planting or 5 months after cover crop incorporation. Although this effect is not sufficient to manage reniform nematode infection on pineapple throughout the pineapple cycle (usually 18 months to fruiting), it suppressed the initial population densities of reniform nematodes. Further studies need to be conducted to search for post-plant treatment for reniform nematode management in pineapple. Continuous increase of NTF population densities for 5 months after sunn hemp incorporation is encouraging for short-term crop production.

8   Production, Standardization, Formulation, Storage, Application, Registration and Safety Problems

In the past, most of the work with antagonists of nematodes has been directed towards searching for potentially useful organisms and assessing the insult ability and efficacy as biological control agents. Some of the nematode-trapping fungi and the opportunistic egg parasite *Paecilomyces lilacinus* are now being used commercially in a few countries but the quantities being produced and distributed are minimal. Organisms such as *Pasteuria penetrans* and *Nematophthora gynophila* have definite biological control potential but they are relatively host-specific and cannot be readily cultured. Consequently, research on these organisms is being concentrated on the development of in vitro culture techniques.

Biological control research has not yet reached the stage where the problems of mass production, standardization, formulation, storage, application and safety have had to be considered in detail, but the time is rapidly approaching when such issues will become major areas of activity. Modern fermentation technology is widely used in the brewing industry and in the production of antibiotics, microbial insecticides and biological herbicides and should be amenable to the mass production of antagonists of nematodes. However, large-scale industrial production requires the development of inexpensive alternatives to the complex media often used to culture nematode antagonists in the laboratory. Ingredients such as peptone, yeast extract and many exotic sugars are generally too expensive for use in commercial fermentations and less expensive sources of nitrogen, carbon and other nutrients must be substituted. Sources of carbohydrate that are suitable for commercial use include glucose, sucrose, starch and hydrolyzed corn products, while potential sources of nitrogen include soybean and cotton seed flour, casein and fish meal.

Biological control agents are generally produced in commercial quantities by one of two fermentation methods. The oldest and perhaps the simplest is solid substrate fermentation, which involves growing the microorganism on the surface of a substrate (e.g. bran) that has been impregnated with nutrients. The substrate may be placed in stationary trays or in drums that rotate continuously or intermittently. Solid substrate fermentation has been used with some success in the production of myco-insecticides and has proved particularly suitable for culturing fungi that do not sporulate in liquid culture. However, for most other applications, it has been largely superseded by submerged culture fermentation, in which micro-organisms are grown in a liquid medium in large, sealed fermentation vessels, with the liquid being kept agitated by the passage of sterile, littered air. Most micro-organisms with potential for development as biological control agents against nematodes are likely to be produced using this method. Submerged culture fermentation is expensive, but has the advantage that parameters such as nutrient level, oxygen concentration, air tlow, incubation temperature and pH can easily be controlled.

However, optimum operating conditions must be determined for each organism being cultured, and this requires a major research effort at the time production is increased to levels beyond those of a small-scale pilot plant.Unmodified fermentation biomass is never likely to be suitable for direct sale as a crop protection product and this material must therefore be formulated into a product that retains its viability in storage, is convenient to use and relatively immune to user abuse. A number of types of formulation, including dusts, granules, wet table powders and liquids have been used in biological crop protection products, but granular formulations are generally considered to be most suitable for micro-organisms that are to be applied to soil. The dry nature of these formulations means that an antagonist of nematodes needs to have the capacity to survive desiccation if it is to be seriously considered for commercial development as a biological control agent. Granular biological products have traditionally been produced by blending the organism with a carrier such as clay or ground corn cobs and alien introducing a material to bind the organism to the carrier, but in recent years there has been considerable interest in encapsulating biological control agents in gallants such as sodium alginate. This process has proved useful for producing granules in the laboratory, but has not yet been scaled-up to the point where it can be used to produce microbial products in commercial quantities.

Although storage, handling and application problems may best be handled by formulating biological control agents in granules, this formulation method will only be suitable for organisms which have considerable powers of spread once they are introduced into soil. Since it will never be possible to incorporate granules evenly into soil and application rates have to he kept to realistic levels, antagonists will have to be able to move from the granule to areas where target nematodes occur. Kerry [8] showed that hyphae of *Verticilium chlamydosporium* grew approximately 1 cm from alginate bran granules, which suggests that such granular formulations may be suitable for this species. However, it is not yet known how the size and distribution of granules in soil affects the performance of the fungus as a biological control agent.

9 Formulation of Microbial Products

Formulation of microbial products in forms which have extended shelf lives and which can be applied to soil using conventional farm equipment is likely to present a major challenge to those interested in commercializing antagonists of nematodes. Organisms which cannot be dried without loss of viability are likely to be difficult to formulate and are unlikely to retain their activity in storage for more than a few months unless expensive storage conditions are employed [8]. The 18-month shelf-life needed by a commercial product as mentioned by Couch and Ignoffo [85] is most likely to be achieved with the endospore-forming bacteria and those fungi which produce thick walled oospores and chlamydospores. However, some other fungi havebeen stored for extended periods as hyphae with little loss of viability [86]. Unfortunately there is little information on the viability of nematode antagonists in long-term storage. When spore-powder preparations of *Pasteuria penetrans*were assessed after ten years, storage at room temperature, spores attached readily to juveniles of *Meloidogyne javanica* as confirmed by Stirling and Wachtel [87]. Although some spores were also able to initiate infection, spore viability appeared to have declined during the storage period. The results of work by Cabanillasi et al. [88] confirmed that it may be possible to develop long-term storage methods for *Paecilomyces lilacinus*, because the viability of formulations on wheat grains or diatomaceous earth granules remained high after storage for eight weeks at 25 ± 2 °C (Fig. 26).

Regardless of the method of fermentation and formulation used, there is always variability in the concentration and viability of organisms in various preparations of a microbial product, and it is imperative that differences in potency between

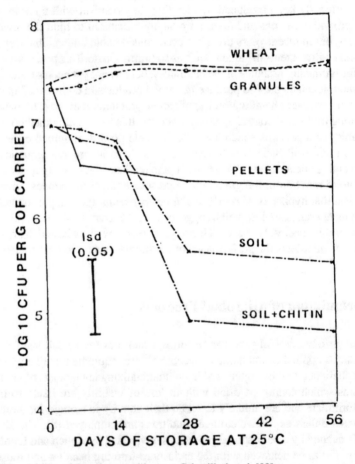

Fig. 26 The viability of *Paecilomyces lilacinus*, Cabanillasi et al. [88]

preparations are measured. Products of constant potency can then be produced and their potency compared with that of other products. In one of the few attempts to address this issue with a microbial nematicides, Stirling and Wachtel [87] developed a method of comparing the potency of different preparations of *Pasteuria penetrans* by modifying the methods used by Burges and Thomson [89] for assaying microbial insecticides. Test preparations were diluted with water; nematodes were added for a standard time and the concentration required for 50% of the nematodes to have spores attached was determined. Although this bioassay has proved useful, it has the disadvantage of being based on the ability of spores to attach to rather than infect nematodes. These parameters are not necessarily always related.

Comparisons of potency are less of a problem with organisms that have readily terminable propgules. For example, where Cabanillas el al. [88] compared the performance of five formulations of *Paecilomyces lilacinus*, the inoculums was standardized by determining the concentration of viable spores in each formulation and then adjusting application rates so that the same number of viable spores was added in each treatment.

In most field and glasshouse tests with biological control agents against nematodes, the antagonist has been cultured on a bulky substrate such as cereal grain and introduced into soil on that substrate in amounts ranging from 1 to 20 t/ha. Since such high application rates are never likely to be suitable lot use in commercial agriculture, it is imperative that more realistic methods of application are devised. Ideally, formulations are required which can be readily delivered through conventional farm equipment designed for applying fertilizers and granular pesticides.Diatomaceous earth granules impregnated with 10% molasses, lignite spillage granules and alginate-clay pellets have proved suitable carriers for the biological control agents developed for use against soil-borne fungi, similar results were obtained by Backman and Rodriguez-Kabana [90], Jones et al. [91], Fravel et al. [92], but such formulation methods have only been tested on a limited scale with antagonists of nematodes.

Commercially prepared liquid and granular formulations of *Arthrobotrys amerospora* failed to control nematodes in the field in Florida as reported by Rhoadcs [93], whereas *Paecilomyces lilacinus* formulated in alginate pellets or diatomaceous earth granules showed promise against *Meloidogyne incognita* in laboratory and micro plot experiments in North Carolina [88]. Although organisms with biological control potential against nematodes have generally been tested by adding them directly to the bulk soil mass, their introduction on seed or planting material may be a more efficient and cost-effective method of application. Such application methods have been widely used in tests with biological control agents against soil-borne pathogens and they warrant further testing with antagonists of nematodes.

Many nematode-susceptible vegetable crops are raised in seedling trays containing sterilized potting mix, and such systems should provide an ideal opportunity to establish potentially useful biological control agents in the rhizosphere before seedlings are transplanted. *Verticilium chlumydosporiuin* was established in field soil after it was grown on a substrate and incorporated into a peat growth medium but there are no other reports of such strategies having been attempted with antagonists of nematodes. However, initial tests in which fungi and bacteria have been applied to seed or to seed pieces for nematode control have yielded encouraging results. Some control of *Meloidogyne incognita* and Globodera. Was obtained by dipping potato tubers in *Paecilomyces lilacinus*, while rhizobacteria applied to sugarbeet seed suppressed early root infection by *Heterodera* sp. However, the colonization of soil and the rhizosphere by organisms applied in this manner requires more de-tailed investigation because showed that competition from the indigenous soil micro flora often prevents seed-inoculated fungi and bacteria from colonizing the rhizosphere more than 2 cm from treated seed.

The transposition of micro-organisms to plants or soil to provide protection against pests and diseases brings biological control into the purview of government health authorities and evidence of safety to users, to consumers of treated produce, to non-target species and to the environment must be established. Such issues have largely been ignored by those interested in developing biological control agents against nematodes, but they will have to be addressed in the near future, because many countries now have mandatory requirements for the registration of biological pesticides [94].

Most parasites and predators of plant parasitic nematodes will biological control potential have a restricted host range and should not pose much olla's threat to non-target organisms. The main non-target species that could be at risk are tile nematodes which prey on other nematodes and those that are used for biological control of insect pests. However, both these groups of beneficial nematodes differ from plant parasitic species in size, distribution in soil, behavior and general biology and therefore may not be affected markedly by biological control agents tiled are developed for plant parasites. The possibility that micro-organisms used for nematode control are also detrimental to plan is perhaps of more concern. Many of tiles bacteria and fungi are intimately associated wilt roots and it will be necessary to establish that they are not plant pathogens before they are widely used for biological control purposes. The only antagonist of nematodes that has caused much *Paecilomyces lilacinus* concern with regard to its mammalian toxicity is the egg-parasitic fungus This opportunistic species causes eye infections and facial lesions in humans and infections in domestic animals as confirmed by Chandler et al. [95], particularly when individuals have suffered physiological stress or injury priority infection or when tile eye tissue lies been weakened by previous surgery [96].

Such reports reaffirm the need for diligent testing of all biological products before release, to ensuretliat they are safe to humans and have a minimal impact on the environment [97]. Since *P. lilacinus* has been mass cultured in many laboratories for more than a decade without causing any obvious problems, the risks involved in working with isolates that parasitizes nematodes may be minimal. Nevertheless, *P. lilacinus* is likely to be categorized as a 'relatively light risk' organism by most governments and initiative lexicological testing is likely to be required before permission to use it lot biological control purposes is granted. Most other nematode antagonists appear to belong in a much lower risk category and it is to be hoped that tile medical, veterinary and plant pathological tests demanded by registration authorities are not as costly as

to prevent or delay the use of these organisms in products. It provides a biological alternative to nematicides.

10 Conclusion and Future Prospects

Most parasites and predators of plant parasitic nematodes will biological control potential have a restricted host range and should not pose much olla's threat to non-target organisms. The main non-target species that could be at risk are tile nematodes which prey on other nematodes and those that are used for biological control of insect pests. However, both these groups of beneficial nematodes differ from plant parasitic species in size, distribution in soil, behavior and general biology and therefore may not be affected markedly by biological control agents tiled are developed for plant parasites. The possibility that micro-organisms used for nematode control are also detrimental to plan is perhaps of more concern. Many of tiles bacteria and fungi are intimately associated wilt roots and it will be necessary to establish that they are not plant pathogens before they are widely used for biological control purposes.

Formulation of microbial products in forms which have extended shelf life and which can be applied to soil using conventional farm equipment is likely to present a major challenge to those interested in commercializing antagonists of nematodes.

Biological control research has not yet reached the stage where the problems of mass production, standardization, formulation, storage, application and safety have had to be considered in detail, but the time is rapidly approaching when such issues will become major areas of activity. Modern fermentation technology is widely used in the brewing industry and in the production of antibiotics, microbial insecticides and biological herbicides and should be amenable to the mass production of antagonists of nematodes. However, large-scale industrial production requires the development of inexpensive alternatives to the complex media often used to culture nematode antagonists in the laboratory. Ingredients such as peptone, yeast extract and many exotic sugars are generally too expensive for use in commercial fermentations and less expensive sources of nitrogen, carbon and other nutrients must be substituted. Sources of carbohydrate that are suitable for commercial use include glucose, sucrose, starch and hydrolyzed corn products, while potential sources of nitrogen include soybean and cotton seed flour, casein and fish meal. Biological control agents are generally produced in commercial quantities by one of two fermentation methods (Solid substrate fermentation or Liquid culture fermentation).

A number of types of formulation, including dusts, granules, wet table powders and liquids have been used in biological crop protection products, but granular formulations are generally considered to be most suitable for micro-organisms that are to be applied to soil. The dry nature of these formulations means that an antagonist of nematodes needs to have the capacity to survive desiccation if it is to be seriously considered for commercial development as a biological control agent. Granular biological products have traditionally been produced by blending the organism with a carrier such as clay or ground corn cobs and alien introducing a material to bind the organism to the carrier, but in recent years there has been considerable interest in encapsulating biological control agents in gallants such as sodium alginate. This process has proved useful for producing granules in the laboratory, but has not yet been scaled-up to the point where it can be used to produce microbial products in commercial quantities. There are many trials to produce the granular formulations for commercial microbial products in future to be applied to soils.

References

1. Kerry BR, Jaffee BA (1997) Fungi as biological control agents for plant parasitic nematodes. In: Esser K, Lemke PA (eds) The mycota, vol 4, pp 204–218. Berlin, Springer
2. Barron GL, Thorne RG (1987) Destruction of nematodes species of Pleurotus. Can J Botany 65:774–778
3. Glawe DA, Stiles CM (1989) Colonization of soybean roots by fungi isolated from cysts of Heterodera glycines. Mycologia 81:797–799
4. Sikora RA (1992) Management of the antagonistic potential in agricultural ecosystems for the biological control of plant parasitic nematodes. Ann Rev Phytopathol 30:245–247
5. Hussey RS, Roncadori RW (1982) Vesicular-arbuscular mycorrhizae may limit nematode activity and improve plant growth. Plant Dis 66:9–14
6. Gray NF (1983) Ecology of nematophagous fungi: distribution and habitat. Annual Rev Appl Biol 102:501–509
7. Barron GL (1977) The nematode-destroying fungi. Can Biol Publishers, Ontaria, Canada, p 140p
8. Kerry B (1988) Fungal parasites of cyst nematodes. Agric Ecosys Environ 24:293–305
9. Timper F, Brodie BB (1993) Infection of *Pratylenchus penetrans* by nematode- pathogenic fungi. J Nematol 25:297–302
10. Jaffee BA, Muldoon AE (1995) Susceptibility of root-knot and cyst nematodes to the nematode-trapping *Monacrosporium ellipsosporum* and *M. cionopagum*. Soil Biol Biochem 27:1083–1090
11. Commandon J, Fonbrune P (1937) Recherches experimentales sur les chapignons predateures de Nematodes du soil. Conditions de formations des organs de capture. Comptes rendus hebdom. Soc. Biologie, Paris T. CXXIX, séance du 19.11.1938, 619–25
12. Harwing J, Scatt PM, Staltez DR, Blanch Field BJ (1979) Toxins of malds from decaying tomato fruit. Appl Environ Microbio 38:267–274
13. Riker AJ, Riker RKS (1936) Introduction to research on plant diseases. John S Swift co., st. Louis, Mo
14. Barnett HL (1955) Illustrated genera of imperfect fungi Ph.D. thesis. Plant pathology Dept, Bacteriology and Entomology. West Virginia Univ, 219p
15. Alexopoulos CJ, Mism CW (1979) Chapter 27. In: Alexapoulos CJ, Mims CW (eds) Introductory mycology. USA, pp 534–572
16. Noweer EMA (2017a) Occurrence and the morphological identity of some antagonistic fungi isolated from soils infested with root-knot nematode. Comm Appl Biol Sci Ghent Univ, 82:261–274
17. Gray NF (1985) Ecology of nematophagous fungi: effect of soil moisture, organic matter, pH and nematode density on distribution. Soil Biol Biochem 17:499–507
18. Cooke RC (1963) Ecological characteristics of nematode-trapping fungi Hyphomycetes. Ann Rev Appl Biolo 52:431–437
19. Jasson HB (1982) Predacity by nematophagous fungi and its relation to the attraction of nematodes. Microbial Ecol 8:233–240
20. Fresenius G (1852) Beitrage zur Mykologie. Heft. 1. Frankfurt am Main, pp 1–38

21. Zopf W (1888) Zur kenntnis der infektionskrankheiten niederer und pflanzen. Nova Acta Leop Carol 42:313–341
22. Drechsler C (1941) Some hyphomycetes parasitic on free- living terricolousnematodes. Phytopathol 31:773–802
23. Galuilina EA (1951) The predatory hyphomycetes from soil of the Turkmenstan. Microbiology (Moscow) 20:489
24. Aboul-Eid HZM (1963) Studies on some aspects of nematode biological control M.Sc. Thesis, Agric Fac Cairo Univ 100p
25. Rao GV, Malek RB (1973) Effects of nematode- trapping fungi on population of *Pratylenchus penetrans*. In: Abstracts from the proceedings 2nd internat congress of plantpathol, st. Paul, MN. No. 0556
26. Mai WF, Bloom JR, Chen TA (1977) Biology and ecology of the plant parasitic nematode *Pratylenchus penetrans*. Pennsylvania State Univ Bull no. 815, University Park
27. Godoy G, Rodriguez- Kabana R, Morgan J (1983) Fungal parasites of *Meloidogyne arenaria* eggs in an Alabama soil. A mycological survey and greenhouse stud. Nematropica 13:201–213
28. Linford MB, Yap F, Oliveria JM (1983) Reduction of soil populations of the root-knot nematode during decomposition of organic matter. Soil Sci 45:127–141
29. Mankau R (1969) Nematicidal activity of *Aspergillus niger*culture filtrates. Phytophathol 59:1170
30. Desai MV, Shah HM, Pillai SN (1973) Effect of *Aspergillus niger* on root-knot nematode *Meloidogyne incognita*. Ind J Nematol 2:210–214
31. Alam MM, Khan MW, Saxena SK (1973) Inhibitory effect of culture filtrates of some rhizosphere fungi of Okra on the mortality and larval hatch of certain plant parasitic nematodes. Ind J Nematol 3:94–98
32. Khan TA, Azam MF, Husain SI (1984) Effect of fungal filtrates of *Aspergillus niger* and *Rhizoctonia solani* on penetration and the plant growth of tomato var. Marglobe. Indian J Nematol 14:106–109
33. RosenzweigWD Premachandran D, Pramer D (1985) Role of trap lectins in the specificity of nematode capture by fungi. Can J Microbiol 31:693–695
34. Niblack TL, Hussey RS (1986) Evaluation of *Arthrobotrys ameraspora* as a biocontrol agent for *Heterodera glycines* on soybean. Plant Dis 70:448–457
35. Jatala P (ed) (1986) Biological control of Nematodes, chapter 26. Department of Neamatology and Entomology International Potato Center, Lima, Peru, pp 303–308
36. Dube B, Smart GC (1987) Biological control of *Meloidogyne incognita* by *Pacilomyces lilacinus* and *Pasturia penetrans*. J Nematol 19:222–227
37. Hewelett TE, Dickson DW, Mitchell DJ, Kannwischer-Mitchell ME (1988) Evaluation of *Paecilomyces lilacinus* as a biocontrol agent of *Meloidogyne incognita* on tobacco. J Nematol 20:578–584
38. Murray DS, Wharton DA (1990) Capture and penetration processes of the free living juveniles of *Trichostrongylus colubriformis* (Nematoda) by the nematophagous fungi, *Arthrobotrys oligospora*. Parasitol 101:93–100
39. Ali HHA (1990) The use of Nematode- trapping fungi and organic amendments to control rootknot nematodes. Bull Agric Fac Cairo Univ 41:1001–1012
40. Voss B, Wyss U (1990) Variation between fungus *Catenaria anguillulae*. Zeit Pflanzenkrankheiten & Pflanzenschutz 94:416–430
41. Mankau R, Ciancio A (1990) Nematode antagonists in a southern California soil developing suppressiveness to *Heterodera schachtii*. Dept Nematology, Cali Univ, Riverside, CA 92521, and Instituto de Nematologia Agraria, CNR, 70126 Bari, Italy
42. Ali HHA, Barakat MIE (1991) Biocontrol and interaction between soil-borne fungi and rootknot nematode associated with Cantaloupe plants. Bull Agric Fac Cairo Univ 42:279–292
43. Hertz BN (1992) Condial traps- a new survival structure of the nematode- trapping fungus *Arthobotrys oligospora*. Mycol Res 96:194–198
44. Hoffmann HS, Sikora RA (1993) Enhancing the bioloigcal control efficacy of nematode- trapping Fungi towards *Heterodera schachtii* with green manure. Zeit Pflanzenk & Pflanzensch 100:1170–1175

45. Park SD, Choo YD, Jung KC, Sim YG, Choi YE (1993) Field application of egg and larval parasitic fungi and chemicals for controlling root-knot nematodes on some medicinal herb. Kor J Appl Entomol 32:105–114
46. Ali EM, Eleraki S, Elgindi AY (1994) Antagonicitic fungi as biological agents for controlling root-knot *Meloidogyne incognita* on tomato. In: Proceedings 2nd International symposium Afro-Asian Society of Nematologists 18–22 Dec 1994
47. Reddy MCM, Sharma NK (1994) Effect of fungal cultures and nematicides on larval penetration, root gall and egg mass counts of *Meloidogyne incognita* on tomato. In: Proceedings 2nd Internat Symposium the Afro- Asian Afro-Asian Society of Nematologists, 18– 22 Dec 1994
48. El-Sawy MA (1994) Advanced studies to control plant parasitic nematodes by non-chemical methods. Ph. D Thesis. Zoology and Nematol Dept, Agric Fac Cairo Univ, 124p
49. Scholler M, Rubner A (1994) Predacious activity of the nematode, destroying fungus *Arthrobotrys oligospora* in dependence of the medium. Microbil Res 149:145–149
50. Stirling GR, Main A (1995) The activity of nematode- trapping fungi following their encapsulation in alginate. Nematologica 41:240–260
51. Reddy PP, Rao MS, Nagesh M (1996) Management of the citrus nematode, *Tylenchulus semipenetrans,* by integration of *Trichoderma harzianum* with oil cakes. Nematol- Mediter 24:265–267
52. Safullah SM (1996) Killing potato cyst nematode males: a possible control strategy. Afro-Asian J Nematol 6:23–28
53. Aboul- Eid HZ, Anter EA, Abdel-Bari NA, Noweer EMA (1997) Effects of different fungal and bacterial treatments on *Meloidogyne incognita* infecting Tomato. Egypt J Agronematol 5:127–144
54. Duponnois R, Mateille T, Ba A (1997) Potential effects of sahelian nematophagous fungi against *Melodiagyne mayagyensis* on tobacco. Ann- du Tabac- Sec-2, 29:61–70
55. Sankaranarayanan C, Hussaini SS, Kumar PS, Prasad RD (1997) Nematicidal effect of fungal filtrates against root- knot nematodes. J Biol Cont 11:37–41
56. Abdel-Bari NA, Aboul-Eid HZ, Anter EA, Noweer EMA (2000) Effects of different fungal filtrates on *Meloidogyne incognita* larvae in laboratory bioassay tests. Egyptian J Agronematol 4:49–70
57. Bandyopadhyay P, Singh KP (2000) Effect of temperature, PH and nematode starvation on induction of rings of *Dactylaria brochopaga.* J Mycol Plant Pathol 30:100–102
58. Al-Shalaby MEM, Noweer EMA (2003) Effects of five plant extracts on the reproduction of root-knot nematode Meloidogyne incognita infested peanut under field condition. J Agric Sci Mans Univ 28:8447–8454
59. Noweer EMA, Al-Shalaby MEM (2003) Effects of some aromatic and medicinal plants as amendments against *Meloidogyne incognita* on peanut under field condition. J Agric Sci Mansoura Univ 28:7437–7445
60. Noweer EMA (2005) Efficacy of the nematode-trapping fungus *Dactylaria brochopaga* and biofertilizers, on controlling the root-knot nematode *Meloidogyne incognita* infecting Tomato. Egypt J Biol Pest Cont 15:135–137
61. Noweer EMA, Hasabo SAA (2005) Effect of different management practices for controlling the root-knot nematode *Meloidogyne incognita* on squash. Egypt J Phytopathol 33:73–81
62. Hasabo SA, Noweer EMA (2005) Management of root-knot nematode *Meloidogyne incognita* on eggplant with some plant extracts. Egypt. J Phytopathol 33:65–72
63. Aboul-Eid HZ, Hasabo SA, Noweer EMA (2006) Effect of a nematode trapping fungus Dactylaria *brochopaga* on *Meloidogyne incognita infesting* olives and coconut palms in Egypt. Internat J Nematol 16:65–70
64. Noweer EMA, El-Wakeil NE (2007) Combination of the entomopathogenic nematode *Heterorhabditis bacteriophora* and the nematode-trapping fungi *Dactylaria brochopaga* and *Arthrobotrys conoides* for controlling *Meloidogyne incognita* in tomato fields. Archiv Phytopathol Plant Prot 40:188–200
65. Noweer EMA, Dawood Mona G (2009) Efficiency of propolis extract on faba bean plants and its role against nematode infection. Comm Appl Biol Sci Ghent Univ 74:593–603

66. Noweer EMA (2013) Effect of the nematode biocide Dbx-1003 in controlling citrus nematode infecting mandarin, and interrelationship with the Co inhabitant fungi. Comm Appl Biol Sci Ghent Univ 78(3):417–423

67. Noweer EMA, Aboul-Eid HZ (2013) Biological control of root-knot nematode *Meloidogyne incognita* infesting cucumber *Cucumis sativus* L. cvs. *Alfa* by the nematode-trapping fungus *Dactylaria brochopaga* under field conditions. Agric Biol J N Am 4:435–440

68. Noweer EMA, Al-Shalaby MEM (2014) Evaluation of Nematophagous fungi *Dactylaria brochopaga* and *Arthrobotrys dactyloides* against *Meloidogyne incognita* infesting peanut plants under field conditions. Agric Biol J N Am 5:193–197

69. Noweer EMA (2014) Effects of some nematode-trapping fungi on the root-knot nematode *Meloidogyne* sp. infesting white bean *Phaseolus vulgaris* and sugar beet *Beta vulgaris* sp. *vulgaris* under field conditions. Agric Biol J N Am 5:209–213

70. Aboul-Eid HZ, Noweer EMA, Ashour NE (2014) Impact of the nematode-trapping fungus, *Dactylaria brochopaga* as a biocontrol agent against *Meloidogyne incognita* infesting Superior grapevine. Egypt J Biol Pest Cont 24:477–482

71. Noweer EMA (2017) A field trial to use the nematode-trapping fungus *Arthrobotrys dactyloides* to control the root-knot nematode *Meloidogyne incognita* infesting bean plants. Comm Appl Biol Sci Ghent Univ 82:275–280

72. De Leij FAAM, Kerry BR (1991) The nematophagous fungus, *Verticillium chlamydosporium* as a potential biocontrol agent for *Meloidogyne arenaria*. Rev Nématolo 14:157–164

73. Kerry BR, De Leij FAAM (1992) Key Factors in the development of fungal agents for the control of cyst and root-knot nematodes. In: Tjamos EC, Papavizas GC, Cook RJ (eds) Biological control of plant diseases: progress and challenges for the future. Plenum Press, New York, pp 139–144

74. Noweer EMA, Al-Shalaby MEM (2009) Effect of *Verticilium chlamydosporium* combined with some organic manure on *Meloidogyne incognita* and other soil micro-organisms on Tomato. Internat J Nematol 19:215–220

75. Noweer EMA (2018) Effect of the nematode-trapping fungus Dactylaria brochopaga and the nematode egg parasitic fungus Verticilium chlamydosporium in controlling citrus Nematode Infesting Mandarin, and Interrelationship with the Co inhabitant Fungi. Internat J Engi Technol 7:19–23

76. Cooke RC, Godfrey BES (1964) A key to the nematode-destroying fungi. Trans Brit Mycol Soc 47:61–74

77. Al-Hazmi AS, Ibrahim AAM, Abdul-Raziq AT (1992) Evaluation of a nematode- encapsulating fungi complex for control of *Meloidogyne javanica* on potato. Pak J Nematol 11:139–149

78. Mankau R (1968) Soil fungistasis and nematophogus fungi phytopathology 52:611–615

79. Persmark L, Yansson H (1997) Nematophagous fungi in the rhizosphere of agriculture crops. FEMS Microbiol Ecol 22:303–312

80. Wang K-H, Sipes BS, Schmitt DP (2001) Suppression of *Rotylenchulus reniformis* by *Crotalaria juncea, Brassica napus,* and *Tagetes erecta*. Nematropica 31:235–249

81. Esser RP, Schubert TS (1983) Fungi that entrap nematodes by mucilaginous droplets borne on glandular cells. Fla. Dept Agric & Consumer Serv, Division of Plant Industry. Nematology Circular no 95

82. Wang K-H, Sipes BS, Schmitt DP (2003) Suppression of *Rotylenchulus reniformis* enhanced by *Crotalaria juncea* amendment in pineapple field soil. Agri Eco Environ 94:197–203

83. Wang K-H, Sipes BS, Schmitt DP (2002) Management of *Rotylenchulus reniformis* in pineapple, *Ananas comosus* by intercycle cover crops. J Nematol 34:106–114

84. Wang K-H, McSorley R (2003) University of Florida, Department of Entomology and Nematology, P.O. Box 110620, Gainesville, FL 32611–0620, USA

85. Couch TC, Ignoffo CM (1981) Formulation of insect pathogens. In: Burges HD (ed) Microbial Control of pests of plant diseases 1970–1980. Academic Press, New York, pp 621–634

86. Papavizas GC, Dunn MT, Lewis JA, Beagle-Ristaino J (1984) Liquid fermentation technology for experimental production of biocontrol fungi. Phytopathol 74:1171–1175

87. Stirling GR, Wachtel MF (1980) Mass production of *Bacillus penetrans* for biological control of root-knot nematodes. Nematol 26:308–312
88. Cabanillas E, Barker KR, Nelson LA (1989) Survival of *Paecilomyces lilacinus* in selected carriers and related effects on *Meloidogyne incognita* on tomato. J Nematol 21:121–130
89. Burges HD, Thomson ME (1971) Standardization and assay of microbial insecticides. In: Burges HD, Hussey NW (eds) Microbial control of insects and mites. Academic Press, New York, pp 591–622
90. Backman PA, Rodriguez-Kabana R (1975) Asystem for growth and delivery of biological control agents to the soil. Phytopathol 65:819–821
91. Jones RW, Pettit RE, Taber RA (1984) Lignite and stillage: carrier and substrate for application of fungal biocontrol agents to soil. Phytopathol 74:1167–1170
92. Fravel DR, Marios JJ, Lumsden RD, Connick WJ Jr (1985) Encapsulation of potential biocontrol agents in an alginate-clay matrix. Phytopathol 75:774–777
93. Rhoades HL (1985) Comparison of fenamiphos and *Arthrobotrys amerospora* for controlling plant nematodes in central Florida. Nematropica 15:1–7
94. Laird M, Lacey LA, Davidson EW (eds) (1990) Safety of microbial insecticides. CRC Press, Boca Raton, Florida, p 259
95. Chandler FW, Kapan W, Ajello L (1980) A colour alas and textbook of the histopathology of mycotic diseases. Wolfe Medical, London
96. Gordon MA, Norton SW (1985) Corneal transplant infection by *Paecilomyces lilacinus*. Sabouraudia. J Med Vet Mycol 23:295–301
97. Sayre RM (1986) Pathogens for biological control of nematodes. Crop Prot 5:268–276

Plant Viral Diseases in Egypt and Their Control

Ahmed Abdelkhalek and Elsayed Hafez

Abstract Plant viruses pose a serious threat to agricultural production and incur enormous costs to growers each year, both directly, in the form of yield and quality loss, and indirectly, in the forms of time and funds spent on scouting and disease management. Virus diseases cause plants crops losses annually in average of US$ 60 billion. Accordingly, viral diseases need to be controlled for the sustainable agriculture and in order to maintain the quality and abundance of food production. Moreover, recently agriculture suffered from different problems such as; the contentious changes in climatic factors globally, increasing the numbers of pathogens per year and appearance of anew pests. Unfortunately, chemical control has negative impact on the environment and on human health as well in addition it creates an imbalance in the microbial biodiversity, which may be unfavorable to the activity of the beneficial organisms and may lead to the development of pathogens-resistant strains. The most important thing is that agricultural sustainability should be supported by eco-friendly approaches such as discovery of new biocontrol agents capable to control the plant viral diseases. To achieve this was inevitable to use the plant growth-promoting microbes as effective biocontrol agents against plant viruses will hold the greatest promise and is considered a pillar of integrated viral diseases management. We argue that the use of growth-promoting microbes will preserve sustainable agriculture as well as a clean environment free from pollution, which will be benefiting for both the farmer and the consumer. So far, there are no such pesticides at the local level, while at the international level there may be one or two products. We succeeded to control some viruses infect potato by using the filtrates of seven *Bacillus spp.* mixed with nanoclay, but the product is still under research and development.

Keywords Plant viruses · Virus control · PGPR · PGPF · Egypt

A. Abdelkhalek (✉) · E. Hafez
Plant Protection and Biomolecular Diagnosis Department, Arid Lands Cultivation Research Institute, City of Scientific Research and Technological Applications, New Borg El Arab, Alexandria, Egypt
e-mail: abdelkhalek2@yahoo.com

© Springer Nature Switzerland AG 2020
N. El-Wakeil et al. (eds.), *Cottage Industry of Biocontrol Agents and Their Applications*, https://doi.org/10.1007/978-3-030-33161-0_13

1 Introduction

Viruses are essentially obligate parasites with fairly simple, unicellular organized only by one type of nucleic acid, either DNA or RNA. Mostly, all types of living organisms including animals, insects, plants, fungi, and bacteria are hosts for viruses [1]. These diseases have existed throughout the history, the archaeological evidence both in Egyptian mummies and in medical texts of readily identifiable viral infections, including genital papillomas and poliomyelitis [2]. In fact, viruses continue to be global threats to public health worldwide by causing AIDS, Chicken pox, Ebola, Hepatitis A, Rabies, Polio, Pneumonia etc. [3]. On the other hand, viruses are a major contributor of plant diseases resulting in losses in agriculture, forestry and productivity of natural ecosystems [4, 5]. It well known that, both cultivated and wild plants are susceptible to be infected by viruses. Meanwhile, the lack of recognition is due to their insidious nature [4, 6], and often the viral disease is less conspicuous than those caused by other plant pathogens [7]. Consequently, viral infection can cause stunting, mosaic pattern, leaf rolling, necrosis, wilting, ringspot and other developmental abnormalities leading to reduction in growth, decrease in vigour by increased predisposition to attack by other pathogens and pests, lesser yields and crop failure and finally decline in quality or market value [6, 8, 9].

In general, viruses are among the most important plant pathogens as near half of the emerging epidemics have a viral etiology [10], in addition, viruses possess some of the properties of living organisms, such as having a genome, which is able to evolve and adapt to the changes of environmental conditions. Normally, some of plant viruses possess single and double strand RNA genomes, single strand DNA, dsDNA; the entire genome covered by coat protein. Moreover, the virus particles are varying in their shape and size for example; isometric or rod-shaped, gemini and etc.

Regarding the virus transmission, viruses can be transmitted between plants either biologically or mechanically. Insects play an important role in virus transmission especially arthropods such as aphids, whiteflies, leaf and other hoppers, others may be transmitted by plant parasitic nematodes and the entophytic fungi. Once inside the plant, the virus replicates in individual cells by modifying and utilizing the plant's replication machinery, and spreads between cells progressively before entering the vascular systems for long-distance systemic spread.

Globally, plant viruses are considered as one of the most problems of food security and they are responsible for huge losses of crop production [11, 12]. Moreover, the world's population is increasing rapidly and world food production needs to be commensurate with the demands of human consumption. So that, viral diseases are needed to be controlled in order to maintain the quality and abundance of food safety. Commonly, controlling viral infections is difficult and it is mainly based on intensive pesticide and insecticide treatments that are usually used to control the spreading of vectors. However, this method is harmful to the environment and it has limited success because it leads for insecticide-resistant populations and in addition to leaving pesticide residues that may affect the ecosystem, soil fertility, underground water and human health [13, 14] leading to serious environmental problems. The use of plant growth-promoting bacteria (PGPR) or plant growth-promoting fungi (PGPF) as biocontrol agents against plant pathogens is becoming more popular in recent years [15]. Many researchers have documented that plant growth-promoting microbes (one or more strains) and they reported that the handling of plant diseases was better than the chemical control under field conditions [16–18]. These growth promoting microorganisms are good contributor for the growth of plants and strengthening the plant immune system resist the virus disease effects [17, 19, 20].

Interestingly, PGPR/PGPF have been widely known for their ability to colonize plant roots and increase plant growth and yield through the uptake of nutrients and the production of growth factors and vitamins. In addition, it can induce systemic resistance or act antagonistically to several soil borne phytopathogens due to the production of siderophores, bacteriosins and antibiotics [21, 22]. Besides its using as anti-microbial agents, there are an increasing number of studies regarding its potential action against viruses. *Bacillus* spp. have been reported to induce antiviral responses against CMV in tomato, pepper and arabidopsis and as well against PVY and PVX in potato plants [23–26]. Moreover, *Pseudomonas* spp. reduced disease severity of TSWV and ToMoV in tomato and of BBTV in banana plants [18, 27, 28].

Raupach et al. [29] used the rhizosphere colorization of some bacteria induced the systemic resistant systemic control for CMV in infected cucumbers and tomatoes. Whenever, Kim et al. [30] used *Acinetobacter* sp. KTB3 for induction of the system control against another viruses existed in Korea. It was reported that, TSWV is a major pathogen for Solanaceae plants; the virus causes cruel losses in tomato crop, which one of the most important crops for the Mediterranean agricultural economy [31]. In particular, filtrates of *Bacillus* spp. bacterial strains were approached as vaccine inoculants, meanwhile, it increase the efficacy of plant growth and induced the field systemic disease protection [32].

Generally, PGPR and PGPF in addition that both are beneficial microorganisms, they protect crops by activating an induced systemic resistance (ISR) via jasmonic acid (JA)-dependent signaling pathway [33].

The main goal of this chapter is to shed light on the plant viral diseases that affect plants and crops in Egypt and all over the world, how these viruses cause great losses. The losses resulted from viral infection affects the production and the crop yields which lead to hunger complained in the third world as a whole. In this chapter we have identified several ways that may be used to eliminate most of these viruses, which are effective as a protection and not a cure. These types of protection depend on the alarming and activating the defense system in the plant to resist the viral entrance and propagation followed by reduction the incidence of infection. We presented several microbes, plant extracts, nanomaterials and etc., that are used to reduce viral infection and ended the chapter with a local Egyptian trail, using bacillus spp. filtrates mixed with nano-clay to protect potato from virus infection. We recommend and acknowledged the usage of the bio-pesticides which considered as a substitution for the chemicals pesticides.

2 Strategies Used in Viral Disease Management

2.1 Protection by Selecting the Resistant Varieties

The use of virus-resistant cultivars is a cheap and effective approach to reduce the economic loss caused by the plant viruses [34]. However; this alternative faces usually many obstacles, such as, the cultivar breakdown resistance which resulted from whether a new mutant of specific virus was emerged and or the environmental conditions affect the plant immune response and it became sensitive for the viral infection [35], in addition the stability of plant resistance will varied according to the virus-host interaction. In another word, it will depend on the ability of the host to resist the newly emerged virulent strains from the virus population.

2.2 Management via Genetically Engineered Plants

The majority of virus-resistant transgenic plants can be considered the result from using a part of the viral sequence in the plant cells leading the cell to be resistant for the viral infection [36]. Among the viral proteins used are replicases, movement proteins, proteases and, most often, coat protein [37, 38]. Generally, the most common viral DNA sequence which heavily used in transgenic plants is CaMV 35S promoter [39], this promoter capable to express the transgenic proteins in GMO. Thus, these transgenic proteins remain stable for a prolonged period based on their potential bioavailability and persistence affecting the soil microorganisms directly [38, 40, 41]. However, some restrictions were noticed through using of GMCs, because it was believed that genetic release from these modified organisms into the other non-transgenic but they are closely related crops [42, 43]. With particular reference to Egypt [44], some varieties of GM sweet potato that are resistant to potato weevils and viral diseases, for example, SPFMV, and potato tuber moth (*Phthorimaea operculella* Zeller) were produced during 1993. However, sweet potato is not yet commercialized to avoid the loss of exporting in the European market [44].

2.3 Viral Disease Control Using Chemically Synthesized Compounds

Chemical control is one of the most used schemes in controlling the majority of plant diseases all over the world, but these pesticides have a serious effect on the environment and the human and plant health. Accordingly, application of synthetic chemicals like fertilizers, fungicides, herbicides, and pesticides, has been reported as non-sustainable and having multiple harmful impacts on both human or animal and plant health as well as environmental well-being [45, 46]. Most of these chemicals are pesticides and physical barriers that used for control of viral vectors [47, 48]. The non-persistently aphid-borne viruses as PVY and SMV are transmitted within seconds or minutes [49, 50], rendering pesticides useless. Under such circumstances, prospective alternatives to the use of chemical or synthetic inputs are environment-friendly microbial inoculants that received widespread acceptance worldwide as biofertilizers, phyto-stimulants, and/or microbial biocontrol agents.

2.4 Biocontrol Strategy

The biocontrol strategy that depends on the using of the microbial agents is fit well in the worldwide trend to produce vegetable crops which are safe for human and animal health [51]. For that reason, the use of plant growth-promoting microbes (PGPMs) as microbial inoculants in agriculture offers considerable advantages, due to their competitive colonization and their abilities to suppress phytopathogens and to enhance plant growth [15]. Therefore, PGPMs hold the prospect of reducing the input of chemical fertilizers, pesticides, and artificial growth regulators, and their inoculation can be regarded as an eco-friendly approach and biotechnological tool for sustainable agricultural applications.

3 Plant Growth-Promoting Bacteria (Pgbr)

A mixture of *Bacillus spp* (MML2501and MML2551), *Pseudomonas aeruginosa* (MML2212) and *Streptomyces fradiae* (MML1042) are formulated and significantly reduced SNV disease up to 51.4% compared to control and eventually improved the yield attributes in field conditions [52], some of the PGPR were used to eradicate the CMV through the induction of ISR on tomato [23]. The biocontrol activity of some bacterial strains *Bacillus* spp, *Kluyvera cryocrescens,* were significantly reduced the mean percentage of symptomatic plants in the field experiments with disease reduction reached from 32 to 58%.

Zehnder et al. [35] conducted a greenhouse screen of PGPR for the potential to elicit ISR against Cucumber mosaic virus (CMV) on tomato. Each of the three strains selected from the 26 tested significantly reduced the mean percentage of symptomatic plants in each of five experiments, with disease incidence ranging from 88 to 98% in the nonbacterized controls and 32–58% for the PGPR-treated plants. A significant reduction, from 50 to 80%, of TSWV incidence was recorded in the various *Bacillus amyloliquefaciens* strain MBI600 application schemes tested. The antiviral effect of MBI600 was proved robust under both plant growth chamber and greenhouse conditions although higher disease reduction scores were achieved under the former, presumably due to a better physiological state of the plants. Moreover, single dry-potting medium amendment prior germination and foliar applications performed equally well to triple drench application against TSWV, offering flexibility in a potential incorporation of MBI600 in integrated pest management programs. Moreover, MBI600 application against PVY resulted in reduced virus accumulation at the early stages of the infection and delay of virus detection in apical leaves [31].

For resistance enhancement, De Meyer et al. [53] enhanced the resistance of tobacco plants against TMV using *Pseudomonas aeruginosa*. While Park et al. [54] and Han et al. [55] induced of systemic resistance against TMV using an antiviral peptide from *Pseudomonas chlororaphis* O6, and an antiviral agent from *Strepcomces noursei* var *xichangensisn*, respectively. Ryu et al. [25, 56] protected *Arabidopsis thaliana* plants against CMV infection using the PGPR *Serratia marcescens* and a mixture of *Bacillus subtilis* GB03 and *B. amyloliquefaciens* IN937a. El-Dougdoug et al. [57] showed that culture filtrate of Streptomyces isolates controls CMV in *Chenopodium amaranticolor*. El-Borollosy and Oraby [58] using crude culture of *Azotobacter chroococcum* induced systemic resistance against CMV in cucumber plants. Khalimi and Suprapta [59] demonstrated that formulated *P. aeruginosa* as; gel, powder, liquid, increased the plant growth and induced resistance against SSV in soybean. Moreover, treatment of tobacco plants with *Bacillus spp*. enhanced the expression of the PR genes, NPR1 and Coi1, and led to increased resistance to CMV [60].

The study performed by Al Shami et al. [61] on the ability of *Frateuria aurantia, Bacillus megaterium* and *Azotobacter chroococcum* in reducing the CMV disease severity revealed that application with single bacteria may resulted in momentous declined in disease severity compared with *Bacillus megaterium* or *Azotobacter chroococcumin*. In contrary, Jetiyanon and Kloepper [62] reported that the mixed treatments with three bacterial species gave the highest reduction in disease severity and increased of free salicylic acid and peroxide activity contained in both CMV-infected and healthy tomato plants. Consequently, application of PGPR isolates viz., *B. amyloliquefaciens* 937b and *B. pumilus* SE-34 reduced ToMoV incidence and disease severity and provide protection of tomatoes against ToMoV under natural conditions [27]. Further, the exploitation of PGPRs was found to be effective to manage the cucumber mosaic virus of tomato and pepper [16], and BBTV in banana [28]. Mann [63] applied cultures of *Bacillus uniflagellatus* and extracts from such cultures to tobacco roots as soil drenches in an attempt to induce systemic resistance to TMV.

On the other hand, systemic acquired resistance (SAR) for virus infections can be induced in plants treated with certain *Streptomyces* strains [64]. *Streptomyces* spp. was the source of many useful and consequently profitable antiviral agents [65]. Many antiviral substances were isolated from *Streptomyces* spp., *i.e.* borrelidin, clindamycin and fattiviracins [66–68]. Foliage treatment with the *Streptomyces* culture filtrates resulted in high reduction of the level of disease severity of CMV infection [69], they got reduction in the diseases incidence ranged between 90 and 85%. The same observation was recorded when the cell-free suspension of *Streptomyces rochei* succeeded to inhibit TMV in leaves on *Datura metel* [70]. Galal and El-Shirbiny [71] enhanced the resistance of *Datura stramonium* plants against PVX using caeseorhodomycin (produced by *S. caeseorhodomyces*). However, Galal [64] mentioned that treatment of cucumber plants with the filtrate of five *Streptomyces* strains, *i.e. S. violatus, S. violaceuisniger, S. aureofaciens, S. nasri* and *Streptomyces* sp., resulted in induced systemic resistance against CMV when applied before virus inoculation than after viral inoculation. He mentioned that soaking cucumber seeds in culture filtrate of the *Streptomyces* strains for two hours gave the highest inhibition of CMV infection.

El-Dougdoug et al. [57] used five identified *Streptomyces* species that they are produce an antiviral component in the culture filtrate, non-phytotoxic and effective in local as well as systematically control of CMV infection. The culture filtrate treated part of the hypersensitive host; *Chenopodium amaranticolor* leaves showed 70.2, 71.4, 74.4, 80 and 82.6% inhibition of the production of local lesions compared to the untreated part of the leaves for *Streptomyces calvus, Streptomyces canarius, Strepo-tomyces vinaceusdrappus, Streptomyces nogalater* and *Streptomyces viridosporus*, respectively. Accordingly, Yassin and Galal [72] reported that the filtrate of some *Actinomycetes* had an inhibitory effect against TNV. Mohamed and Galal [65] found that mixing each isolate of *Streptomyces* spp. with *Potato virus Y* inoculum completely inhibited the inducing of necrotic local lesions produced on *Chenopodium amaranticolor*. Mixture of the *Streptomyces* isolates with the crude sap on infected source of TMV reduced the number of necrotic local lesions formed on *Datura metel* leaves [73, 74]. Bio-formulations of mixtures of the rhizobacterial isolate *Pseudomonas fluorescens* and endophytic *Bacillus spp.* at the time of planting and during third, fifth and seventh month after planting of banana were effective in reducing the incidence of BBTV under green-house (80%) and field conditions (52%) [28].

In a greenhouse experiment, two PGPR strains (*Pseudomonas fluorescens* FB11 and a *Rhizobium leguminosarum* bv. *viceae* FBG05 were examined either singly or mixed as inducer for the plant immune system to resist the BYMV virus infected *faba bean* [75]. The results demonstrated that each *Pseudomonas* and *Rhizobium* singly showed a significant reduction in both percent disease incidence and virus concentration.

Shoman et al. [76] study the capacity of foliage treatment of *Bacillus globisporus, Candida glabrata, Pseudomonas fluorescens,* and *Streptomyces gibsonii* to protect *Phaseolus vulgaris* plants from TNV infection. The results show that the culture filtrates of the four-rhizosphere microbial isolates reduced TNV incidence in bean plants. However, the filtrate of the two strains *P. fluorescens* and *S. gibsonii* when approached on the plant reduced the TNV infection with percentage ranged from 92 to 97, respectively. Lee and Ryu [24] reported that leaf-colonizing *Bacillus amyloliquefaciens* strain 5B6 that isolated from a cherry tree leaf was protected *Nicotiana benthamiana* and pepper plants against CMV.

Sofy et al. [77] showed that *R. leguminosarum* bv. *Viceae* reduced 55% of BYMV infectivity and decreased the level of disease severity with 36.6% related to infected faba bean plants. Megahed et al. [78] reported that *Bacillus circulans*, and *Pseudomonas fluorescens harzianum* had the ability to reduce the mean number of local lesions for ToMV infection on *Datura metel* plant. *P. fluorescens* showed the best treatment on reduction of local lesions numbers 49.16, 57.66 and 58.47% of microbial liquid culture, microbial cells or spores and microbial culture filtrate, respectively. While *B. circulans* reductions of local lesions numbers were 42.29, 46.83 and 47.35%

of microbial liquid culture, microbial cells or spores and microbial culture filtrate, respectively.

Pre-inoculation of the soil with the actinomycetes, *Streptomyces pactum* Act12, agent resulted in reduction of the TYLCV virus incidence in the tomato plants and increase the crop yield [79]. Pretreated of tomato plants with *Enterobacter asburiae* BQ9 had increased fresh mass and delayed the appearance of TYLCV symptoms for 7 days [80]. Moreover, the symptoms that developed on leaves were milder and less distinct than control. At 30 days after inoculation, the biocontrol efficacy of TYLCV reached 58.7%. However, after 45 days the biocontrol efficacy decreased to 42% but still provided significant disease reduction. The same observation was obtained when the potato seeds were coated with *Bacillus vallismortis* strain EXTN-1, a significant decrease of disease severity caused by PVY and PVX in potato plants compared to those in the untreated control [26].

Al-Ani et al. [81] study the activity of *Pseudomonas fluorescens* and Rhodotorula sp. to protect potato plants against PVY disease development under field conditions, and the results showed that the treatment of PVY-infected tubers with *P. fluorescens* and *Rhodotorula* sp revealed that such treatment increased the plant resistance against viral infection a combined with incensement in the plant growth and its dry weight. Maurhofer et al. [82] reported that growth of *Nicotiana glutinosa*, *N. tabacum* 'Xanthi nc' and *N. tabacum* 'Burley 63' in soil previously inoculated with *Pseudomonas fluorescens* strain CHA0 resulted in a significant reduction in lesion number, diameter and area after infection with TNV compared with TNV control. After 6 weeks of infection, all the plants tested showed resistance in leaves to infection with the virus. Kandan et al. [18] isolated some strains of *Pseudomonas fluorescens* from the rhizosphere of tomato and studied their activity against TSWV. A maximum disease reduction of 84% was observed in tomato plants treated with a mixture of three *P. fluorescens* strains (CoP-1, CoT-1, CHAO) followed by strain CHAO alone, and the strain mixture of CoT-1/CHAO compared to the untreated control plants. Ranasinghe et al. [83] used some *Pseudomonas* spp. (i.e. *P. fluorescens*, *P. putida*, *P. aeruginosa*, *P. taiwanensis* and *Bacillus* spp. isolates for the management of PRSV. The application of the bacterial isolates either by seed or by root dip method reduced the severity of PRSV symptoms on papaya leaves and fruits and increased root and shoot dry weight and plant growth. The details of some applications of PGPB as biocontrol agents for plant viral infections were listed in Table 1.

4   Plant Growth-Promoting Fungi (PGPF)

PGPF are a class of non-pathogenic, soil-borne, filamentous fungi that confer beneficial effects on plants [84]. Moreover, it was observed that *Fusarium equiseti* as PGPF have high capability to induce the plant systemic resistance against cucumber diseases [85]. Additionally, The PGPF, *Penicillium chrysogenum*, and its dry mycelium, have been reported as potential inducers of protection from fungal or viral diseases

Table 1 Plant growth-promoting bacteria (PGPB) used in plant viral biocontrol

Virus	PGPB	Tested plant	Reference
BBTV	*Pseudomonas fluorescens, Bacillus* sp.	Banana	[28]
BYMV	*Pseudomonas fluoresceni*	Faba bean	[75]
	R. leguminosarum bv. *Viceae*		[75, 77]
CMV	*Serratia marcescens*	*Arabidopsis thaliana*	[25]
	Bacillus subtilis, B. amyloliquefaciens		[56]
	Streptomyces spp., *S. calvus, S. canarius, S. vinaceusdrappus, S. nogalater, S. viridosporus*	*Chenopodium amaranticolor*	[57]
	Azotobacter chroococcum	Cucumber	[58]
	S. griseorebens, S. cavourensis		[69]
	Streptomyces spp.,*S. aureofaciens , S. nasri , S. violaceuisniger, S. violatus*		[64]
	11 mixture bacteria		[62]
	Bacillus amyloliquefaciens	Pepper, *Nicotiana benthamiana*	[24]
	Bacillus spp.	Tobacco	[60]
CMV	*Azotobacter chroococcum, Bacillus megateriu, Frateuria aurantia*	Tomato	[61]
	Bacillus amyloliquefaciens, Bacillus pumilus, Bacillus subtilus, Kluyvera cryocrescens		[23]
PRSVD	*Bacillus* spp., *P. aeruginosa, P. fluorescens, P. putida, P. taiwanensis*	Papaya	[83]
PVX	*S. caeseorhodomyces*	*Datura stramonium*	[71]
PVY	*Streptomyces* spp.	*Chenopodium amaranticolor*	[65]
	Bacillus vallismortis	Potato	[26]
	Pseudomonas fluorescens, Rhodotorula sp.	Potato	[81]
	Bacillus amyloliquefaciens	Tomato	[31]
SNV	*Bacillus Licheniformis Bacillus* sp. *Pseudomonas aeruginosa Streptomyces fradiae*	Sunflower	[52]
SSV	*P. aeruginosa*	Soybean	[59]

(continued)

Table 1 (continued)

Virus	PGPB	Tested plant	Reference
TMV	*Streptomyces rochei*	*Datura metel*	[70]
	Streptomyces spp.		[73, 74]
	Pseudomonas chlororaphis	Tobacco	[54]
	S. noursei var *xichangensisn*		[55]
	Pseudomonas aeruginosa		[53]
	Bacillus uniflagellatus		[63]
TNV	*Actinomycetes* sp.		[72]
	Pseudomonas fluorescens	*Nicotiana glutinosa, N. tabacum* 'Xanthi nc', *N. tabacum* 'Burley 63'	[82]
	Bacillus globisporus, Candida glabrata, Pseudomonas fluorescens, Streptomyces gibsonii	*Phaseolus vulgaris*	[76]
ToMV	*Bacillus circulans, P. fluorescens harzianum*	*Datura metel*	[78]
ToMoV	*B. amyloliquefaciens, B. pumilus*	Tomato	[27]
TSWV	*Bacillus amyloliquefaciens*	Tomato	[31]
	Pseudomonas fluorescens		[18]
TYLC	*Enterobacter asburiae*	Tomato	[80]
	Streptomyces pactum		[79]

in various crops [33]. It was reported that the water extract of dry mycelium of *Penicillium chrysogenum* protects tobacco from TMV infection through activating the synthesis of secondary metabolites and the expression of defense-related genes [33].

It was demonstrated that the pre-application of leaf colonizing yeast, *Pseudozyma churashimaensis* strain RGJ1, on pepper leaves reduced disease symptoms after infection of pepper with several viruses under field conditions [86]. Moreover, the quantification of naturally occurring CMV, BBWV, PepMoV and PMMoV by virus-specific primer-based qRT-PCR demonstrated that viral-mediated disease symptoms were significantly reduced by the foliar pre-application. *Penicillium simplicissimum* elicited the ISR against the cucumber mosaic virus in *Arabidopsis thaliana* and tobacco by multiple defense pathways including the salicylic acid signaling pathway [87]. Elsharkawy et al. [88] speculated that prior treatments with *Fusarium equiseti* alone or combined inoculation of *Glomus mosseae* induced SAR and protected cucumber plants against CMV for more than 21 day post inoculation (dpi).

Under pot and field conditions, application of PGPF, *Phoma sp.* GS8-3, and its culture filtrate led to reduction of CMV severity and viral accumulation in tobacco and cucumber plants [89]. In addition, the treatment of cucumber plants with culture filtrate of the *Penicillium simplicissimum* GP17-2 were significantly reduced the severity of PRSV and its accumulation in cucumber leaves through increased defense mechanism against PRSV [90]. It was postulated that a variety of plant systemic resistance could be brought by some *Trichoderma spp*. agansit some plant pathogens such as; fungi, bacteria and viruses [91]. Subsequently, roots treatment with culture filtrate of *Trichoderma asperellum* SKT-1 one-day before the challenged inoculation with CMV significantly reduced the disease severity and decreased CMV titer in arabidopsis plants relative to the controls plants [92].

The foliar spray of leaf-colonizing yeast, *Pseudozyma churashimaensis* strain RGJ1, at 108 cfu/mL conferred significant protection against CMV, PepMoV, PMMoV and BBWV under field conditions [86]. They observed significantly fewer viral symptoms such as mosaic leaf, shoestring patterns, or leaf shape deformation on yeast-treated plants than on water-treated controls.

Kolase and Sawant [93] reported that some *Trichoderma* spp succeeded to increase the *Nicotiana glutinosa* plant resistance system against TMV infection. Megahed et al. [78] study the effect of *Trichoderma harzianum* on ToMV infection. *T. harzianum* reduced the mean number of ToMV local lesions on *Datura metel* plants. The percentage reduction of mean local lesions numbers was 25.66, 19.58 and 33.38% of microbial liquid culture, microbial cells or spores and microbial culture filtrate, respectively. Tobacco plants treated with barley grain inoculums of *Penicillium sp*. GP16-2 or its culture filtrate showed significant decline in CMV severity and concentration in comparison with control plants. At 14 dpi, plants grown in soils amended with the BGI of GP16-2 displayed a dramatic reduction in CMV symptoms in comparison with the non-treated control plants. Moreover, roots treatment with culture filtrate 1-day before CMV inoculation significantly reduced disease severity in tobacco plants relative to the controls [94]. Also, Vitti et al. [95] study the ability of *Trichoderma harzianum*, strain T-22 to control CMV in tomato plants at three different treatments. Among three months old tomato plants, the treatment with T22 always led to a significant modulation of symptoms, in all three conditions (PD, PE, and PF) and a significant decrease in virus level in tomato. Thus, the early treatment is able to induce systemic defense responses against CMV and led to a systemic resistance of tomato. The details of some applications of PGPF as biocontrol agents for plant viral infections were listed in Tables 1 and 2.

5 The Case Study: Biocontrol of Potato Leaf Roll Virus (PLRV)

The PLRV virus particles were detected by the PCR using the coat protein primers on the Almond potato tubers before starting the experiment. It was observed that about 30% of tubers were infected with the *Potato leaf roll virus*. On the other hand, the overnight culture filtrates of five isolates of *bacillus* spp were mixed with nanocaly particles (Egypt Bentonite and Derivatives Co. - EBDC, new Borg El Arab, Alexandria, Egypt). The Potato tubers were soaked for two hours in the mixture (1 L of filtrate to 39 L of H_2O mixed with one Kgm on nano-clay). After the two

Table 2 Plant growth-promoting fungi (PGPF) used in plant viral biocontrol

Virus	PGPF	Tested plant	Reference
BBWV	*Pseudozyma churashimaensis*	Pepper	[86]
CMV	*Penicillium simplicissimum*	*Arabidopsis thaliana*	[87]
	Trichoderma asperellum		[92]
	Phoma sp. GS8-3	Cucumber	[89]
	Fusarium equiseti, Glomus mosseae		[88]
	Pseudozyma churashimaensis	Pepper	[86]
	Penicillium spp.	Tobacco	[94]
	Phoma spp.		[89]
	Trichoderma harzianum	Tomato	[95]
PepMoV	*Pseudozyma churashimaensis*	Pepper	[86]
PMMoV			
PMV			
PRSV	*Penicillium simplicissimum*	Cucumber	[90]
TMV	*Penicillium chrysogenum*	Tobacco	[33]
	Trichoderma harzianum, T. longisporum, T. viride.	Tomato, *Nicotiana glutinosa*	[93]
ToMV	*Trichoderma harzianum*	*Datura metel*	[78]

hours of dipping (Fig. 1), the treated tubers were planted in soil. The experiment was designed as treated field (five acres) and non-treated field (1 acre) in Kafr El-Sheikh Governorate in the last season of 2018. The results showed that after two months of planting, plant leaves were free of any viral symptoms, while the non-treated field showed many viruses diseases symptoms, among which the *Potato leaf roll virus*. Later, the field was sprayed with the mixture two times in the third month and the beginning of forth month. In the end of the experiment it was observed that, no viral symptoms was noticed in the treated plants, the number of the tubers was ranged from 5 to 8 tubers in each plant and the size of the tuber was more larger when compared with the non-treated plants (Fig. 2). It can conclude that our product (proposed) will be approached on large farmer and the results should be compared with the other commercialized pesticides in the market. The list of plant virus names and their abbreviations used in this chapter was listed in Table 3

6 Conclusion

As we mentioned above, there have been many attempts to manufacture viral biocontrol agents, whether bacterial or fungal, and how this was applied in the laboratory, green houses and in the open field. All these attempts were mostly international but

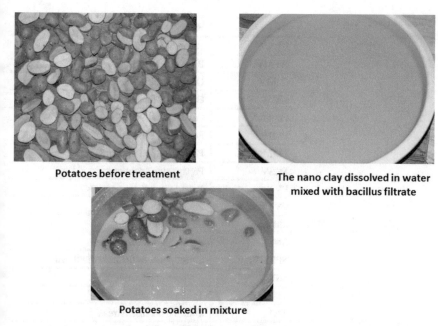

Potatoes before treatment

The nano clay dissolved in water mixed with bacillus filtrate

Potatoes soaked in mixture

Fig. 1 The treatment process of the Almond potato tubers in the nano-clay mixed with the bacillus culture filtrate

The open filed trail: untreated plants

The open filed trail: treated plants

Fig. 2 Open filed experiment: the three above pictures showed the plants without treatment contains some viruses symptoms, the below picture showed the plant free viruses as a results of the treatment

Table 3 List of virus abbreviation

BBTV	Banana bunchy top virus
BBWV	Broad bean wilt virus
BYMV	Bean yellow mosaic virus
CaMV	Cauliflower mosaic virus
CMV	Cucumber mosaic virus
PepMoV	Pepper mottle virus
PLRV	Potato leaf roll virus
PMMoV	Pepper mild mottle virus
PRSV	Papaya ring spot virus
PVX	Potato virus X
PVY	Potato virus Y
SMV	Soybean mosaic virus
SNV	Sunflower necrosis virus
SPFMV	Sweet potatoes feathery mottle virus
SSV	Soybean stunt virus
TMV	Tobacco mosaic virus
TNV	Tobacco necrosis virus
ToMoV	Tomato mottle virus
ToMV	Tomato mosaic virus
TSWV	Tomato spot wilt virus
TYLCV	Tomato yellow leaf curl virus

one or two attempts were local and even wasn't reach the field application yet. We would like to say that the antiviral products are still in their early stages, and the market for these bio antiviral has only a few numbers, which do not exceed the fingers of one hand. Work in this area is still open until humanity can rid itself of the bad effects of pesticides and the huge losses caused by plant viruses. Bio-anti viruses have been internationally and locally accepted by senior and small farmers as well as in scientific media. In fact, there is no evidence so far that there is a substance whether naturally synthesized or manufactured can eliminate the plant virus within the plant host cell. But all the available bio/compounds which have antiviral activity resulted from their capability to stimulate the plant immune response .

References

1. Fenner F, Maurin J (1976) The classification and nomenclature of viruses. Arch Virol 51:141–149
2. Bryant JL (2008) Animal models in virology. In: Sourcebook of models for biomedical research. Springer, pp 557–563

3. Willey J (2008) Prescott, Harley, and Klein's Microbiology-7th international. In: Willey JW, Sherwood LM, Woolverton CJ (eds). McGraw-Hill Higher Education, New York [etc.]
4. Strange RN, Scott PR (2005) Plant disease: a threat to global food security. Annu Rev Phytopathol 43:83–116
5. Gergerich RC, Dolja VV (2006) Introduction to plant viruses, the invisible foe. Plant Health Instruct. https://doi.org/10.1094/PHI-I-2006-0414-01
6. Hull R (2014) Plant virology, 5th edn. Elsevier, London, United Kingdom
7. Bisnieks M, Kvarnheden A, Turka I, Sigvald R (2006) Occurrence of *barley yellow dwarf virus* and *cereal yellow dwarf virus* in pasture grasses and spring cereals in Latvia. Acta Agric Scand Sect B Soil Plant Sci 56:171–178
8. Hafez E, El-Morsi A, El-Shahaby O, Abdelkhalek A (2014) Occurrence of *iris yellow spot virus* from onion crops in Egypt. Virus Disease 25:455–459
9. Abdelkhalek A, Sanan-Mishra N (2019) Differential expression profiles of tomato miRNAs induced by *Tobacco Mosaic Virus*. J Agr Sci Tech 21:475–485
10. Lewsey M, Palukaitis P, Carr JP (2009) Plant—virus interactions: defence and counter-defence. In: Parker J (ed) Molecular aspects of plant disease resistance. Wiley-Blackwell, Oxford, pp 134–176
11. Abdelkhalek A, Eldessoky D, Hafez E (2018) Polyphenolic genes expression pattern and their role in viral resistance in tomato plant infected with *Tobacco mosaic virus*. Biosci Res 15:3349–3356
12. Abdelkhalek A, ElMorsi A, AlShehaby O, Sanan-Mishra N, Hafez E (2018) Identification of genes differentially expressed in *Iris Yellow Spot Virus* infected onion. Phytopathologia Mediterranea 57:334–340
13. Arias-Estévez M, Lopez-Periago E, Martinez-Carballo E, Simal-Gandara J, Mejuto JC, Garcia-Rio L (2008) The mobility and degradation of pesticides in soils and the pollution of ground water resources. Agric Ecosyst Environ 123:247–260
14. Nayak SK, Dash B, Baliyarsingh B (2018) Microbial remediation of persistent agro-chemicals by soil bacteria: an overview. In: Patra J, Das G, Shin HS (eds) Microbial Biotechnol. Springer, Singapore
15. Abdel-Gayed M, Abo-Zaid G, Matar S, Hafez E (2019) Fermentation, formulation and evaluation of PGPR Bacillus subtilis isolate as a bioagent for reducing occurrence of peanut soil-borne diseases. J Integr Agric. https://doi.org/10.1016/S2095-3119(19)62578-5
16. Murphy JF, Reddy MS, Ryu CM, Kloepper JW, Li R (2003) Rhizobacteria mediated growth promotion of tomato leads to protection against cucumber mosaic virus. Phytopathology 93:1301–1307
17. Gray EJ, Smith DL (2005) Intracellular and extracellular PGPR: commensalities and distinctions in the plant-bacterium signaling processes. Soil Biol Biochem 37:395–412
18. Kandan A, Ramaiah M, Vasanthi VJ, Radjacommare R, Nandakumar R, Ramanathan A, Samiyappan R (2005) Use of Pseudomonas fluorescens based formulations for management of *tomato spot wilt virus* (TSWV) and enhanced yield in tomato. Biocontrol Sci Tech 15:553–569
19. Jetiyanon K, Fowler WD, Kloepper JW (2003) Broad spectrum protection against several pathogens by PGPR mixtures under field conditions. Plant Dis 87:1390–1394
20. Hass D, Defago G (2005) Biological control of soil-borne pathogens by fluorescent pseudomonads. Nat Rev Microbiol 3:307
21. Matar S, El-Kazzaz S, Wagih E, El-Diwany A, Moustafa H, Abo-Zaid G, Abd-Elsalam HE, Hafez E (2009) Antagonistic and inhibitory effect of *Bacillus subtilis* against certain plant pathogenic fungi. Biotechnology 8:53–61
22. Beneduzi A, Ambrosini A, Passaglia LMP (2012) Plant growth-promoting rhizobacteria (PGPR): Their potential as antagonists and biocontrol agents. Genet Mol Biol 35:1044–1051
23. Zehnder GW, Yao C, Murphy JF, Sikora ER, Kloepper JW (2000) Induction of resistance in tomato against *Cucumber mosaic cucumovirus* by plant growth-promoting rhizobacteria. Biocontrol 45:127–137
24. Lee GH, Ryu CM (2016) Spraying of leaf-colonizing *Bacillus amyloliquefaciens* protects pepper from *Cucumber mosaic virus*. Plant Dis 100:2099–2105

25. Ryu CM, Murphy JF, Mysore KS, Kloepper JW (2004) Plant growth-promoting rhizobacteria systemically protect *Arabidopsis thaliana* against *Cucumber mosaic virus* by a salicylic acid and NPR1-independent and jasmonic acid-dependent signaling pathway. Plant J 39:381–392
26. Park KS, Paul D, Ryu KR, Kim EY, Kim YK (2006) *Bacillus vallismortis* strain EXTN-1 mediated systemic resistance against *Potato Virus X* and Y (PVX & PVY) in the field. Plant Pathol J 22:360–363
27. Murphy JF, Zehnder GW, Schuster DJ, Sikora EJ, Polstan JE, Kloepper JW (2000) Plant growth-promoting rhizobacterial mediated protection in tomato against *Tomato mottle virus*. Plant Dis 84:779–784
28. Harish S, Kavino M, Kumar N, Balasubramanian P, Samiyappan R (2009) Induction of defense-related proteins by mixtures of plant growth promoting endophytic bacteria against *Banana bunchy top virus*. Biol Control 51:16–25
29. Raupach GS, Liu L, Murphy JF, Tuzun S, Kloepper JW (1996) Induced systemic resistance in cucumber and tomato against *Cucumber mosaic cucmovirus* using plant growth-promoting rhizobacteria (PGPR). Plant Dis 80:891–894
30. Kim YS, Hwang EI, Jeong-Hun O, Kim KS, Ryu MH, Yeo WH (2004) Inhibitory effects of *Acinetobacter* sp. KTB3 on infection of *Tobacco mosaic virus* in tobacco plants. Plant Pathol J 20:293–296
31. Beris D, Theologidis I, Skandalis N, Vassilakos N (2018) *Bacillus amyloliquefaciens* strain MBI600 induces salicylic acid dependent resistance in tomato plants against *Tomato spotted wilt virus* and *Potato virus Y*. Scientific Reports 8:10320
32. Kloepper JW, Ryu CM, Zhang S (2004) Induced systemic resistance and promotion of plant growth by *Bacillus* spp. Phytopathology 94(11):1259–1266
33. Zhong Y, Peng J-j, Chen Z-z, Xie H, Luo D, Dai J-r, Yan F, Wang J-g, Dong H-z, Chen S-y (2015) Dry mycelium of *Penicillium chrysogenum* activates defense responses and restricts the spread of *Tobacco Mosaic Virus* in tobacco. Physiol Mol Plant Pathol 92:28–37
34. Cerqueira-Silva CB, Moreira CN, Figueira AR, Corrêa RX, Oliveira AC (2008) Detection of a resistance gradient to *Passion fruit woodiness virus* and selection of 'yellow' passion fruit plants under field conditions. Genet Mol Res 7:1209–1216
35. Lecoq H, Moury B, Desbiez C, Palloix A, Pitrat M (2004) Durable viral resistance in plants through conventional approaches: a challenge. Virus Res 100:31–39
36. Prins M, Laimer M, Noris E, Schubert J, Wassenegger M, Tepfer M (2008) Strategies for antiviral resistance in transgenic plants. Mol Plant Pathol 9:73–83
37. Tepfer M (2002) Risk assessment of virus-resistant transgenic plants. Annu Rev Phytopathol 40:467–491
38. Giovannetti M, Sbrana C, Turrini A (2005) The impact of genetically modified crops on soil microbial communities. Riv Biol 98:393–417
39. Ho MW, Ryan A, Cummins J (1999) *Cauliflower mosaic viral* promoter- a recipe for disaster. Microb Ecol Health Dis 11:194–197
40. Saxena D, Flores S, Stotzky G (1999) Insecticidal toxin in root exudates from *Bt* corn. Nature 402:480
41. Zwahlen C, Hilbeck A, Gugerli P, Nentwig W (2003) Degradation of the Cry1Ab protein within transgenic *Bacillus thuringiensis* corn tissue in the field. Mol Ecol 12:765–775
42. Mercer KL, Wainwright JD (2008) Gene flow from transgenic maize to landraces in Mexico: an analysis. Agri Ecosyst Environ 123:109–115
43. Prakash D, Verma S, Bhatia R, Tiwary BN (2011) Risks and precautions of genetically modified organisms. ISRN Ecol 369573:13. https://doi.org/10.5402/2011/369573
44. Eicher CK, Maredia K, Sithole-Niang I (2006) Crop biotechnology and the African farmer. Food Policy 31(6):504–527
45. Franks A, Ryan RP, Abbas A, Mark GL, O'Gara F (2006) Molecular tools for studying plant growth promoting rhizobacteria. In: Cooper JE, Rao JR (eds) Molecular approaches to soil rhizosphere and plant microorganisms analysis. Biddes Ltd Kings, Lynn, pp 116–131
46. Glick BR (2014) Bacteria with ACC deaminase can promote plant growth and help to feed the world. Microbiol Res 169:30–39

47. Hilje L, Costa HS, Stansly PA (2001) Cultural practices for managing Bemisia tabaci and associated viral diseases. Crop Prot 20:801–812
48. Palumbo JC, Horowitz AR, Prabhaker N (2001) Insecticidal control and resistance management for *Bemisia tabaci*. Crop Prot 20:739–766
49. Satapathy MK (1998) Chemical control of insect and nematode vectors of plant viruses. In: Hadidi A, Khetarpal RK, Koganezawa H (eds) Plant virus disease control. APS Press, St. Paul, MN, USA. pp 188–195
50. Fereres A (2000) Barrier crops as a cultural control measure of non-persistently transmitted aphid-borne viruses. Virus Res 71:221–231
51. Prasad RD, Rangeshwaran R (2000) Effect of soil application of a granular formulation of *Trichoderma harzianum* on *Rhizoctonia solani* incited seed rot and damping-off of chickpea. J Mycol Plant Pathol 30:216–220
52. Srinivasan K, Mathivanan N (2009) Biological control of *sunflower necrosis virus* disease with powder and liquid formulations of plant growth promoting microbial consortia under field conditions. Biol Control 51:395–402
53. De Meyer G, Audenaert K, Höfte M (1999) *Pseudomonas aeruginosa 7NSK2*-induced systemic resistance in tobacco depends on in planta salicylic acid accumulation but is not associated with PR1a expression. Eur J Plant Pathol 105:513–517
54. Park JY, Yang SY, Kim YC, Kim JC, Le Dang Q, Kim JJ, Kim IS (2012) Antiviral peptide from *Pseudomonas chlororaphis* O6 against Tobacco Mosaic Virus (TMV). J Korean Soc Appl Biol 55:89–94
55. Han Y, Luo Y, Qin S, Xi L, Wan B, Du L (2014) Induction of systemic resistance against *Tobacco Mosaic Virus* by Ningnanmycin in tobacco. Pestic Biochem Phys 111:14–18
56. Ryu C, Murphy JF, Reddy M, Kloepper JW (2007) A two-strain mixture of rhizobacteria elicits induction of systemic resistance against Pseudomonas syringae and *Cucumber mosaic virus* coupled to promotion of plant growth on Arabidopsis thaliana. J Microbiol Biotechnol 17:280
57. El-Dougdoug KhA, Ghaly MF, Taha MA (2012) Biological control of *Cucumber Mosaic Virus* by certain local streptomyces isolates: inhibitory effects of selected five Egyptian isolates. Int J Virol 8:151–164
58. El-Borollosy AM, Oraby MM (2012) Induced systemic resistance against Cucumber mosaic cucumovirus and promotion of cucumber growth by some plant growth-promoting rhizobacteria. Ann Agric Sci 57:91–97
59. Khalimi K, Suprapta DN (2011) Induction of plant resistance against Soybean stunt virus using some formulations of Pseudomonas aeruginosa. J ISSAAS Int Soc Southeast Asian Agric Sci 17:98–105
60. Wang S, Wu H, Qiao J, Ma L, Liu J, Xia Y, Gao X (2009) Molecular mechanism of plant growth promotion and induced systemic resistance to *Tobacco mosaic virus* by Bacillus spp. J Microbiol Biotechnol 19:1250–1258
61. Al Shami R, Ismail I, Hammad Y (2017) Effect of three species of rhizobacteria (PGPR) in stimulating systemic resistance on tomato plants against *Cucumber Mosaic Virus* (CMV). SSRG-IJAES 4(6):11–16
62. Jetiyanon K, Kloepper JW (2002) Mixtures of plant growth-promoting rhizobacteria for induction of systemic resistance against multiple plant diseases. Biol Control 24:285–291
63. Mann EW (1965) Inhibition of *tobacco mosaic virus* by a bacterial extract. Phytopathology 59:658–662
64. Galal AM (2006) Induction of systemic acquired resistance in cucumber plant against Cucumber mosaic *cucumovirus* by local *Streptomyces* strains. Pl Pathol J 5:343–349
65. Sonya HM, Galal AM (2005) Identification and antiviral activities of some halotoletant *Streptomycetes* isolated from Qaroonlake. Int J Agric Biol 7:747–753
66. Ghaly MF, Awny AM, Galal AM, Askora A (2005) Characterization and action of antiphytoviral agent produced by certain *Streptomyces* species against *Zucchini yellow mosaic virus*. Egypt. J Biotechnol 19:209–223
67. Bhikshapathi DVRN, Krishna DR, Kishan V (2010) Anti-HIV, antitubercular and mutagenic activities of borrelidin. Indian J Biotechnol 9:265–270

68. Chaudhary HS, Soni B, Shrivastava AR, Shrivastava S (2013) Diversity and versatility of *Actinomycetes* and its role in antibiotic production. J Appl Pharm Sci 3:883–894
69. Shafie RM, Hamed AH, El-Sharkawy HHA (2016) Inducing systemic resistance against *Cucumber Mosaic Cucumovirus* using *Streptomyces* spp. Egypt J Phytopathol 44:127–142
70. Mansour FA, Soweha HE, Desouki SSAS, Mohamadin AH (1988) Studies on the antiviral activity of some bacterial isolates belonging to Streptomycetes. Egypt J Bot 31:167–183
71. Galal AM, El-Sherbieny SA (1995) Antiphytoviral activity of caesearhodomycin isolated from Streptomyces caesius var. Egypt Fac Educ, Ain Shams Univ 20:121–128
72. Yassin MH, Galal AM (1998) Antiphytoviral potentialities of some fungi and *actinomycete* isolates against kidney bean plants infected with TNV. In: 1st Conf Protec Egypt, pp 156–161
73. Ara I, Bukhari NA, Aref NM, Shinwari MMA, Bakir MA (2012) Antiviral activities of *Streptomycetes* against *Tobacco mosaic virus* (TMV) in datura plant: evaluation of different organic compounds in their metabolites. African J Biotechnol 11:2130–2138
74. Mohamed SH, Omran WM, Abdel-Salam MS, Sheri ASA, Sadik AS (2012) Isolation and identification of some halotolerant actinomycetes having antagonistic activities against some plant pathogens (*i.e. Tobacco mosaic virus, Aspergillus* sp. & *Fusarium* sp.) from soil of Taif governorate KSA. Pak J Biotechnol 9:1–12
75. Elbadry M, Taha RM, Eldougdoug KA, Gamal-Eldin H (2006) Induction of systemic resistance in faba bean (*Vicia faba* L.) to *bean yellow mosaic potyvirus* (BYMV) via seed bacterization with plant growth promoting rhizobacteria. J Plant Dis Protect 113(6):247–251
76. Shoman SA, Abd-Allah NA, El-Baz AF (2003) Induction of resistance to *Tobacco necrosis virus* in bean plants by certain microbial isolates. Egypt J Biol 5:10–18
77. Sofy AR, Attia MS, Sharaf AMA, El-Dougdoug KhA (2014) Potential impacts of seed bacterization or salix extract in faba bean for enhancing protection against bean yellow mosaic disease. Nat Sci 12(10):67–82
78. Megahed AA, El-Dougdoug KA, Othman BA, Lashin SM, Ibrahim MA, Sofy AR (2013) Induction of resistance in tomato plants against *tomato mosaic tobamovirus* using beneficial microbial isolates. Pak J Biol Sci 16:385–390
79. Li Y, Guo Q, Li Y, Sun Y, Xue Q, Lai H (2019) Streptomyces pactum Act12 controls tomato yellow leaf curl virus disease and alters rhizosphere microbial communities. Biol Fertil Soils 55:149–169
80. Li H, Ding X, Wang C, Ke H, Wu Z, Wang Y, Liu H, Guo J (2016) Control of *tomato yellow leaf curl virus* disease by Enterobacter asburiae BQ9 as a result of priming plant resistance in tomatoes. Turk J Biol 40:150–159
81. Al-Ani AR, Adhab AM, Matny NO (2013) Management of *potato virus Y* (PVY) in potato by some biocontrol agent under field condition. Int J Microbiol Mycol 1(1):1–6
82. Maurhofer M, Hase C, Meuwly Ph, Métraux J-P, Défago G (1994) Induction of systemic resistance of tobacco to *tobacco necrosis virus* by the root colonizing *Pseudomonas fluorescens* strain CHA0: influence of the gacA gene and of pyoverdine production. Phytopathology 84:139–146
83. Ranasinghe C, De Costa1 DM, Basnayake BMVS, Gunasekera DM, Priyadharshani S, Navagamuwa NVR (2018) Potential of Rhizobacterial *Pseudomonas* and *Bacillus* spp. to *Manage Papaya Ringspot Virus* Disease of Papaya (*Carica papaya* (L.). Trop Agric Res 29(4):271–283
84. Hyakumachi M (1994) Plant-growth-promoting fungi from turfgrass rhizosphere with potential for disease suppression. Soil Microorg 44:53–68
85. Macia-Vicente JG, Jansson HB, Talbot NJ, Lopez-Llorca LV (2009) Real-time PCR quantification and live-cell imaging of endophytic colonization of barley (Hordeum vulgare) roots by Fusarium equiseti and Pochonia chlamydosporia. New Phytol 182:213–228
86. Lee G, Lee S-H, Kim KM, Ryu C-M (2017) Foliar application of the leaf colonizing yeast *Pseudozyma churashimaensis* elicits systemic defense of pepper against bacterial and viral pathogens. Sci Rep 7:39432
87. Elsharkawy M, Shimizu M, Takahashi H, Hyakumachi M (2012) Induction of systemic resistance against *Cucumber mosaic virus* by Penicillium simplicissimum GP17-2 in Arabidopsis and tobacco. Plant Pathol 61:964–976

88. Elsharkawy MM, Shimizu M, Takahashi H, Hyakumachi M (2012) The plant growth-promoting fungus *Fusarium equiseti* and the arbuscular mycorrhizal fungus *Glomus mosseae* induce systemic resistance against *Cucumber mosaic virus* in cucumber plants. Plant Soil 361:397–409
89. Elsharkawy MM, Suga H, Shimizu M (2018) Systemic resistance induced by Phoma sp. GS8–3 and nanosilica against *Cucumber mosaic virus*. Environ Sci Pollut Res. https://doi.org/10.1007/s11356-018-3321-3
90. Elsharkawy MM, Mousa KM (2015) Induction of systemic resistance against *Papaya ring spot virus* (PRSV) and its vector Myzus persicae by *Penicillium simplicissimum* GP17-2 and silica (Sio2) nanopowder. Int J Pest Manag 61:353–358
91. Harman GE, Howell CR, Viterbo A, Chet I, Lorito M (2004) Trichoderma species-opportunistic avirulent plant symbionts. Nat Rev 2:43–56
92. Elsharkawy MM, Shimizu M, Takahashi H, Ozaki K, Hyakumachi M (2013) Induction of systemic resistance against *Cucumber mosaic virus* in Arabidopsis thaliana by *Trichoderma asperellum* SKT-1. Plant Pathol J 29:193–200
93. Kolase SV, Sawant DM (2007) Isolation and efficacy of antiviral principles from Trichoderma spp. against *Tobacco Mosaic Virus* (TMV) on tomato. J Maharashtra Agric Univ 32:108–110
94. Elsharkawy MM, Abass JM, Kamel SM, Hyakumachi M (2017) The plant growth promoting fungus *Penicillium* sp. GP16-2 enhances the growth and confers protection against *Cucumber mosaic virus* in tobacco. J Virol Sci 1:145–154
95. Vitti A, Pellegrini E, Nali C, Lovelli S, Sofo A, Valerio M, Scopa A, Nuzzaci M (2016) *Trichoderma harzianum* T-22 Induces Systemic Resistance in Tomato Infected by *Cucumber mosaic virus*. Front Plant Sci 7:1520

Bio-products Against Abiotic Factors

Biochemical Indicators and Biofertilizer Application for Diagnosis and Allevation Micronutrient Deficiency in Plant

Zeinab A. Salama and Magdi T. Abdelhamid

Abstract Deficiencies in micronutrients are well established causal factors for sub-optimal plant production. Therefore, the first option is to apply micronutrient-containing fertilizers, both chemical and organic, and application method should be based on soil nutrient management guidelines along with crop types. It is often misleading to identify nutrient constraints based on morphological symptoms or in combination with leaf/soil analysis, particularly with regard to remedying the nutritional problems of a standing crop. The objective of this chapter is to identify and diagnose common symptoms of plant nutrient deficiency and to understand how to use chemical indicators and bio fertilizer to identify and alleviate deficiencies of micronutrients. The possibilities of using biochemical markers and application of biofertilizers to diagnose and mitigate deficiencies in micronutrients are presented. Foliar analysis is a useful tool for detecting deficiencies of micronutrients before macroscopic symptoms occur in plants. This work will therefore focus on the other diagnostic tools used to evaluate micronutrient deficiencies that include soil analysis, plant-growth response (in annual plants), and visual symptom observation. More biochemical indicators recently have been used as early detectors of deficiencies in micronutrients i.e. for iron (Fe), include peroxidase and catalase, and zinc (Zn) include carbonic anhydrase, superoxide dismutase, protein electrophoresis, and isozymes. The roles of the necessary micronutrients for normal plant growth need to have a deep knowledge of the redox system mechanisms. The biochemical markers for the deficiency of micronutrients in diagnosis and the role and benefits of application of bio fertilizers are presented in detail and discussed in this chapter in full. One more aim of this chapter is to summarize the updating knowledge of biochemical markers and the role of biofertilizers that would play a key role in soil productivity and sustainability as well as protecting the environment as environmentally friendly and cost-effective inputs for farmers.

Z. A. Salama
Plant Biochemistry Department, National Research Centre, EL Bohouth St, Dokki, Giza 12622, Egypt

M. T. Abdelhamid (✉)
Botany Department, National Research Centre, EL Bohouth St, Dokki, Giza 12622, Egypt
e-mail: magdi.abdelhamid@yahoo.com

© Springer Nature Switzerland AG 2020
N. El-Wakeil et al. (eds.), *Cottage Industry of Biocontrol Agents and Their Applications*, https://doi.org/10.1007/978-3-030-33161-0_14

Keywords Micronutrient deficiency · Biochemical markers · Bio fertilizer · Crop productivity

1 Introduction

Micronutrients like iron (Fe), copper (Cu), zinc (Zn), manganese (Mn), and boron (B) are important for an abundance of physiological functions in plant growth, development, and oxidative stress response. At low concentrations in electron transport and antioxidant systems, micronutrients are desired for cellular structures and protein stabilization [1].

Micronutrients are metalloprotein cofactors that involve storage and transportation, signal transduction, enzymatic reactions, and other functions. Micronutrients penetrate biomembranes either through unspecified ion carriers or through passive diffusion, metal efflux, or heavy metal carrying ATPases. The bioavailability of others may be affected by declining levels of one micronutrient [2]. Homeostasis of micronutrients must be strictly regulated to sustain cellular processes and prevent oxidative stress.

Micronutrient bioavailability depends on several factors. Hydrous Fe and Mn oxides induce precipitation and specific adsorption reactions to the solubility of Zn and Cu [3]. Ordinary agricultural practices for instance intensive agriculture, monoculture and acid soil liming, exhausting micronutrients from the soil, or reducing bioavailability. Therefore, the deficiency of micronutrients in agricultural soils is widespread. In addition, it has been shown that extensive glyphosate use interferes with Fe, Zn, and Mn uptake [4, 5]. Climate is another factor affecting micronutrient bioavailability. Drought, heavy rainfall or waterlogging cause either micronutrient deficiency or toxicity [6, 7]. These events become more frequent and affect crop production due to global warming and climate change.

Last century studies focused on morphological symptoms, physiological functions and the transport of micronutrients (e.g., [8, 9]). The rapid development of omics technologies and the decoding of crop genomes are currently promoting a view of molecular mechanisms in micronutrient deficiencies and toxicity in more biological systems [10, 11]. Bottom-up and top-down strategies with gel-free and gel-based proteomics have been used to explore the availability of plant nutrients. Fe homeostasis has been a major focus, whereas only a few studies have been published for other micronutrients like Cu, Zn, Mn, and B. The degree of adequacy of different elements in plant nutrition is measured by the appearance of symptoms of deficiency or by the plant tissue elemental analysis [12].

Metabolic processes in root and shoot levels are affected in such micronutrient deficiencies. Many authors therefore appreciate the use of distinct physiological and biochemical responses for monitoring the nutritional status of different plant species in root and leaves of different plant species. Recently, proton release, Fe reduction mechanism, root exudate release, release of phytosiderophores compounds (PS) in grasses, lipid peroxidation as well as physiologically active iron and zinc, pigment content, enzyme activity and isozymes have been suggested as diagnostic parameters for early diagnosis of micronutrient deficiencies in different plant species. In this chapter, we epitomize sources and factors that influence the behavior and availability of micronutrients in soils, possible approaches to control the availability of soil micronutrients, and present physiological and biochemical changes in plants, and then explain some strategies to increase the availability of micronutrients in soil for plant uptake.

2   Sources and Factors Affecting Soil Micronutrients

The soil is made up of a variety of parent materials like minerals and organic matter and has a range of micronutrients depending on their composition. When dissolved in soil solution in ion or chelating forms, micronutrients become available to plants. Nevertheless, the dissolved micronutrients undertake a rapid reaction with compounds such as phosphates and carbonates to form chemical precipitates, or they may interact with clay, other mineral complexes, and organic matter resulting in plant micronutrient unavailability [13]. The availability of micronutrients in soils is arranged under the influence of different edaphic and biological factors such as pH, redox potential, interaction with coexisting ions, organic matter dynamics, and soil microbiology. In igneous rocks, concentrations of Fe, Mn and Cu are generally found to be higher than in sedimentary rocks such as calcareous and dolomite, an increase in the order of basic (mafic), intermediate compared to acid (felsic) rocks. Micronutrient concentrations in soils also showed wide ranges affected by soil formation/degradation processes and parent material types. Because of their low mobility compared to other major elements such as calcium (Ca), magnesium (Mg), potassium (K) and silicon (Si) [14], micronutrients are usually concentrated from rock to soil during rock weathering and soil formation processes [14].

3   Regular Behavior of Micronutrients in Soils

In previous works (for example [15, 16]), soil micronutrient chemistry and its status are well summarized. Ferro magnesium (olivine), which releases Fe by weathering, is the widespread primary mineral of Fe. The released Fe precipitates like ferric oxides and hydroxides.

Fe concentration in soil solution is mainly governed by Fe^{3+}. Fe oxide solubility such as hematite (Fe_2O_3), goethite (FeO[OH]), and magnetite (Fe_3O_4) are very low compared to other micronutrient minerals. In aerated soil conditions, the predominant form is Fe^{3+} in the soil solution, which is very low compared to other micronutrient cations such as Mn, Cu and Zn (around 10–9–10–20 M Fe^{3+} at 4 to 8 pH).

Mn^{2+} ranges from 0.01 to 1.0 mg L^{-1} and increases below -200 mV under submerged conditions under low pH and redox potential, is the common form of manganese in a soil solution. Similar to Fe, reducing the potential for redox in acid soil results in Mn toxicity, which is rich in total Mn. On the contrary, by liming in acid soils, an increase in pH precipitates Mn^{2+} as MnO_2 and decreases solution and Mn^{2+} exchangeable. In soils, $ZnFe_2O_4$ (franklinite) can account for Zn^{2+}'s solubility depicted in soil by the Zn available. Depending on Fe^{3+} activity, the concentration of Zn^{2+} may be affected by franklinite solubility. For instance, $Fe(OH)_3$ (amorphous) depresses the solubility of franklinites, while crystalline Fe (III) oxides such as lower Fe^{3+} magnetite or goethite allow superior levels of Zn^{2+} balance in soils. While $ZnOH^+$ is more common than this pH, in a soil solution below pH 7.7, the predominant Zn species is Zn^{2+}.

The major soil complexes are $CuSO_4$ and $CuCO_3$. At lower pH, SO_4^{2-} concentration is the main controlling factor in the availability of soil Cu, whereas at high pH, and partial CO_2 (g) pressure are the main controlling factors. Activities are equal to $10-2.36$ M of SO_4^{2-}, $CuSO_4$, and Cu^{2+}.

The soil solution is expected to have a maximum Mo concentration of $10-3.68$ M at 10% moisture. MoO_4^{2-} is the major species in solution in soil with pH greater than 4.24. This ion form MoO_4^{2-} is considered the available Mo for plant and its concentration in soils is very close to $PbMoO_4$ (wulfenite). Pb^{2+} concentration affects its solubility that in order controlled by concentration of phosphate and other parameters such as pH.

Numerous soluble organic compounds, produced by microbial activity and secreted from plant root, are capable of solubilizing (chelate) certain micronutrients, such as Fe, Mn, and others [17]. Organic acids like citric, oxalic, malonic, malic and tartaric acids are common compounds of natural chelate. Synthetic compounds like EDTA, DTPA, etc. are also used as effective micronutrient fertilizers to make chelates with micronutrient. The complexes contribute significantly to micronutrient solubility in soils. In addition, the factors described above, the movement of micronutrients in soil is influenced by soil texture and soil moisture conditions. Soil texture, such as sandy soil, for example, increases leaching of available micronutrients and dry soil conditions generally reduce movement due to lower soil solution diffusion and mass flow of available micronutrients.

4   Symptoms of Micronutrient Deficiencies

It has been reported that micronutrient deficiencies are associated with high soils pH, and calcareous soils with low organic matter. Deficiencies in micronutrients have different cases and are classified as follows by El-Fouly [18] as follows:

1. Apparent deficiency: symptoms of deficiency can be identified in the field on the leaves.
2. Hidden deficiency: no symptoms of deficiency can be seen on plant leaves although the deficiency may be severe, in this case hidden deficiencies can be recognized through leaf analysis.

3. Deficiency in one micronutrient: under Egyptian conditions, a very rare case.
4. Deficiency in more than micronutrient: It is very common in Egypt, where crops mainly suffer from deficiencies in Zn, Mn, Fe and Cu to some extent.
5. True deficiency: due to the lack of one more nutrients in total. This case in Egypt is not common.
6. Induced deficiency: due to one or more agro-ecological factors that affect the nutrient availability and the plant's ability to absorb it. This is a common case in Egypt.

4.1 Fe Deficiency

Iron deficiency in the younger leaves always begins. Chlorosis is interveinal in most species and it is often possible to observe a fine reticulate pattern (Fig. 1). The youngest leaves can be white and completely free of chlorophyll [20]. Iron plays a major role in plant photosynthetic and respiratory reactions. Fe deficiency reduces chlorophyll production and is characterized by interveinal chlorosis with strong differentiation between veins and chlorotic areas in young leaves (Fig. 1). As the deficiency develops, the entire leaf turns whitish-yellow and progresses to necrosis, resulting in the plant's slow growth. The Fe deficient field, if viewed from a distance, displays irregularly shaped yellow areas, particularly where the subsoil is exposed to the surface [21].

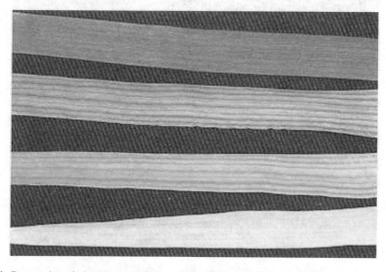

Fig. 1 Progression of wheat leaves deficiency in Fe. Normal top leaf; middle leaves with interveinal chlorosis with prominent green veins; and fully chlorotic bottom leaf [19]

4.2 Zn Deficiency

The most characteristic visible symptoms of zinc deficiency in Dicotyledon are stunted growth due to internode shortness (reset) and a dramatic decrease in leaf size (small leaf). The shoot apexes die (die-back) under severe zinc deficiency. These symptoms are often combined in the interveinal areas of the leaf with chlorosis, and these are pale green, yellow, or white areas (Fig. 2). Chlorotic bands form in the monocots on either side of the leaf's midrib [20]. Plants require zinc to produce growth hormones, and internode elongation which is particularly important. As noted earlier, Zn has intermediate mobility in the plant and symptoms will appear in the middle leaves initially. Zn deficient leaves show interveinal chlorosis, particularly midway between the margin and midrib, resulting in a stripping effect; mottling may also occur (Fig. 2). Chlorotic areas can be pale green, yellow, or even white. Severe Zn deficiency leads to leaves turning gray-white and prematurely falling or dying. Because Zn plays a prominent role in internode elongation, Zn deficient plants are generally severely stunted. The flowering and seed set is also poor in affected plants. Specific crop symptoms include smaller alfalfa leaves, gray or bronze banding leaves and reduced production of tiller in small grains and abnormal grain formation [22]. Zn deficiency generally does not affect fields uniformly and where topsoil has been removed usually occurs deficient areas [21]. Forage Zn deficiencies have been shown to reduce reproductive efficiency.

Fig. 2 Zn deficiency showing interveinal striped chlorosis [23]

Fig. 3 Wheat Cu deficiency: severely impaired (left), moderately impaired (centre), unimpaired (right). Deficient wheat shows poor production and filling of melanosis [23]

4.3 Copper Deficiency

The typical visible symptoms of cereal Cu deficiency are stunted growth distortion of young leaves, apical meristem necrosis, and "White tip" leaf bleaching (Fig. 3). Improved tiller formation in cereals and auxiliary shoots in dicotyledons are secondary symptoms caused by apical meristem necrosis. Also characteristic of Cu deficient plants is wilting in young leaves [20]. The production of chlorophyll, respiration and protein synthesis requires copper. In younger leaves, Cu deficient plants show chlorosis, stunted growth, delayed maturity (excessively late tillering in grain crops), and sometimes melanosis (brown discolouration). Grain production and filling are often poor in cereals, and under severe deficiencies, grain heads may not even form (Fig. 3). Cu deficient plants are susceptible to increased disease, especially ergot (a fungus that causes reduced yield and grain quality [24]. The onset of symptoms caused by disease may confuse the identification display of chlorosis in younger leaves, stunted growth, delayed maturity (excessively late tillering of deficient symptoms of Cu in grain. The most sensitive crops to Cu deficiency are winter and spring wheat [24].

4.4 Manganese (Mn) Deficiency

The most sensitive of cell organelles to Mn deficiency is chloroplasts (plant organelles where photosynthesis occurs) [25]. As a result, interveinal chlorosis in young leaves is a common symptom of Mn deficiency (Fig. 4). There is no sharp difference between veins and interveinal areas, however, unlike Fe, but rather a more diffuse chlorotic effect. Two well-known Mn deficiencies in arable crops are gray speck in the marsh spot of oats and peas. White wheat streak and interveinal brown spot of barley are also symptoms of Mn deficiency [26]. Visual observation can be limited by different factors as a diagnostic tool, including hidden hunger and pseudo-deficiencies, and soil or plant testing will be required to verify nutrient stress. Nevertheless, field visual symptom assessment is an low-cost and fast method for identifying potential crop nutrient deficiencies or toxicity, and learning to recognize symptoms and their causes is an important skill in managing and correcting soil fertility and crop production problems.

Fig. 4 Mn wheat deficient with interveinal chlorosis [23]

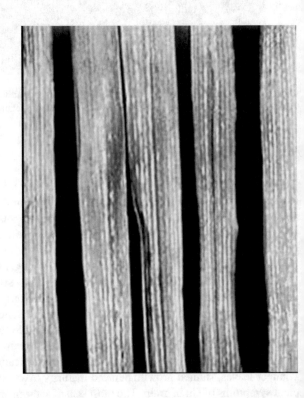

5   Physiological and Biochemical of Micronutrients Changes in Plant

5.1   Dicots and Non-grasses Monocots (Strategy 1)

Higher plant root cells contain several redox systems associated with plasmalemma that can mediate the transportation of electrons from certain cytosolic donors (NADH or NADPH) to different extracellular electron acceptors such as ferric chelates, ferricyanide. Plasma membrane redox-systems are supposed to be involved in regulating the transport of ions, proton flux, and membrane energization in plant roots [27]. Dicots and non-grass monocots show root system adaptive responses to avoid Fe deficiency (Fig. 5). These responses are (1) increased net proton excretion, and (2) reductase-bound plasma membrane. The relative importance of the two mechanisms between plant species seems to differ considerably. Major plant nutrient transport systems are sensitive to nutritional conditions and are repressed or controlled by an adequate supply of specific ion. Depression of ion transport systems in root cells can be caused by deficiency or plant growth demand, allowing a limited physiological range to maintain internal levels [28]. Stimulating the activity of ferric chelate reductase, which is an important plasma-membrane that fulfills the reduction of Fe

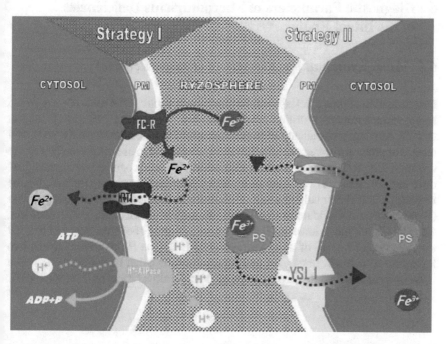

Fig. 5 Schematic representation of the iron-taking mechanisms of Strategy I and Strategy II. IRT1, iron-regulated transporter, YSL1, yellow strip such as 1, PS, phytosiderophores, PM, and plasma membrane [35]

(III) in the cell, surface and the immediate increase in leaf content Fe and chlorophyll following these responses suggests that all these factors act in concert, not independently, to assist in the absorption and transportation of Fe to the tops of the plant.

5.2   Grasses (Strategy II)

In grasses for the acquisition of Fe^{+3} and to some extent other micronutrient cations in the rizosphere, a particular mechanism exists, this mechanism is characterized by two components (Fig. 5), as follows: (1) Iron deficiency (also Zn, Cu, Mn and some Ca and Mg) induced increased release of compounds of phytosiderophores (PS), and (2) a highly efficient Fe^{+3} PS uptake system, which is activated further under Fe deficiency. Both mechanism components are in the apical root zones. Phytosiderophores (PS) chelate and solubilize Fe in high pH and high bicarbonate soils, while bicarbonate inhibits Fe solubilization by releasing H+ ions and enhancing Fe root reduction capacity (Strategy I) [29].

6   Diagnostic Parameters of Micronutrients Deficiencies in the Intact Roots

6.1   Acidification of the Rizosphere (Proton Release)

Determination of proton release and Fe reduction capacity in the roots of peas grown under various mineral nutrition conditions (FNS − Fe) and (FNS + Fe), support that Fe deficiency stimulates both mechanisms: Fe reduction capacity and plant release of protons (Strategy 1) [30]. It was recommended to use the proton H^+ release determination technique as a useful diagnostic tool to recognize plant genotypes' differential efficacy to (−Fe) deficiency stress. The same procedure for studying the response of subclover cultivars to Fe deficiency was adopted. Whilst koala cultivar (Fe resistant) and Karridale cultivar (susceptible) were grown under (−Fe) treatment, Karridale had a much lower rate of release of H+ and (+Fe) treatment led to a net negative rate of release of hydrogen ions (H^+) [31]. The H^+ proton release in two lupine cultivars has been reduced and it has been shown that Fe deficiency has been found to promote the release of protons by Giza 1 cultivar roots on 7 days old [32].

6.2 Iron Reduction Mechanism—Iron Reductase Enzyme Activity

It is recognized that the main inducing factor of Fe deficiency in dicotyledonous species is high concentration of bicarbonate in the soil solution. The stress associated with Fe deficiency is well documented [33]. Nongraminate monocots and dicots (plants of Strategy I) induce the activity of ferric chelate reductase (FCR) on the root cell membrane, an enzyme involved in Fe^{3+} to Fe^{2+} conversion. The function of Strategy II is observed in monocots dissolving ferric compounds by exuding mugineic acid from the root. Strategy I plants have many complex functions compared to Strategy II plants which involve not only FCR but also extrusion of protons [34].

Reduction of iron (Fe^{3+}) was used to predict resistance of Fe deficiency in soybean [36], while roots of tomato and soybean were documented in the release of H^+ ions, reducing agents and reducing Fe^{3+} to Fe^{2+} [37]. Furthermore, iron stress results in branching of root [38, 39] and exudation of ferric chelate compound [40], which also chelates ferric ions. Cucumber grows thin lateral roots [41], whereas red clover has a different pattern of root branching [39]. Many types of phenols are released from the roots of Fe deficient plant, i.e. in alfalfa [42], and red clover [39]. Fe chlorosis is considered a major problem in the nutrition of plants in citrus trees. The reactions of "Newhall" orange plants to iron deficiency in the presence or absence of $CaCO_3$, where the presence of $CaCO_3$ resulted in a significant increase in FC-R activity (about 2.5-fold) for Fe at concentrations of 0.0 and 5 μM compared to plants grown with Fe concentration of 10 and 20 μM [43]. It was reported that neither the reduction of Fe (III) by roots nor its induction by Fe deficiency are unique root reduction characteristics, chelated Cu (II) and chelated Fe of roots can be reduced by Fe deficiency on the root surface of peas [30].

6.3 The Release of Phytosiderophore (PS)

Grass roots have a different mechanism for physiological responses as reported by Takagi [44]. Grass roots stressed with deficiency of Fe and Zn release phytosiderophores compounds (PS) that are efficiently chelate and mobilize Zn^{+3} and Fe^{+3} [28]. The main sites for releasing phytosiderophores (PS) and absorption of Fe are apical root zones. The release of phytosiderophores (PS) from roots is considered among the physiological response mechanisms as an important mechanism for detecting deficiencies of Fe and Zn in various plant species. It has been shown that under Zn and Fe deficiency stress the synthesis and release of (PS) at Fe and Zn deficiency is significantly enhanced. The determination of PS release was suggested as a signal of the differential effectiveness of the different strategy (II) genotypes for Fe and Zn [45].

Various morphological and physiological plant factors have been studied to understand the basis under Zn deficiency of genotypical diversity. Furthermore, genotypical differences in Zn deficiency may be attributed to variations in phytosiderophores root release. Phytosiderophores release is a well-known phenomenon that take places not only under Fe deficiency but also under Zn deficiency [46, 47], and involves the mobilization of Zn in the rhizosphere [48] and from the walls of the root cells [49]. The release rate of phytosiderophores under Fe deficiency was closely associated with the differences in sensitivity to Fe deficiency between and within cereal species [50, 51].

6.4 Membrane Integrity

Zinc is needed to maintain biomembrane integrity. It may be linked to phospholipid and sulfhydryl membrane groups consisting of iron tetrahedral complexes with polypeptide chain cysteine residues, thus protecting membrane lipids and proteins from oxidative damage [52]. Root cell membrane permeability is enhanced in plants with Zn deficiency and exudation of low molecular weight solutes such as K^+ and NO_3, sugars and amino acids are enhanced. The exudation of solutes from a number of plant species with and without Zn deficiency was investigated and found that the root exudation of amino acids, sugar and phenolic in all species was increased. In addition, root exudate measurements in different plants have been suggested to be useful for checking zinc nutrition status [49].

6.5 Lipid Peroxidation

The cell membrane is a bilayer of lipids with saturated and unsaturated fatty acids within the structure. The protection against oxidation of membrane lipids and protein depends on the supply of Zn. Under Zn stress, lipid peroxidation was greatly increased by more than 143% compared to a corresponding value in a microsomal fraction of *Phaseolus* plant roots under normal Zn level [53]. As a result, an increase in the rate of lipid peroxidation can be used as a diagnostic tool for stress with Zn deficiency.

7 Diagnostic Parameters of Micronutrient Deficiencies in the Plant Leaves

7.1 Iron

7.1.1 Visual Chlorosis Score (Vs)

Deficiency diagnosis of Fe and evaluation of genotype efficiency under field conditions were determined using visual score. It was found that ranking cultivars by Vs would be most efficient because much less labor and time is required. Five Egyptian maize hybrids were used to compare maize response to Fe deficiency and evaluate the methods regularly used to evaluate the appearance of chlorosis [54]. Scores of visual chlorosis are i.e. chlorophyll a, chlorophyll a + b, or active iron. The mentioned criteria could be used to evaluate Fe chlorosis, whilst leaf total Fe concentration is not a suitable marker for Fe chlorosis assessment [54].

7.1.2 N-Tester (Chlorophyll Meter)

It is very useful to use chemical analysis of leaf samples to monitor the nutritional status of field crops to meet their needs at the right times. Laboratory analysis is a destructive measurement requiring effort, time, and chemicals. Predicting nutritional status directly in the field by the N-tested pocket instrument (Minolta, Japan) is cheaper, faster and more appreciated for farmers' use [55]. The possibility of predicting Fe status in leaves of various fruit crops using N-tester (chlorophyll meter) has been studied [55]. They found that it was not possible to measure Fe-status using N-tester under field conditions [55].

7.1.3 Chlorophyll Content

Several authors found that the content of chlorophyll was correlated with active Fe^{+2}. These results suggest that chlorophyll is related quantitatively to chloroplast's Fe content [53]. El-Baz et al. [53] reported that chlorophyll a and chlorophyll a + b were significantly reduced when the iron supply to *Phaseolus* plants decreased. These findings support the use of changes in chlorophyll as a biochemical criterion for Fe deficiency assessment.

7.1.4 Photosynthesis Process

When iron deficiency gets worse, PII activity also drops drastically and becomes much harder to restore. It was found that leaves of orange plants grown in the nutrient solution in the absence of Fe had lower oxygen evolution rates in both the presence or absence of $CaCO_3$ compared with 10 μM Fe [43]. A minor decline in efficiency of photosystem II was observed in plants grown without Fe and in the presence of $CaCO_3$ [43].

7.1.5 Physiological Active Iron (Fe^{+2})

The degree of adequacy of different elements in plant nutrition is measured by symptom appearance or elementary tissue analysis [12]. Several investigators have made an assessment of the iron status of plants by determining their active form, which is useful as measuring the total amount of micronutrient in plant tissues and they found in all cases does not correspond to the physiological status of the plant and the biological efficacy of the element [56]. El-Baz et al. [56] Tested two Fe^{+2} extractants (1.5% O-pH and 1.5 N HCl). He found high correlations between Fe supply and Fe extracted by O-phenanthroline method (O-ph). In *Phaseolus* plant, the high efficiency of an O-ph determination method of Fe^{+2}, is a more promising technique for assessing Fe status in various plants, while when grown at high concentration of $CaCO_3$, some plant species have limited capacity to adaptive processes [56]. It was found that the symptoms of Fe chlorosis do not appear when grapevine grew in (−Fe) for 2 weeks, although the concentration of total and active iron in leaves decreases. The decrease was significantly in active iron in the presence of $CacO_3$ due to inhibited root acquisition by Fe [57].

7.1.6 Enzyme Activities

Another approach is based on the hypothesis that under stress conditions the enzyme's limiting activity provides a quick and sensitive indicator of micronutrient deficiencies. The method's advantage is due to the high sensitivity and enzymatic analysis specificity. Fe is the component of a number of well characterized enzymes that iron performs as a metal component in redox reactions or as a bridge between enzymes and substrates, the main cause of metabolism alterations. The functional analysis of nutrients is based on the examination of certain molecular compounds related to their functional activity. There may be specific reference to the study of enzymatic activities directly affected by the metabolic activity of the nutrient. In the early diagnosis of mineral deficiencies in lemon trees, aconitase proved to be as accurate as peroxidase, a specific Fe-metalloenzyme, or even more [58].

Therefore, in view of this background, efforts have been made to analyze the prospective value of various biochemical indicators appropriate for the diagnosis of citrus mineral deficiencies as a feasible substitute to other popular techniques of diagnosis such as soil or leaf analysis. It is often misleading to identify nutrient constraints based on morphological symptoms or alternatively in combination with leaf/soil analysis, especially with regard to the remediation of standing crop nutrition problems. El-Baz et al. [53] reported that catalase activity in bean plants is declining under conditions of iron deficiency. In addition, the effect of Fe deficiency on enzyme activity (peroxidase and catalase) in young maize leaves grown at various levels of Fe. Enzyme activity reduced due to Fe deficiency as the impact of Fe deficiency on enzyme apoprotein synthesis [59]. The concentration of catalase and peroxidase enzyme activity suggested in this concept to be used as sensitive and precise parameters for early diagnosis of Fe deficiency.

7.1.7   Isozymes

Recently, isozymes have been used as genetic markers in various plant species for early evaluation of nutrient efficiency. A relationship between plant nutritional status and isozymes was found. Lake in available cellular iron may decrease the synthesis of peroxidase isozyme, Fe-efficient maize cultivar Alice, iron inefficient mutant yellow strip (ysl) and six additional local hi bird lines were grown in hydroponics with 0.1 mM Fe-EDTA and 0.01 mM to establish a genetic assay for early diagnosis of POD isozyme-based Fe deficiency [60]. Fe-efficient maize lines may have an additional band (C_3) reflecting the Fe-nutritional status of maize lines or C_3 associated with Fe (III) PS uptake [60].

7.2   Zinc

7.2.1   Extractable Zinc

In the leaves, total zinc is not always a reliable indicator for the plant's diagnosis of zinc deficiency. In contrast, El-Baz et al. [56] reported that HCl's use of extractable Zn as a diagnostic tool for zinc deficiency in the upper and lower leaves is of small value as the total zinc content.

7.2.2   Photosynthesis and Carboxylation Enzyme Activities

Another approach to Zn deficiency diagnosis is based on combining zinc with several metabolic processes (i.e. photosynthesis, carbon metabolism, carboxylated carbohydrate-related enzymes, and protein synthesis) [52]. Nutrient deficiency inhibits photosynthesis in higher plants. Low Zn conditions reduced photosynthesis compared to the adequate level provided with Zn, this can be attributed to the role of zinc in regulating the activity of electron transport, carbon metabolism and carboxylated enzyme modification [32]. In maize (C_4) and chickpea (C_3) plants, the relationships between Zn deficiency stress and photosynthesis, CO_2 fixation, and carboxylated enzyme activity were studied [32]. Zinc deficiency in maize plants resulted in a sharp reduction in chlorophyll content, whereas CO_2 fixation rates showed low values for both plants (-Zn level) [32]. In leaves with increasing zinc deficiency, a sharp decline in CA activity is the most evident of the changes occurring in the CO_2 assimilation pathway enzyme activity. The main carboxylating enzyme in C_3 plants is RuBP-carboxylase and may represent 50% of the total soluble protein of the leaf. RuBP was reduced by 22 and 34% for maize and chickpea, respectively under Zn deficiency. In maize plants, PEP-carboxylase was nearly 5 times higher than chickpea grown without Zn [32].

7.2.3 Antioxidants Enzyme Activity

Zn ions are involved in enzymatic cell defense against free radical damage. The influence of zinc nutritional status in bean leaves grown in nutrient solution for 15 days on superoxide hydrogen peroxide scavenging enzymes has been studied [61]. Zn deficiency was found to roughly decrease protein and SOD activity. Ascorbate peroxidase, glutathione reductase, and catalase except guaiacol peroxidase activity, within 72 h or re-supplying Zn to deficient plants, the enzyme activity reached sufficient plants at the Zn level [61].

7.2.4 Protein Synthesis and SDS-PAGE Protein

In terms of protein synthesis, it was reported that zinc deficiency decreased the ribosomal content followed by a reduction in protein synthesis when investigating the effects of Zn on the nitrogen metabolism of meristematic tissues in tobacco plants [62]. Zn deficiency was found to significantly reduce the ribosomal content and the protein composition remained nearly unchanged. Deficiencies in micronutrients can modulate antioxidant enzyme activity [63]. Yu et al. [63] found that SOD enzyme concentration measured by a newly developed capillary electrophoresis technique in wheat genotypes plant tissues was related to the early stage of Zn deficiency tolerance of wheat to Zn deficiency.

7.3 Copper

Cu is an essential element in the metabolism of plants, it is needed in small quantities. Cu is a constituent of a large number when Cu proteins and enzymes are essential in plant metabolism for their function.

7.3.1 Photosynthesis

Cu deficiency has a direct impact on energy metabolism because it affects the synthesis of Cu-containing electron carriers like plastocyanine as a component of the electron transport chain of PSI. There is therefore a close association between the leaf copper concentration and the plastocyanin under Cu deficiency, resulting in a drastic decrease in activity, consequently a reduction in photosynthesis and respiration [52].

7.3.2　Cupric Chelate Reductase Activity

Copper as an essential micronutrient ion is necessary for growth and it is possible to include the reduction step at the plasmalemma level by depolarizing the membrane potential or cupric reductase as a standard reductase associated with the regulation of other redox systems in plant roots [30]. Babalakova et al. [30] studied the influence of 10 and 20 μmol Cu as $CuCl_2$ on pea grew hydroponically under different mineral nutritional conditions on the activity of Cupric chelate reductase in intact pea roots (Strategy I) and wheat (Strategy II) plants (Fig. 5).

7.3.3　Proton Release and Enzyme Activities

Copper interferes with multiple enzymes, mainly those involved in redox system mechanisms, and ATPase activity (throughout proton release determination), which was found to be a good tool for Cu deficiency diagnosis. The effect of various copper ion concentrations in the root growth medium varied from (0.002 to 1.25 ppm) on plasma membrane activity in rice plants (as measured throughout the release of protons) [64]. Fernando et al. [64] observed that with 1.25 ppm Cu treatment, root proton extrusion increased 4 folds. This could be explained by the increase in H-ATP-use plasma membrane activity that pumps protons from the protoplast to the space of the cell wall, inducing growth enhancement. Salama [65] supported the previous results and the determination of proton release mechanism by roots of *Pisum sativum* grown with two levels of Cu (zero-0.2 mmol) and activities of SOD and AO antioxidant enzymes as valuable parameters for early diagnosis of Cu deficiency.

8　Strategies to Increase the Availability of Micronutrients in Soil for Plant Uptake

Implementing either chemical or organic fertilizers is an appropriate approach for preserving and improving the soil's biological and physicochemical properties, as well as providing essential elements for plant growth. Micronutrient shortages not only reduce crop yields, but also reduce crop concentration of micronutrients. Zn and Fe are the most studied elements among the micronutrients. Zn-containing fertilizers in Turkey have been reported as a rapid and effective technique for enriching Zn cereal content [66]. The application of urea-enriched Zn concentration to 3% has been reported to increase grain yield and rice grain Zn concentration in India by 23 and 56% respectively [67]. Iron is the second element most studied, but it is more difficult to improve the agronomic concentration of Fe in grain than with Zn because Fe is readily precipitated into insoluble forms of soil. For example, a greenhouse experiment with Zn and Fe application on wheat showed increased concentrations of Zn grain, whereas concentrations of Fe were not improved effectively [66].

The other micronutrients were also studied and guidelines on applications of micronutrients for crop types were developed [16]. Dobermann and Fairhurst [68] provided detailed guidance on rice nutrient management. Micronutrient application in the form of sulfates or organic chelates is the ordinary method for mitigating micronutrient deficiencies. Furthermore, there is a recent technology to improve the availability of fertilizer nutrients, which makes fertilizers in the form of nano-size with the initial rapid release and long-term slow release [69]. Although the most common method is the direct application of micronutrients to soils, the availability could be reduced as they react with soil minerals and organic matter [70].

In contrast to chemical fertilizer, the adoption of organic amendment n techniques was promoted as the most feasible and sustainable approach to reinstate soil fertility. One of the most common organic fertilization techniques is the use of manure from farmyards (FYM) as a source of organic matter, nitrogen and micronutrients to compensate the low content of organic matter in soils [71]. Micronutrient amendments are also possible from animal compost and sewage sludge. Despite its soil benefits, organic fertilizer can not be considered a short-term substitute for chemical fertilization due to the gradual release of nutrients into the soil as organic matter breaks down and the process is based on many environmental factors. A combination of inorganic fertilizers is therefore advisable in this regard [72]. Continuous use of these organic materials, however, is likely to accumulate micronutrients in soils.

Foliar fertilization is another agronomic approach that is widely used to correct micronutrient deficiencies. This technique is considered an effective short-term approach to increase the mineral content of rice grains [73]. Although foliar fertilization is considered a nutrient corrective approach in line with several studies, it can not be a substitute for chemical fertilization [74] or organic fertilizers, which is a common approach to maintaining and improving the biological and physicochemical properties of soils and providing essential elements for plant growth. In addition to reducing crop yield, micronutrient deficiency also reduces crop concentration of micronutrients.

Besides to the application of micronutrient in soil to increase its availability for plant uptake, a foliar spraying strategy with micronutrients has been used intensively for plant canopies as a foliar application to compensate its deficiencies in plant micronutrients, thereby increasing its growth and productivity [75–77]. Reda et al. [75], for example, reported that salt-stressed faba bean plants treated with B or Zn and the combination of both had significantly increased photosynthetic pigments, osmotic compounds i.e. soluble sugar, proline, amino acid and phenolic content in the leaves, which in turn helped faba bean plants to increase its tolerance to irrigation diluted seawater salinity. Additionally, Zn and B treatments minimized the unfavorable effects of salt stress on faba bean seed yield and even increased their contents of total carbohydrate, starch, protein, phenolics and flavonoids compounds. Boron foliar spray at 5 ppm was found to be highly effective in maximizing the ratios of $K^+:Na^+$ and $Ca^{2+}:Na^+$ and reducing the risk of salinity in faba bean plants [76]. Furthermore, Mohamed et al. [77] reported that foliar spray with $ZnSO_4$ and $FeSO_4$'s with two concentrations 1 or 3 g L^{-1} assisted faba bean plants by producing antioxidant enzymes to overcome the deficiency of these minerals. They concluded that improved faba bean growth through adequate Fe and Zn foliar spray is likely to be a promising strategy to improve faba bean plants.

9 Conclusions

The availability of micronutrients in soils is associated with the type of parent materials and the environment in which soils are formed. Cu and Mo's availability in soils appears to be more related to soil organic matter content. Predicting behavior of micronutrients in soils is not secure. The first choice, however, is to use fertilizers containing micronutrients, both chemical and organic. Application method must be conducted based on soil nutrient management guidelines along with crop types. Increased use of modern biotechnology techniques, such as gene transformation, will lead to a promising prospect of higher nutritional stress resistance. There are already transgenic plants with enhanced nutrient efficiency such as zinc (Zn), boron (B) and copper (Cu). Further plant stress research will also enhance the knowledge of mechanisms of adaptation. This will then stimulate breeding for the efficiency of nutrients and the tolerance of stress.

Acknowledgements The authors Dr. Zeinab Salama and Dr. Magdi Abdelhamid, would like to express their gratitude to Dr. Mohamed M. El-Fouly, Professor of Plant Nutrition, National Research Centre, Egypt, and their colleagues at Department of Plant Biochemistry, and Department of Botany, National Research Centre, Egypt for their valuable contribution in some phases of this study.

References

1. O'Neil MA, Ishiim T, Albersheimm P, Darvill AG (2004) Rhamnogalacturonan II: structure and function of a borate cross-linked cell wall pectic polysaccharide. Ann Rev Plant Biol 55:109–139
2. Patterson J, Ford K, Cassin A, Natera S, Bacic A (2007) Increased abundance of protein involved in phytosiderophore production in boron-tolerant barley. Plant Physiol 144:1612–1631
3. Rieuwerts JS, Thornton I, Farago ME, Ashmore MR (1998) Factors influencing metal bioavailability in soils: preliminary investigations for the development of a critical loads approach for metals. Chem Spec Bioavailab 10:61–75
4. Tsui MT, Wang WX, Chu LM (2005) Influence of glyphosate and its formulation (roundup) on the toxicity and bioavailability of metals to Ceriodaphnia dubia. Environ Pollut 138:59–68
5. Eker S, Ozturk L, Yazici A, Erenoglu B, Romheld V, Cakmak I (2006) Foliar-applied glyphosate substantially reduced uptake and transport of iron and manganese in sunflower (Helianthus annuus L.) plants. J Agric Food Chem 54:10019–10025

6. Steffens D, Hutsch BW, Eschholz T, Lošak T, Schubert S (2005) Waterlogging may inhibit plant growth primarily by nutrient deficiency rather than nutrient toxicity. Plant Soil Environ 51:545–552
7. Waraich EA, Ahmad R, Ashraf MY (2011) Role of mineral nutrition in alleviation of drought stress in plants. Aust J Crop Sci 5:764–777
8. Brown PH, Bellaloui N, Wimmer MA, Bassil ES, Ruiz J, Hu H, Pfeffer H, Dannel F, Romheld V (2002) Boron in plant biology. Plant Biol 4:203–223
9. Kobayashi T, Nishizawa NK (2012) Iron uptake, translocation, and regulation in higher plants. Annu Rev Plant Biol 63:131–152
10. Yan X, Wu P, Ling H, Xu G, Xu F, Zhang Q (2006) Plant nutrigenomics in China: an overview. Ann Bot 98:473–482
11. Ahsan N, Lee DG, Lee SH, Kang KY, Lee JJ, Kim PJ, Yoon HS, Kim PJ, Lee BH (2007) Excess copper-induced physiological and proteomic changes in germinating rice seeds. Chemosphere 67:1182–1193
12. El-Fouly MM, El-Baz FK, Youssef AM, Salama ZA (1998) Carbonic anhydrase, aldolase, and catalase activities as affected by spraying different concentrations and forms of zinc and iron on faba bean and wheat. Egypt J Sci 22:1–11
13. Yin Y, Impellitteri CA, You SJ, Allen HE (2002) The importance of organic matter distribution and extract soil: solution ratio on the desorption of heavy metals from soils. Sci Total Environ 287:107–119
14. Bowen HJM (1979) Environmental chemistry of the elements. Academic Press, London
15. Lindsay WL (2001) Chemical equilibria in soils. The Blackburn Press, New Jersey
16. Halvin JL, Tisdale SL, Nelson WL, Beaton JD (2014) Soil fertility and fertilizer: an introduction to nutrient management, 8th edn. Pearson Education, New Jersey
17. Neumann G, Römheld V (2012) Rhizosphere chemistry in relation to plant nutrition. In: Marschner P (ed) Mineral nutrition of higher plants, 3rd edn. Academic Press, London, pp 347–368
18. El-Fouly MM (1987) Use of micronutrient under practical conditions in Egypt. In: El-Fouly et al (ed.) Proceedings of symposium "Application of special fertilizers" Alex, Egypt, 21–23.02.1986, pp 71–86
19. Grundon NJ (1987) Hungry crops: a guide to nutrient deficiencies in field crops. Queensland Government, Brisbane, Australia, p 246
20. Bergmann W (1986) Ernaehrungsstorungen bei Kulturpflanzen: Visuelle and analytische Dignose. VEB Gustar Fishers Verlage, Jena, Germany, p 306
21. Follett RH, Westfall DG (1992) Identifying and correcting zinc and iron deficiency in field crops. Colorado State University Cooperative Extension. Service in action no 545
22. Wiese MV (1993) Wheat and other small grains. In: Nutrient deficiencies and toxicities in crop plants. APS Press, St. Paul, Minnesota, p 202
23. McCauly A, Jones C, Jacobsen J (2009) Plant nutrient functions and deficiency and toxicity symptoms. In: Nutrient management module, no 9, p 16. Montana State University, Bozeman, MT, USA
24. Solberg E, Evans I, Penny D (1999) Copper deficiency: diagnosis and correction. Government of Alberta Agriculture and Rural Development. Agdex 532-3
25. Mengel K, Kirkby EA (2001) Principles of plant nutrition. Kluwer Academic Publishers, Netherlands, p 849
26. Jacobsen JS, Jasper CD (1991, Feb) Diagnosis of nutrient deficiencies in alfalfa and wheat. EB 43. Montana State University Extension, Bozeman, Montana
27. Bienfait HF (1988) Mechanisms in Fe-efficiency reactions of a higher plant. J Plant Nutr 11:605–629
28. Marschner H, Roemheld V (1996) Root-induced changes in the availability of micronutrients in the rhizosphere. In: Waisel Y, Eshel A (eds) Plant roots the hidden half. Marcel Dekker, New York, pp 557–579
29. von Wiren N, Mori S, Marschner H, Roemheld V (1994) Iron-inefficiency in the maize mutant ys1 (Zea mays L. cv. Yellow-stripe) is caused by a defect in uptake of iron phytosiderophores. Plant Physiol 106:71–77

30. Babalakova NK, Traykova D, Matsumoto H (1993) The reaction of H+ transporting activity membrane potential forming by tonoplast ATPase proton pump of barley roots after uptake of Cu^{2+}. Biol Biochime 46:117–120

31. Wei L, Loeppert RH, Ocumpauch WR (1998) Analysis of iron-deficiency undocked hydrogen release by plant roots using chemical equilibrium and pH-stat methods. J Plant Nutr 21:1539–1549

32. Salama ZA, Laszova GN, Stoinova ZG, Popova LP (2002) Effect of zinc deficiency on photosynthesis in maize and chick-pea plants. CR Acad Bulg Sci 55:65–68

33. Brown JC, Jolley VD (1989) Plant metabolic responses to iron deficiency stress. BioSci 39:546–551

34. Brown JC, Ambler JE (1974) Iron stress in tomato (Lycopersicon esculentum). I. Sites of Fe reduction, absorption and transport. Physiol Plant 31:221–224

35. Vigani G, Donnini, S, Zocchi G (2015) Metabolic adjustment under Fe deficiency in roots of dicotyledonous plants. Chapter Nova, 1–24

36. Jolley VD, Fairbanks DJ, Stevens WB, Terry RE, Orf JH (1992) Root iron-reduction capacity for genotypic evaluation of iron efficiency in soybean. J Plant Nutr 15:1679–1690

37. Camp SD, Jolley VD, Brown JC (1987) Comparative evaluation of factors involved in Fe stress response in tomato and soybean. J Plant Nutr 4:423–442

38. Hagström J, James WM, Skene KR (2001) A comparison of structure, development and function in cluster roots of *Lupinus albus* under phosphate and iron stress. Plant Soil 232:81–90

39. Jin CW, Chen WW, Meng ZB, Zheng SJ (2008) Iron deficiency-induced increase of root branching contributes to the enhanced root ferric chelate reductase activity. J Integr Plant Biol 50:1557–1562

40. Noguchi A, Yoshihara T, Ichihara A, Sugihara S, Koshino M, Kojima M, Masaoka Y (1994) Ferric phosphate-dissolving compound, alfa furan, from alfalfa (*Medicago sativa*) in response to iron deficiency-stress. Biosci Biotechnol Biochem 58:2312–2313

41. Dell'Orto M, Santi S, De Nisi P, Cesco S, Varanini Z, Zocchi G, Pinton R (2000) Development of Fe-deficiency responses in cucumber (*Cucumis sativus* L.) roots: involvement of plasma membrane H^+ -ATPase activity. J Exp Bot 51:695–701

42. Koshino H, Masaoka Y, Ichihara A (1993) A benzofuran derivative released by Fe-deficient Medicago sativa. Phytochem 33:1075–1077

43. Pestana M, David M, de Varennes A, Abadia J, Faria AE (2001) Responses of "Newhall" orange trees to Iron deficiency in hydroponic: effect of leaf chlorophyll photosynthetic efficiency, and root ferric chelate reductase activity. J Plant Nutr 24:1609–1620

44. Takagi S (1976) Naturally occurring iron-chelating compounds in oat-and rice-root washings 1. Activity measurements and preliminary characterization. Soil Sci Plant Nutr 22:423–433

45. Erenouglu B, Eker S, Cakmak L, Derici R, Roemheld V (2000) Effect of iron and zinc deficiency of release of phytosiderophores in barley cultivars differing in zinc efficiency. J Plant Nutr 23:1645–1656

46. Zhang F, Roemheld V, Marschner H (1989) Effect of zinc deficiency in wheat on the release of zinc and iron mobilizing exudates. Z Pflanzenernaehr Bodenk 152:205–210

47. Cakmak I, Gulut K, Marschner H, Graham RD (1994) Effect of zinc and iron deficiency on phytosiderophore release in wheat genotypes differing in zinc efficiency. J Plant Nutr 17:1–17

48. Treeby M, Marschner H, Romheld V (1989) Mobilization of iron and other micronutrient cations from a calcareous soil by plant-borne, microbial, and synthetic metal chelators. Plant Soil 114:217–226

49. Zhang F, Romheld V, Marschner H (1991) The release of zinc mobilizing root exudates in different plant species as affected by zinc nutritional status. J Plant Nutr 14:675–686

50. Takagi S, Nomoto K, Takemoto T (1984) Physiological aspect of mugineic acid, a possible phytosiderophore of graminaceous plants. J Plant Nutr 7:469–477

51. Marschner H, Römheld V, Kissel M (1986) Different strategies in higher plants in mobilization and uptake of iron. J Plant Nutr 9:695–713

52. Marschner H (1986) Mineral nutrition of higher plants. Academic Press Inc., London, p 674

53. El-Baz FK, Salama ZA, Mohamed AA (1996) Evaluation of catalase, carbonic (CA), Aldolase activities and chlorophyll as indicators for Fe and Zn deficiency in snap bean (*Phosphorus Vulgaris*) and faba bean (*Vicia Faba*) plants. J Agric Sci Mans Univ 21:2569–2581

54. El-Bendary AA, Abou El-Nour EAA, El-Sayed AA (1999) Responses of maize hybrids to Fe-stress in calcareous soil. Alex J Agric Res 44:181–190

55. Shaaban MM, El-Sayed AA, Abou El-Nour EAA (1999) Predicting nitrogen magnesium and iron nutritional status in some crops a portable. Sci Hortic 82:339–348

56. El-Baz FK, El-Monde EA, Salama ZA, Mohamed AA (1998) Determination of Fe^{2+} and soluble zinc as biochemical indicators for the diagnosis of iron and zinc deficiency in snap bean *Phaseolus Vulgaris* and fava bean *Vicia faba* plants. Egypt J Physiol Sci 22:25–39

57. Nikolic M, Kastori R (2000) Effect of bicarbonate and Fe supply of Fe nutrition of grapevine. J Plant Nutr 23:1619–1627

58. Garcia AL, Galindo L, Sanchez-Blanco MJ, Torrecillas A (1990) Peroxidase assay using 3, 3′, 5, 5′ tetramethylbenzidine as H-donor for rapid diagnosis of the iron deficiency in citrus. Sci Hortic 92:251–255

59. Nenova V, Stoyanov I (1995) Physiological and biochemical changes in young maize plants under iron deficiency: 2, catalase, peroxidase, and nitrate reductase activities in leaves. J Plant Nutr 18:2081–2091

60. El-Bendary AA, Mabrouk Y, El-Metainy A (1998) Peroxidase isozyme variants as genetic markers for early evaluation of Fe-efficiency and Fe-nutritional status in maize lines. Field Crops Res 59:181–185

61. Cakmak I, Marschner H (1993) Effect of zinc nutritional status on activities of superoxide-radical and hydrogen peroxide scavenging enzymes in bean leaves. In: Barrow NJ (ed) Plant nutrition, from genetic engineering to field practice, pp 133–137

62. Obata H, Umebayashi M (1988) Effect of zinc deficiency or protein synthesis in cultured tobacco plant cells. Soil Sci Plant Nutr 34:351–357

63. Yu Q, Worth C, Rengel Z (1999) Using capillary electrophoresis to measure Cu/Zn-SOD concentration in leaves of wheat genotypes differing in tolerance to Zn deficiency. Plant Sci 143:231–239

64. Fernando C, Lidon C, Fernando SH (1993) Effects of copper toxicity on growth and uptake and translocation of metals in rice plants. J Plant Nutr 16:1449–1464

65. Salama ZA (2001) Diagnosis of copper deficiency through growth, nutrient uptake and some biochemical reactions in Pisum sativum L. Pakistan J Biolog Sci 4:1299–1302

66. Cakmak I (2009) Enrichment of fertilizers with zinc: an excellent investment for humanity and crop production in India. J Trace Elem Med Biol 23:281–289

67. Shivay YS, Kumar D, Prasad R (2008) Effect of zinc-enriched urea on productivity, zinc uptake and efficiency of an aromatic rice-wheat cropping system. Nutr Cycle Agroecosyst 81:229–243

68. Dobermann A, Fairhurst T (2000) Rice: nutrient disorders and nutrient management. Potash & Phosphate Institute (PPI), Potash & Phosphate Institute of Canada (PPIC), and International Rice Research Institute (IRRI), Singapore and Los Baños, Philipines

69. Dimkpa CO, McLean JE, Britt DW, Anderson AJ (2012) Bioactivity and biomodification of Ag, ZnO and CuO nanoparticles with relevance to plant performance in agriculture. Ind Biotechnol 8:344–357

70. Khoshgoftarmanesh AH, Schulin R, Chaney RL, Daneshbakhsh B, Afyuni M (2010) Micronutrient-efficient genotypes for crop yield and nutritional quality in sustainable agriculture. Agron Sustain Dev 30:83–107

71. Gao M, Che FC, Wei CF, Xie DT, Yang JH (2000) Effect of long-term application of manures on forms of Fe, Mn, Cu and Zn in purple paddy soil. Plant Nutr Fertil Sci 6:11–17

72. IRRI (2012) Using organic materials and manures. Available from www.knowledgebank. irri.org/training/factsheets/nutrient-management/item/using-organic-materials-and-manures. Accessed on 9 Aug 2017

73. He W, Shohag MJ, Wei Y, Feng Y, Yang X (2013) Iron concentration, bioavailability, and nutritional quality of polished rice affected by different forms of foliar iron fertilizer. Food Chem 141:4122–4126

74. Fageria NK, Barbosa FMP, Moreira A, Guimaraes CM (2009) Foliar fertilization of crop plants. J Plant Nutr 32:1044–1064
75. Reda F, Abdelhamid MT, El-Lethy SR (2014) The role of Zn and B for improving *Vicia faba* L. tolerance to salinity stress. Middle East J Agric Res 3:707–714
76. Hellal FA, El Sayed SAA, Zewainy RM, Abdelhamid M (2015) Interactive effects of calcium and boron application on nutrient content, growth and yield of faba bean irrigated by saline water. Int J Plant Soil Sci 4:288–296
77. Mohamed HI, Elsherbiny EA, Abdelhamid MT (2016) The changes induced on physiological and biochemical responses of *Vicia faba* plants to foliar application with zinc and iron. Gesunde Pflanzen 68:201–212

Conclusions

Conclusions and Recommendations of Biological Control Industry

Nabil El-Wakeil, Mahmoud Saleh and Mohamed Abu-hashim

Abstract Nowadays, there is a big gap between plant nutrients produced and the required nutrients. The agricultural production increase should be attained to face the population increase in a system that conserve the environment, humankind and limits the insecticide uses and unattractive chemicals in agriculture. This chapter summarizes the key biological control industry challenges. This chapter spotlights on the sustainable biological control strategies of the agricultural environment that was documented in this book. Finally, four main contribution areas were identified which include; parasitoids, predacious insects and mites for managing insects and mites, Microorganisms for controlling insect pests, Biocontrol products for plant diseases management, and Bio-products against abiotic factors and micronutrient deficiency. Thus, recommendations and conclusions would be built on scientist and researcher visions added regarding to their research findings. In addition, this chapter includes information on a set of conclusions and recommendations to direct future research toward industry of biological control, which is one of the main tactical strategies of the Egyptian economy and environment.

Nowadays, there is a big gap between plant nutrients produced and the required nutrients. The agricultural production increase should be attained to face the population increase in a system that conserve the environment, humankind and limits the insecticide uses and unattractive chemicals in agriculture. This chapter summarizes the key biological control industry challenges.

This chapter spotlights on the sustainable biological control strategies of the agricultural environment that was documented in this book. Finally, four main contribution areas were identified which include; parasitoids, predacious insects and mites for managing insects and mites, Microorganisms for controlling insect pests, Biocontrol products for plant diseases management, and Bio-products against abiotic factors and micronutrient deficiency. Thus, recommendations and conclusions would be built on

N. El-Wakeil (✉) · M. Saleh
Pests and Plant Protection Department, National Research Centre (NRC), Cairo, Egypt
e-mail: nabil.elwakeil@yahoo.com

M. Abu-hashim
Department of Soil Sciences, Agriculture Faculty, Zagazig University, Zagazig, Egypt

© Springer Nature Switzerland AG 2020
N. El-Wakeil et al. (eds.), *Cottage Industry of Biocontrol Agents and Their Applications*, https://doi.org/10.1007/978-3-030-33161-0_15

scientist and researcher visions added regarding to their research findings. In addition, this chapter includes information on a set of conclusions and recommendations to direct future research toward industry of biological control, which is one of the main tactical strategies of the Egyptian economy and environment.

In last decades, there are some challenges which Egyptian producers faced them due to unprofessional conduct of natural resource and environmental degradation. Egypt is facing unmatched resource catastrophes particularly in water, and food. These impacts incorporate the degradation of soil fertility, water hardness, the increment in the dangerous residue, and advancement of resistance in insects. On the other hand, this agricultural production would be conducted in an ecological strategy which reduces the use of harmful insecticides.

Therefore, we try in this chapter to present a general idea of biological control industry and its significance for Egyptian farmers generally and especially for the researchers and students. In manipulative industry of biological control system, it is essential to give due kindness to the characteristics of various biocontrol agents used, which causes to be the resulting agricultural production system sustainable. So, the intention of this book is to improve and discuss the following main points:

- Parasitoids, predacious insects and mites for managing insects and mites,
- Microorganisms for controlling insect pests,
- Biocontrol products for plant diseases management,
- Bio-products against abiotic factors and micronutrient deficiency.

A brief overview of the important findings of the recent published studies on mass production of biocontrol agents and the main conclusions will be mentioned below. As well as the main recommendations for farmers, students, researchers and decision-makers will be presented.

The following are the main major studied points

1. Searching and studying the abundance of different biocontrol agents would be an economic method to have an enough biological control agents, which play a vital role in increasing the organic and clean agricultural production.
2. Using natural enemies and organic fertilizers to help in solving the pollution problem as trails for solve food shortage to face the growing population in Egypt.
3. Integrated different biocontrol agents, biopesticides and biofertilizers for organic and sustainable agriculture as natural products to increase organic agriculture in Egypt.
4. Climate and weather can significantly affect the developmental time and distribution of insect pests and associated natural enemies.
5. Factors affect mass production and field application of biocontrol agents.
6. Artificial fertilizers may help to increase the agricultural production, but it also can cause the emission of greenhouse gases, which threaten the environment [1–3].
7. The other potential ways are to how to mass produce the biocontrol agents, parasitoids, insect and mite predators, pathogens and biological control products of plant diseases as well as biofertilizers against micro nutrients deficiency [4].

8. Most of these strategies help to rise productivity of crops, vegetables as well as oil crops and grains, which become a requirement to minimize the gap between agricultural production and human consumption.

1 Part I: Parasitoids, Predacious Insects and Mites as Bioagents

In the first part, the authors of these book chapters are active and working in biological control institutions using parasitoids, predatory insects and mites for managing various insect and mite pests.

1.1 Chapter 1. Egg Parasitoids

Although the authors mentioned in the first chapter that egg parasitoids are predominant over the world, but selection of the strain/ecotype plays a crucial role in managing their host insects [5]. Tritrophic interactions (among plant-insect and parasitoid) should be known before egg parasitoid releases in a target cropping system. The volatile compounds which are released by plants in response to herbivore feeding, attracted many of egg parasitoids. Additionally being highly detectable and reliable indicators of herbivore presence, herbivore-induced plant volatiles may convey herbivore-specific information that allows parasitoids to discriminate closely-related herbivore species [6].

Egg parasitoids could be stored for about one week at 8–10 °C in the refrigerator without adversely affecting their mergence and parasitism efficiency [7]. Biological control using egg parasitoids is one of the most important strategies which will be the development of extension support to deliver the product to the user. Therefore, egg parasitoids are a promising biocontrol agent for agricultural insect pest's mainly Lepidopteran insects.

1.2 Chapter 2. Larval Parasitoids

The larval parasitoids come in the second importance level after egg parasitoids. Therefore, combining larval parasitoids with other biocontrol agents seems to be well suited to protect many crops from insect infestations; conversely, sometimes few species will probably be ineffective in some crop species, then the producers would use the other biocontrol agents. *Habrobracon hebetor* seems to be well suitable to some stored product insects, for example, it could use to find *Plodia interpunctella* larvae in damaged packages or storages. Therefore, the combination of releasing both egg parasitoids with larval parasitoids (*i.e.*, *Trichogramma evanescens* and *B. hebetor*) should provide the best control over the long term [8, 9]. However, it is important to realize that unlike pest mortality resulting from egg parasitism by *T. evanescens* and larval mortality caused by *B. hebetor* will avoid future infestation. For adequate management of corn borers to occur, releases both parasitoid wasps should probably be applied, as early as possible, so that the parasitoids have a better chance of managing pest populations before they reach the economic levels.

The authors of this book chapter confirmed that potential of other biocontrol species should be studied for combining them with other active agents, detect their efficiency on the primary hosts, and evaluate them in biological control and IPM programs in commercial tomato plantations [10]. The compatibility of the two bio-control groups for insect pest control could be easily managed if mass releases of parasitoids have been applied 48 h or more after for example EPNs applications [11].

1.3   Chapter 3. Aphid Parasitoids

Another group of biocontrol agents is aphid parasitoids which infect the aphids; aphids suck the plant sab and affect negatively the plant productivity and its quality. The author of this chapter mentioned that aphids have an economic importance, which disturb the agriculture development and production. This parasitoid group plays a significant role in controlling different aphid species in various crops and vegetables mainly cabbage [12–14].

The author of this chapter highlights the following conclusions:

1. *Diaeretialla rapae* is an important primary parasitoid of a wide range of aphid species and is considered a promising biological control agent against cabbage aphids.
2. Freshly formed mummies of *D. rapae* could be conserved at 5 °C in cold storage for 2 months.
3. The author confirmed that *L. fabarum* and *D. rapae* could be recommended as biological control candidates against this aphid species under Egyptian conditions.
4. It is to be recommended to consider aphid parasitoids in IPM program designed for the control of *B. brassicae* to decrease the environmental pollution by using the traditional insecticides.

1.4   Chapter 4. Insect Predators

Using insect predators in programs of biological control might be a foremost part of ecological agriculture. The authors mentioned that ca. 27 species of insect predators listed in that book chapter have been considered by Egyptian biocontrol researchers as reported by Saleh et al. [2]. Those predators prey most all the economic agricultural insect pests of field crops, vegetables, fruit orchards, ornamentals and greenhouse crops. The authors had tried to focus in this chapter on richness of predators in our environment and the challenges of mass rearing and field application of the most significant predators.

As declared in Chap. 4 the biocontrol production is influenced by many factors, such as infrastructure, personnel, tools, transportation and market, beside presence of the mentioned predators in the local environment and tries of mass rearing and releasing [15–19]. In Egypt, almost of these elements are found, what is still need is the support of decision makers in private and public sectors for having the infrastructure as well as some equipment. The author confirmed that the producers are required to increase their consciousness to avoid the risks of agricultural chemicals and the need to the biocontrol for preserving the environment in Egypt.

1.5   Chapter 5. Predatory Mites

Mites of the family Tetranychidae are strictly phytophagous and are represented in all region of Egypt. Spider mites can undoubtedly cause severe crop loss [20]. The authors of Chap. 5 stated that predatory mites could keep its potential efficiency after long-term rearing on alternative food to control natural prey. The effective factitious food supply sufficient nutritional requirements to the predators and lead to continuous production of high quality progeny with no reverse effect on the predator performance [21]. For example, *Neoseiulus barkeri* is predatory mite which distributed worldwide and has been used to manage pests in plastic houses since the method of its mass-rearing were developed. Hansen [22] reared *N. barkeri* on storage mites (*Acarus* spp.) in large numbers at a relatively small cost and was utilized to successfully control *T. tabaci*, on cucumber plants in glasshouses. *Amblyseius swirskii* has been proved to be an effective biological control agent against whiteflies, thrips, broad mites in several plastic houses crops [23–25].

Another objective of this chapter was to explore the potential of *E. kuehniella* eggs on development and reproduction of different life stages of *A. swirskii* for five successive generations [26]. Moreover, the study reported that *A. swirskii* did not lose their power to control its natural prey first instars of *F. occidentalis* after six tested generations [25]. Only the predation rates of the predators of generation six were slightly lower than the predation rates of the predators of generation one. The authors concluded that twenty five phytoseiid, three stigmaeid, six laelapid, ten ascid, five melicharid, and only one Rhodacarid species namely *Protogamasellopsis denticus* were recorded in Egypt. Also, Predatory phytoseiid mites are successfully used as bio-control agents in controlling of phytophagous mites, thrips and whitefly. *Agistemus exsertus* is one of the most important species of the family Stigmaeidae and able to attack pests of the families Tetranychidae and Eriophyidae as well as scale insects and whitefly in field crops and orchards.

The predatory mites played an important role for inhabiting mites either in the soil or on vegetation as biological control agents. This chapter concentrated on the method of mass production of aerial and soil predacious mites that can be reared traditionally by introducing natural or by providing factitious foods. It was explained in detail the rearing system taking into consideration the cost, advantage and disadvantage.

2 Part II: Microorganisms as Biocontrol Agents

In the second part of this book, role of microorganisms were reviewed and discussed in four book chapters dealing with using entomopathogenic bacteria, fungi, nematodes and viruses as shown in the followings:

2.1 Chapter 6. Bacillus Thuringiensis

The author of this book chapter confirmed that *Bacillus thuringiensis* is a desirable biocontrol agent for pest control and ideal for use in Egypt and other developing countries because its possible low production cost and lack of toxicity for other organisms. The possibility of increasing regional production using inexpensive material and agro-industrial byproducts is a particularly attractive option. Novel approaches of *B. thuringiensis* could be adopted to enhance the potency of *B.t.* preparations by using feeding stimulants and safe and cheap chemical additives leading to biochemical reactions in the midgut of the treated insects [27–30]. These additives were able to increase the potency and to extend spectrum level of activity of *B.t.* preparations by many folds. It was found that *B.t.* can affect the various developmental stages of lepidopteran insects other than the larvae.

Field application of the local *B. thuringiensis* product against lepidopteran insects in cotton, soybean, vegetables and some other field crops indicate that the yield was almost equal to that obtained with economically profitable strategies [28, 29]. The author mentioned that develop the GM crops is on-going in a number of research institutions and universities in Egypt, but the developed plants did not reach the stage of commercial release due to lack of national legislation on Biotech crops. However the approval for commercialization of genetically modified (GM) *B.t.* corn hybrid was made for a short period.

2.2 Chapter 7. Entomopathogenic Fungi

The entomopathogenic fungi (EPFs) have played a significant role in the history of microbial control of insects. Entomopathogenic fungi were the first to be recognized as microbial diseases in insects [31].

The authors reported that EPFs are one of the most promising agents for the biological control of insects, where it permits the cost of production and preservation of public health. As the fungus *Beauveria bassiana* successfully used against different insects as well as the fungus *Metarhizium anisopliae* and especially the rank of sheaths wings also *Verticillium lecanii* against insects sucking mouth parts [32, 33]. For mass production of EPFs, the authors stated that mass production of Entomopathogenic Fungi in various forms of liquid and powder as well as the body of nanoparticles through laboratories and factories are discussed. The dissemination of this means of promising in Integrated Pest Management (IPM) programs. Spreading awareness among the farmers about the importance of using this method as one of the most important means of biological control of pests as summarized in the following points:

1. The entomopathogenic fungi were isolated from soil samples using the *Galleria*-bait technique based on Zimmermann [34].
2. *Beauveria bassiana*, commonly known as white muscardine fungus attacks a wide range of immature and adult insects.
3. *Metarhizium anisopliae* a green muscardine fungus is reported to infect 200 species of insects and arthropods.
4. Both of these entomopathogenic fungi are soil borne and widely distributed.
5. The mortality percentage of *O. surinmensis*, *Tuta absoluta* and *Bemisia tabaci* treated with *B. bassiana* and *M. anisopliae* differed significantly among different concentrations [35].

2.3   Chapter 8. Entomopathogenic Nematodes

After discussing role of bacteria and fungi in controlling insect pests, we move to discuss role EPNs in insect management. The authors of that chapter explained the attributes that EPNs have made them excellent insect biocontrol agents as they can kill their insect host within 48 h; have a wide host range; have the ability to move searching for their hosts either in plants or in the soils.

Nematodes have been commercially developed to achieve progress in developing large-scale production and application technology which has led to the expanded use of EPNs either in In vivo or in vitro solid culture [36–38]. The authors confirmed that EPNs have been commercially produced by numerous companies in large liquid fermentation tanks in different industrialized countries. However, this technique requires greater funds investment and an advanced level of technical proficiency. They also mentioned there are different biotic and abiotic factors can affect efficiency of EPN application. Nematodes can be suppressed a diversity of economically significant insect pests in different habitats [39, 40].

The authors recommended that additional technological advancements are desired to expand and develop the market potential of the nematode-based biopesticides.

Isolation of additional species and selective breeding are required for proper classification, for biodiversity studies. This will also contribute to enhance the economic value of EPNs in biological control. Improved efficiency in nematode applications can be supported through improved formulation.

2.4 Chapter 9. Insect Viruses

The last chapter in this group is using insect viruses for controlling insect pests. Insect viruses as a biological control agent represent an important component in IPM programs, mainly because it is specific and safe for the environment for Lepidopteran, Dipteran, Orthopteran and Colepoteran insects [41].

The authors of this book chapter mentioned that specificity and the production of secondary inoculum make baculoviruses and other insect viruses attractive alternatives to chemicals insecticides and ideal components of Insect. In addition, the use of insect viruses as bioinsecticides is compatible with many other components of biological control agents. In addition, the fact that insect viruses are unable to infect mammals, including humans, makes them very safe to handle and attractive candidates as alterative biopesticides to avoid the use of the harmful pesticides. Therefore, the future for the continuous use of the insect viruses will depend on the success to overcome these limitations. The development of formulation which include protectant materials against UV could increase the sustainability of the viral product that can tolerate the UV effect; therefore increase the virus persistence [42, 43].

The use of recombinant bocaviruses that include the deletion of virus genes that delay the virus killing (e.g. the deletion of the ecdysteroid UDP-glucosyltransferase (*egt*) gene) or the expression of toxins that accelerate the killing effect has been developed for some viruses. However there are several challenges facing the large-scale production of these viruses as fast killing of the host affect negatively the amount of produced virus from infected host [44, 45]. Finally, the authors confirmed that using the correct virus (or a mixture of virus) strains in the biopesticides to overcome the development of resistant against the virus in the host population might help to face the resistance challenges [46]. The success in facing the abomination limitation will shape the future viruses use as biopesticides to control the major insect pests.

3 Part III: Biocontrol Products for Managing Plant Diseases

In the third part, the editors are moving to other pest group (plant diseases), which were caused by microorganisms. In this section, using biological control products to manage plant diseases are reviewed.

3.1 Chapter 10. Phyto-Pathogenic Bacteria

This chapter covered the biological control strategies for controlling bacterial plant pathogens in the field. For achieving the best control of bacterial plant diseases, identifying the plant pathogens must be correct identified by known the pathogen's life cycle and how it relates to the cycle of disease development [47].

The author stated that for developing a suitable management program that attacks the pathogen at the weakest point in it life cycle should be applied. The development of alternative approaches to control pathogens of crops utilizing bio-control agents is necessary to reduce risk pesticides [48, 49]. To success of biological control, the author mentioned that bio-control agent must be grown very fast, the environment is favorable for their growth and development and it must be applied at pre-planting or prior to the onset of disease.

The bio-control agents had different mechanisms such as antibiosis, competition, hyperparasitism, cell-wall degrading enzymes and induction of systemic resistance which play an important role for controlling bacterial plant diseases, where inducing systemic resistance protects plants not only against the attacking pathogen, but against other types of pathogens [50]. Therefore, the biological control successfully applied for controlling many bacterial plant diseases as safe alternative tools replace the chemical bactericides which had mammalian toxicity and environmental pollution.

3.2 Chapter 11. Fungal Plant Diseases

In this book chapter, the authors discussed importance of fungal diseases and their effects on the agricultural production as well as using biological control agents for controlling the fungal plant pathogens. They mentioned that biocontrol has many advantages in relation to soil fertility, plant, animal and human health [51, 52]. The authors also tried to highlight the most important biological control practices either in Egypt or over the world [53]. It is found that throughout the last decades, the attention to biological control of economic crops has increased from both the government and the researchers starting from the ordinary application of biocontrol agents in contact directly to the soil and in form of gelatin capsules to insertion of the resistance genes in the plant and produce what we know today GM plants (genetically modified plants) [54]. Finally, the authors of this chapter confirmed that the biological control of different diseases becomes common due to the awareness of farmers about the benefits of biocontrol applications in Egypt.

3.3   Chapter 12. Plant Parasitic Nematodes

Nematodes are invertebrate roundworms, which comprise one of the largest and most diverse groups of multicultural organisms in existence and they spend at least some part of their lives in soil. The author mentioned that parasites, predators and micro-organisms used for nematode biocontrol are effective and environment friend. Many of bacteria and fungi are intimately associated wilt roots and it will be necessary to establish that they are not plant pathogens before they are widely used for biological control purposes.

The author confirmed that some formulation of microbial products in forms which have extended shelf life and which can be applied to soil using conventional farm equipment is likely to present a major challenge to those interested in commercializing antagonists of nematodes [55]. A number of types of formulation, including dusts, granules, wet table powders and liquids have been used in biological crop protection products, but granular formulations are generally considered to be most suitable for micro-organisms that are to be applied to soil. There are many trials to produce the granular formulations for commercial microbial products in future to be applied to soils [56].

One of the important problem faces the mass production of biocontrol agents of parasitic nematodes in Egypt is standardization, formulation, storage, application and safety have had to be considered in detail, but the time is rapidly approaching when such issues will become major areas of activity. Egyptian researchers intended to transfer the modern fermentation technology which is widely used in the brewing industry and in the production of antibiotics and microbial insecticides that should be agreeable to the mass production of antagonists of nematodes. The author reported that the large-scale industrial production requires the development of inexpensive alternatives to the complex media often used to culture nematode antagonists in the laboratory [57, 58]. Biological control agents are generally produced in commercial quantities by one of two fermentation methods (Solid substrate fermentation or liquid culture fermentation).

3.4   Chapter 13. Plant Viral Diseases

Viruses are among the most important plant pathogens as near half of the emerging epidemics have a viral etiology [59]. Regarding the virus transmission, viruses can be transmitted between plants either biologically or mechanically [60]. The authors of this chapter mentioned that there are many trials to manufacture viral biocontrol agents, whether bacterial or fungal, and how this would be applied in the green houses and in the open fields. All these attempts have been mostly international but one or two attempts were local and even wasn't reach the field application yet. The authors reported that the antiviral products are still in their early stages, and the market for these bio antiviral has only a few numbers, which do not exceed the fingers of one hand [61].

Work in this area is still need more efforts and research until humanity can rid itself of the bad effects of pesticides and the huge yield losses caused by plant viruses [62]. Bio-anti viruses have been internationally and locally accepted by farmers and investment companies as well as in scientific media. In fact, there is no evidence so far that there is a substance whether naturally synthesized or manufactured can eliminate the plant virus within the plant host cell [63]. Nevertheless all the available bio/compounds which have antiviral activity resulted from their capability to stimulate the plant immune response.

4 Part IV: Bio-Products Against Abiotic Factors

In the fourth part of the book which contains only one book chapter entitled "Biochemical indicators and biofertilizer application for diagnosis and alleviation micronutrient deficiency in plant"; the authors tried to find the best strategy to solve the micronutrient deficiency. The editors suggested this theme, after controlling various species of insects, mites and plant diseases biologically to complete the whole story of biological agents or products to sustain the organic or ecological friendly agricultural production.

4.1 Chapter 14. Biochemical Indicators of Micronutrient Deficiency

Micronutrients are very important for an abundance of physiological functions in plant growth, development, and oxidative stress response. In case of low concentrations in electron transport and antioxidant systems, micronutrients are desired for cellular structures and protein stabilization [64]. The possibilities of using biochemical markers and application of biofertilizers to diagnose and mitigate deficiencies in micronutrients are discussed and reviewed in this book chapter.

The authors confirmed that availability of micronutrients in soils is associated with the type of parent materials and the environment in which soils are formed. Cu and Mo's availability in soils appears to be more related to soil organic matter content. The authors reported that predicting behavior of micronutrients in soils is not secure [65]. The first choice to solve this problem, is to use fertilizers containing micronutrients, both chemical and organic. Application method must be conducted based on soil nutrient management guidelines along with crop types [66]. Finally, the authors mentioned that increase using the modern biotechnology techniques, such as gene transformation, will lead to a promising prospect of higher nutritional stress resistance. There are already transgenic plants with enhanced nutrient efficiency such as zinc (Zn), boron (B) and copper (Cu). Further plant stress research will also enhance the knowledge of mechanisms of adaptation. This will then stimulate breeding for the efficiency of nutrients and the tolerance of stress.

5 Recommendations

The main aspects of biological control industry is the ability to mass production and mass field application as well as to acclimate to future challenges and prospects. We argue that supportable arrangements need built-in flexibility to accomplish this goal. Throughout this book, the editorial board noted some areas that could be explored to further improvement. Based on the authors' chapters, the following recommendations could be mentioned for future researchers in exceeding the scope of this book.

Egyptian organic agriculture has a comparative advantage in terms of production dates and quality of European market countries. The future application of sustainable agriculture approach will help in the following:

- Significantly, the collaboration between biology, chemistry, and pest management researchers is essential for standardization of mass rearing of beneficial insects and mites.
- The majority of findings of factors which affect toward production or development of commercial biocontrol products.
- Several factors delay such advancement either in Egypt including lack of (a) real links between industry and research institutions, (b) adequate investments for scientists to prepare their products commercially, (d) components that safe to the environment, (e) effective natural components as pesticides, (f) Governmental regulations on the use of natural products in pest control programs.
- There is a growing consciousness toward the hazards of chemicals and the need for the biological control for conserving the environment and realizing the organic agriculture in Egypt.
- Egypt has a large market of biological products represented mainly by the agricultural investment companies, which export these organic products of vegetables and fruit to Arab and European countries.
- Biological control infrastructure, mass production, and field application technologies would help to have clean crops, vegetables, and fruits to reduce the economic cost and to minimize environmental and health risks.

6 Conclusion

The different biological control strategies would be used in the appropriate routines and against certain insect pests in large areas, the proposed approach might be highly effective and environmentally acceptable. Furthermore, if this kind of biological control could be mass produced and mass applied in conjunction with other control strategies, this could result in significant and important synergistic effects on pest population suppression. Egypt has a great market of biological products represented mostly by the agricultural private sector which distribute these bio- or organic products of vegetables and fruit to Arab and European countries. Climate change is expected to have a negative or positive effect on the short and long-term diversity of pest's abundance, pest's-host plant interactions, and an abundance of natural enemies. Further biocontrol strategies research would improve the knowledge of mechanisms of adaptation.

The proposed aims of this book mass production of biocontrol agent with cost-effective, which are considered as an essential to increasing food supply sustainably and ecologically in Egypt. Integrated biopesticides and biofertilizers for organic agriculture. Finally, it is expected to have an effect on the Egyptian economy due to the impact on agricultural economic crops. This will motivate breeding for resistance of insects, mites and plant diseases as well as to solve problem of nutrient deficiency for IPM and Integrated Crop Management (ICM) programs for keeping the agroecosystem clean and safe.

References

1. Negm AM, Abu-hashim M (eds) (2019) Sustainability of agricultural environment in Egypt: part II—soil-water-plant nexus. In: The handbook of environmental chemistry, vol 77, pp 3–30. Springer International Publishing, Part II © Springer Nature Switzerland. https://doi.org/10.1007/698_2017_76. ISBN 978-3-319-95356-4
2. Saleh MME, El-Wakeil NE, Elbehery H, Gaafar N, Fahim S (2019) Biological pest control for sustainable agriculture in Egypt. In: Negm AM, Abu-hashim M (eds) Sustainability of agricultural environment in Egypt: Part II. The handbook of environmental chemistry, vol 77, pp 145–188. © Springer Nature Switzerland AG 2019—Soil-Water-Plant Nexus, Springer Publisher. https://doi.org/10.1007/978-3-319-95357-1. ISBN 978-3-319-95356-4
3. Ajmal M, Hafiza IA, Rashid S, Asna A, Muniba T, Muhammad ZM, Aneesa A (2018) Biofertilizer as an alternative for chemical fertilizers. Nawaz Sharif Medical College, University of Gujrat, Gujrat, Punjab
4. Nour-El-deen MA, Abo-zid AE, Azouz HA (2014) Use of some environmentally safe materials as alternatives to the chemical pesticides in controlling *Polyphagotarsonimus lauts* (banks) mite & *Myzus persica* (Koch) aphid which attack potatoes crop. Middle East J Agric Res 3:32–41
5. Kumar P, Shenhmar M, Brar KS (2004) Field evaluation of trichogrammatids for the control of *Helicoverpa armigera* (Hübner) on tomato. Biol Cont 18:45–50
6. De Moraes CM, Lewis WJ, Tumlinson JH (2000) Examining plant-parasitoid interactions in tritrophic systems. An Soc Entomol Bras 29:189–203
7. Khosa SS, Brar KS (2000) Effect of storage on the emergence and parasitization efficiency of laboratory reared and field collected population of *Trichogramma chilonis* Ishii. Biol Cont 14:71–74
8. Grieshop JG, Flinn PW, Nechols JR (2006) Biological control of Indian meal moth on finished stored products using egg and larval parasitoids. J Econ Entomol 99:1080–1084
9. Briggs CJ, Latto J (2001) Interactions between the egg and larval parasitoids of a gall-forming midge and their impact on the host. Ecol Entomol 26:109–116

10. Lacey LA, Unruh TR, Headrick HL (2003) Interactions of two idiobiont parasitoids (Ichneu-monidae) of codling moth (Tortricidae) with the entomopathogenic nematode *Steinernema carpocapsae* (Steinernematidae). J Inverteb Pathol 83:230–239

11. Batalla-Carrera L, Morton A, García-del-Pino F (2010) Efficacy of entomopathogenic nema-todes against the tomato leafminer *Tuta absoluta* in laboratory and greenhouse conditions. Biocontrol 55:523–530

12. Nematollahi MR, Fathipour Y, Talebi AA, Karimzadeh J, Zalucki MP (2014) Parasitoid- and hyperparasitoid-mediated seasonal dynamics of the cabbage aphid (Hemiptera: Aphididae). Environ Entomol 43:1542–1551

13. Saleh AAA, Desuky WMH, Hashem HHA, Gatwary WGT (2009) Evaluation the role of aphid parasitoid *Diaeretiella rapae* (M´ Intosh) (Hymenoptera: Aphidiidae) on cabbage aphid *Brevicoryne brassicae* L. (Homoptera: Aphididae) in Sharkia district. Egypt J Biol Pest Cont 19:151–155

14. Nematollahi MR, Fathipour Y, Talebi AA, Karimzadeh J, Zalucki MP (2014) Parasitoid and hyperparasitoid-mediated seasonal dynamics of the cabbage aphid (Hemiptera: Aphididae). Environ Entomol 43:1542–1551

15. Holling CS (1961) Principles of insect predation. Annu Rev Entomol 6:163–182

16. Holling CS (1959) Some characteristics of simple types of predation and parasitism. Can Entomol 91:385–398

17. Zheng Y, Km Daane, Hagen KS, Mittler TE (1993) Influence of larval food consumption on the fecundity of the lacewing *Chrysoperla camea*. Ent Exp Appl 67:9–14

18. Lopez-Arroyo JI, Tauber CA, Tauber MJ (2000) Storage of lacewing eggs: post-storage hatch-ing and quality of subsequent larvae and adults. Biol Cont 18:165–171

19. Gaafar N (2002) Effects of some Neem products on *Helicoverpa armigera* and their natural enemies *Trichogramma* spp. and *Chrysoperla carnea*. MSc. Agric Fac., Goerge August Univ., Goettingen, Germany

20. Gerson U (2008) The Tenuipalpidae: an under-explored family of plant feeding mites. Syst Appl Acarol 2:83–101

21. Cohen AC (2004) Insect diets: science and technology. CRC Press, Boca Raton

22. Hansen LS (1988) Control of *Thrips tabaci* (Thysanoptera: Thripidae) on glasshouse cucum-ber using large introductions of predatory mites *Amblyseius barkeri* (Acarina: Phytoseiidae). Entomophaga 33:33–42

23. van Maanen R, Vila E, Janssen A (2010) Biological control of broad mite (*Polyphagotarsone-mus latus*) with the generalist predator *Amblyseius swirskii*. Exp Appl Acarol 52:29–34

24. Nomikou M, Janssen A, Schraag R, Sabelis MW (2002) Phytoseiid predators suppress popula-tions of *Bemisia tabaci* on cucumber plants with alternative food. Exp Appl Acarol 27:57–68

25. Messelink GJ, van Steenpaal SE, Ramakers PM (2006) Evaluation of phytoseiid predators for control of western flower thrips on greenhouse cucumber. Biocontrol 51:753–768

26. Nguyen DT, Vangansbeke D, De Clercq P (2014) Artificial and factitious foods support the development and reproduction of the predatory mite *Amblyseius swirskii*. Exp Appl Acarol 2:181–194

27. Salama HS, Foda S, Sharaby A (1984) Novel biochemical avenues for enhancing *Bacillus thuringiensis* endotoxin potency against *Spodoptera litoralis*. Entomophaga 29:171–178

28. Salama HS, Foda S, Sharaby A (1986) Possible extension of the activity spectrum of *Bacillus thuringiensis* strains through chemical additives. J Appl Ent 101:304–313

29. Salama HS (1993) Enhancement of *Bacillus thuringiensis* for field application. In: Morris O, Rached E, Salama HS (eds) The biopesticide *Bacillus thuringiensis* and its application in developing countries. Al-Ahram Press, Cairo, pp 105–116

30. Osman GEH, Already R, Assaeedi ASA, Organji SR El-Ghareeb D, Abulreesh HH, Althubiani AS (2015) Bioinsecticide *Bacillus thuringiensis* a comprehensive review. Egypt J Biol Pest Control 25:271–288

31. Shah PA, Pell JK (2003) Entomopathogenic fungi as biological control agents. Appl Microbiol Biotech 61:413–423

32. Abdel-Raheem MA (2011) Impact of entomopathogenic fungi on Cabbage Aphids, *Brevicoryne brassica*. Egypt Bull NRC 36:53–62
33. Abdel-Raheem et al (2011) Effect of entomopathogenic fungi on the green stink bug, *Nezara viridula* L. in sugar beet. Bull NRC 36:145–152
34. Zimmermann (1986) The *Galleria* bait method for detection of entomopathogenic fungi in soil. J Appl Entomol 102(2):213–215
35. Sabbour M, Abdel-Raheem M (2015) Efficacy of *Beauveria brongniartii* and *Nomuraea rileyi* against the potato tuber moth, *Phthorimaea operculella* (zeller). Am J Innovative Res Appl Sci 1(6):197–202
36. Metwally HM, Hafez GA, Hussein MA, Hussein MA, Salem HA, Saleh MME (2012) Low cost artificial diet for rearing the greater wax moth, *Galleria mellonella* L. (Lepidoptera: Pyralidae) as a host for entomopathogenic nematodes. Egypt J Biol Pest Cont 22:15–17
37. Shapiro-Ilan DI, Lewis EE, Behle RW, McGuire MR (2001) Formulation of entomopathogenic nematode-infected-cadavers. J Invert Pathol 78:17–23
38. Shapiro-Ilan DI, Morales-Ramos JA, Rojas MG, Tedders WL (2010) Effects of a novel ento-mopathogenic nematode–infected host formulation on cadaver integrity, nematode yield, and suppression of *Diaprepes abbreviatus* and *Aethina tumida* under controlled conditions. J Invert Pathol 103:103–108
39. Saleh MEE, Metwally HM, Mahmoud YA (2018) Potential of the entomopathogenic nema-tode, *Heterorhabditis marelatus*, isolate in controlling the peach fruit fly, *Bactrocera zonata* (Saunders) (Tiphritidae). Egypt J Biol Pest Control 28(22). https://doi.org/10.1186/s41938-018-0029-0
40. Moawad SS, Saleh MME, Metwally HM, Ebadah IM, Mahmoud YA (2018) Protective and curative treatments of entomopathogenic nematodes against the potato tuber moth, *Phthori-maea operculella* (Zell.). Biosci Res 15:2602–2610
41. Lacey LA, Grzywacz D, Shapiro-Ilan DI, Frutos R, Brownbridge M, Goettel MS (2015) Insect pathogens as biological control agents: back to the future. J Invertebr Pathol 132:1–41
42. Dougherty EM, Guthrie KP, Shapiro M (1996) Optical brighteners provide baculovirus activity enhancement and UV radiation protection. Biol Cont 7:71–74
43. Petrik DT, Iseli A, Montelone BA, Van Etten JL, Clem RJ (2003) Improving baculovirus resistance to UV inactivation: increased virulence resulting from expression of a DNA repair enzyme. J Invertebr Pathol 82:50–56
44. Sun X (2015) History and current status of development and use of viral insecticides in China. Viruses 7:306–319
45. Sun X, Peng H (2007) Recent advances in biological control of pest insects by using viruses in China. Virol Sin 22:158–162
46. Eberle KE, Asser-Kaiser S, Sayed SM, Nguyen HT, Jehle JA (2008) Overcoming the resistance of codling moth against conventional *Cydia pomonella* granulovirus (CpGV-M) by a new isolate CpGV-I12. J Inverteb Pathol 98:293–298
47. Lwin M, Ranamukhaarachchi SL (2006) Development of biological control of *Ralstonia solanacearum* through antagonistic microbial populations. Int J Agric Biol 8:1560–8530
48. Verdier V, Vera Cruz C, Leach JE (2011) Controlling rice bacterial blight in Africa: needs and prospects. J Biotechnol 159:320–328
49. Ramanamma CH, Santoshkumari M (2017) Biological control of blight of rice using RR8 rhizosphere bacteria. Int J Current Microbiol Appl Sci (Special Issue-5):124–128
50. Mello MRF, Silveira EB, Viana IO, Guerra ML, Mariano RLR (2011) Use of antibiotics and yeasts for controlling Chinese cabbage soft rot. Hortic Bras 29:78–83
51. Brundrett MC (2009) Mycorrhizal associations and other means of nutrition of vascular plants: understanding the global diversity of host plants by resolving conflicting information and developing reliable means of diagnosis. Plant Soil 320(1–2):37–77
52. Chen M, Arato M, Borghi L et al (2018) Beneficial services of Arbuscular Mycorrhizal fungi—from ecology to application. Front Plant Sci 9:1270
53. Barros DCM, Fonseca ICB, Balbi-Peña MIP et al (2015) Biocontrol of *Sclerotinia sclerotiorum* and white mold of soybean using saprobic fungi from semi-arid areas of Northeastern Brazil. Summa Phytopathologica 41(4):251–255

54. Wang SM, Liang Y, Shen T et al (2016) Biological characteristics of *Streptomyces albospinus* CT205 and its biocontrol potential against cucumber Fusarium wilt. Biocontrol Sci Techn 26(7):951–963
55. Noweer EMA, Hasabo SAA (2005) Effect of different management practices for controlling the root-knot nematode *Meloidogyne incognita* on squash. Egypt J Phytopathol 33:73–81
56. Rhoades HL (1985) Comparison of fenamiphos and *Arthrobotrys amerospora* for controlling plant nematodes in central Florida. Nematropica 15:1–7
57. Noweer EMA, Al-Shalaby MEM (2009) Effect of *Verticilium chlamydosporium* combined with some organic manure on *Meloidogyne incognita* and other soil micro-organisms on tomato. Int J Nematol 19:215–220
58. Noweer EMA (2018) Effect of the nematode-trapping fungus Dactylaria brochopaga and the nematode egg parasitic fungus Verticilium chlamydosporium in controlling citrus nematode infesting mandarin, and interrelationship with the Co inhabitant fungi. Int J Eng Technol 7:19–23
59. Lewsey M, Palukaitis P, Carr JP (2009) Plant—virus interactions: defence and counter-defence. In: Parker J (ed) Molecular aspects of plant disease resistance. Wiley-Blackwell, Oxford, pp 134–176
60. Fereres A (2000) Barrier crops as a cultural control measure of non-persistently transmitted aphid-borne viruses. Virus Res 71:221–231
61. Prins M, Laimer M, Noris E, Schubert J, Wassenegger M, Tepfer M (2008) Strategies for antiviral resistance in transgenic plants. Mol Plant Pathol 9:73–83
62. Abdelkhalek A, ElMorsi A, AlShehaby O, Sanan-Mishra N, Hafez E (2018) Identification of genes differentially expressed in *Iris Yellow Spot Virus* infected onion. Phytopathol Mediterr 57:334–340
63. Mohamed SH, Galal AM (2005) Identification and antiviral activities of some halotolerant *Streptomycetes* isolated from Qaroonlake. Int J Agric Biol 7:747–753
64. O'Neil MA, Ishiim T, Albersheimm P, Darvill AG (2004) Rhamnogalacturonan II: structure and function of a borate cross-linked cell wall pectic polysaccharide. Ann Rev Plant Biol 55:109–139
65. El-Baz FK, El-Monde EA, Salama ZA, Mohamed AA (1998) Determination of Fe^{2+} and soluble zinc as biochemical indicators for the diagnosis of iron and zinc deficiency *Phaseolus Vulgaris* and *Vicia faba* plants. Egypt J Physiol Sci 22:25–39
66. McCauly A, Jones C, Jacobsen J (2009) Plant nutrient functions and deficiency and toxicity symptoms. In: Nutrient management module, no 9, p 16. Montana State University, Bozeman, MT, USA

Printed in the United States
By Bookmasters